NATO ASI Series

Advanced Science Institutes Series

A series presenting the results of activities sponsored by the NATO Science Committee, which aims at the dissemination of advanced scientific and technological knowledge, with a view to strengthening links between scientific communities.

The Series is published by an international board of publishers in conjunction with the NATO Scientific Affairs Division.

A Life Sciences B Physics	Plenum Publishing Corporation London and New York
C Mathematical and Physical Sciences D Behavioural and Social Sciences E Applied Sciences	Kluwer Academic Publishers Dordrecht, Boston and London
F Computer and Systems Sciences G Ecological Sciences H Cell Biology I Global Environmental Change	Springer-Verlag Berlin Heidelberg New York Barcelona Budapest Hong Kong London Milan Paris Santa Clara Singapore Tokyo

Partnership Sub-Series

1. Disarmament Technologies	Kluwer Academic Publishers
2. Environment	Springer-Verlag / Kluwer Academic Publishers
3. High Technology	Kluwer Academic Publishers
4. Science and Technology Policy	Kluwer Academic Publishers
5. Computer Networking	Kluwer Academic Publishers

The Partnership Sub-Series incorporates activities undertaken in collaboration with NATO's Cooperation Partners, the countries of the CIS and Central and Eastern Europe, in Priority Areas of concern to those countries.

NATO-PCO Database

The electronic index to the NATO ASI Series provides full bibliographical references (with keywords and/or abstracts) to about 50 000 contributions from international scientists published in all sections of the NATO ASI Series. Access to the NATO-PCO Database is possible via the CD-ROM "NATO Science & Technology Disk" with user-friendly retrieval software in English, French and German (© WTV GmbH and DATAWARE Technologies Inc. 1992).

The CD-ROM can be ordered through any member of the Board of Publishers or through NATO-PCO, B-3090 Overijse, Belgium.

Series F: Computer and Systems Sciences, Vol. 161

Springer-Verlag Berlin Heidelberg GmbH

Computational Methods in Mechanical Systems

Mechanism Analysis, Synthesis, and Optimization

Edited by

Jorge Angeles

Department of Mechanical Engineering
and McGill Centre for Intelligent Machines
McGill University, 817 Sherbrooke Street W.
Montreal, Quebec, Canada H3A 2K6

Evtim Zakhariev

Institute of Mechanics
Bulgarian Academy of Sciences
Acad. G. Bonchev Street, bl. 4
BG-1113 Sofia, Bulgaria

Springer
Published in cooperation with NATO Scientific Affairs Division

Proceedings of the NATO Advanced Study Institute on Computational Methods in Mechanisms, held near Varna, Bulgaria, June 16–28, 1997

Library of Congress Cataloging-in-Publication Data

Computational methods in mechanical systems : mechanism analysis, synthesis and optimization / Jorge Angeles, Evtim Zakhariev.
p. cm. -- (NATO ASI series. Series F, Computer and systems sciences ; vol. 161)
Includes bibliographical references and index.

1. Machinery, Kinematics of. 2. Machinery, Dynamics of. 3. Engineering mathematics. I. Angeles, Jorge, 1943- . II. Zakhariev, Evtim, 1951- . III. Series.
TJ175.C66 1998
621.8'15--dc21 98-10329
CIP

CR Subject Classification (1991): J.6, I.6

ISBN 978-3-642-08369-3 ISBN 978-3-662-03729-4 (eBook)
DOI 10.1007/978-3-662-03729-4

Originally published by Springer-Verlag Berlin Heidelberg New York 1998.
Softcover reprint of the hardcover 1st edition 1998

Typesetting: Camera-ready by authors
Printed on acid-free paper
SPIN: 10648745 45/3142 - 5 4 3 2 1 0

Preface

The chapters of this book summarize the lectures delivered during the NATO Advanced Study Institute (ASI) on *Computational Methods in Mechanisms*, that took place in the Sts. Constantin and Elena Resort, near Varna, on the Bulgarian Coast of the Black Sea, June 16–28, 1997. The purpose of the ASI was to bring together leading researchers in the area of mechanical systems at large, with special emphasis in the computational issues around their analysis, synthesis, and optimization, during two weeks of lectures and discussion. A total of 89 participants from 23 countries played an active role during the lectures and sessions of contributed papers. Many of the latter are being currently reviewed for publication in specialized journals.

The subject of the book is *mechanical systems*, i.e., systems composed of rigid and flexible bodies, coupled by mechanical means so as to constrain their various bodies in a goal-oriented manner, usually driven under computer control. Applications of the discipline are thus of the most varied nature, ranging from transportation systems to biomedical devices. Under normal operation conditions, the constitutive bodies of a mechanical system can be considered to be rigid, the rigidity property then easing dramatically the analysis of the kinematics and dynamics of the system at hand. Examples of these systems are the suspension of a terrestrial vehicle negotiating a curve at speeds within the allowed or recommended limits and the links of multiaxis industrial robots performing conventional pick-and-place operations. Many an application, however, calls for an analysis that considers the flexibility of the various links, as is the case in light devices like the solar panels of a satellite undergoing a reorientation maneuver, or the body of a high-speed train experiencing a collision. The correct operation of systems like those mentioned above, which are set to work under increasingly demanding conditions, calls for increasingly detailed and accurate analyses both at the design stage and when driven under computer control. Hence the need to review the state of the art in areas pertaining to the mechanics of multibody systems.

The book is intended to cover a broad scope of multibody-system mechanics, from the kinematics fundamentals to applications such as motion-simulator design and crashworthiness evaluation by simulation means. We have thus divided the book into three parts: Kinematics of Mechanical Systems; Dynamics and Control of Rigid-Body Systems; and Dynamics of Flexible Multibody Systems. With this organization of the book, the reader should be able to find quickly the chapter of his or her own interest.

The book is, in a way, a follow-up of that containing the lectures of the NATO ASI on *Computer-Aided Analysis of Rigid and Flexible Mechanical Systems*, held in Tróia, Portugal, from June 27 to July 9, 1993, and published by Kluwer Academic Publishers in 1994. Besides including techniques and software developed in the last four years since that ASI, the present book includes a greater component of kinematics than the previous one. In fact, seven chapters out of a total of sixteen are devoted to this discipline.

The reader will find review chapters on topics such as optimization of multi-dof mechanisms, with classical and modern techniques, and on the dynamics of mechanical systems with flexible bodies. Besides, chapters on elimination techniques based on computer-algebra software in the realm of computational kinematics, as well as on design techniques, are included. Advanced software for design and simulation is introduced in some chapters.

The sponsors of the ASI on Computational Methods in Mechanisms are cited below. Herewith we acknowledge dutifully their support.

- NATO Scientific and Environmental Affairs Division;
- Ministry of Defense of Bulgaria;
- Ministry of Foreign Affairs of Bulgaria;
- National Science Fund MM 526/95, Ministry of Education, Science and Technology of Bulgaria;
- U.S. National Science Foundation;
- Institute of Mechanics, Bulgarian Academy of Sciences;
- Department of Mechanical Engineering, McGill University; and
- Centre for Intelligent Machines, McGill University.

The Organizing Committee of the NATO ASI was composed of:

- Prof. Jorge Angeles, McGill University, Montreal, Que., Canada
- Prof. Nikolai Denchev, Director of the Research Institute of the Ministry of Defense of Bulgaria;
- Prof. Clément Gosselin, Laval University, Quebec City, Que., Canada;
- Prof. Edward J. Haug, The University of Iowa, Iowa City, IN, USA;
- Prof. Bojan M. Petkov, Assistant Minister of Research, Science, and Technology of Bulgaria;
- Professor Bernard Roth, Stanford University, Stanford, CA, USA; and
- Dr. Evtim V. Zakhariev, Academy of Sciences of Bulgaria, Sofia.

All members of the Organizing Committee played a key role both in the organization of the ASI and in the editing of the book.

Finally, the assistance of Ms. Irène Cartier and Ms. ChinPun Teng in the editing of this book is herewith gratefully acknowledged. The technical assistance of Dr. Hans Woessner and Mr. J. Andrew Ross, of Springer-Verlag Heidelberg, was crucial in the production of this work.

Jorge Angeles and Evtim Zakhariev, December 1997

Table of Contents

Part II. Dynamics and Control of Rigid-Body Systems

Part III. Dynamics of Flexible Multibody Systems

List of Contributors

J. A. C. Ambrósio
IDMEC - Polo do I.S.T.
Av. Rovisco Pais
P-1096 Lisboa Codex
Portugal
D2060@beta.ist.utl.pt

J. Angeles
Department of Mechanical
Engineering & McGill Centre
for Intelligent Machines
McGill University
817 Sherbrooke St. W.
Montreal, Quebec
Canada H3A 2K6
angeles@cim.mcgill.ca

H. Brauchli
Institut für Mechanik
HG F38.1
ETH Zentrum
CH-8092 Zurich
Switzerland
brauchli@ifm.mavt.ethz.ch

P. Chedmail
IRCyN
Unité mixte de Recherche C.N.R.S.
UMR 6597
Ecole centrale de Nantes
1, rue de la Noë
BP 92101
F-44321 Nantes Cedex 3, France
Patrick.Chedmail@lan.ec-nantes.fr

C. M. Gosselin
Laval University
Mechanical Engineering Dept.
Ste.-Foy, Quebec
Canada G1K 7P4
gosselin@gmc.ulaval.ca

E. J. Haug
CAD-Research Engineering
2136 EB
The University of Iowa
Iowa City, IA 52242
USA
haug@ccad.uiowa.edu

L. K. Lilov
Sofia University
Department of Mathematics
and Informatics
James Bouchier Street 5
BG-1126 Sofia
Bulgaria
lilov@fmi.uni-sofia.bg

J. M. McCarthy
Department of Mechanical
Engineering
University of California
Irvine, CA 92717
USA
jmmccart@uci.edu

B. Ravani
Department of Mechanical Engineering
University of California, Davis
Davis, CA 95616
USA
bravani@ucdavis.edu

B. Roth
Dept. of Mechanical Engineering - Design Division
Stanford University
Stanford, CA 94305
USA
roth@flamingo.Stanford.EDU

W. Schiehlen
Institut B für Mechanik
Pfaffenwaldring 9
Universität Stuttgart
D-7000 Stuttgart 80
Germany
gmb@mechb1.fertigungstechnik.uni-stuttgart.de

R. Schwertassek
Institute for Robotics and System Dynamics
DLR Oberpfaffenhoffen
D-82234 Wessling
Germany
Richard.Schwertassek@dlr.de

O. I. Sivertsen
Dept. of Machine Design
The Norwegian Institute of Tech.
N-7034 Trondheim
Norway
oles@protek.unit.no

C. Woernle
Institut für allgemeinen Maschinenbau
FB Maschinenbau und Schiffstechnik
Universität Rostock
D-18051 Rostock
Germany
woernle@agr.uni-rostock.de

E. Zakhariev
Institute of Mechanics
Bulgarian Academy of Sciences
Acad. G. Bonchev Street, bl. 4
BG-1113 Sofia
Bulgaria
evtimvz@bgcict.acad.bg

PART I
Kinematics of Mechanical Systems

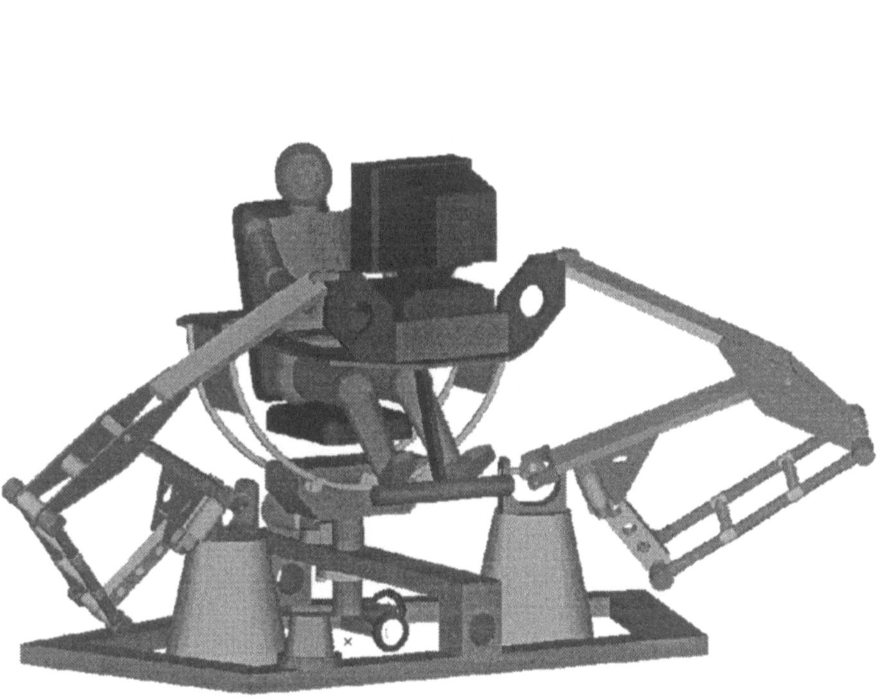

The Application of Dual Algebra to Kinematic Analysis

Jorge Angeles

Department of Mechanical Engineering & McGill Centre for Intelligent Machines, McGill University, 817 Sherbrooke St. W., Montreal, Quebec, Canada H3A 2K6
angeles@cim.mcgill.ca

Summary. The algebra of dual quantities is applied in this chapter to kinematic analysis. The fundamental problem of determining the parameters of both the finite screw and the instant screw of a rigid body under general motion is used to illustrate the power of dual algebra in kinematics. It is shown that the calculation of these parameters reduces essentially to a sum of the diagonal entries and to the differences of the off-diagonal entries of the associated dual matrix.

Keywords. Dual algebra, kinematic analysis, screw parameters.

1. Introduction

The aim of this chapter is to outline the applications of dual algebra to kinematic analysis. To this end, the algebra of dual scalars, vectors, and matrices is first recalled. The applications included here refer to the computation of the parameters of the screw of a rigid body between two finitely-separated positions and of the instant screw. However, the applications of dual numbers go beyond that in kinematics. Indeed, the well-known *Principle of Transference* (Dimentberg 1965; Bottema and Roth 1978; Rico Martínez and Duffy 1993) has been found extremely useful in spatial kinematics, since it allows the derivation of spatial kinematic relations by simply *dualizing* the corresponding relations of spherical kinematics.

Dual numbers were first proposed by Clifford (1873), their first applications to kinematics being attributed to both Kotel'nikov (1895) and Study (1903). A comprehensive analysis of dual numbers and their applications to the kinematic analysis of spatial linkages was conducted by Yang (1963) and Yang and Freudenstein (1964). Bottema and Roth (1978) include a treatment of theoretical kinematics using dual numbers. More recently, Agrawal (1987) reported on the application of dual quaternions to spatial kinematics, while Pradeep, Yoder, and Mukundan (1989) used the dual-matrix exponential in the analysis of robotic manipulators. Shoham and Brodsky (1993, 1994) have proposed a dual inertia operator for the dynamical analysis of mechanical systems. A comprehensive introduction to dual quaternions is to be found in (McCarthy 1990), while an abstract treatment is found in (Chevallier 1991.)

2. Definitions

A *dual number* $\hat{a}$ is defined as the sum of a *primal* part a, and a *dual* part a_0, namely,

$$\hat{a} = a + \epsilon a_0 \ , \tag{1}$$

where ϵ is the dual unity, which verifies $\epsilon \neq 0$, $\epsilon^2 = 0$, while a and a_0 are real numbers, the former being the *primal part* of $\hat{a}$, the latter its *dual part*. Actually, dual numbers with complex parts can be equally defined (Cheng and Thompson 1996). For the purposes of this chapter, real numbers will suffice.

If $a_0 = 0$, $\hat{a}$ is called a *real number*, or, correspondingly, a *complex number*; if $a = 0$, $\hat{a}$ is called a *pure dual number*; and if neither is zero $\hat{a}$ is called a *proper dual number*.

Let $\hat{b} = b + b_0$ be another dual number. Equality, addition, multiplication, and division are defined, respectively, as

$$\hat{a} = \hat{b} \quad - \quad a = b, \quad a_0 = b_0 \tag{2a}$$

$$\hat{a} + \hat{b} \quad = \quad (a + b) + \epsilon(a_0 + b_0) \tag{2b}$$

$$\hat{a}\hat{b} \quad = \quad ab + \epsilon(ab_0 + a_0 b) \tag{2c}$$

$$\frac{\hat{a}}{\hat{b}} \quad = \quad \frac{a}{b} - \epsilon\left(\frac{ab_0 - a_0 b}{b^2}\right), \quad b \neq 0 \ . \tag{2d}$$

From eq.(2d) it is apparent that the division by a pure dual number is not defined. Hence, dual numbers do not form a *field* in the algebraic sense; they do form a *ring* (Simmons, 1963).

All formal operations involving dual numbers are identical to those of ordinary algebra, while taking into account that $\epsilon^2 = \epsilon^3 = \cdots = 0$. Therefore, the series expansion of the *analytic function* $f(\hat{x})$ of a dual argument $\hat{x}$ is given by

$$f(\hat{x}) = f(x + \epsilon x_0) = f(x) + \epsilon x_0 \frac{\mathrm{d}f(x)}{\mathrm{d}x} \ . \tag{3}$$

As a direct consequence of eq.(3), we have the expression below for the exponential of a dual number $\hat{x}$:

$$\mathrm{e}^{\hat{x}} = \mathrm{e}^{x} + \epsilon\, x_0 \mathrm{e}^{x} = \mathrm{e}^{x}(1 + \epsilon\, x_0) \ , \tag{4}$$

and hence, *the dual exponential cannot be a pure dual number.*

The *dual angle* $\hat{\theta}$ between two skew lines $\mathcal{L}_1$ and $\mathcal{L}_2$, introduced by Study (1903), is defined as

$$\hat{\theta} = \theta + \epsilon s \ , \tag{5}$$

where θ and s are, respectively, the twist angle and the distance between the two lines. The *dual trigonometric functions* of the dual angle $\hat{\theta}$ are derived directly from eq.(3), namely,

$$\cos\hat{\theta} = \cos\theta - \epsilon s \sin\theta, \quad \sin\hat{\theta} = \sin\theta + \epsilon s \cos\theta, \quad \tan\hat{\theta} = \tan\theta + \epsilon s \sec^2\theta \,. \tag{6}$$

Moreover, all identities for ordinary trigonometry hold for dual angles. Likewise, the square root of a dual number can be readily found by a straightforward application of eq.(3), namely,

$$\sqrt{\hat{x}} = \sqrt{x} + \epsilon \frac{x_0}{2\sqrt{x}} \,, \tag{7}$$

A *dual vector* $\hat{\mathbf{a}}$ is defined as the sum of a primal vector part $\mathbf{a}$, and a dual vector part $\mathbf{a}_0$, namely,

$$\hat{\mathbf{a}} = \mathbf{a} + \epsilon \mathbf{a}_0 \,, \tag{8}$$

where both $\mathbf{a}$ and $\mathbf{a}_0$ are Cartesian, 3-dimensional vectors. Henceforth, all vectors are assumed to be of this kind. Further, let $\hat{\mathbf{a}}$ and $\hat{\mathbf{b}}$ be two dual vectors and $\hat{c}$ be a dual scalar. The concepts of dual-vector equality, multiplication of a dual vector by a dual scalar, inner product and vector product of two dual vectors are defined below:

$$\hat{\mathbf{a}} = \hat{\mathbf{b}} \quad - \quad \mathbf{a} = \mathbf{b} \quad \text{and} \quad \mathbf{a}_0 = \mathbf{b}_0 \,; \tag{9a}$$

$$\hat{c}\hat{\mathbf{a}} = c\mathbf{a} + \epsilon(c_0\mathbf{a} + c\mathbf{a}_0) \,; \tag{9b}$$

$$\hat{\mathbf{a}} \cdot \hat{\mathbf{b}} = \mathbf{a} \cdot \mathbf{b} + \epsilon(\mathbf{a} \cdot \mathbf{b}_0 + \mathbf{a}_0 \cdot \mathbf{b}) \,; \tag{9c}$$

$$\hat{\mathbf{a}} \times \hat{\mathbf{b}} = \mathbf{a} \times \mathbf{b} + \epsilon(\mathbf{a} \times \mathbf{b}_0 + \mathbf{a}_0 \times \mathbf{b}) \,. \tag{9d}$$

In particular, when $\hat{\mathbf{b}} = \hat{\mathbf{a}}$, eq.(9c) leads to the *Euclidean norm* of the dual vector $\hat{\mathbf{a}}$, i.e.,

$$\|\hat{\mathbf{a}}\|^2 = \|\mathbf{a}\|^2 + \epsilon\, 2\mathbf{a} \cdot \mathbf{a}_0 \,. \tag{9e}$$

Furthermore, the six *normalized Plücker coordinates* of a line $\mathcal{L}$ passing through a point P of position vector $\mathbf{p}$ and parallel to the unit vector $\mathbf{e}$ are given by the pair $(\mathbf{e}, \mathbf{p} \times \mathbf{e})$, where the product $\mathbf{e}_0 \equiv \mathbf{p} \times \mathbf{e}$ denotes the *moment* of the line. The foregoing coordinates can be represented by a *dual unit vector* $\hat{\mathbf{e}}^*$, whose six real components in $\mathbf{e}$ and $\mathbf{e}_0$ are the Plücker coordinates of $\mathcal{L}$, namely,

$$\hat{\mathbf{e}}^* = \mathbf{e} + \epsilon \mathbf{e}_0, \quad \text{with} \quad \|\mathbf{e}\| = 1 \quad \text{and} \quad \mathbf{e} \cdot \mathbf{e}_0 = 0 \,. \tag{10}$$

The reader is invited to verify the results summarized below:

Lemma 2.1. *For* $\hat{\mathbf{e}}^* \equiv \mathbf{e} + \epsilon \mathbf{e}_0$ *and* $\hat{\mathbf{f}}^* \equiv \mathbf{f} + \epsilon \mathbf{f}_0$ *defined as two dual unit vectors representing lines* $\mathcal{L}$ *and* $\mathcal{M}$*, respectively, we have:*

(*i*) *If* $\hat{\mathbf{e}}^* \times \hat{\mathbf{f}}^*$ *is a pure dual vector, then* $\mathcal{L}$ *and* $\mathcal{M}$ *are parallel;*
(*ii*) *if* $\hat{\mathbf{e}}^* \cdot \hat{\mathbf{f}}^*$ *is a pure dual number, then* $\mathcal{L}$ *and* $\mathcal{M}$ *are perpendicular;*
(*iii*) $\mathcal{L}$ *and* $\mathcal{M}$ *are coincident if and only if* $\hat{\mathbf{e}}^* \times \hat{\mathbf{f}}^* = \mathbf{0}$*; and*
(*iv*) $\mathcal{L}$ *and* $\mathcal{M}$ *intersect at right angles if and only if* $\hat{\mathbf{e}}^* \cdot \hat{\mathbf{f}}^* = 0$.

Dual matrices can be defined likewise, i.e., if $\mathbf{A}$ and $\mathbf{A}_0$ are two real $n \times n$ matrices, then the dual $n \times n$ matrix $\hat{\mathbf{A}}$ is defined as

$$\hat{\mathbf{A}} \equiv \mathbf{A} + \epsilon \mathbf{A}_0 . \tag{11}$$

We will work with 3×3 matrices in connection with dual vectors, but the above definition can be applied to any square matrices, which is the reason why n has been left arbitrary. Equality, multiplication by a dual scalar, and multiplication by a dual vector are defined as in the foregoing cases. Moreover, matrix multiplication is defined correspondingly, but then the order of multiplication must be respected. We thus have that, if $\hat{\mathbf{A}}$ and $\hat{\mathbf{B}}$ are two $n \times n$ dual matrices, with their primal and dual parts self-understood, then

$$\hat{\mathbf{A}}\hat{\mathbf{B}} = \mathbf{AB} + \epsilon(\mathbf{AB}_0 + \mathbf{A}_0\mathbf{B}) . \tag{12}$$

Therefore, matrix $\hat{\mathbf{A}}$ is real if $\mathbf{A}_0 = \mathbf{O}$, where $\mathbf{O}$ denotes the $n \times n$ zero matrix; if $\mathbf{A} = \mathbf{O}$, then $\hat{\mathbf{A}}$ is called a *pure dual matrix*. Moreover, as we shall see below, a square dual matrix admits an inverse if and only if its primal part is nonsingular.

Now we can define the inverse of a dual matrix, if this is nonsingular. Indeed, it suffices to make $\hat{\mathbf{B}} = \hat{\mathbf{A}}^{-1}$ in eq.(12) and the right-hand side of this equation equal to the $n \times n$ identity matrix, $\mathbf{1}$, thereby obtaining two matrix equations that allow us to find the primal and the dual parts of $\hat{\mathbf{A}}^{-1}$, namely,

$$\mathbf{AB} = \mathbf{1}, \quad \mathbf{AB}_0 + \mathbf{A}_0\mathbf{B} = \mathbf{O} ,$$

whence

$$\mathbf{B} = \mathbf{A}^{-1}, \quad \mathbf{B}_0 = -\mathbf{A}^{-1}\mathbf{A}_0\mathbf{A}^{-1} ,$$

which are defined because $\mathbf{A}$ is invertible by hypothesis, and hence, for any nonsingular dual matrix $\hat{\mathbf{A}}$,

$$\hat{\mathbf{A}}^{-1} = \mathbf{A}^{-1} - \epsilon \mathbf{A}^{-1}\mathbf{A}_0\mathbf{A}^{-1} . \tag{13}$$

Note the striking similarity of the dual part of the foregoing expression with the time-derivative of the inverse of $\mathbf{A}(t)$, namely,

$$\frac{\mathrm{d}}{\mathrm{d}t}[\mathbf{A}^{-1}(t)] = -\mathbf{A}^{-1}(t)\dot{\mathbf{A}}(t)\mathbf{A}^{-1}(t) .$$

In order to find an expression for the determinant of an $n \times n$ dual matrix, we need to recall the general expression for the dual function defined in eq.(3). However, that expression has to be adapted to a dual-matrix argument, which leads to

$$f(\hat{\mathbf{A}}) = f(\mathbf{A}) + \epsilon \operatorname{tr}\left[\mathbf{A}_0 \left(\frac{\mathrm{d}f}{\mathrm{d}\hat{\mathbf{A}}}\right)^T\right]\Bigg|_{\hat{\mathbf{A}}=\mathbf{A}} . \tag{14}$$

In particular, when $f(\hat{\mathbf{A}}) = \det(\hat{\mathbf{A}})$, we have, recalling the formula for the derivative of the determinant with respect to its matrix argument (Angeles 1982), for any $n \times n$ matrix $\mathbf{X}$,

$$\frac{\mathrm{d}}{\mathrm{d}\mathbf{X}}[\det(\mathbf{X})] = \det(\mathbf{X})\mathbf{X}^{-T} \, ,$$

where $\mathbf{X}^{-T}$ denotes the transpose of the inverse of $\mathbf{X}$ or, equivalently, the transpose of $\mathbf{X}^{-1}$. Therefore,

$$\mathrm{tr}\left[\mathbf{A}_0\left(\frac{\mathrm{d}f}{\mathrm{d}\hat{\mathbf{A}}}\right)^T\right]\Bigg|_{\hat{\mathbf{A}}=\mathbf{A}} = \det(\mathbf{A})\mathrm{tr}(\mathbf{A}_0\mathbf{A}^{-1}) \, ,$$

and hence,

$$\det(\hat{\mathbf{A}}) = \det(\mathbf{A})[1 + \epsilon\, \mathrm{tr}(\mathbf{A}_0\mathbf{A}^{-1})] \, . \tag{15}$$

Now we can define the eigenvalue problem for the dual matrix $\hat{\mathbf{A}}$ defined above. Let $\hat{\lambda}$ and $\hat{\mathbf{e}}$ be a dual eigenvalue and a dual (unit) eigenvector of $\hat{\mathbf{A}}$, respectively. Then,

$$\hat{\mathbf{A}}\hat{\mathbf{e}} = \hat{\lambda}\hat{\mathbf{e}}, \quad \|\hat{\mathbf{e}}\| = 1 \, . \tag{16a}$$

For the foregoing linear homogeneous equation to admit a nontrivial solution, we must have

$$\det(\hat{\lambda}\mathbf{1} - \hat{\mathbf{A}}) = 0 \, , \tag{16b}$$

which yields an nth-order dual polynomial in the dual number $\hat{\lambda}$. Its n dual roots, real and complex, constitute the n *dual eigenvalues* of $\hat{\mathbf{A}}$. Note that, associated with each dual eigenvalue $\hat{\lambda}_i$, a corresponding *dual (unit) eigenvector* $\hat{\mathbf{e}}_i^*$ is defined, for $i = 1, 2, \ldots, n$. Moreover, if we recall eq.(4), we can write

$$\mathrm{e}^{\hat{\mathbf{A}}} = \mathrm{e}^{\mathbf{A}} + \epsilon\, \mathbf{A}_0\mathrm{e}^{\mathbf{A}} \, . \tag{17}$$

Upon expansion, the foregoing expression can be cast in the form

$$\mathrm{e}^{\hat{\mathbf{A}}} = (\mathbf{1} + \epsilon\, \mathbf{A}_0)\mathrm{e}^{\mathbf{A}} \neq \mathrm{e}^{\mathbf{A}}(\mathbf{1} + \epsilon\, \mathbf{A}_0) \, , \tag{18}$$

the inequality arising because, in general, $\mathbf{A}$ and $\mathbf{A}_0$ do not commute. They do so only in the case in which they share the same set of eigenvectors. A special case in which the two matrices share the same set of eigenvectors is when one matrix is an *analytic function* of the other. More formally, we have

Lemma 2.2. *If* $\mathbf{F}$ *is an analytic matrix function of matrix* $\mathbf{A}$*, then the two matrices*

(i) share the same set of eigenvectors, and
(ii) commute under multiplication.

Typical examples of analytic matrix functions are $\mathbf{F} = \mathbf{A}^N$ and $\mathbf{F} = \mathrm{e}^{\mathbf{A}}$, for an integer N.

3. Fundamentals of Rigid-Body Kinematics

We review in this section some basic facts from rigid-body kinematics. For the sake of conciseness, some proofs are not given, but the pertinent references are cited whenever necessary.

3.1 Finite Displacements

A rigid body is understood as a particular case of the continuum with the special feature that, under any given motion, *any* two points of the rigid body remain equidistant. A rigid body is available through a *configuration* or *pose* that will be denoted by $\mathcal{B}$. Whenever a *reference configuration* is needed, this will be labelled $\mathcal{B}^0$. Moreover, the position vector of a point P of the body in configuration $\mathcal{B}$ will be denoted by $\mathbf{p}$, that in $\mathcal{B}^0$ being denoted correspondingly by $\mathbf{p}^0$.

A rigid-body motion leaving a point O of the body fixed is called a *pure rotation*, and is represented by a *proper orthogonal matrix* $\mathbf{Q}$, i.e., $\mathbf{Q}$ verifies the two properties below:

$$\mathbf{Q}\mathbf{Q}^T = \mathbf{1}, \quad \det(\mathbf{Q}) = +1 . \tag{19}$$

According to Euler's Theorem (Euler 1776), a pure rotation leaves a set of points of the body immutable, this set lying on a line $\mathcal{L}$, which is termed the *axis of rotation*. If we draw the perpendicular from an arbitrary point P of the body to $\mathcal{L}$ and denote its intersection with $\mathcal{L}$ by P', the angle ϕ between $P'P^0$ and $P'P$, where, according to our convention, P^0 denotes the point P in the reference configuration $\mathcal{B}^0$ of the body, is called the *angle of rotation*. Note that a direction must be specified along this line to define the sign of the angle. Furthermore, the direction of the line is specified by the unit vector $\mathbf{e}$. We term $\mathbf{e}$ and ϕ the *natural invariants* of $\mathbf{Q}$.

As a result of Euler's Theorem, the rotation $\mathbf{Q}$ can be represented in terms of its natural invariants. This representation takes the form

$$\mathbf{Q} = \mathbf{e}\mathbf{e}^T + \cos\phi(\mathbf{1} - \mathbf{e}\mathbf{e}^T) + \sin\phi\mathbf{E} , \tag{20}$$

where $\mathbf{E}$ denotes the *cross-product matrix* of $\mathbf{e}$, i.e., for any 3-dimensional vector $\mathbf{v}$,

$$\mathbf{e} \times \mathbf{v} = \mathbf{E}\mathbf{v} .$$

As a result of the foregoing definition, $\mathbf{E}$ is skew-symmetric, i.e., $\mathbf{E} = -\mathbf{E}^T$ and, moreover, it has the properties below:

$$\mathbf{E}^{2k+1} = (-1)^k\mathbf{E}, \quad \mathbf{E}^{2k} = (-1)^k(\mathbf{1} - \mathbf{e}\mathbf{e}^T), \quad \text{for} \quad k = 1, 2, \ldots$$

By virtue of the foregoing properties of the cross-product matrix $\mathbf{E}$ of $\mathbf{e}$, the rotation matrix $\mathbf{Q}$ can be written in the alternative form

$$\mathbf{Q} = \mathbf{1} + \sin\phi\mathbf{E} + (1 - \cos\phi)\mathbf{E}^2 . \tag{21}$$

Now, if we recall the *Cayley-Hamilton Theorem* (Halmos 1974), we can realize that the right-hand side of the foregoing equation is nothing but the exponential of $\phi\mathbf{E}$, i.e.,

$$\mathbf{Q} = e^{\phi\mathbf{E}} , \tag{22}$$

which is the *exponential form of the rotation matrix.* Now it is a simple matter to obtain the eigenvalues of the rotation matrix if we first notice that one eigenvalue of $\mathbf{E}$ is 0, the other eigenvalues being readily derived as $\pm$i, where i is the imaginary unit, i.e., $\mathrm{i} \equiv \sqrt{-1}$. Therefore, if $\mathbf{Q}$ is the exponential of $\phi\mathbf{E}$, then the eigenvalues of $\mathbf{Q}$ are the exponentials of the eigenvalues of $\phi\mathbf{E}$:

$$\lambda_1 = e^0 = 1, \quad \lambda_{2,3} = e^{\pm \mathrm{i}\phi} = \cos\phi \pm \mathrm{i}\sin\phi . \tag{23}$$

Moreover, we recall below the *Cartesian decomposition* of an $n \times n$ matrix $\mathbf{A}$, namely,

$$\mathbf{A} = \mathbf{A}_s + \mathbf{A}_{ss} , \tag{24a}$$

where $\mathbf{A}_s$ is symmetric and $\mathbf{A}_{ss}$ is skew-symmetric. These matrices are given by

$$\mathbf{A}_s \equiv \frac{1}{2}(\mathbf{A} + \mathbf{A}^T), \quad \mathbf{A}_{ss} \equiv \frac{1}{2}(\mathbf{A} - \mathbf{A}^T) . \tag{24b}$$

Any 3×3 skew-symmetric matrix is fully defined by three scalars, which means that such a matrix can then be made isomorphic to a 3-dimensional vector. Indeed, let $\mathbf{S}$ be a 3×3 skew-symmetric matrix and $\mathbf{v}$ be an arbitrary 3-dimensional vector. Then, we have

$$\mathbf{S}\mathbf{v} \equiv \mathbf{s} \times \mathbf{v} . \tag{25}$$

When the above items are expressed in a given coordinate frame $\mathcal{F}$, the components of $\mathbf{S}$, indicated as $\{ s_{i,j} \}_{i,j=1}^3$, and of $\mathbf{s}$, indicated as $\{ s_i \}_1^3$, bear the relations below:

$$\mathbf{S} = \begin{bmatrix} 0 & -s_3 & s_2 \\ s_3 & 0 & -s_1 \\ -s_2 & s_1 & 0 \end{bmatrix} , \quad \mathbf{s} = \frac{1}{2} \begin{bmatrix} s_{32} - s_{23} \\ s_{13} - s_{31} \\ s_{21} - s_{12} \end{bmatrix} . \tag{26}$$

In general, we define the *axial vector* of an arbitrary 3×3 matrix $\mathbf{A}$ in terms of the difference of its off-diagonal entries, as appearing in eq.(26) for the entries of matrix $\mathbf{S}$. Apparently, the axial vector of any 3×3 matrix is identical to that of its skew-symmetric component; this vector, represented as $\mathbf{a} \equiv \mathrm{vect}(\mathbf{A})$, is the *vector linear invariant* of $\mathbf{A}$. The *scalar linear invariant* of the same matrix is its trace, $\mathrm{tr}(\mathbf{A})$. With this notation, note that

$$\frac{1}{2}(\mathbf{A} - \mathbf{A}^T)\mathbf{v} = \mathbf{a} \times \mathbf{v} .$$

Further, with reference to Fig. 1, let A and P be two points of a rigid body, which is shown in its reference and its current configurations.

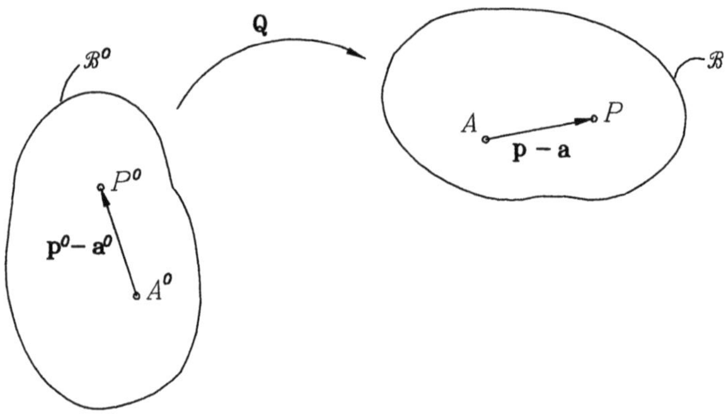

Fig. 1. Displacements of two points of a rigid body in two finitely-separated configurations

We can regard vector $\mathbf{p} - \mathbf{a}$ as the image of $\mathbf{p}^0 - \mathbf{a}^0$ under the rotation $\mathbf{Q}$, namely,

$$\mathbf{p} - \mathbf{a} = \mathbf{Q}(\mathbf{p}^0 - \mathbf{a}^0) , \tag{27}$$

whence an expression for $\mathbf{p}$ can be derived as

$$\mathbf{p} = \mathbf{a} + \mathbf{Q}(\mathbf{p}^0 - \mathbf{a}^0) . \tag{28}$$

Furthermore, the *displacement* $\mathbf{d}_A$ of A is defined as the difference $\mathbf{a}-\mathbf{a}^0$, with a similar definition for the displacement $\mathbf{d}_P$ of P. From the above equation, it is now apparent that a linear relation between the two displacements follows:

$$\mathbf{d}_P = \mathbf{d}_A + (\mathbf{Q} - \mathbf{1})(\mathbf{p}^0 - \mathbf{a}^0) . \tag{29}$$

Therefore,

Theorem 3.1. *The displacements of all the points of a rigid body have identical projections onto the axis of the concomitant rotation.*

The proof of the foregoing result follows upon dot-multiplying both sides of eq.(29) by $\mathbf{e}$:

$$\mathbf{e} \cdot \mathbf{d}_P = \mathbf{e} \cdot \mathbf{d}_A .$$

From the previous result it is apparent that $\|\mathbf{d}_P\|$ can attain infinitely large values, depending on $\|\mathbf{p}^0 - \mathbf{a}^0\|$, but, in general, $\mathbf{d}_P$ does not vanish. Hence, a minimum of $\|\mathbf{d}_P\|$ can be found, a result summarized in the Mozzi-Chasles Theorem (Mozzi 1763; Chasles 1830). This theorem states that the points of $\mathcal{B}$ of minimum-norm displacement lie in a line $\mathcal{M}$ that is parallel to the axis of the rotation represented by matrix $\mathbf{Q}$, the minimum-norm displacement being a vector parallel to the same axis. If we recall that $\mathbf{e}$ and ϕ denote the natural invariants of $\mathbf{Q}$, then the position vector $\mathbf{p}^*$ of the point P^* of $\mathcal{M}$ lying closest to the origin O is given by (Angeles 1997)

$$\mathbf{p}^* = \frac{(\mathbf{Q}-\mathbf{1})^T(\mathbf{Q}\mathbf{a}^0-\mathbf{a})}{2(1-\cos\phi)}, \quad \text{for} \quad \phi \neq 0\,, \tag{30}$$

the special case in which $\phi = 0$ corresponding to a *pure translation*, whereby all points of $\mathcal{B}$ undergo identical displacements. In this case, then, the axis $\mathcal{M}$ is indeterminate, because all points of the body can be thought of as undergoing minimum-norm displacements. Henceforth, line $\mathcal{M}$ will be termed the *Mozzi-Chasles axis.* Note that the Plücker coordinates of the Mozzi-Chasles axis are given by $\mathbf{e}$ and $\mathbf{e}_0 \equiv \mathbf{p}^* \times \mathbf{e}$. We shall denote with $\mathbf{d}^*$ the minimum-norm displacement, which can be represented in the form

$$\mathbf{d}^* = d^*\mathbf{e}, \quad d^* = \mathbf{d}_P \cdot \mathbf{e}\,. \tag{31}$$

Therefore, the body under study can be regarded as undergoing, from $\mathcal{B}^0$ to $\mathcal{B}$, a screw motion, as if the body were rigidly attached to the bolt of a screw of axis $\mathcal{M}$ and *pitch* p given by

$$p = \frac{d^*}{\phi} = \frac{\mathbf{e}\cdot\mathbf{d}_P}{\phi}\,. \tag{32}$$

We list below further results:

Lemma 3.1. *Let A and P be two points of a rigid body undergoing a general motion from a reference pose $\mathcal{B}^0$ to a current pose $\mathcal{B}$. Then, under the notation adopted above, the difference $\mathbf{p}-\mathbf{Q}\mathbf{p}^0$ remains constant and is denoted by $\mathbf{d}$, i.e.,*

$$\mathbf{p}-\mathbf{Q}\mathbf{p}^0 = \mathbf{a}-\mathbf{Q}\mathbf{a}^0 = \mathbf{d}\,. \tag{33}$$

Proof: If we recall eq.(28) and substitute the expression therein for $\mathbf{p}$ in the difference $\mathbf{p}-\mathbf{Q}\mathbf{p}^0$, we obtain

$$\mathbf{p}-\mathbf{Q}\mathbf{p}^0 = \mathbf{a}+\mathbf{Q}(\mathbf{p}^0-\mathbf{a}^0)-\mathbf{Q}\mathbf{p}^0 = \mathbf{a}-\mathbf{Q}\mathbf{a}^0 = \mathbf{d}\,,$$

thereby completing the intended proof.

Note that the kinematic interpretation of $\mathbf{d}$ follows directly from eq.(33): $\mathbf{d}$ represents the displacement of the point of $\mathcal{B}$ that coincides with the origin in the reference pose $\mathcal{B}^0$.

The geometric interpretation of the foregoing lemma is given in Fig. 2. What this figure indicates is that the pose $\mathcal{B}$ can be attained from $\mathcal{B}^0$ in two stages: (a) first, the body is given a rotation $\mathbf{Q}$ about the origin O, that takes the body to the intermediate pose $\mathcal{B}'$; (b) then, from $\mathcal{B}'$, the body is given a pure translation of displacement $\mathbf{d}$ that takes the body into $\mathcal{B}$.

Therefore, eq.(30) for the position vector of the point of the Mozzi-Chasles axis lying closest to the origin can be expressed in terms of vector $\mathbf{d}$ as

$$\mathbf{p}^* = \frac{(\mathbf{1}-\mathbf{Q})^T\mathbf{d}}{2(1-\cos\phi)}, \quad \text{for} \quad \phi \neq 0\,. \tag{34}$$

Note that, in general, $\mathbf{d}$ is not of minimum norm. Additionally, $\mathbf{d}$ is origin-dependent, and hence, is not an invariant of the motion under study. Now, if

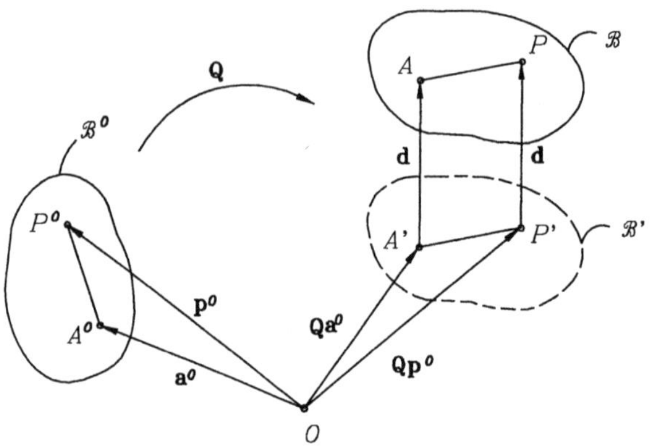

Fig. 2. Geometric interpretation of Lemma 3.1

we choose the origin on the Mozzi-Chasles axis $\mathcal{M}$, then we have the layout of Fig. 3, and vector $\mathbf{d}$ becomes a multiple of $\mathbf{e}$, namely, $\mathbf{d} = d^*\mathbf{e}$.

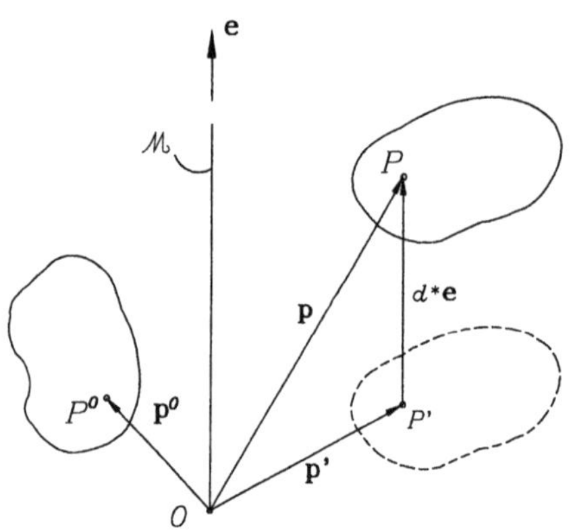

Fig. 3. Rigid-body displacement with origin on the Mozzi-Chasles axis

We can now express the Plücker coordinates of a line $\mathcal{L}$ of a rigid body $\mathcal{B}$ in terms of those of the line in its reference configuration $\mathcal{L}^0$ (Bottema and Roth 1978; Pradeep, Yoder, and Mukundan 1989), as shown in Fig. 4. To this end, we let $\mathbf{f}$ be the unit vector parallel to $\mathcal{L}$ and P be a point of $\mathcal{L}$, and arrange the Plücker coordinates of $\mathcal{L}^0$ and $\mathcal{L}$ in the 6-dimensional arrays $\boldsymbol{\lambda}^0$ and $\boldsymbol{\lambda}$, respectively, defined as

$$\boldsymbol{\lambda}^0 \equiv \begin{bmatrix} \mathbf{f}^0 \\ \mathbf{p}^0 \times \mathbf{f}^0 \end{bmatrix}, \quad \boldsymbol{\lambda} \equiv \begin{bmatrix} \mathbf{f} \\ \mathbf{p} \times \mathbf{f} \end{bmatrix}. \tag{35}$$

We thus have

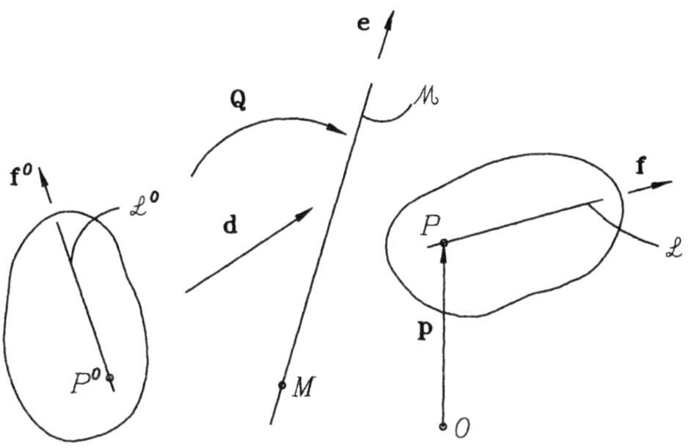

Fig. 4. The reference and the current configurations of a body and one of its lines

$$\mathbf{f} = \mathbf{Q}\mathbf{f}^0, \quad \mathbf{p} = \mathbf{Q}\mathbf{p}^0 + \mathbf{d}\,,$$

and hence,

$$\mathbf{p} \times \mathbf{f} = (\mathbf{Q}\mathbf{p}^0 + \mathbf{d}) \times \mathbf{Q}\mathbf{f}^0 = (\mathbf{Q}\mathbf{p}^0) \times \mathbf{Q}\mathbf{f} + \mathbf{d} \times \mathbf{Q}\mathbf{f}^0\,.$$

Now, the first term of the rightmost-hand side of the above equation can be simplified upon noticing that the cross product of two rotated vectors is identical to the rotated cross product. Furthermore, the second term of the same side can be expressed in terms of $\mathbf{D}$, the cross-product matrix of $\mathbf{d}$, thereby obtaining

$$\mathbf{p} \times \mathbf{f} = \mathbf{Q}(\mathbf{p}^0 \times \mathbf{f}^0) + \mathbf{D}\mathbf{Q}\mathbf{f}^0\,.$$

Upon substituting the foregoing expressions for $\mathbf{f}$ and $\mathbf{p} \times \mathbf{f}$ into eq.(35), we obtain

$$\boldsymbol{\lambda} = \begin{bmatrix} \mathbf{Q}\mathbf{f}^0 \\ \mathbf{D}\mathbf{Q}\mathbf{f}^0 + \mathbf{Q}(\mathbf{p}^0 \times \mathbf{f}^0) \end{bmatrix}\,,$$

which can be readily cast in the form of a linear transformation of $\boldsymbol{\lambda}^0$, i.e.,

$$\begin{bmatrix} \mathbf{f} \\ \mathbf{p} \times \mathbf{f} \end{bmatrix} = \begin{bmatrix} \mathbf{Q} & \mathbf{O} \\ \mathbf{D}\mathbf{Q} & \mathbf{Q} \end{bmatrix} \begin{bmatrix} \mathbf{f}^0 \\ \mathbf{p}^0 \times \mathbf{f}^0 \end{bmatrix}\,, \tag{36a}$$

where $\mathbf{O}$ denotes the 3×3 zero matrix.

As the reader can readily verify, the inverse relation of eq.(36a) takes the form

$$\begin{bmatrix} \mathbf{f}^0 \\ \mathbf{p}^0 \times \mathbf{f}^0 \end{bmatrix} = \begin{bmatrix} \mathbf{Q}^T & \mathbf{O} \\ -\mathbf{Q}^T\mathbf{D} & \mathbf{Q}^T \end{bmatrix} \begin{bmatrix} \mathbf{f} \\ \mathbf{p} \times \mathbf{f} \end{bmatrix}\,. \tag{36b}$$

By inspection of eq.(36a), and recalling the dual-unit-vector representation of a line, as given in eq.(10), we can realize that the dual unit vector of $\mathcal{L}$ can be expressed as the image of the dual unit vector of $\mathcal{L}^0$ upon a linear transformation given by a dual matrix $\hat{\mathbf{Q}}$. Moreover, the dual matrix of interest can be readily derived from the real matrix of eq.(36a). Indeed, it can be realized from Section 2 that the difference between the primal and the dual parts of a dual quantity is that the units of the dual part are those of the primal part times units of length. Hence, the primal part of the dual matrix sought is bound to be $\mathbf{Q}$, which is dimensionless, the corresponding dual part being $\mathbf{DQ}$, which has units of length. A plausible form of the matrix sought is, then,

$$\hat{\mathbf{Q}} = \mathbf{Q} + \epsilon\,\mathbf{DQ}\,. \tag{37}$$

The correctness of the above expression can be readily realized. Indeed, let $\hat{\mathbf{f}}^* = \mathbf{f} + \epsilon\,\mathbf{p} \times \mathbf{f}$ and $\hat{\mathbf{f}}^{0*} = \mathbf{f}^0 + \epsilon\,\mathbf{p}^0 \times \mathbf{f}^0$ be the dual unit vectors of $\mathcal{L}$ and $\mathcal{L}^0$, respectively. Then upon performing the product $\hat{\mathbf{Q}}\hat{\mathbf{f}}^{0*}$, we note that the product is rightfully $\hat{\mathbf{f}}^*$, i.e., $\hat{\mathbf{f}}^* = \hat{\mathbf{Q}}\hat{\mathbf{f}}^{0*}$. In the derivations below, we will need expressions for the vector and scalar linear invariants of the product of two matrices, one of which is skew-symmetric. These expressions are derived in detail in (Angeles 1997). For quick reference, we recall these relations below:

Theorem 3.2. *Let both* $\mathbf{A}$ *and* $\mathbf{S}$ *be* 3×3 *matrices, the former arbitrary, the latter skew-symmetric. Then,*

$$\operatorname{vect}(\,\mathbf{SA}\,) = \frac{1}{2}[\operatorname{tr}(\mathbf{A})\mathbf{1} - \mathbf{A}]\mathbf{s}\,, \tag{38}$$

where $\mathbf{s} \equiv \operatorname{vect}(\,\mathbf{S}\,)$.

Now, as a direct consequence of the above result, we have

Corollary 3.1. *If* $\mathbf{A}$ *in Theorem 3.2 is skew-symmetric, then the axial vector of the product* $\mathbf{SA}$ *reduces to*

$$\operatorname{vect}(\,\mathbf{SA}\,) = -\frac{1}{2}\mathbf{As} = -\frac{1}{2}\mathbf{a} \times \mathbf{s}\,, \tag{39}$$

where $\mathbf{a} \equiv \operatorname{vect}(\mathbf{A})$.

Moreover,

Theorem 3.3. *Let* $\mathbf{A}$*,* $\mathbf{S}$*, and* $\mathbf{s}$ *be defined as in Theorem 3.2. Then,*

$$\operatorname{tr}(\,\mathbf{SA}\,) = -2\mathbf{s} \cdot [\operatorname{vect}(\,\mathbf{A}\,)]\,. \tag{40}$$

Furthermore, we prove now that $\hat{\mathbf{Q}}$ is proper orthogonal. Indeed, orthogonality can be proven by performing the product $\hat{\mathbf{Q}}\hat{\mathbf{Q}}^T$ and noticing that this product yields the 3×3 identity matrix, i.e., $\hat{\mathbf{Q}}\hat{\mathbf{Q}}^T = \mathbf{1}$. Proper orthogonality is proven, in turn, upon application of formula (15) to matrix $\hat{\mathbf{Q}}$, as given by eq.(37), namely,

$$\det(\hat{\mathbf{Q}}) = \det(\mathbf{Q})[1 + \epsilon\operatorname{tr}(\mathbf{DQQ}^{-1})] = \det(\mathbf{Q})[1 + \epsilon\operatorname{tr}(\mathbf{D})] = 1\,,$$

thus completing the proof.

The exponential form of the dual rotation matrix can be obtained if we note that the exponential of a pure dual number $\hat{x} = \epsilon x_0$ reduces to

$$e^{\epsilon x_0} = 1 + \epsilon x_0 . \tag{41}$$

On the other hand, we can write

$$\hat{\mathbf{Q}} = (\mathbf{1} + \epsilon \mathbf{D})\mathbf{Q} . \tag{42}$$

In analogy with eq.(41), the foregoing expression takes the form

$$\hat{\mathbf{Q}} = e^{\epsilon \mathbf{D}}\mathbf{Q} .$$

Furthermore, if we recall the exponential form of $\mathbf{Q}$, as given in eq.(22), the foregoing expression simplifies to

$$\hat{\mathbf{Q}} = e^{\epsilon \mathbf{D}} e^{\phi \mathbf{E}} . \tag{43}$$

However, since $\mathbf{D}$ and $\mathbf{E}$ are unrelated, they do not share the same set of eigenvectors, and hence, they do not commute under multiplication, the foregoing expression thus not being further reducible to one single exponential. Nevertheless, if the origin is placed on the Mozzi-Chasles axis, as depicted in Fig. 3, then the dual rotation matrix becomes

$$\hat{\mathbf{Q}} = \mathbf{Q} + \epsilon\, d^* \mathbf{E}\mathbf{Q} , \tag{44}$$

where $d^*\mathbf{E}$ is, apparently, the cross-product matrix of vector $d^*\mathbf{e}$. Furthermore, the exponential form of the dual rotation matrix, eq.(43), then simplifies to $\hat{\mathbf{Q}} = e^{(\phi + \epsilon\, d^*)\mathbf{E}}$ or, if we let $\hat{\phi} = \phi + \epsilon\, d^*$, then we can write $\hat{\mathbf{Q}} = e^{\hat{\phi}\mathbf{E}}$.

3.2 Velocity Analysis

Upon differentiation with respect to time of both sides of eq.(27), we obtain

$$\dot{\mathbf{p}} - \dot{\mathbf{a}} = \dot{\mathbf{Q}}(\mathbf{p}^0 - \mathbf{a}^0) ,$$

and, if we solve for $(\mathbf{p}^0 - \mathbf{a}^0)$ from the equation mentioned above, we obtain

$$\dot{\mathbf{p}} - \dot{\mathbf{a}} = \dot{\mathbf{Q}}\mathbf{Q}^T(\mathbf{p} - \mathbf{a}) , \tag{45}$$

where $\dot{\mathbf{Q}}\mathbf{Q}^T$ is defined as the angular-velocity matrix of the motion under study, and is represented as $\mathbf{\Omega}$, namely,

$$\mathbf{\Omega} \equiv \dot{\mathbf{Q}}\mathbf{Q}^T . \tag{46a}$$

It can be readily proven that the foregoing matrix is skew-symmetric, i.e.,

$$\mathbf{\Omega}^T = -\mathbf{\Omega} . \tag{46b}$$

Moreover, the axial vector of $\mathbf{\Omega}$ is the angular-velocity vector $\boldsymbol{\omega}$:

$$\boldsymbol{\omega} = \mathrm{vect}(\mathbf{\Omega}) . \tag{46c}$$

We can now write eq.(45) in the form

$$\dot{\mathbf{p}} = \dot{\mathbf{a}} + \mathbf{\Omega}(\mathbf{p} - \mathbf{a}) = \dot{\mathbf{a}} + \boldsymbol{\omega} \times (\mathbf{p} - \mathbf{a}) \,, \tag{47}$$

whence,

$$\dot{\mathbf{p}} - \boldsymbol{\omega} \times \mathbf{p} = \dot{\mathbf{a}} - \boldsymbol{\omega} \times \mathbf{a} \equiv \mathbf{v}^0 = \text{const} \,. \tag{48}$$

Therefore, the difference $\dot{\mathbf{p}} - \boldsymbol{\omega} \times \mathbf{p}$ is the same for all points of a rigid body. The kinematic interpretation of this quantity is straightforward: If we rewrite $\mathbf{v}^0$ in the form $\mathbf{v}^0 = \dot{\mathbf{p}} + \boldsymbol{\omega} \times (-\mathbf{p})$, then we can readily realize that, $-\mathbf{p}$ being the vector directed from point P of the rigid body to the origin O, $\mathbf{v}^0$ represents the velocity of the point of the body that coincides instantaneously with the origin. Furthermore, we express $\mathbf{d}$, as given by eq.(33), in terms of the position vector of an arbitrary point P, $\mathbf{p}$, thus obtaining

$$\mathbf{d} = \mathbf{p} - \mathbf{Q}\mathbf{p}^0 \,. \tag{49}$$

Upon differentiation of the two sides of the above expression with respect to time, we obtain

$$\dot{\mathbf{d}} = \dot{\mathbf{p}} - \dot{\mathbf{Q}}\mathbf{p}^0 \,,$$

which can be readily expressed in terms of the current value of the position vector of P, by solving for $\mathbf{p}^0$ from eq.(49), namely,

$$\dot{\mathbf{d}} = \dot{\mathbf{p}} - \mathbf{\Omega}(\mathbf{p} - \mathbf{d}) \quad \text{or} \quad \dot{\mathbf{d}} - \boldsymbol{\omega} \times \mathbf{d} = \dot{\mathbf{p}} - \boldsymbol{\omega} \times \mathbf{p} \,, \tag{50}$$

and hence, the difference $\dot{\mathbf{d}} - \boldsymbol{\omega} \times \mathbf{d}$ is identical to the difference $\dot{\mathbf{p}} - \boldsymbol{\omega} \times \mathbf{p}$, i.e.,

$$\dot{\mathbf{d}} - \boldsymbol{\omega} \times \mathbf{d} = \mathbf{v}^0 \,. \tag{51}$$

Furthermore, upon dot-multiplying the two sides of eq.(48) by $\boldsymbol{\omega}$, we obtain an interesting result, namely,

$$\boldsymbol{\omega} \cdot \dot{\mathbf{p}} = \boldsymbol{\omega} \cdot \dot{\mathbf{a}} \,, \tag{52}$$

and hence,

Theorem 3.4. *The velocities of all points of a rigid body have the same projection onto the angular-velocity vector of the motion under study.*

Similar to the Mozzi-Chasles Theorem, we have now

Theorem 3.5. *Given a rigid body $\mathcal{B}$ under general motion, a set of its points, on a line $\mathcal{L}$, undergoes the identical minimum-magnitude velocity $\mathbf{v}^*$ parallel to the angular velocity $\boldsymbol{\omega}$.*

The Plücker coordinates of line $\mathcal{L}$, grouped in the 6-dimensional array $\boldsymbol{\lambda}$, are given as

$$\boldsymbol{\lambda} \equiv \begin{bmatrix} \mathbf{f} \\ \boldsymbol{\pi} \times \mathbf{f} \end{bmatrix}, \quad \mathbf{f} \equiv \frac{\boldsymbol{\omega}}{\|\boldsymbol{\omega}\|}, \quad \boldsymbol{\pi} \equiv \frac{\boldsymbol{\omega} \times \mathbf{v}^0}{\|\boldsymbol{\omega}\|^2} \,, \tag{53}$$

where $\mathbf{v}^0$ was previously introduced as the velocity of the point of $\mathcal{B}$ that coincides instantaneously with the origin. Line $\mathcal{L}$ is termed the *instant screw axis*–ISA, for brevity.

Thus, the instantaneous motion of $\mathcal{B}$ is defined by a screw of axis $\mathcal{L}$ and pitch p', given by

$$p' = \frac{\dot{\mathbf{p}} \cdot \boldsymbol{\omega}}{\|\boldsymbol{\omega}\|^2}, \tag{54}$$

where $\dot{\mathbf{p}}$ is the velocity of an arbitrary point P of $\mathcal{B}$, the product $\dot{\mathbf{p}} \cdot \boldsymbol{\omega}$ being constant by virtue of Theorem 3.4. A proof of the foregoing results is available in (Angeles 1997).

3.3 The Linear Invariants of the Dual Rotation Matrix

We start by recalling the *linear invariants* of the real rotation matrix (Angeles 1997). These are defined as

$$\mathbf{q} \equiv \text{vect}(\mathbf{Q}) = (\sin\phi)\mathbf{e}, \quad q_0 \equiv \frac{\text{tr}(\mathbf{Q}) - 1}{2} = \cos\phi\,. \tag{55a}$$

Note that the linear invariants of any 3×3 matrix can be obtained from simple differences of its off-diagonal entries and sums of its diagonal entries. Once the foregoing linear invariants are calculated, the natural invariants can be obtained uniquely as indicated below: First, note that the sign of $\mathbf{e}$ can be changed without affecting $\mathbf{q}$ if the sign of ϕ is changed accordingly, which means that the sign of ϕ–or that of $\mathbf{e}$, for that matter–is undefined. In order to define this sign uniquely, we will adopt a positive sign for $\sin\phi$, which means that ϕ is assumed, henceforth, to lie in the interval $0 < \phi \leq \pi$.

We can thus obtain the inverse relations of eq.(55a) in the form

$$\mathbf{e} = \frac{\mathbf{q}}{\|\mathbf{q}\|}, \quad \phi = \arctan\left(\frac{\|\mathbf{q}\|}{q_0}\right), \quad \mathbf{q} \neq \mathbf{0}\,, \tag{55b}$$

the case $\mathbf{q} = \mathbf{0}$ being handled separately. Indeed, $\mathbf{q}$ vanishes under two cases: (a) $\phi = 0$, in which case the body undergoes a pure translation and the axis of rotation is obviously undefined; and (b) $\phi = \pi$, in which case $\mathbf{Q}$ is symmetric and takes the form

$$\text{For} \quad \phi = \pi: \quad \mathbf{Q} = -\mathbf{1} + 2\mathbf{e}\mathbf{e}^T\,, \tag{55c}$$

whence the natural invariants become apparent and can be readily extracted from $\mathbf{Q}$.

Similar to the linear invariants of the real rotation matrix, in the dual case we have

$$\hat{\mathbf{q}} \equiv \text{vect}(\hat{\mathbf{Q}}), \quad \hat{q}_0 \equiv \frac{\text{tr}(\hat{\mathbf{Q}}) - 1}{2}\,. \tag{56}$$

Expressions for the foregoing quantities in terms of the motion parameters are derived below; in the sequel, we also derive expressions for the *dual natural invariants* in terms of the same parameters. We start by expanding the vector linear invariant:

$$\mathrm{vect}(\hat{\mathbf{Q}}) = \mathrm{vect}(\mathbf{Q} + \epsilon\,\mathbf{DQ}) = \mathrm{vect}(\mathbf{Q}) + \epsilon\,\mathrm{vect}(\mathbf{DQ}) \,. \tag{57a}$$

But, by virtue of eq.(20),

$$\mathrm{vect}(\mathbf{Q}) = (\sin\phi)\mathbf{e} \,. \tag{57b}$$

Furthermore, the second term of the rightmost-hand side of eq.(57a) can be readily calculated if we recall Theorem 3.2, with $\mathbf{d} \equiv \mathrm{vect}(\mathbf{D})$:

$$\mathrm{vect}(\mathbf{DQ}) = \frac{1}{2}[\mathrm{tr}(\mathbf{Q})\mathbf{1} - \mathbf{Q}]\mathbf{d} \,. \tag{57c}$$

Now, if we recall expression (20), we obtain

$$\mathrm{tr}(\mathbf{Q})\mathbf{1} - \mathbf{Q} = (1+\cos\phi)\mathbf{1} - \sin\phi\mathbf{E} - (1-\cos\phi)\mathbf{e}\mathbf{e}^T \,.$$

Upon substitution of the foregoing expression into eq.(57c), the desired expression for vect($\mathbf{DQ}$) is readily derived, namely,

$$\mathrm{vect}(\mathbf{DQ}) = \frac{1}{2}[(1+\cos\phi)\mathbf{d} - \sin\phi\mathbf{e}\times\mathbf{d} - (1-\cos\phi)(\mathbf{e}\cdot\mathbf{d})\mathbf{e}] \,, \tag{57d}$$

and hence,

$$\hat{\mathbf{q}} = (\sin\phi)\mathbf{e} + \epsilon\frac{1}{2}[(\cos\phi)(\mathbf{e}\cdot\mathbf{d})\mathbf{e} + (1+\cos\phi)\mathbf{d} + (\sin\phi)\mathbf{d}\times\mathbf{e} - (\mathbf{e}\cdot\mathbf{d})\mathbf{e}]. \tag{57e}$$

On the other hand, the position vector $\mathbf{p}^*$ of the Mozzi-Chasles axis, given by eq.(34), can be expressed as

$$\mathbf{p}^* = \frac{1}{2}\frac{\sin\phi}{1-\cos\phi}\mathbf{e}\times\mathbf{d} + \frac{1}{2}\mathbf{d} - \frac{1}{2}(\mathbf{e}\cdot\mathbf{d})\mathbf{e} \,, \tag{58a}$$

and hence,

$$\mathbf{p}^*\times\mathbf{e} = \frac{1}{2}\frac{\sin\phi}{1-\cos\phi}\mathbf{d} - \frac{1}{2}\frac{\sin\phi}{1-\cos\phi}(\mathbf{e}\cdot\mathbf{d})\mathbf{e} + \frac{1}{2}\mathbf{d}\times\mathbf{e} \,. \tag{58b}$$

Moreover, let us recall the identity

$$\frac{1+\cos\phi}{\sin\phi} = \frac{\sin\phi}{1-\cos\phi} \,, \tag{58c}$$

which allows us to rewrite eq.(58b) in the form

$$\mathbf{p}^*\times\mathbf{e} = \frac{1}{2}\frac{1+\cos\phi}{\sin\phi}\mathbf{d} - \frac{1}{2}\frac{1+\cos\phi}{\sin\phi}(\mathbf{e}\cdot\mathbf{d})\mathbf{e} + \frac{1}{2}\mathbf{d}\times\mathbf{e} \,, \tag{58d}$$

whence,

$$(\sin\phi)\mathbf{p}^*\times\mathbf{e} = \frac{1}{2}[(1+\cos\phi)\mathbf{d} - (1+\cos\phi)(\mathbf{e}\cdot\mathbf{d})\mathbf{e} + (\sin\phi)\mathbf{d}\times\mathbf{e}] \,,$$

and $\hat{\mathbf{q}}$ takes the form

$$\hat{\mathbf{q}} = (\sin\phi)\mathbf{e} + \epsilon\,[(\cos\phi)(\mathbf{e}\cdot\mathbf{d})\mathbf{e} + (\sin\phi)\mathbf{p}^* \times \mathbf{e}]\,. \tag{59}$$

If we now recall eqs.(31) and (32), $\mathbf{d}\cdot\mathbf{e} \equiv d^* = p\phi$, while $\mathbf{p}^* \times \mathbf{e}$ is the moment of the associated Mozzi-Chasles axis, $\mathbf{e}_0$, and hence, eq.(59) becomes

$$\hat{\mathbf{q}} = (\sin\phi)\mathbf{e} + \epsilon\,[(\cos\phi)p\phi\mathbf{e} + (\sin\phi)\mathbf{e}_0]\,. \tag{60}$$

Thus, $\hat{\mathbf{q}}$ can be further simplified to

$$\hat{\mathbf{q}} = \hat{\mathbf{e}}^* \sin\hat{\phi}, \quad \hat{\phi} \equiv \phi(1+\epsilon p)\,, \tag{61}$$

where $\hat{\mathbf{e}}^*$ is the dual unit vector representing the Mozzi-Chasles axis, i.e., $\hat{\mathbf{e}}^* = \mathbf{e} + \epsilon\,\mathbf{e}_0$.

Now, such as in the real case, we can calculate the dual natural invariants of the motion under study in terms of the foregoing dual linear invariants. We do this by mimicking eqs.(55b), namely,

$$\hat{\mathbf{e}}^* = \frac{\hat{\mathbf{q}}}{||\hat{\mathbf{q}}||}, \quad \hat{\phi} = \arctan\left(\frac{||\hat{\mathbf{q}}||}{\hat{q}_0}\right), \quad ||\hat{\mathbf{q}}|| \neq 0\,, \tag{62}$$

where $||\hat{\mathbf{q}}||$ is calculated from eq.(9e), which gives $||\hat{\mathbf{q}}||^2$, the square root of the latter then following from eq.(7), thus obtaining

$$||\hat{\mathbf{q}}|| = \sin\hat{\phi} = \sin\phi + \epsilon\,(\mathbf{e}\cdot\mathbf{d})\cos\phi\,, \tag{63}$$

and hence, upon simplification,

$$\hat{\mathbf{e}}^* = \mathbf{e} + \epsilon\,\mathbf{p}^* \times \mathbf{e} = \mathbf{e} + \epsilon\,\mathbf{e}_0\,, \tag{64}$$

which is rightfully the dual unit vector of the Mozzi-Chasles axis. Furthermore,

$$\mathrm{tr}(\hat{\mathbf{Q}}) = \mathrm{tr}(\mathbf{Q}) + \epsilon\,\mathrm{tr}(\mathbf{DQ})\,, \tag{65a}$$

where, from Theorem 3.3, $\mathrm{tr}(\mathbf{DQ})$ turns out to be

$$\mathrm{tr}(\mathbf{DQ}) = -2[\mathrm{vect}(\mathbf{Q})]\cdot\mathbf{d} = -2\sin\phi(\mathbf{e}\cdot\mathbf{d})\,, \tag{65b}$$

whence,

$$\mathrm{tr}(\hat{\mathbf{Q}}) = 1 + 2\cos\phi - \epsilon\,2(\sin\phi)\mathbf{e}\cdot\mathbf{d}\,, \tag{65c}$$

and so, from the second of eqs.(56),

$$\hat{q}_0 \equiv \cos\hat{\phi} = \cos\phi - \epsilon\,(\sin\phi)(\mathbf{e}\cdot\mathbf{d})\,,$$

which, by virtue of eqs.(31), leads to

$$\hat{q}_0 = \cos\phi - \epsilon\,(\sin\phi)d^*, \quad \hat{\phi} = \phi + \epsilon\,d^* = \phi(1+\epsilon p)\,. \tag{65d}$$

In summary, the dual angle of the dual rotation under study comprises the angle of rotation of $\mathbf{Q}$ in its primal part and the axial component of the displacement of all points of the moving body onto the Mozzi-Chasles axis. Upon comparison of the dual angle between two lines, as given in eq.(5), with the dual angle of rotation $\hat{\phi}$, it is then apparent that the primal part of the latter plays the role of the angle between two lines, while the corresponding dual part plays the role of the distance s between those lines. It is noteworthy that a pure rotation has a dual angle of rotation that is real, while a pure translation has an angle of rotation that is a pure dual number.

Example 1: Determination of the screw parameters of a rigid-body motion. We take here an example of (Angeles 1997): The cube of Fig. 5 is displaced from configuration $A^0B^0 \ldots H^0$ into configuration $AB \ldots H$. Find the Plücker coordinates of the Mozzi-Chasles axis of the motion undergone by the cube.

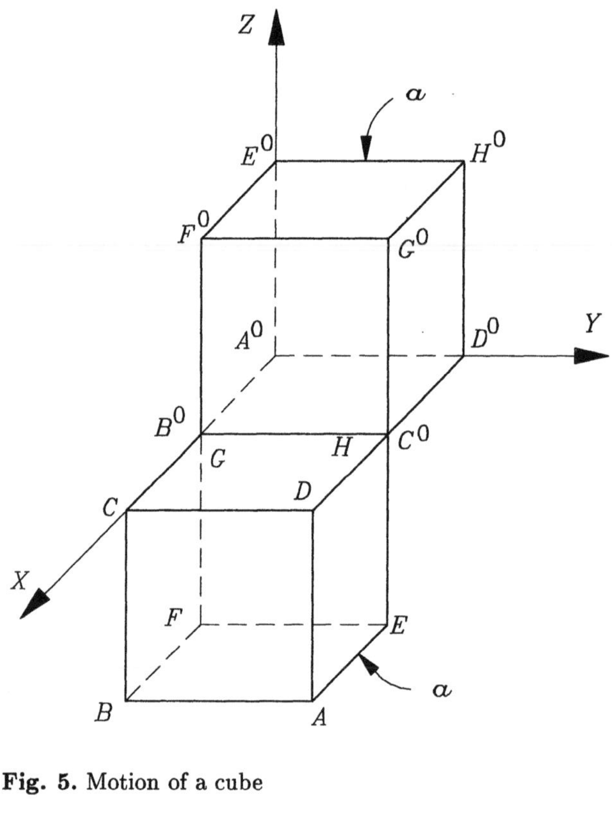

Fig. 5. Motion of a cube

Solution: We start by constructing $\hat{\mathbf{Q}}$: $\hat{\mathbf{Q}} \equiv [\,\hat{\mathbf{i}}^* \;\; \hat{\mathbf{j}}^* \;\; \hat{\mathbf{k}}^*\,]$, where $\hat{\mathbf{i}}^*$, $\hat{\mathbf{j}}^*$, and $\hat{\mathbf{k}}^*$ are the dual unit vectors of lines AB, AD, and AE, respectively. These lines are, in turn, the images of lines A^0B^0, A^0D^0, and A^0E^0 under the rigid-body motion at hand. The dual unit vectors of the latter are denoted by $\hat{\mathbf{i}}^{0*}$, $\hat{\mathbf{j}}^{0*}$, and $\hat{\mathbf{k}}^{0*}$, respectively, and are parallel to the X, Y, and Z axes of the figure. We thus have

$$\hat{\mathbf{i}}^* = -\mathbf{j}^0 + \epsilon\,\mathbf{a}\times(-\mathbf{j}^0), \quad \hat{\mathbf{j}}^* = \mathbf{k}^0 + \epsilon\,\mathbf{a}\times\mathbf{k}^0, \quad \hat{\mathbf{k}}^* = -\mathbf{i}^0 + \epsilon\,\mathbf{a}\times(-\mathbf{i}^0)\,,$$

where $\mathbf{a}$ is the position vector of A, and is given by

$$\mathbf{a} = [2 \quad 1 \quad -1]^T a\,.$$

Hence,

$$\begin{aligned}
\hat{\mathbf{i}}^* &= -\mathbf{j}^0 + \epsilon\,a(-\mathbf{i}^0 - 2\mathbf{k}^0)\\
\hat{\mathbf{j}}^* &= \mathbf{k}^0 + \epsilon\,a(\mathbf{i}^0 - 2\mathbf{j}^0)\\
\hat{\mathbf{k}}^* &= -\mathbf{i}^0 + \epsilon\,a(\mathbf{j}^0 + \mathbf{k}^0)
\end{aligned}$$

Therefore,

$$\hat{\mathbf{Q}} = \begin{bmatrix} -\,\epsilon\,a & +\,\epsilon\,a & -1\\ -1 & -\,\epsilon\,2a & +\,\epsilon\,a\\ -\,\epsilon\,2a & 1 & +\,\epsilon\,a \end{bmatrix},$$

whence,

$$\text{vect}(\hat{\mathbf{Q}}) = \frac{1}{2}\begin{bmatrix} 1-\epsilon\,a\\ -1+\epsilon\,2a\\ -1-\epsilon\,a \end{bmatrix}, \quad \text{tr}(\hat{\mathbf{Q}}) = -\,\epsilon\,(2a)\,,$$

and

$$\|\text{vect}(\hat{\mathbf{Q}})\|^2 = \|\frac{1}{2}\begin{bmatrix}1\\-1\\-1\end{bmatrix}\|^2 + \epsilon\,2\frac{1}{2}[1 \quad -1 \quad -1]\begin{bmatrix}-1\\2\\-1\end{bmatrix}\frac{a}{2} = \frac{3}{4} - \epsilon\,a\,.$$

Thus,

$$\|\text{vect}(\hat{\mathbf{Q}})\| = \frac{\sqrt{3}}{2} + \epsilon\,\frac{-a}{\sqrt{3}} = \frac{\sqrt{3}}{2} - \epsilon\,\frac{\sqrt{3}}{3}a\,.$$

Therefore, the unit dual vector representing the Mozzi-Chasles axis of the motion at hand, $\hat{\mathbf{e}}^*$, is given by $\hat{\mathbf{e}}^* = \text{vect}(\hat{\mathbf{Q}})/\|\text{vect}(\hat{\mathbf{Q}})\|$, i.e.,

$$\hat{\mathbf{e}}^* = \frac{1}{\sqrt{3}/2}\frac{1}{2}\begin{bmatrix}1\\-1\\-1\end{bmatrix} - \epsilon\,\frac{a}{3/4}\left(\frac{1}{2}\begin{bmatrix}1\\-1\\-1\end{bmatrix}\frac{-\sqrt{3}}{3} - \frac{1}{2}\begin{bmatrix}-1\\2\\-1\end{bmatrix}\frac{\sqrt{3}}{2}\right).$$

After various stages of simplification, the foregoing expression reduces to

$$\hat{\mathbf{e}}^* = \frac{\sqrt{3}}{3}\begin{bmatrix}1\\-1\\-1\end{bmatrix} + \epsilon\,\frac{\sqrt{3}}{9}\begin{bmatrix}-1\\4\\-5\end{bmatrix}a\,.$$

Thus, the Mozzi-Chasles axis is parallel to the unit vector $\mathbf{e}$, which is given by the primal part of $\hat{\mathbf{e}}$, while the dual part of the same dual unit vector represents the moment of the Mozzi-Chasles axis, from which the position vector $\mathbf{p}^*$ of P^*, the point of the Mozzi-Chasles axis closest to the origin, is readily found as

$$\mathbf{p}^* = \mathbf{e}\times\mathbf{e}_0 = \frac{a}{3}[3 \quad 2 \quad 1]^T\,.$$

3.4 The Dual Euler-Rodrigues Parameters of a Rigid-Body Motion

We first recall the definition of the Euler-Rodrigues parameters of a pure rotation, which are isomorphic to the *quaternion* of the rotation (Hamilton 1844). These are most naturally introduced as the linear invariants of the square root of the rotation at hand, and represented, paralleling the definition of the linear invariants, as

$$\mathbf{r} \equiv \text{vect}(\sqrt{\mathbf{Q}}), \quad r_0 \equiv \frac{\text{tr}(\sqrt{\mathbf{Q}}) - 1}{2}, \tag{66}$$

the *proper orthogonal* square root of $\mathbf{Q}$ being given as (Angeles 1997):

$$\sqrt{\mathbf{Q}} = \mathbf{1} + \sin\left(\frac{\phi}{2}\right)\mathbf{E} + \left[1 - \cos\left(\frac{\phi}{2}\right)\right])\mathbf{E}^2 . \tag{67}$$

The *dual Euler-Rodrigues parameters* of a rigid-body motion are thus defined as

$$\hat{\mathbf{r}} \equiv \text{vect}(\sqrt{\hat{\mathbf{Q}}}), \quad \hat{r}_0 \equiv \frac{\text{tr}(\sqrt{\hat{\mathbf{Q}}}) - 1}{2} . \tag{68}$$

Below we derive an expression for $\sqrt{\hat{\mathbf{Q}}}$. Prior to this, we introduce a relation that will prove useful:

Lemma 3.2. *Let* $\mathbf{a}$ *and* $\mathbf{b}$ *be arbitrary 3-dimensional vectors, and* $\mathbf{c} \equiv \mathbf{a} \times \mathbf{b}$. *The cross-product matrix* $\mathbf{C}$ *of* $\mathbf{c}$ *is given by*

$$\mathbf{C} = \mathbf{b}\mathbf{a}^T - \mathbf{a}\mathbf{b}^T . \tag{69}$$

Proof: This follows by noticing that, for any 3-dimensional vector $\mathbf{u}$,

$$\mathbf{c} \times \mathbf{u} = (\mathbf{a} \times \mathbf{b}) \times \mathbf{u} = \mathbf{b}(\mathbf{a}^T\mathbf{u}) - \mathbf{a}(\mathbf{b}^T\mathbf{u}) ,$$

which readily leads to

$$\mathbf{C}\mathbf{u} = (\mathbf{b}\mathbf{a}^T - \mathbf{a}\mathbf{b}^T)\mathbf{u} ,$$

thereby completing the proof.

Now we proceed to determine $\sqrt{\hat{\mathbf{Q}}}$. To this end, we regard the motion at hand, from a reference configuration $\mathcal{B}^0$ to a current configuration $\mathcal{B}$, as consisting of a rotation $\mathbf{Q}$ about the origin O followed by a translation $\mathbf{d}$. Then, this motion is decomposed into two parts, as shown in Fig. 6: First, the body is rotated about the origin O by a rotation $\sqrt{\mathbf{Q}}$ and a translation $\mathbf{d}_s$; then, from the configuration $\mathcal{B}'$ thus attained, the body is given a new rotation $\sqrt{\mathbf{Q}}$ about O as well, followed by the same translation $\mathbf{d}_s$.

It is apparent that, from the general expression for the dual rotation matrix, eq.(42), $\sqrt{\hat{\mathbf{Q}}}$ can be represented as

$$\sqrt{\hat{\mathbf{Q}}} = (\mathbf{1} + \epsilon \mathbf{D}_s)\sqrt{\mathbf{Q}} \,, \tag{70}$$

the calculation of $\sqrt{\hat{\mathbf{Q}}}$ thus reducing to that of the skew-symmetric matrix $\mathbf{D}_s$, which is the cross-product matrix of $\mathbf{d}_s$. This matrix is calculated below in terms of $\sqrt{\mathbf{Q}}$ and $\mathbf{D}$. We thus have

$$\mathbf{p}^2 = \sqrt{\mathbf{Q}}\mathbf{p}^0 + \mathbf{d}_s \,, \tag{71}$$

$$\mathbf{p}^4 = \sqrt{\mathbf{Q}}\mathbf{p}^2 + \mathbf{d}_s = \mathbf{Q}\mathbf{p}^0 + (\mathbf{1} + \sqrt{\mathbf{Q}})\mathbf{d}_s \,. \tag{72}$$

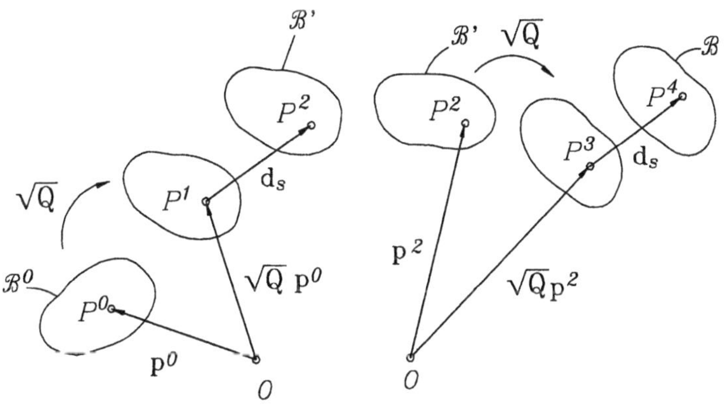

Fig. 6. Decomposition of the motion of a rigid body

But $\mathbf{p}^4$ is the position vector of point P in $\mathcal{B}$, which can be attained by a rotation $\mathbf{Q}$ about O followed by a translation $\mathbf{d}$, i.e.,

$$\mathbf{p}^4 = \mathbf{Q}\mathbf{p}^0 + \mathbf{d} \,. \tag{73}$$

Upon comparing the right-hand sides of eqs.(72) and (73), we obtain

$$(\mathbf{1} + \sqrt{\mathbf{Q}})\mathbf{d}_s = \mathbf{d} \,,$$

whence,

$$\mathbf{d}_s = (\mathbf{1} + \sqrt{\mathbf{Q}})^{-1}\mathbf{d} \,. \tag{74}$$

An expression for the above inverse can be derived if we realize that this inverse is an analytic function of $\sqrt{\mathbf{Q}}$, which is, in turn, an analytic function of $\mathbf{Q}$. We can thus conclude that by virtue of the Cayley-Hamilton Theorem, invoked when deriving the exponential form of the rotation matrix in eq.(22), the inverse sought must be a linear combination of the first three powers of $\mathbf{E}$: $\mathbf{E}^0 \equiv \mathbf{1}$, $\mathbf{E}$, and $\mathbf{E}^2$, namely,

$$(\mathbf{1} + \sqrt{\mathbf{Q}})^{-1} = \alpha\mathbf{1} + \beta\mathbf{E} + \gamma\mathbf{E}^2 \,, \tag{75}$$

where α, β, and γ are to be determined. To this end, we write

$$(\mathbf{1}+\sqrt{\mathbf{Q}})(\alpha\mathbf{1}+\beta\mathbf{E}+\gamma\mathbf{E}^2)=\mathbf{1}\,.$$

If we now substitute in the above equation the expression for $\sqrt{\mathbf{Q}}$ displayed in eq.(67), we obtain three equations for the three unknowns α, β, and γ, from which it is a simple matter to solve for these unknowns, namely,

$$\alpha=\frac{1}{2},\quad \beta=-\frac{\sin(\phi/2)}{2[1+\cos(\phi/2)]},\quad \gamma=0\,, \tag{76}$$

the inverse sought thus taking the form

$$(\mathbf{1}+\sqrt{\mathbf{Q}})^{-1}=\frac{1}{2}\left[\mathbf{1}-\frac{\sin(\phi/2)}{1+\cos(\phi/2)}\mathbf{E}\right]\,. \tag{77}$$

Therefore, eq.(74) yields

$$\mathbf{d}_s=(\mathbf{1}+\sqrt{\mathbf{Q}})^{-1}\mathbf{d}=\frac{1}{2}\left[\mathbf{1}-\frac{\sin(\phi/2)}{1+\cos(\phi/2)}\mathbf{E}\right]\mathbf{d}\,,$$

i.e.,

$$\mathbf{d}_s=\frac{1}{2}\left[\mathbf{d}-\frac{\sin(\phi/2)}{1+\cos(\phi/2)}\mathbf{e}\times\mathbf{d}\right]\,. \tag{78}$$

Thus, $\mathbf{D}_s$ is the cross-product matrix of the sum of two vectors, and hence, $\mathbf{D}_s$ reduces to the sum of the corresponding cross-product matrices. The cross-product matrix of the first term of the right-hand side of the foregoing equation is apparently proportional to $\mathbf{D}$, that of the second term being proportional to the cross-product matrix of $\mathbf{e}\times\mathbf{d}$. The latter can be readily obtained by application of Lemma 3.2, which leads to

$$\mathbf{D}_s=\frac{1}{2}\left[\mathbf{D}-\frac{\sin(\phi/2)}{1+\cos(\phi/2)}(\mathbf{d}\mathbf{e}^T-\mathbf{e}\mathbf{d}^T)\right]\,. \tag{79}$$

Hence, upon substitution of eq.(79) into eq.(70), we obtain

$$\sqrt{\hat{\mathbf{Q}}}=\left\{\mathbf{1}+\epsilon\frac{1}{2}\left[\mathbf{D}-\frac{\sin(\phi/2)}{1+\cos(\phi/2)}(\mathbf{d}\mathbf{e}^T-\mathbf{e}\mathbf{d}^T)\right]\right\}\sqrt{\mathbf{Q}}\,. \tag{80}$$

Now, the linear invariants of $\sqrt{\hat{\mathbf{Q}}}$ are

$$\operatorname{vect}(\sqrt{\hat{\mathbf{Q}}})=\operatorname{vect}(\sqrt{\mathbf{Q}})+\epsilon\operatorname{vect}(\mathbf{D}_s\sqrt{\mathbf{Q}}) \tag{81a}$$

and

$$\operatorname{tr}(\sqrt{\hat{\mathbf{Q}}})=\operatorname{tr}(\sqrt{\mathbf{Q}})+\epsilon\operatorname{tr}(\mathbf{D}_s\sqrt{\mathbf{Q}})\,. \tag{81b}$$

An expression for $\operatorname{vect}(\sqrt{\mathbf{Q}})$, appearing in the first term of $\operatorname{vect}(\sqrt{\hat{\mathbf{Q}}})$, can be obtained from eq.(67), namely,

$$\operatorname{vect}(\sqrt{\mathbf{Q}})=\sin\left(\frac{\phi}{2}\right)\operatorname{vect}(\mathbf{E})=\sin\left(\frac{\phi}{2}\right)\mathbf{e}\,, \tag{82}$$

while an expression for the second term of the right-hand side of eq.(81b) is obtained by application of Theorem 3.2:

$$\operatorname{vect}(\mathbf{D}_s\sqrt{\mathbf{Q}}) = \frac{1}{2}[\operatorname{tr}(\sqrt{\mathbf{Q}})\mathbf{1} - \sqrt{\mathbf{Q}}]\mathbf{d}_s ,$$

which can be further expanded without intermediate lengthy derivations if we realize that the above expression is the counterpart of that appearing in eq.(57c); the latter is expanded in eq.(57d). Thus, all we need now is mimic eq.(57d), if with ϕ and $\mathbf{d}$ substituted by their counterparts $\phi/2$ and $\mathbf{d}_s$, respectively, i.e.,

$$\operatorname{vect}(\mathbf{D}_s\sqrt{\mathbf{Q}}) = \frac{1}{2}\left\{\left[1+\cos\left(\frac{\phi}{2}\right)\right]\mathbf{d}_s - \sin\left(\frac{\phi}{2}\right)\mathbf{e}\times\mathbf{d}_s - \left[1-\cos\left(\frac{\phi}{2}\right)\right](\mathbf{e}\cdot\mathbf{d}_s)\mathbf{e}\right\} . \tag{83}$$

If we now simplify the above expression for $\operatorname{vect}(\mathbf{D}_s\sqrt{\mathbf{Q}})$, and substitute the simplified expression into eq.(81a), along with eq.(82), we obtain the desired expression for $\hat{\mathbf{r}}$. Note that the latter is defined in eq.(68), and hence,

$$\hat{\mathbf{r}} = \sin\left(\frac{\phi}{2}\right)\mathbf{e} + \epsilon\left[\cos\left(\frac{\phi}{2}\right)p_s\frac{\phi}{2}\mathbf{e} + \sin\left(\frac{\phi}{2}\right)\mathbf{e}_0\right] , \tag{84}$$

where p_s is the pitch associated with the motion represented by $\sqrt{\hat{\mathbf{Q}}}$, namely,

$$p_s\frac{\phi}{2} \equiv \mathbf{d}_s\cdot\mathbf{e} - \frac{1}{2}\mathbf{d}\cdot\mathbf{e} , \tag{85}$$

where we have recalled the expression for $\mathbf{d}_s$ displayed in eq.(78). Similar to eq.(61), then, the dual vector of the Euler-Rodrigues parameters is given by

$$\hat{\mathbf{r}} = \hat{\mathbf{e}}^*\sin\left(\frac{\hat{\phi}}{2}\right), \quad \hat{\phi} \equiv \phi + \epsilon\, d_s^*, \quad d_s^* \equiv \mathbf{d}_s\cdot\mathbf{e} . \tag{86}$$

The scalar of the Euler-Rodrigues parameters under study, $\hat{r}_0$, is now found in terms of the trace of $\sqrt{\hat{\mathbf{Q}}}$, which is displayed in eq.(81b). In that equation,

$$\operatorname{tr}(\sqrt{\mathbf{Q}}) = 1 + 2\cos\left(\frac{\phi}{2}\right) ,$$

the dual part of the right-hand side of eq.(81b) being calculated by application of Theorem 3.3:

$$\operatorname{tr}(\mathbf{D}_s\sqrt{\mathbf{Q}}) = -2\mathbf{d}_s\cdot\operatorname{vect}(\sqrt{\mathbf{Q}}) = -2\mathbf{d}_s\cdot\mathbf{e}\sin\left(\frac{\phi}{2}\right)$$

or, in terms of the corresponding pitch p_s,

$$\operatorname{tr}(\mathbf{D}_s\sqrt{\mathbf{Q}}) = -2p_s\sin\left(\frac{\phi}{2}\right) .$$

Therefore,

$$\mathrm{tr}(\sqrt{\hat{\mathbf{Q}}}) = 1 + 2\cos\left(\frac{\phi}{2}\right) - \epsilon\, 2p_s \sin\left(\frac{\phi}{2}\right) ,$$

and hence,

$$\hat{r}_0 = \cos\left(\frac{\phi}{2}\right) - \epsilon\, p_s \sin\left(\frac{\phi}{2}\right) = \cos\left(\frac{\hat{\phi}}{2}\right) , \tag{87}$$

which is the counterpart of the second of eqs.(55a). The set ($\hat{\mathbf{r}}$, $\hat{r}_0$) constitutes the *dual quaternion* of the motion under study (McCarthy 1990).

4. The Dual Angular Velocity

Similar to the angular-velocity matrix $\mathbf{\Omega}$ introduced in eq.(46a), the *dual angular velocity matrix* $\hat{\mathbf{\Omega}}$ is defined as

$$\hat{\mathbf{\Omega}} \equiv \dot{\hat{\mathbf{Q}}}\hat{\mathbf{Q}}^T . \tag{88}$$

Now we differentiate with respect to time the expression for $\hat{\mathbf{Q}}$ introduced in eq.(42), which yields

$$\dot{\hat{\mathbf{Q}}} = (\mathbf{1} + \epsilon\,\mathbf{D})\dot{\mathbf{Q}} + \epsilon\dot{\mathbf{D}}\mathbf{Q} .$$

Upon substitution of the above expression for $\dot{\hat{\mathbf{Q}}}$ and of the expression for $\hat{\mathbf{Q}}$ of eq.(42) into eq.(88), we obtain

$$\hat{\mathbf{\Omega}} = \mathbf{\Omega} + \epsilon\,(\mathbf{D}\mathbf{\Omega} - \mathbf{\Omega}\mathbf{D} + \dot{\mathbf{D}}) . \tag{89}$$

The *dual angular-velocity vector* $\hat{\omega}$ of the motion under study is then obtained as the axial vector of the foregoing expression, namely,

$$\hat{\omega} = \mathrm{vect}(\hat{\mathbf{\Omega}}) = \omega + \epsilon\,[\mathrm{vect}(\mathbf{D}\mathbf{\Omega} - \mathbf{\Omega}\mathbf{D}) + \dot{\mathbf{d}}] , \tag{90}$$

with $\dot{\mathbf{d}}$ being the time-derivative of vector $\mathbf{d}$, introduced in eq.(33). Thus, in order to determine $\hat{\omega}$, all we need is the axial vector of the difference $\mathbf{D}\mathbf{\Omega} - \mathbf{\Omega}\mathbf{D}$. An expression for this difference can be obtained in various manners, one of which is outlined below: First, note that this difference is skew-symmetric, and hence,

$$\mathrm{vect}(\mathbf{D}\mathbf{\Omega} - \mathbf{\Omega}\mathbf{D}) = 2\,\mathrm{vect}(\mathbf{D}\mathbf{\Omega}) .$$

Further, the vector of $\mathbf{D}\mathbf{\Omega}$ is computed by means of Corollary 3.1, eq.(39), upon substituting $\mathbf{A}$ by $\mathbf{\Omega}$ in that expression. Thus,

$$\mathrm{vect}(\mathbf{D}\mathbf{\Omega}) = -\frac{1}{2}\omega \times \mathbf{d} . \tag{91}$$

Therefore,

$$\hat{\omega} = \omega + \epsilon(\dot{\mathbf{d}} - \omega \times \mathbf{d}) , \tag{92}$$

and, if we recall eq.(51), the foregoing expression takes the alternative form

$$\hat{\omega} = \omega + \epsilon \mathbf{v}^0 . \tag{93}$$

In consequence, the dual angular velocity is the dual representation of the *twist* $\mathbf{t}$ of $\mathcal{B}$, defined as the 6-dimensional array

$$\mathbf{t} \equiv \begin{bmatrix} \omega \\ \mathbf{v}^0 \end{bmatrix} . \tag{94}$$

We can therefore find the angular velocity vector and the moment of the ISA about the given origin–i.e., the ***instant screw parameters*** of the motion at hand–if we are given enough information as to allow us to compute $\hat{\omega}$. The information required to determine the screw parameters of the motion under study can be given as the position and velocity vectors of three non-collinear points of a rigid body (Angeles 1997). However, note that the dual rotation matrix was obtained in Example 1 in terms of the dual unit vectors representing three mutually orthogonal lines. Also note that, by virtue of Lemma 2.1, the three lines of Example 1 were chosen concurrent and mutually orthogonal.

Now, in order to find the instant-screw parameters of interest, we need the time-derivatives of the dual unit vectors representing three concurrent, mutually orthogonal lines, but all we have at our disposal is the position and velocity vectors of three non-collinear points. Nevertheless, once we know three non-collinear points of a rigid body, say A, B, and C, along with their velocities, it is possible to find the position and velocity vectors of three pairs of points defining a triad of concurrent, mutually orthogonal lines, an issue that falls beyond the scope of this chapter. Rather than discussing the problem at hand in its fullest generality, we limit ourselves to the special case in which the position vector $\mathbf{p}$ of a point P of the rigid body under study can be determined so that the three lines PA, PB, and PC are mutually orthogonal. Further, we let the position vectors of the three given points be $\mathbf{a}$, $\mathbf{b}$, and $\mathbf{c}$. Thus, point P of the body in this case forms a rectangular trihedron with vertex at P and edges PA, PB, and PC. We can thus express $\mathbf{p}$ as a nonlinear function of the three position vectors $\mathbf{a}$, $\mathbf{b}$, and $\mathbf{c}$:

$$\mathbf{p} = \mathbf{p}(\mathbf{a}, \mathbf{b}, \mathbf{c}) . \tag{95}$$

Moreover, the velocity of point P, $\dot{\mathbf{p}}$, can be calculated now as a linear combination of the velocities of the three given points, by straightforward differentiation of the foregoing expression, namely,

$$\dot{\mathbf{p}} = \mathbf{P}_a\dot{\mathbf{a}} + \mathbf{P}_b\dot{\mathbf{b}} + \mathbf{P}_c\dot{\mathbf{c}} , \tag{96}$$

where $\mathbf{P}_a$, $\mathbf{P}_b$, and $\mathbf{P}_c$ denote the partial derivatives of $\mathbf{p}$ with respect to $\mathbf{a}$, $\mathbf{b}$, and $\mathbf{c}$, respectively. Once the position and the velocity vectors of point P are known, it is possible to determine the time-rates of change of the dual unit vectors representing the three lines PA, PB and PC, as described below.

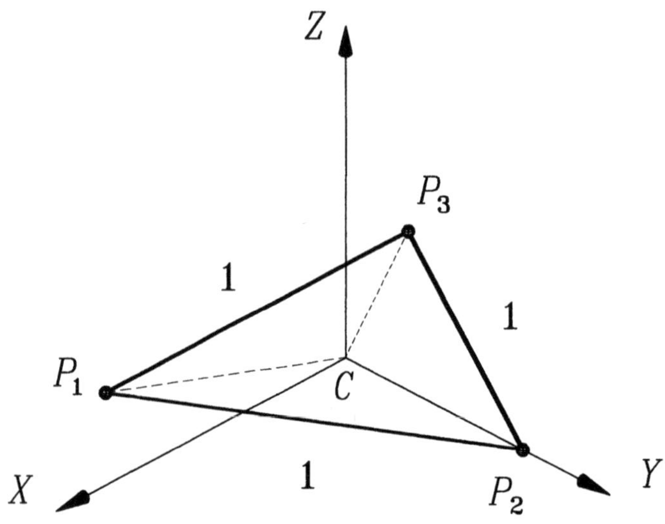

Fig. 7. A rigid triangular plate undergoing a motion given by the velocity of its vertices

Let $\hat{\mathbf{e}}^*$ denote the dual unit vector representing the line determined by points A and P, its primary and dual parts, $\mathbf{e}$ and $\mathbf{e}_0$, being given by

$$\mathbf{e} = \frac{\mathbf{a}-\mathbf{p}}{||\mathbf{a}-\mathbf{p}||}, \quad \mathbf{e}_0 = \mathbf{p} \times \frac{\mathbf{a}-\mathbf{p}}{||\mathbf{a}-\mathbf{p}||} . \tag{97}$$

Straightforward differentiation of the foregoing expressions with respect to time leads to

$$\begin{aligned} \dot{\mathbf{e}} &= \frac{1}{||\mathbf{a}-\mathbf{p}||}\left(\dot{\mathbf{a}} - \dot{\mathbf{p}} - \mathbf{e}\frac{\mathrm{d}}{\mathrm{dt}}||\mathbf{a}-\mathbf{p}||\right), \\ \dot{\mathbf{e}}_0 &= \dot{\mathbf{p}} \times \frac{\mathbf{a}-\mathbf{p}}{||\mathbf{a}-\mathbf{p}||} + \mathbf{p} \times \frac{1}{||\mathbf{a}-\mathbf{p}||}\left(\dot{\mathbf{a}} - \dot{\mathbf{p}} - \mathbf{e}\frac{\mathrm{d}}{\mathrm{dt}}||\mathbf{a}-\mathbf{p}||\right) . \end{aligned}$$

Upon simplification, we obtain the desired expression for $\dot{\hat{\mathbf{e}}}^*$, namely,

$$\dot{\hat{\mathbf{e}}}^* = \frac{1}{||\mathbf{a}-\mathbf{p}||}\left[\dot{\mathbf{a}} - \dot{\mathbf{p}} + \epsilon(\mathbf{p}\dot{\mathbf{a}} + \dot{\mathbf{p}} \times \mathbf{a})\right] . \tag{98}$$

Therefore, knowing the velocity of two points of a line, we can determine the time-rate of change of the dual unit vector representing the line. The foregoing idea is best illustrated with the aid of the example included below.

Example 2: Determination of the ISA of a rigid-body motion. For comparison purposes, we take an example from (Angeles 1997): The three vertices of the equilateral triangular plate of Fig. 7, which lie in the X-Y plane, $\{P_i\}_1^3$, have the position vectors $\{\mathbf{p}_i\}_1^3$. Moreover, the origin of the coordinate frame X, Y, Z lies at the centroid C of the triangle, and the velocities of the foregoing points, $\{\dot{\mathbf{p}}_i\}_1^3$, are given in this coordinate frame as

$$\dot{\mathbf{p}}_1 = \frac{4-\sqrt{2}}{4}\begin{bmatrix}0\\0\\1\end{bmatrix}, \quad \dot{\mathbf{p}}_2 = \frac{4-\sqrt{3}}{4}\begin{bmatrix}0\\0\\1\end{bmatrix}, \quad \dot{\mathbf{p}}_3 = \frac{4+\sqrt{2}}{4}\begin{bmatrix}0\\0\\1\end{bmatrix}.$$

With the above information, compute the instant-screw parameters of the motion under study.

Solution: Since the centroid C of the triangle coincides with that of the three given points, we have $\mathbf{c} = \mathbf{0}$, where $\mathbf{c}$ is the position vector of C. Moreover,

$$\mathbf{p}_1 = \begin{bmatrix}1/2\\-\sqrt{3}/6\\0\end{bmatrix}, \quad \mathbf{p}_2 = \begin{bmatrix}0\\\sqrt{3}/3\\0\end{bmatrix}, \quad \mathbf{p}_3 = \begin{bmatrix}-1/2\\-\sqrt{3}/6\\0\end{bmatrix}.$$

First and foremost, we have to verify the compatibility of the data. To do this, we calculate the component of the relative velocities of two given points onto the line that they define. It can be readily shown that the data are compatible, and hence, the motion is possible. Next, we obtain the position vector of the point P that, along with $\{P_i\}_1^3$, forms an orthogonal trihedron. It is not difficult to realize that the position vector of point P can be expressed as[1]

$$\mathbf{p} = \mathbf{c} + \frac{\sqrt{2}}{3}(\mathbf{p}_2 - \mathbf{p}_1) \times (\mathbf{p}_3 - \mathbf{p}_1),$$

and hence,

$$\dot{\mathbf{p}} = \dot{\mathbf{c}} + \frac{\sqrt{2}}{3}[(\mathbf{p}_3 - \mathbf{p}_2) \times \dot{\mathbf{p}}_1 + (\mathbf{p}_1 - \mathbf{p}_3) \times \dot{\mathbf{p}}_2 + (\mathbf{p}_2 - \mathbf{p}_1) \times \dot{\mathbf{p}}_3],$$

with the numerical values of $\mathbf{p}$ and $\dot{\mathbf{p}}$ given below:

$$\mathbf{p} = \frac{\sqrt{6}}{6}\begin{bmatrix}0\\0\\1\end{bmatrix}, \quad \dot{\mathbf{p}} = \frac{1}{12}\begin{bmatrix}2\sqrt{3}\\\sqrt{6}\\12-\sqrt{3}\end{bmatrix}.$$

Now, let $\hat{\mathbf{e}}_i^*$ denote the dual unit vector representing the line that passes through P and P_i, i.e.,

$$\hat{\mathbf{e}}_i^* = \frac{1}{\|\mathbf{p}_i - \mathbf{p}\|}[\mathbf{p}_i - \mathbf{p} + \epsilon\mathbf{p} \times \mathbf{p}_i],$$

where

$$\|\mathbf{p}_i - \mathbf{p}\| = \frac{\sqrt{2}}{2}, \quad i = 1, 2, 3.$$

[1] Although $\mathbf{c} = \mathbf{0}$ in this case, $\dot{\mathbf{c}} \neq \mathbf{0}$, and hence, $\mathbf{c}$ must be written explicitly in the expression for $\mathbf{p}$.

Next, the three foregoing dual unit vectors are stored columnwise in the dual rotation matrix $\hat{\mathbf{Q}}$, i.e.,

$$\hat{\mathbf{Q}} = [\hat{\mathbf{e}}_1^* \quad \hat{\mathbf{e}}_2^* \quad \hat{\mathbf{e}}_3^*] \; .$$

Upon substitution of the numerical values of these vectors into the above expression, we obtain

$$\hat{\mathbf{Q}} = \frac{\sqrt{12}}{12}\begin{bmatrix} 6+\epsilon 2 & -\epsilon 2\sqrt{2} & -6+\epsilon\sqrt{2} \\ -2\sqrt{3}+\epsilon\sqrt{6} & 4\sqrt{3} & -2\sqrt{3}-\epsilon\sqrt{6} \\ -2\sqrt{6} & -2\sqrt{6} & -2\sqrt{6} \end{bmatrix} .$$

Likewise, the time derivative of $\hat{\mathbf{Q}}$ is computed as

$$\begin{aligned}\dot{\hat{\mathbf{Q}}} &= \frac{\sqrt{2}}{24}\left(\begin{bmatrix} -4\sqrt{3} & -4\sqrt{3} & -4\sqrt{3} \\ -2\sqrt{6} & -2\sqrt{6} & -2\sqrt{6} \\ -6\sqrt{2}+2\sqrt{3} & -4\sqrt{3} & 6\sqrt{2}+2\sqrt{3} \end{bmatrix}\right. \\ &\quad \left. +\,\epsilon \begin{bmatrix} -1+4\sqrt{3} & 2-8\sqrt{3} & -1+4\sqrt{3} \\ 12-\sqrt{3} & 0 & -12+\sqrt{3} \\ -(2+\sqrt{6}) & 4 & -2+\sqrt{6} \end{bmatrix}\right) .\end{aligned}$$

Therefore,

$$\hat{\mathbf{\Omega}} = \dot{\hat{\mathbf{Q}}}\hat{\mathbf{Q}}^T = \frac{1}{12}\begin{bmatrix} 0 & -\epsilon(12-\sqrt{3}) & 6\sqrt{2} \\ +\epsilon(12-\sqrt{3}) & 0 & 6 \\ -6\sqrt{2} & -6 & 0 \end{bmatrix} ,$$

which, as expected, is a dual skew-symmetric matrix. Hence,

$$\hat{\boldsymbol{\omega}} = \text{vect}(\hat{\mathbf{\Omega}}) = \frac{1}{2}\begin{bmatrix} -1 \\ \sqrt{2} \\ 0 \end{bmatrix} + \epsilon\,\frac{12-\sqrt{3}}{12}\begin{bmatrix} 0 \\ 0 \\ 1 \end{bmatrix} ,$$

from which we can readily identify

$$\boldsymbol{\omega} = \frac{1}{2}\begin{bmatrix} -1 \\ \sqrt{2} \\ 0 \end{bmatrix} , \quad \mathbf{v}^0 = \frac{12-\sqrt{3}}{12}\begin{bmatrix} 0 \\ 0 \\ 1 \end{bmatrix} .$$

Furthermore, the position vector $\boldsymbol{\pi}^*$ of the point P^* of the ISA lying closest to the origin can be obtained from $\mathbf{v}^0$. Indeed, let $\mathbf{v}^*$ be the velocity of P^*, which thus allows us to write

$$\mathbf{v}^0 = \mathbf{v}^* + \boldsymbol{\omega} \times (-\mathbf{p}^*) = \mathbf{v}^* + \mathbf{p}^* \times \boldsymbol{\omega} \, .$$

Upon cross-multiplying the two sides of the foregoing expression by $\boldsymbol{\omega}$, we obtain

$$\mathbf{v}^0 \times \boldsymbol{\omega} = \mathbf{v}^* \times \boldsymbol{\omega} + (\mathbf{p}^* \times \boldsymbol{\omega}) \times \boldsymbol{\omega} \, ,$$

whose first term of the right-hand side vanishes because $\mathbf{v}^*$ and $\boldsymbol{\omega}$ are parallel. Therefore,

$$\mathbf{v}^0 \times \boldsymbol{\omega} = (\mathbf{p}^* \times \boldsymbol{\omega}) \times \boldsymbol{\omega} = (\mathbf{p}^* \cdot \boldsymbol{\omega})\boldsymbol{\omega} - \|\boldsymbol{\omega}\|^2\mathbf{p}^* \, .$$

The first term of the rightmost-hand side of the foregoing equation vanishes because $\mathbf{p}^*$ being the position vector of the point of the ISA that lies closest to the origin, and the ISA being parallel to $\boldsymbol{\omega}$, these two vectors are orthogonal. We can thus solve for $\mathbf{p}^*$ from the above expression, which yields

$$\mathbf{p}^* = -\frac{\mathbf{v}^0 \times \boldsymbol{\omega}}{\|\boldsymbol{\omega}\|^2} \, .$$

The quantities involved in the foregoing expression are now evaluated:

$$-\mathbf{v}^0 \times \boldsymbol{\omega} = \boldsymbol{\omega} \times \mathbf{v}^0 = \frac{12-\sqrt{3}}{24}\begin{bmatrix} \sqrt{2} \\ 1 \\ 0 \end{bmatrix}, \quad \|\boldsymbol{\omega}\|^2 = \frac{3}{4} \, .$$

Finally, $\mathbf{p}^* = \{[(12-\sqrt{3})]/18\}[\sqrt{2} \;\; 1 \;\; 0]^T$, which coincides with the results reported in (Angeles 1997), obtained by another method.

5. Conclusions

We revisited dual algebra in the context of kinematic analysis, which led us to a straightforward introduction of dual quaternions. In the process, we showed that the parameters of both the finite screw and the instant screw of a rigid-body motion can be computed from the sum of the diagonal and the difference of the off-diagonal entries of the dual rotation and, correspondingly, the dual angular-velocity matrices.

Acknowledgements

The work reported here was possible under the support of both Canada's Natural Sciences and Engineering Research Council and Quebec's Fonds pour la formation de chercheurs et l'aide à la recherche. NATO SEAD's financial support is gratefully acknowledged.

References

Agrawal, O.P. (1987): Hamilton operators and dual-number-quaternions in spatial kinematics. Mech. Mach. Theory **22** (6), 569–575

Angeles, J. (1982): Spatial Kinematic Chains. Analysis, Synthesis, Optimization. Springer-Verlag, Berlin

Angeles, J. (1997): Fundamentals of Robotic Mechanical Systems. Theory, Methods, and Algorithms. Springer-Verlag, New York

Bottema, O. and Roth, B. (1978): Theoretical Kinematics. North-Holland Publishers Co., Amsterdam

Cheng, H.H. and Thompson, S. (1996): Dual polynomials and complex dual numbers for analysis of spatial mechanisms. Proc. 1996 ASME Design Engineering Technical Conference and Computers in Engineering Conference. August 18–22, Irvine, California. 96-DETC/MECH-1221, CD-ROM ISBN 0-7918-1232-4

Chevallier, D.P. (1991): Lie algebras, modules, dual quaternions and algebraic methods in kinematics. Mechanism and Machine Theory **26** 6, 613–627

Clifford, W.K. (1873): Preliminary sketch of bi-quaternions. Proc. London Math. Soc. **4**, 381–395

Dimentberg, F.M. (1965): The Screw Calculus and Its Applications in Mechanics. Izdat. Nauka. Moscow. English Translation: AD680993, Clearinghouse for Federal and Scientific Technical Information

Euler, L. (1776): Nova methodus motum corporum rigidorum determinandi. Novii Comentarii Academiæ Scientiarum Petropolitanæ **20** (1775); 1776, 208–238 = *Opera Omnia* (2) 9, 99–125

Halmos, P. (1974): Finite-Dimensional Vector Spaces. Springer-Verlag, New York

Hamilton, W.R. (1844): On quaternions: or a new system of imaginaries in algebra. Phil. Mag., 3rd. ser. 2, 489–495

Kotel'nikov, A.P. (1895): Screw calculus and some of its applications to geometry and mechanics. Annals of The Imperial University of Kazan

McCarthy, J.M. (1990): An Introduction to Theoretical Kinematics. The MIT Press, Cambridge (MA)

Pradeep, A.K., Yoder, P.J., and Mukundan, R. (1989): On the use of dual matrix esponentials in robot kinematics. The Int. J. Robotics Res. **8** 5, 57–66

Rico Martínez, J.M. and Duffy, J. (1994): The Principle of Transference: History, statement and proof. Mechanism and Machine Theory **28** 1, 165–177

Shoham, M. and Brodsky, V. (1993): Analysis of mechanisms by the dual inertia operator. In Angeles, J., Hommel, G., and Kovács, P. (editors): Computational Kinematics. Kluwer Academic Publishers. Dordrecht, 129–138

Shoham, M. and Brodsky, V. (1994): The dual inertia operator and its application to robot dynamics. ASME J. Mechanical Design **116**, 1089–1095

Simmons, G.F. (1963): Introduction to Topology and Modern Analysis. McGraw-Hill Book Co., New York.

Study, E. (1903): Geometrie der Dynamen. Leipzig

Yang, A.T. (1963): Application of Quaternion Algebra and Dual Numbers to the Analysis of Spatial Mechanisms. Doctoral Dissertation, Columbia University, New York. No. 64-2803 (University Microfilm, Ann Arbor, Michigan)

Yang, A.T. and Freudenstein, F. (1964): Application of dual-number quaternion algebra to the analysis of spatial mechanisms. J. of Applied Mechanics **31**, 300–308

Formulation and Solution for the Direct and Inverse Kinematics Problems for Mechanisms and Mechatronics Systems

James Nielsen and Bernard Roth

Design Division, Department of Mechanical Engineering, Stanford University
Stanford, CA. 94305 USA.
nielsen@leland.stanford.edu, roth@flamingo.stanford.edu

Summary. The formulation of a kinematic analysis problem often leads to a set of nonlinear equations. There are a variety of both numerical and algebraic techniques available to solve such systems of equations, and to give bounds on the number of solutions. In this chapter, the formulations of the direct and inverse kinematics of the most important mechanisms used today are reviewed, along with a variety of solution techniques. The reviewed solution methods include polynomial continuation, Gröbner basis, and resultants. The relative merits of these techniques are discussed, and different problem formulations are compared.

Keywords. Inverse kinematics, direct kinematics, Bezout bound, BKK bound, polynomial continuation, Gröbner basis, dialytic elimination, resultant, Dixon determinants

1. Introduction

Sets of nonlinear equations are frequently encountered during the design and analysis of mechanical systems. For example, the kinematic analysis of linkages and robotic manipulators leads naturally to such equations. For sets of linear equations, there is a universally applicable and recognizably superior solution procedure, but for sets of nonlinear equations no such procedure is known; instead, there are a variety of techniques used to predict bounds on the number of solutions and to find those solutions.

In this chapter, after first reviewing the formulation of kinematic synthesis, direct kinematics, and inverse kinematics problems, several methods will be presented for the solution of nonlinear equation sets. Techniques for bounding the number of solutions will also be presented. Examples will be included to show how to apply the theory and to formulate the problems for analysis of the most important mechanisms used in today's mechatronic systems.

2. Problem Context

Though we are interested in solving a wide variety of kinematics problems, several problem classes can be considered prototypical of the problems encountered in kinematic analysis. Specifically, linkage synthesis problems, inverse kinematics problems, and direct kinematics problems are the three most common categories of nonlinear problems in kinematics. Studying the formulation and solution of these three problems will help us to understand not only the sources of nonlinearities in kinematics, but also the implications of those nonlinearities with respect to solution methods.

In linkage synthesis problems, the designer seeks the linkage dimensions and parameters which will guide a rigid body through a series of positions and orientations. For example, we may seek a four-bar linkage which guides a body through a specified set of positions, perhaps with restrictions on the locations of the ground pivots. Such a problem leads to a set of nonlinear equations, which can be solved to find all possible four-bar linkages meeting the restrictions and requirements.

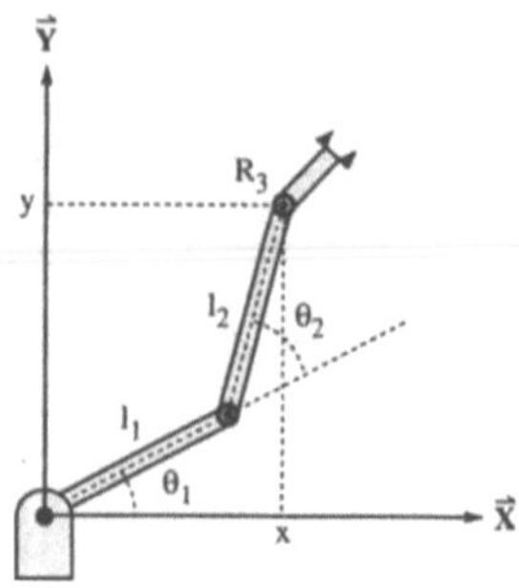

Fig. 2.1. Planar three-link serial mechanism

The inverse kinematics problem, while generally trivial for in-parallel mechanisms, can be very challenging for serial mechanisms. Given the dimensions of a three-revolute joint mechanism, shown in Fig. 2.1, and given the position and orientation of the end effector, we seek the actuation angles θ_1 and θ_2. The position of the third revolute joint R_3 can be easily calculated from the position and orientation of the end-effector; to find θ_1 and θ_2, we write the Cartesian projections of the loop equation:

$$\begin{aligned} l_1 \cos\theta_1 + l_2 \cos(\theta_1 + \theta_2) - x &= 0 \\ l_1 \sin\theta_1 + l_2 \sin(\theta_1 + \theta_2) - y &= 0 \end{aligned} \tag{2.1}$$

This set of equations (which are polynomials in sine and cosine) can be solved simultaneously for θ_1 and θ_2, thus finding all possible configurations of the mechanism for a given position and orientation of the end-effector. In this case, there will be two solutions, corresponding to the elbow-up and elbow-down positions of the planar robotic arm.

Most research into nonlinear equation solving techniques has focused on algebraic polynomials; it will therefore be useful to convert the sine-cosine polynomials in Eq. (2.1) to algebraic polynomials. The tangent-of-the-half-angle substitution accomplishes this task: we have $\sin\theta_i = \frac{2x_i}{1+x_i^2}$ and $\cos\theta_i = \frac{1-x_i^2}{1+x_i^2}$, where $x_i = \tan\frac{\theta}{2}$. After introducing this substitution, the denominators may be cleared to obtain an algebraic polynomial. Another possible conversion technique is to introduce separate variables for $\sin\theta_i$ and $\cos\theta_i$, with the additional restriction $\sin^2\theta_i + \cos^2\theta_i = 1$.

Given a full set of actuation parameters, the direct kinematics problem seeks the position and orientation of the end-effector. The actuation parameters could be, for example, angles for revolute joints driven by motors, or link lengths for variable-length links controlled by linear actuators. Direct kinematics problems are also called assembly mode problems. For serial mechanisms, the direct kinematics problem is trivial because the relative position and orientation of each link is dependent on the previous link. However, for parallel mechanisms, where the relative position and orientation of certain links depend on more than one other link, the direct kinematics problem leads to a set of nonlinear equations.

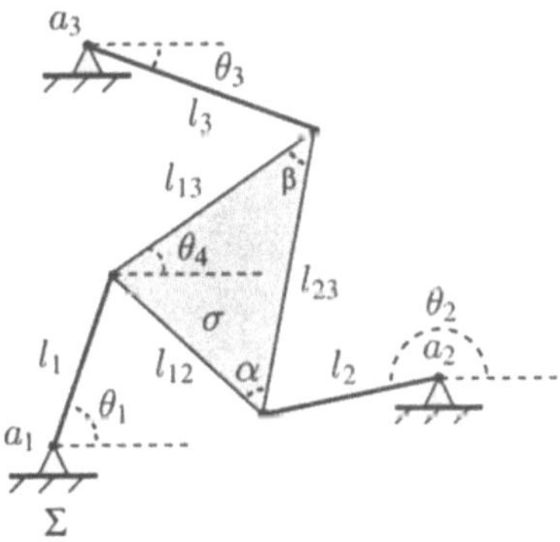

Fig. 2.2. Planar five-bar structure

Consider the planar linkage shown in Fig. 2.2. We are given three points a_1, a_2, and a_3 in the fixed coordinate system Σ, and dimensions l_{12}, l_{13}, and l_{23} of a moving platform. The actuation parameters are l_1, l_2, and l_3; for a given set of these parameters, we seek the position and orientation of the moving platform (the end-effector for this mechanism). Again, we begin by writing loop equations for the system. There are four unknown angles, θ_i, $i = 1, 2, 3, 4$, and two loop equations with Cartesian projections:

$$\begin{aligned}
l_3\cos\theta_3 - l_{23}\cos(\theta_4+\beta) - l_2\cos\theta_2 &= a_{2_x} - a_{3_x}\\
l_3\sin\theta_3 + l_{23}\sin(\theta_4+\beta) + l_2\sin\theta_2 &= a_{3_y} - a_{2_y}\\
l_1\cos\theta_1 + l_{12}\cos(\pi-\alpha-\beta-\theta_4) - l_2\cos\theta_2 &= a_{2_x} - a_{1_x}\\
l_1\sin\theta_1 - l_{12}\sin(\pi-\alpha-\beta-\theta_4) - l_2\sin\theta_2 &= a_{2_y} - a_{1_y}
\end{aligned} \quad (2.2)$$

where (a_{i_x}, a_{i_y}) are the coordinates of a_i. We have four sine-cosine polynomials in four unknowns; this set of nonlinear equations can be solved for θ_i to find the position and orientation of the end-effector.

As for most problems in kinematics, different formulations are possible for the direct kinematics of the planar mechanism described above. Husty (1996) has described a solution method which relies on a kinematic mapping of displacements in the plane to three-dimensional space. Bottema and Roth (1979) developed this mapping in detail. Using isotropic coordinates, Wampler (1996) presented a formulation for all planar direct kinematics problems, which in the case of this particular five-bar structure leads to four bilinear equations in four unknowns.

There is a spatial analogue to the previous problem; the Stewart-Gough platform shown in Fig. 2.3 has fixed and moving platforms of arbitrary geometry, with attached coordinate systems Σ and σ respectively. The moving platform has a full six-degrees-of-freedom in space, due to the six extensible legs with length l_i. To solve the direct kinematics of the mechanism, one must find the position and orientation of σ with respect to Σ for a given set of leg lengths l_i. This problem has attracted a great deal of research effort, and many solutions have been presented for different cases of the platform geometry.

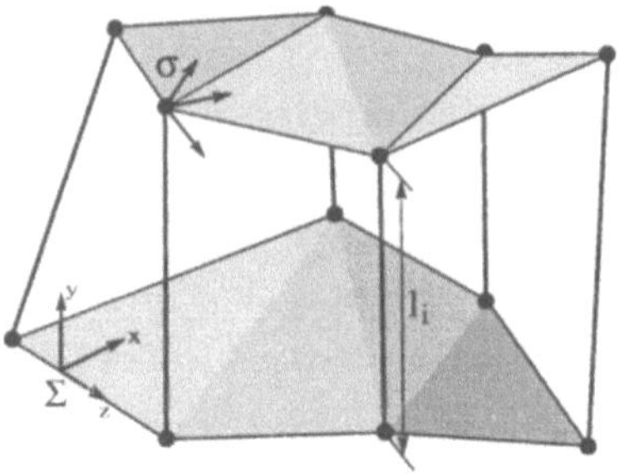

Fig. 2.3. General Stewart-Gough Platform

In formulating the direct kinematics of the Stewart-Gough platform, one must choose a representation for the orientation of σ with respect to Σ from the many ways available to represent spatial rotations. The first confirmation that the number of solutions to this problem is 40 used Euler angles to represent rotation and was accomplished numerically using Polynomial Continuation (Raghavan 1993). Quaternions (or, equivalently, Soma coordinates or Rodrigues parameters) have proven very useful in subsequent algebraic proofs of the solution number (Mourrain 1993, Wampler 1994), as well as in developing the first algorithm which reduces the problem to a 40th degree univariate polynomial (Husty 1996).

3. Number of Solutions

In contrast to linear equations, sets of nonlinear equations generally have multiple solutions. In kinematics, the set of solutions may correspond, for example, to different poses of a mechanism, or different possible design parameters of a mechanism. Given a particular set of nonlinear equations, the number of finite solutions cannot be stated without solving the problem, but several different bounds can be calculated to find a limit on the possible number of finite solutions. Consider the following set of two algebraic polynomial equations in two unknowns (note that any sine-cosine polynomial can be converted to an algebraic polynomial):

$$\begin{aligned} f_1: \quad & 3x_1^2x_2^2 + 2x_1 + 9 = 0 \\ f_2: \quad & 6x_1x_2 + x_1 + 8 = 0 \end{aligned} \tag{3.1}$$

This set of equations will be used to illustrate the different bounds which may be calculated, as well as the commonly-used solution methods.

3.1 Bezout Number

The oldest and best-known bound on the number of finite solutions to a system of polynomial equations is provided by Bezout's theorem, which states that the number of solutions, including asymptotic solutions at infinity, is equal to the Bezout number, which is the total degree $\prod_{j=1}^{n} d_j$ of the n polynomials, where d_j is the degree of the jth polynomial. The polynomial degree is determined by the degree of its highest-order term. Though this bound is quite easy to calculate, it is generally not a tight bound on the number of finite solutions, since all asymptotic solutions at infinity are included.

For Eqs. (3.1), f_1 has degree 4, while f_2 has degree 2. The total degree of the system, then, is $2 \times 4 = 8$. The Bezout bound on the number of finite solutions is therefore 8.

3.2 Multihomogeneous Bezout Number

If one views a set of equations as separately homogeneous in groups of variables, then a different Bezout number can be developed which often reduces the number of solutions at infinity (see Wampler et al. 1990). A suitable choice of variable groupings can lead to a significantly tighter bound than the 1-homogeneous Bezout bound described in Sec. 3.1. Suppose we are given n non-homogeneous equations in n unknowns. To begin, we divide the n variables into m variable groups $\{x_{11}, \ldots, x_{1k_1}\}$, $\{x_{21}, \ldots, x_{2k_2}\}$, ..., $\{x_{m1}, \ldots, x_{mk_m}\}$, where k_j is the number of variables in group j. If the degree of equation l with respect to variable group j is defined to be d_{jl}, then the multihomogeneous Bezout number is the coefficient of $\prod_{j=1}^{m} \alpha_j^{k_j}$ in the product $\prod_{l=1}^{n} \left(\sum_{j=1}^{m} d_{jl}\alpha_j \right)$. This multihomogeneous Bezout number provides an upper bound on the number of finite solutions. Obviously, the value of the bound depends on the variable grouping selected. It is easy to show that the multihomogeneous Bezout bound reduces to the 1-homogeneous Bezout bound when $m = 1$.

For Eqs. (3.1), one possible selection of variable groupings is $\{x_1\}$ and $\{x_2\}$. With this choice, the multihomogeneous Bezout number is the coefficient of $\alpha_1\alpha_2$ in the product $(2\alpha_1 + 2\alpha_2)(\alpha_1 + \alpha_2)$, which is 4. The upper limit on the number of finite solutions is therefore 4; note that this bound is tighter than the 1-homogeneous Bezout bound, which gave an upper limit of 8 finite solutions. The only other choice of variable groupings for this problem is $\{x_1, x_2\}$, which corresponds to the 1-homogeneous Bezout number.

3.3 BKK Bound

The tightest known bound on the number of finite solutions to a nonlinear system of equations is the BKK bound, based on work by Bernstein (1975), Khovanskii (1978), and Kushnirenko (1975). This bound is based on a remarkable and unexpected connection between combinatorial geometry and the intersection of polynomial equations. The BKK bound has proven to be an impressively tight bound for many kinematics problems, and much effort has been made to adapt classical solution methods to account for this new result (Emiris 1993b, Emiris 1994b, Li et al. 1996).

Some preliminary definitions are required before describing the computation of the BKK bound. The exponent vector of a given term $x_1^{e_1}x_2^{e_2}\cdots x_n^{e_n}$ is the vector $(e_1, e_2, \ldots, e_n)$. Associated with a given polynomial f_i is the set of all exponent vectors for its terms, which is called the support of the polynomial. This terminology is taken from the work of Emiris (1993b, 1994b). The Newton polytope of f_i is the convex hull of the support of f_i, that is the smallest convex polyhedron which contains all of the exponent vectors in the support.

Two useful concepts from combinatorial geometry are the Minkowski Sum and the mixed volume function. For sets A_1 and A_2 in $\mathbb{R}^n$, the Minkowski Sum $A_1 + A_2$ is $\{a_1 + a_2 \mid a_1 \in A_1, a_2 \in A_2\}$. If A_1 and A_2 are convex polytopes, then so is their Minkowski Sum $A_1 + A_2$. The mixed volume of a collection of convex polytopes is a unique, real-valued function which is defined by the requirements of multilinearity with respect to Minkowski addition and scalar multiplication. The mixed volume for n polytopes A_i is given by the formula

$$\sum_{I \subset \{1,\ldots,n\}} (-1)^{n-|I|} Vol\left(\sum_{i \in I} A_i\right) \tag{3.2}$$

where I ranges over all subsets of $\{1, \ldots, n\}$, $|I|$ denotes the cardinality of I, and the second sum represents Minkowski addition.

The BKK bound can be stated as follows: an upper bound on the number of finite solutions to a set of polynomial equations f_i is given by the mixed volume of the Newton polytopes A_i corresponding to the polynomial supports. Equality holds for general coefficients (Bernstein 1975); in fact, only the coefficients corresponding to the Newton polytope vertices need be general for equality to hold (Canny and Rojas 1991). For Eqns. (3.1), the Newton polytopes A_1 and A_2 corresponding to f_1 and f_2, respectively, are shown in Fig. 3.1, as well as the Newton polytope for the Minkowski sum $A_1 + A_2$. To find a bound on the number of solutions, we apply the BKK formula to this system:

$$(-1)^{2-|\{1\}|}Vol(A_1)+(-1)^{2-|\{2\}|}Vol(A_2)+(-1)^{2-|\{1,2\}|}Vol(A_1+A_2) \quad (3.3)$$

Using the volumes of the polytopes in Fig. 3.1, the BKK bound is $-1 - \frac{1}{2} + 3\frac{1}{2} = 2$. This bound is less than both the Bezout number (8) and the Multihomogeneous Bezout number (4) for this system. The BKK bound is the tightest bound, but it is also computationally the most complex, particularly for large systems. There are, however, software packages available to perform the computation (Verschelde 1996, Emiris 1993a).

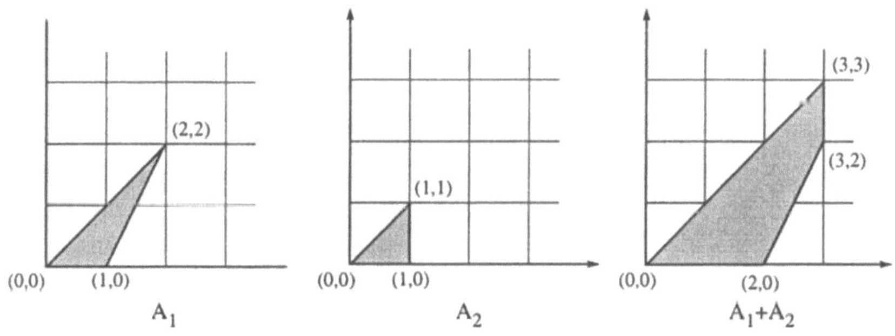

Fig. 3.1. Newton polytopes for calculation of BKK bound

4. Solution Methods

A variety of solution methods have been developed for solving sets of nonlinear polynomial equations. These methods all have strengths and weaknesses, and the applicability of each method is highly dependent on the particular problem being addressed. The following sections will describe polynomial continuation and resultant methods in detail. The Gröbner Basis method is also briefly reviewed. Only polynomial continuation is a purely numerical method; Gröbner Basis is an iterative algebraic variable elimination technique, while the resultant method is an algebraic technique capable of eliminating all but one variable in a single step.

4.1 Polynomial Continuation

For kinematic analysis, we are interested mainly in solution methods which give all possible solutions to a particular set of nonlinear equations. While numerical methods, such as Newton-Raphson, converge to a single solution, the polynomial continuation method is a numerical procedure which can find all solutions to a given problem. In its original form, this technique was known as the *Bootstrap Method*, as developed by Roth and Freudenstein (1963). Wampler et al. (1990) have provided a thorough overview of the procedure in its current highly-developed state.

The basic premise is that small perturbations in the coefficients of a system lead to small changes in the solutions. Given this fact, the idea is to begin with a "start" system whose solutions are known, then transform it gradually to the system whose solutions are sought. During this transformation, all solutions are tracked so that in the end all the solutions to the final system are found.

There is a strong correspondence between the bounds described in Sec. 3. and the choice of a start system. For 1-homogeneous systems, the following system suffices:

$$x_j^{d_j} - 1 = 0, \qquad (j = 1, 2, \ldots, n) \tag{4.1}$$

where d_j is the degree of equation j. Note that there are $\prod_{j=1}^{n} d_j$ distinct solutions to this system. We now require a schedule, also called a homotopy, for transforming the start system $G(x)$ to the final system $F(x)$. The following system, parameterized by t, will perform this transformation:

$$H(x,t) = (1-t)e^{i\Theta}G(x) + tF(x) \tag{4.2}$$

where Θ is a random real number. When $t = 0$, the solutions of $H(x,t)$ are the same as the start system $G(x)$; when $t = 1$, the solutions of $H(x,t)$ are the same as the final system $F(x)$, whose solutions are sought. As t is varied smoothly from 0 to 1, the solutions may be obtained using a Newton-Raphson procedure at each step of the transformation, with the results from the previous step as an initial guess. Not all paths will necessarily converge to a finite solution; some may go to infinity. It is guaranteed that all solutions to $F(x)$ will be found. The factor $e^{i\Theta}$ in Eq. (4.2) is included for numerical stability, to avoid Jacobian singularities which would cause the Newton-Raphson iteration to fail during certain intermediate values of t.

Consider once again the system given by Eqs. (3.1). Following Eq. (4.1), we have as a start system

$$\begin{aligned} g_1: &\quad x_1^4 - 1 = 0 \\ g_2: &\quad x_2^2 - 1 = 0 \end{aligned} \tag{4.3}$$

which has the eight solutions $(1,1)$, $(1,-1)$, $(-1,1)$, $(-1,-1)$, $(i,1)$, $(i,-1)$, $(-i,1)$, and $(-i,-1)$, where $i = \sqrt{-1}$. Following the homotopy given by

Eq. (4.2), the eight paths corresponding to these solutions are tracked as t varies from 0 to 1. Six paths diverge to infinity, while two paths converge to the solutions $(-35.1, -0.129)$ and $(-4.90, 0.105)$. Note that the BKK bound for this system was 2, so in this case that bound was exact.

If a multihomogeneous start system is used, then the number of paths that must be tracked is equal to the multihomogeneous Bezout number, which may reduce the computational burden. If the degree of equation l with respect to variable group j is d_{jl}, then the corresponding start equation is given as a product of factors $\prod_{j=1}^{m} f_{jl}(x_{j1}, x_{j2}, \ldots, x_{jk_j})$, where the degree of f_{jl} is d_{jl}. For the example problem given by Eqns. (3.1), if we use variable groupings $\{x_1\}$ and $\{x_2\}$, then the following start system will suffice:

$$\begin{aligned} g_1 : &\quad (x_1^2 - 1)\,(x_2^2 - 1) = 0 \\ g_2 : &\quad (x_1 - 2)\,(x_2 - 2) = 0 \end{aligned} \tag{4.4}$$

Recognizing that one factor from each equation must vanish for a solution to exist, the solutions to this start system are $(1, 2)$, $(-1, 2)$, $(2, 1)$, and $(2, -1)$. The paths of these four start solutions may then be tracked as the system is transformed using the standard homotopy of Eq. (4.2).

Other procedures reduce the number of paths still further. There has been some effort to develop polynomial continuation methods which exploit the BKK bound (Li et al. 1996, Verschelde 1994). It is also possible to solve a general case of a specific problem, then use that solution as a start system for new instances of the same problem. Using this procedure, called Coefficient Continuation, only a number of paths equal to the number of finite solutions for the problem need be tracked, which may reduce computational burden considerably. If the coefficients themselves are functions of some parameters, such as link lengths for kinematic analysis, then Parameter Polynomial Continuation may be used. A general case of the problem is solved, then this solution is used as a start system and the coefficient parameters are transformed to solve for a specific instance of the problem.

Perhaps the greatest strengths of polynomial continuation are its ability to solve very large systems, and the fact that the procedure itself need not be modified for different polynomial systems. It is also virtually guaranteed to find all solutions to a system, assuming there are no numerical anomalies (which can usually be handled by the path-tracking algorithm). For these reasons, polynomial continuation has been the tool that enabled the original solutions of many long-standing kinematics problems. Tsai and Morgan (1985) first showed that the inverse kinematics of the general 6-R serial manipulator has 16 solutions using polynomial continuation; Raghavan (1993) used the method to show that the direct kinematics of the general Stewart-Gough platform has 40 solutions. Also, the nine-point path synthesis problem for four-bar linkages was shown by polynomial continuation to have 1442 non-degenerate solutions (Wampler et al. 1992). This latter result relied on a problem formulation which represents the links as vectors in the complex plane. Using this formulation reduces the multihomogeneous Bezout bound relative to other formulations, which means fewer paths needed to be followed during continuation. For spatial problems, quaternion formulations tend to lead to a similar reduction in solution bounds.

The disadvantage of polynomial continuation is mainly its speed. For many kinematics problems, the number of paths which must be tracked can be large enough that the continuation calculation is too slow for real-time control or other applications where speed is important. Much progress in execution time has been made, though. For example, all 40 solutions of the general Stewart-Gough platform direct kinematics problem can be calculated in 14 seconds on an IBM RS/6000 (Wampler 1994). Moreover, the continuation algorithm is naturally amenable to parallel computation, since the paths are independent. Computer packages for performing polynomial continuation are publicly available (Rheinboldt and Burkardt 1991, Verschelde 1996).

4.2 Gröbner Basis

The Gröbner basis technique is an iterative algebraic variable elimination technique. The basic elimination procedure resembles Gaussian elimination in many ways. First, as in Gaussian elimination, an ordering for the unknowns is chosen, then an algorithm is used to eliminate the unknowns sequentially, eventually achieving a triangular system. For Gaussian elimination, the triangular system means that the last equation has only one unknown, and each subsequent linear equation adds a single unknown. For Gröbner basis, the last equation is a univariate polynomial, and each subsequent equation adds at most one new variable, although the equation may not be linear in that variable. The univariate polynomial may be solved to find all possible values of one unknown, and the other equations will yield the values of the other variables for each solution.

Unlike Gaussian elimination, however, the choice of ordering for the polynomial terms in a nonlinear equations is not obvious. It turns out that the choice of a lexicographic ordering, which resembles the ordering of words in a dictionary, will always lead to a triangular Gröbner basis. For a lexicographic ordering, $t_1 = x_1^{\alpha_1} x_2^{\alpha_2} \cdots x_n^{\alpha_n}$ is less than $t_2 = x_1^{\beta_1} x_2^{\beta_2} \cdots x_n^{\beta_n}$ if the leftmost non-zero entry in the difference of the exponent vectors (i.e., $(\alpha_1 - \beta_1, \alpha_2 - \beta_2, \ldots, \alpha_n - \beta_n)$) is negative. Buchberger (1976) first developed an elimination procedure to find the Gröbner basis of a set of polynomial equations; an overview of the algorithm is provided by Raghavan and Roth (1995). The basic idea is to eliminate the highest-ordered terms in a given set by adding multiples of the other equations in the set, a process known as reduction.

The polynomials in the triangular set given by the Gröbner basis algorithm generate the same polynomial ideal as the original set, and are thus different "bases" for that ideal. The ideal generated by polynomial equations $f_1, f_2, \ldots, f_n$ in the variables $x_1, x_2, \ldots, x_m$ is defined to be the set of all elements of the form $f_1g_1 + f_2g_2 + \cdots + f_ng_n$, where g_i are arbitrary polynomials in the variables x_i. From this definition, it is clear that all polynomial sets which generate the same polynomial ideal have the same set of zeroes. Thus, the solutions to the triangular basis are the same as the solutions to the original system.

The disadvantage of the Gröbner basis technique is that the Buchberger algorithm may generate a large number of complex intermediate polynomials before converging to the Gröbner basis. As a result, computation time may be prohibitively long. Also, the complexity of a given problem is unpredictable. Nevertheless, the technique has proven useful in kinematic analysis, most notably in establishing the number of solutions for the general case of the Stewart-Gough platform direct kinematics problem (Mourrain 1993), as well as special cases where platform legs are required to share pivot locations (Lazard and Faugère 1995).

Most modern computer algebra systems include implementations of the Gröbner basis algorithm (Char et al. 1992, Wolfram 1991), and specialized implementations are available as well (Chauvin and Faugère 1994).

4.3 Resultant Methods

The resultant of n non-homogeneous polynomial equations in $n-1$ variables is a homogeneous function of the coefficients, whose vanishing is a necessary and sufficient condition for the existence of a solution to the system of equations. The basic resultant construction procedure is best illustrated by an example. Eqns. (3.1) are repeated for convenience:

$$\begin{aligned} 3x_1^2x_2^2 + 2x_1 + 9 &= 0 \\ 6x_1x_2 + x_1 + 8 &= 0 \end{aligned} \tag{4.5}$$

These are two equations in the two variables x_1 and x_2. To solve the system by resultant calculation, we begin by rewriting the equations with one of the variables included in the coefficient field, or "suppressed." Suppressing x_2, Eqns. (4.5) become two equations in one variable:

$$\begin{aligned} (3x_2^2)x_1^2 + (2)x_1 + (9)1 &= 0 \\ (6x_2 + 1)x_1 + (8)1 &= 0 \end{aligned} \tag{4.6}$$

where the coefficients are in parentheses and the constant 1 has been treated as an unknown. This step seems counterintuitive but the logic becomes more clear if the equations are now viewed as linear in the unknowns $\{x_1^2, x_1, 1\}$. With this view, there are two equations in three unknowns, and only one more equation is needed to be able to solve the system. Note that the polynomial terms x_1^2 and x_1 are treated as separate linear variables in this analysis. These unsuppressed polynomial terms are called power products.

The extra equation may be obtained by multiplying the second equation above by x_1, which yields the following augmented set of equations:

$$\begin{aligned}(3x_2^2)x_1^2 + (2)x_1 + (9)1 &= 0\\ (6x_2 + 1)x_1 + (8)1 &= 0\\ (6x_2 + 1)x_1^2 + (8)x_1 &= 0\end{aligned} \tag{4.7}$$

This set of equations may be rewritten in matrix form:

$$\begin{bmatrix} 3x_2^2 & 2 & 9 \\ 0 & 6x_2+1 & 8 \\ 6x_2+1 & 8 & 0 \end{bmatrix} \begin{bmatrix} x_1^2 \\ x_1 \\ 1 \end{bmatrix} = \begin{bmatrix} 0 \\ 0 \\ 0 \end{bmatrix} \tag{4.8}$$

With the equations in matrix form, the rationale for treating 1 as an unknown becomes clear. If the matrix of coefficients in Eq. (4.8) were invertible, then both sides of the equation could be multiplied by that inverse, which would yield $1 = 0$ for the final equation. This is a contradiction, which implies that the matrix is not invertible, and therefore has a determinant equal to 0. This determinant is the resultant of the system, a univariate polynomial in the suppressed variable x_2. Finding the zeroes of this polynomial yields all values of x_2 for which a solution of the original system exists. For Eq. (4.8), the determinant of the matrix of coefficients is

$$g = -516x_2^2 - 12x_2 + 7, \tag{4.9}$$

and the zeroes of g are $x_2 = -0.129$ and $x_2 = 0.105$, which are in agreement with the solutions found by polynomial continuation in Sec. 4.1.

The variable elimination procedure outlined above, where a number of equations equal to the number of power products is obtained by multiplication of equations by the variables, is known generally as dialytic elimination. It has been known since the 19th century that a resultant for two equations may be formed by the above procedure (Salmon 1885). Because only one variable remains unsuppressed for two-equation problems, it is straightforward to specify the multiplying terms for the elimination procedure. The result is the classical Sylvester resultant of two binary forms. Note that this procedure could be used to solve the input-output problem for any one-loop mechanism in the plane, with revolute and slider joints. Single loop mechanisms in the plane, including four-bar mechanisms, lead to two equations in two unknowns. Efficient implementations of the binary resultant are available in most modern computer algebra systems (Char et al. 1992, Wolfram 1991).

4.3.1 Resultants of Multivariate Problems. For more than two equations, the set of multiplying terms necessary to create a multivariate resultant is difficult to predict. The challenge is finding a determinant whose vanishing is a necessary and sufficient condition for the existence of a solution, and which is not identically equal to zero due to linear dependencies among the rows of the matrix. If the vanishing of the determinant is a necessary but not a sufficient condition for the existence of a solution, then there could be extraneous solutions in the univariate polynomial obtained from the determinant. For kinematics problems, which tend to lead to sparse polynomials, there is unfortunately no systematic procedure for finding the optimal set of multiplying terms.

One of the earliest attempts to develop a generalized procedure for creating a multivariate resultant was due to Macaulay (1916). Suppose we are given n homogeneous polynomials $f_1, f_2, \ldots, f_n$ in n variables $x_1, x_2, \ldots, x_n$. Let $d_M = 1 + \sum_{i=1}^{n} d_i$, where d_i is the degree of equation f_i. Macaulay's construction gives a square matrix A of size $\binom{d_M+n-1}{d_M}$. For general coefficients, the determinant of this square matrix is not identically zero, and the disappearance of the determinant is a necessary but not sufficient condition for the existence of a solution. The resultant may be recovered as the ratio of the determinant of A to the determinant of a submatrix of A. Unfortunately, for kinematics problems either the determinant of A or the submatrix of A are often identically zero. Also, the Macaulay construction often leads to matrices which are of unreasonable size for even relatively simple problems.

More recently, Sturmfels and Zelevinski (1994) have presented a significant advance in resultant construction techniques for certain multihomogeneous systems. The formula for the resultant is said to be of the Sylvester type because the resultant is obtained from a single determinant rather than a ratio of determinants as in Macaulay's construction. The following development is taken from (Sturmfels and Zelevinsky 1994). Suppose we are given $l+1$ generic polynomials which are separately homogeneous in r groups of variables, where the kth group consists of $l_k + 1$ variables $\mathbf{x}_k = \{x_{k0}, x_{k1}, \ldots, x_{kl_k}\}$. The degree of the kth variable group is d_k, and $l = l_1 + l_2 + \cdots + l_r$.

If $l_k = 1$ or $d_k = 1$ for $k = 1, 2, \ldots, r$, then there are at least $r!$ different Sylvester type formulas for the resultant. These resultant formulas are indexed by permutations π of $\{1, 2, \ldots, r\}$. For a given permutation, to find the resultant matrix each equation in the multihomogeneous set is multiplied by a set of terms determined by the multiindex $(m_1, m_2, \ldots, m_r)$ which is obtained by the following rule:

$$m_k = (d_k - 1)l_k + d_k \sum_{j:\pi(j)<\pi(k)} l_j \tag{4.10}$$

To construct the matrix, we multiply each of the equations by all terms of degree m_k in the variable groups $\mathbf{x}_k$. The result is an equal number of equations and power products, which leads to a square matrix whose determinant is the resultant for the system.

For example, consider the following three-equation polynomial system in the variables x_1, x_2, and x_3:

$$\begin{aligned} 3x_1x_2x_3 + 5x_1x_3 + 2x_2 + 6 &= 0 \\ 6x_1x_2 + 6x_1 - x_2x_3 - 5 &= 0 \\ 7x_1x_2x_3^2 - 8x_1 + 2x_2 - 3 &= 0 \end{aligned} \tag{4.11}$$

After suppressing the variable x_3, we homogenize the equations separately with respect to x_1 and x_2, using w_1 and w_2 as the homogenizing variables. The result is a set of three multihomogeneous equations:

$$\begin{aligned}(3x_3)x_1x_2 + (5x_3)x_1w_2 + (2)w_1x_2 + (6)w_1w_2 = 0\\ (6)x_1x_2 + (6)x_1w_2 + (-x_3)w_1x_2 + (-5)w_1w_2 = 0\\ (7x_3^2)x_1x_2 + (-8)x_1w_2 + (2)w_1x_2 + (-3)w_1w_2 = 0\end{aligned} \tag{4.12}$$

There are two sets of variable groups: $\mathbf{x}_1 = \{x_1, w_1\}$ is of degree 1, as is $\mathbf{x}_2 = \{x_2, w_2\}$. Following the notation above, $l = 2$, $r = 2$, $l_1 = 1$, $d_1 = 1$, $l_2 = 1$, and $d_2 = 1$. There are two possible multihomogeneous resultants, indexed by the permutations of $\{1, 2\}$. For the permutation $\{1, 2\}$, we have $m_1 = (1-1)1 + 1(0) = 0$ and $m_2 = (1-1)1 + 1(1) = 1$. To find the resultant, then, we multiply each equation in Eqns. (4.12) by all terms of degree 1 in the variable group $\{x_2, w_2\}$. The augmented set of equations may be written in matrix form as

$$\begin{bmatrix} 0 & 0 & 3x_3 & 2 & 5x_3 & 6 \\ 0 & 0 & 6 & -x_3 & 6 & -5 \\ 0 & 0 & 7x_3^2 & 2 & -8 & -3 \\ 3x_3 & 2 & 5x_3 & 6 & 0 & 0 \\ 6 & -x_3 & 6 & -5 & 0 & 0 \\ 7x_3^2 & 2 & -8 & -3 & 0 & 0 \end{bmatrix} \begin{bmatrix} x_1x_2^2 \\ w_1x_2^2 \\ x_1x_2w_2 \\ w_1x_2w_2 \\ x_1w_2^2 \\ w_1w_2^2 \end{bmatrix} = \begin{bmatrix} 0 \\ 0 \\ 0 \\ 0 \\ 0 \\ 0 \end{bmatrix} \tag{4.13}$$

and the determinant of the matrix of coefficients gives the resultant of the system. In this case, the determinant is a sixth-degree polynomial in x_3, whose vanishing is a necessary and sufficient condition for the existence of a solution to Eqns. (4.11).

Given the multihomogeneous resultant theory, it is advantageous to seek problem formulations which meet the criteria for the existence of a Sylvester type resultant. Several recent solutions to kinematic synthesis problems have fallen under multihomogeneous resultant theory (Nielsen and Roth 1995, Innocenti 1994). Unfortunately, the existence of a Sylvester type resultant does not guarantee that the determinant will not be identically zero for some kinematics problems, because the equation coefficients may be interrelated or special in some sense.

There has been some effort to develop resultants which take the BKK bound into account (Emiris 1993b, Gelfand et al. 1994). An entity known as the sparse mixed resultant, which is a function of the coefficients of a set of $n + 1$ polynomials, is determined by the Newton polytopes associated with those polynomials. As a consequence of Bernstein's work (1975), this resultant must be separately homogeneous in the coefficients of each polynomial f_i, and its degree in the coefficients of a specific polynomial f_i is the mixed volume of the other n polynomials. Algorithms to construct the sparse mixed resultant have been suggested (Emiris 1994a, Emiris 1994b), although there is currently no known procedure that will yield the sparse mixed resultant for all cases.

4.3.2 Resultant Applications in Kinematics. One of the most successful applications of resultant theory to kinematics was the solution of the inverse kinematics problem for serial manipulators (Lee and Liang 1988, Raghavan and Roth 1990). Resultant-based solution methods for 6-R inverse kinematics problems are capable of finding all 16 solutions to a general problem in 11 milliseconds on an IBM RS/6000 computer (Manocha and Canny 1994). This solution illustrates the greatest strength of resultant methods: computation time. If a resultant-based solution method can be found for a particular problem, it normally leads to much faster computation times than polynomial continuation or Gröbner basis methods. Resultant methods can also give greater insight into a problem than a purely numerical solution, as demonstrated by the work of Mavroidis and Roth (1995) in analyzing overconstrained mechanisms. The main disadvantage of resultant-based methods is the challenge of finding an appropriate multivariate resultant for a particular problem.

The application of resultant theory to serial mechanisms has benefited greatly from some unique properties associated with the inverse kinematics equations for such mechanisms. Certain functions of the equations themselves (which belong to the same ideal as the original equations) are composed of the same power products as the original equations. This unexpected result makes the resultant calculation possible, since the new equations are linearly independent of the original set. An example of this property may be found in Eqns. (2.1) for the inverse kinematics of a three-revolute joint serial mechanism in the plane. Those equations may be rewritten as

$$\begin{aligned} l_1 \cos\theta_1 + l_2 \cos\theta_1 \cos\theta_2 - l_2 \sin\theta_1 \sin\theta_2 &= x \\ l_1 \sin\theta_1 + l_2 \sin\theta_1 \cos\theta_2 + l_2 \cos\theta_1 \sin\theta_2 &= y \end{aligned} \tag{4.14}$$

and the sum of the squares of these equations may be simplified to

$$l_1^2 + l_2^2 + 2l_1 l_2 \cos\theta_2 = x^2 + y^2 \tag{4.15}$$

This equation is already a univariate equation in θ_2, so solving it will give all values of θ_2 for which a solution to the original system of equations exists. Notice that despite having squared and added the original equations, the simplified equation is linear in $\cos\theta_2$, as are Eqns. (4.14). No new power products were created.

Resultant methods have also been used extensively for the analysis of in-parallel mechanisms. Numerous authors have used resultant methods to solve the direct kinematics problems for Stewart platforms with some special geometric constraints, such as concentric spherical joints. The first algorithm to solve the general case of the Stewart platform direct kinematics problem, and obtain a 40th degree univariate polynomial, uses resultants as well (Husty 1996). This solution procedure uses a formulation based on Soma coordinates, and applies the Sylvester resultant of binary forms repeatedly, along with polynomial greatest-common-divisor calculations to eliminate extraneous solutions. The Soma coordinate formulation helps to reduce the size of the intermediate polynomials in this solution algorithm.

4.3.3 Implementation Details. Instead of expanding the determinant obtained by a dialytic elimination procedure to obtain a univariate polynomial, it is possible to find all roots of that polynomial from an eigenvalue problem. This calculation is much easier to implement in practice, and robust eigenvalue routines are readily available (Anderson et al. 1992).

Suppose that before dialytic elimination, the final matrix equation may be written as

$$[A_n x^n + A_{n-1} x^{n-1} + \cdots + A_0]\mathbf{y} = 0 \tag{4.16}$$

where the suppressed variable is x, and $\mathbf{y}$ is a vector of unsuppressed power products, including 1. The following generalized eigenvalue problem may be solved to find all values of x for which the determinant of the above matrix vanishes:

$$\begin{bmatrix} I & 0 & 0 & & \\ 0 & I & 0 & \cdots & \\ 0 & 0 & I & & \\ & \vdots & & \ddots & \\ A_{n-1} & A_{n-2} & A_{n-3} & \cdots & A_0 \end{bmatrix} \begin{bmatrix} x^{n-1}\mathbf{y} \\ x^{n-2}\mathbf{y} \\ \vdots \\ \vdots \\ \mathbf{y} \end{bmatrix} =$$

$$x \begin{bmatrix} 0 & I & 0 & \cdots & \\ 0 & 0 & I & & \\ 0 & 0 & 0 & \ddots & \\ & \vdots & & & I \\ -A_n & 0 & 0 & \cdots & 0 \end{bmatrix} \begin{bmatrix} x^{n-1}\mathbf{y} \\ x^{n-2}\mathbf{y} \\ \vdots \\ \vdots \\ \mathbf{y} \end{bmatrix} \tag{4.17}$$

This is a generalized eigenvalue problem of the form $G\mathbf{z} = xH\mathbf{z}$, or equivalently $G\mathbf{z} - xH\mathbf{z} = 0$. If H is invertible, then we have $H^{-1}G\mathbf{z} - x\mathbf{z} = 0$, which is a regular eigenvalue problem. Otherwise, the generalized problem $G\mathbf{z} - xH\mathbf{z} = 0$ may be solved using a QZ algorithm (Golub and Van Loan 1989). To find the roots of the univariate polynomial which would be obtained by expanding the determinant of an $m \times m$ matrix which contains a suppressed variable of degree n, then, a generalized eigenvalue problem of size mn may be solved. Note also that the corresponding eigenvectors contain the values for all power products, from which one may find the values of the unsuppressed variables in the problem.

4.3.4 Dixon Determinants. There is an alternative procedure for finding the resultant of a system of equations based on work done by Bezout, reformulated by Cayley and later extended to bivariate cases by Dixon (1908). Suppose we are given $n+1$ polynomials f_j in n variables $\{x_1, x_2, \ldots, x_n\}$. Consider the following determinental equation for Δ:

$$\Delta = \begin{vmatrix} f_1(x_1,x_2,\ldots,x_n) & f_2(x_1,x_2,\ldots,x_n) & \cdots & f_{n+1}(x_1,x_2,\ldots,x_n) \\ f_1(a_1,x_2,\ldots,x_n) & f_2(a_1,x_2,\ldots,x_n) & \cdots & f_{n+1}(a_1,x_2,\ldots,x_n) \\ f_1(a_1,a_2,\ldots,x_n) & f_2(a_1,a_2,\ldots,x_n) & \cdots & f_{n+1}(a_1,a_2,\ldots,x_n) \\ \vdots & \vdots & & \vdots \\ f_1(a_1,a_2,\ldots,a_n) & f_2(a_1,a_2,\ldots,a_n) & \cdots & f_{n+1}(a_1,a_2,\ldots,a_n) \end{vmatrix} \tag{4.18}$$

where the a_i are new variables which replace the x_i sequentially in the rows of the matrix. Dixon observed that the above determinant will vanish if $x_i = a_i$ for any i. Therefore, Δ must have as a factor the polynomial $(x_1 - a_1)(x_2 - a_2)\cdots(x_n - a_n)$. The Dixon polynomial δ is defined as

$$\delta = \frac{\Delta}{(x_1 - a_1)(x_2 - a_2)\cdots(x_n - a_n)} \tag{4.19}$$

Moreover, the determinant Δ must vanish at any solution $(x_1, x_2, \ldots, x_n)$ of the original system. As a result, all coefficients of the polynomial δ when considered to be a polynomial in the a_i, $i = 1, 2, \ldots, n$, must vanish at any solution as well. This requirement leads to a set of equations g_j in x_i which are a basis for the same polynomial ideal as the original set of equations. The equations g_j may be written as $M\mathbf{y} = 0$, where $\mathbf{y}$ is a vector of power products in the variables x_i. For a set of generic polynomials f_i which are all of degree m, it can be shown that the matrix M is square. The determinant of this matrix is the resultant of the system.

For example, consider the Eqns. (3.1), which may be rewritten with x_2 suppressed as

$$\begin{aligned} f_1: &\quad (3x_2^2)x_1^2 + (2)x_1 + (9)1 = 0 \\ f_2: &\quad (6x_2 + 1)x_1 + (8)1 = 0 \end{aligned} \tag{4.20}$$

This is $n + 1$ equations in n unknowns, where $n = 1$, so we can follow the procedure outlined above to find the resultant. We have

$$\Delta = \begin{vmatrix} (3x_2^2)x_1^2 + (2)x_1 + (9) & (6x_2 + 1)x_1 + (8) \\ (3x_2^2)a_1^2 + (2)a_1 + (9) & (6x_2 + 1)a_1 + (8) \end{vmatrix} \tag{4.21}$$

Expanding the determinant and factoring out $(x_1 - a_1)$ gives

$$\delta = \left(18x_1x_2^3 + 3x_1x_2^2 + 24x_2^2\right)a_1 + \left(24x_1x_2^2 - 54x_2 + 7\right)1 \tag{4.22}$$

Setting the coefficients of a_1 and 1 to zero gives two equations in x_1 and the suppressed variable x_2. These equations may be written as:

$$\begin{bmatrix} 3x_2^2 + 18x_2^3 & 24x_2^2 \\ 24x_2^2 & 7 - 54x_2 \end{bmatrix} \begin{bmatrix} x_1 \\ 1 \end{bmatrix} = \begin{bmatrix} 0 \\ 0 \end{bmatrix} \tag{4.23}$$

and the determinant of the above matrix gives the resultant of the system: $-1548x_2^4 - 36x_2^3 + 21x_2^2$. Comparing this resultant with the Sylvester resultant of binary forms given in Eq. (4.9), there is an extraneous factor $3x_2^2$. This fact points to one weakness in the Dixon formulation, which is that it assumes that the input polynomials are of equal degree. If they are not, then there can be extraneous factors in the determinant. For two equations, the extraneous factor is ${I_{f_1}}^{degree(f_1)-degree(f_2)}$, where I_{f_1} is the coefficient of the highest-degree term in f_1 (Kapur and Lakshman 1992). Another problem with the Dixon formulation is that for a set of polynomials with differing degrees the matrix M is not guaranteed to be square.

Despite its weaknesses, the Dixon determinant performs well in practice. Recent interest has focused on its ability to take advantage of sparsity in polynomials. Kapur and Saxena (1996) have shown that the size of matrices created by the Dixon formulation compares favorably with the size of those created by the current algorithms for finding sparse resultants. Nevertheless, specific applications to kinematics problems have been very limited.

5. Conclusions

Regardless of the method of problem formulation for a kinematic synthesis, direct kinematics, or inverse kinematics problem, the result is a set of nonlinear equations. Though the number of solutions to such a problem is never known until the nonlinear equations are actually solved, the BKK bound provides a remarkably tight bound on the number of solutions for most kinematics problems. The Bezout and multihomogeneous Bezout numbers are somewhat looser bounds, but are computationally easier to calculate.

Polynomial continuation is applicable to any set of nonlinear equations, but the computation time is better for formulations that lead to a low BKK or multihomogeneous Bezout bound. Gröbner basis solution methods have proven useful for establishing the number of solutions to problems, but algorithmic complexity has limited their usefulness in an industrial setting. Lastly, resultant methods can lead to computationally fast solutions, but a formula for the resultant is difficult to find for most problems. Finding a problem formulation which is amenable to elimination by resultants is often a matter of trial-and-error, but if such a formulation can be found then the computational complexity of solving a given problem may be reduced considerably versus polynomial continuation or Gröbner basis methods.

References

Anderson, E., Bai, Z., Bischof, C., Demmel, J., Dongarra, J., Du Cros, J., Greenbaum, A., Hammarling, S., Mckenney, A., Ostrouchov, S., and Sorensen, D., 1992, *LAPACK users' guide*, Society for Industrial and Applied Mathematics, Philadelphia.

Bernstein, D., 1975, The number of roots of a system of equations, *Funct. Anal. and Appl.*, Vol. 9, pp. 183–185.

Bottema, O., and Roth, B., 1979, *Theoretical Kinematics*, pp. 393–445, North-Holland (reprinted by Dover Publications, NY, 1990).

Buchberger, B., 1976, A Theoretical Basis for the Reduction of Polynomials to Canonical Form, *ACM SIGSAM Bull.*, Vol. 10, pp. 19–29.

Canny, J., and Rojas, J., 1991, An Optimal Condition for Determining the Exact Number of Roots of a Polynomial system, *Proceedings ACM International Symposium on Symbolic and Algebraic Computation*, Bonn, pp. 96–102.

Char, B., Geddes, K., Gonnet, G., Leong, B., Monagan, M., and Watt, S., 1992, *First Leaves: A Tutorial Introduction to Maple V*, Springer-Verlag, Berlin.

Chauvin, C., and Faugère, J.-C., 1994, Basic user's manual of the Faugère's GB package (available at ftp://posso.ibp.fr/pub/softwares/GB/).

Dixon, A. L., 1908, The Eliminant of Three Quantics in Two Independent Variables, *Proceedings London Mathematical Society*, Volume 6, Series 2, pp. 468–478.

Emiris, I., 1993a, Mymix (computer program available on the Internet at ftp://robotics.eecs.berkeley.edu/pub/MixedVolume/).

Emiris, I., 1993b, A Practical Method for the Sparse Resultant, *Proceedings ACM International Symposium on Symbolic and Algebraic Computation*, Kiev, pp. 183–192.

Emiris, I., 1994a, Polynomial System Solving Package (computer program available at ftp://robotics.eecs.berkeley.edu/pub/emiris/res_solver/).

Emiris, I., 1994b, *Sparse Elimination and Applications in Kinematics*, Ph.D. Thesis, University of California at Berkeley.

Gelfand, I., Kapranov, M., and Zelevinsky, A., 1994, *Discriminants, Resultants, and Multidimensional Determinants*, Birkhäuser, Boston.

Golub, G. H., and Van Loan, C. F., 1989, *Matrix Computations*, The Johns Hopkins University Press, Baltimore, 2nd edn.

Husty, M., 1996, An Algorithm for Solving the Direct Kinematics of General Stewart-Gough Platforms, *Mech. Mach. Theory*, Vol. 31, pp. 365–379.

Innocenti, C., 1994, Polynomial Solution of the Spatial Burmester Problem, *Mechanism Synthesis and Analysis*, edited by Pennock, G. R., New York, Volume DE-70, pp. 161–166. The American Society of Mechanical Engineers, New York.

Kapur, D., and Lakshman, Y. N., 1992, Elimination Methods: an Introduction, *Symbolic and Numerical Computation for Artificial Intelligence*, edited by Donald, B. R., Kapur, D., and Mundy, J. L., Chapter 2, pp. 45–87. Academic Press, London.

Kapur, D., and Saxena, T., 1996, Sparsity Considerations in Dixon Resultants, *Conference Proceedings of the Annual ACM Symposium on Theory of Computing 1996*, New York, pp. 184–191. ACM.

Khovanskii, A., 1978, Newton Polyhedra and the Genus of Complete Intersections, *Funktsional'nyi Analiz i Ego Prilozheniya*, Vol. 12, pp. 51–61.

Kushnirenko, A. G., 1975, The Newton Polyhedron and the Number of Solutions of a System of k Equations in k Unknowns, *Uspekhi Mat. Nauk.*, Vol. 30, pp. 266–267.

Lazard, D., and Faugère, J.-C., 1995, The Combinatorial Classes of Parallel Manipulators, *Mech. Mach. Theory*, Vol. 30, pp. 765–776.

Lee, H., and Liang, C., 1988, Displacement Analysis of the General Spatial 7-Link 7R Mechanism, *Mechanism and Machine Theory*, Vol. 15, pp. 219–226.

Li, T., Wang, T., and Wang, X., 1996, Random Product Homotopy with Minimal BKK Bound, *The Mathematics of Numerical Analysis*, Volume 32 of *Lectures in Applied Mathematics*, pp. 503–512. American Mathematical Society.

Macaulay, F. S., 1916, *The Algebraic Theory of Modular Systems*, Volume 19 of *Cambridge Tracts in Mathematics and Mathematical Physics*, Cambridge University Press, Cambridge, UK.

Manocha, D., and Canny, J., 1994, Efficient Inverse Kinematics for General 6R Manipulators, *IEEE Trans. on Robotics and Automation*, Vol. 10, pp. 648–657.

Mavroidis, C., and Roth, B., 1995, Analysis of Overconstrained Mechanisms, *ASME Journal of Mechanical Design*, Vol. 117, pp. 69–74.

Mourrain, B., 1993, The 40 Generic Positions of a Parallel Robot, *Proceedings ACM International Symposium on Symbolic and Algebraic Computation*, Kiev, pp. 173–182.

Nielsen, J., and Roth, B., 1995, Elimination Methods for Spatial Synthesis, *Computational Kinematics '95*, edited by Merlet, J. P., and Ravani, B., Volume 40 of *Solid Mechanics and its Applications*, pp. 51–62. Kluwer Academic Publishers.

Raghavan, M., 1993, The Stewart Platform of General Geometry Has 40 Configurations, *ASME Journal of Mechanical Design*, Vol. 115, pp. 277–282.

Raghavan, M., and Roth, B., 1990, Kinematic Analysis of the 6R Manipulator of General Geometry, *Robotics Research, The Fifth International Symposium*, edited by Miura, H., and Arimoto, S., pp. 263–270. MIT Press, Cambridge, MA.

Raghavan, M., and Roth, B., 1995, Solving Polynomial Systems for the Kinematic Analysis and Synthesis of Mechanisms and Robot Manipulators, *Special 50th Anniversary Design Issue, Trans. ASME*, Vol. 117, pp. 71–79.

Rheinboldt, W., and Burkardt, J., 1991, PITCON (FORTRAN subroutine library available at ftp://netlib.bell-labs.com/netlib/contin/).

Roth, B., and Freudenstein, F., 1963, Synthesis of Path-Generating Mechanisms by Numerical Methods, *ASME Journal of Engineering for Industry*, Vol. 85, pp. 298–307.

Salmon, G., 1885, *Lessons Introductory to the Modern Higher Algebra*, pp. 79–83, Hodges, Figgis, and Co., Dublin, 4th edn.

Sturmfels, B., and Zelevinsky, A., 1994, Multigraded Resultants of Sylvester Type, *J. Algebra*, Vol. 163, pp. 115–127.

Tsai, L.-W., and Morgan, A., 1985, Solving the Kinematics of the Most General Six- and Five-Degree-of-Freedom Manipulators by Continuation Methods, *ASME Journal of Mechanisms, Transmissions, and Automation in Design*, Vol. 107, pp. 189–200.

Verschelde, J., 1994, Homotopies Exploiting Newton Polytopes for Solving Sparse Polynomial Systems, *SIAM J. Numer. Anal.*, Vol. 31, pp. 915–930.

Verschelde, J., 1996, Polynomial Homotopy Continuation: a portable Ada software package (documentation for a computer program which may be obtained at ftp://ftp.cs.kuleuven.ac.be/pub/NumAnal-ApplMath/PHC/).

Wampler, C., 1994, Forward Displacement Analysis of General Six-in-Parallel SPS (Stewart) Platform Manipulators using Soma Coordinates, GM R & D-8179.

Wampler, C., 1996, Isotropic Coordinates, Circularity, and Bezout Numbers: Planar Kinematics from a New Perspective, GM R & D-8511 (also in the *Proceedings of the 1996 ASME Design Engineering Technical Conference*).

Wampler, C., Morgan, A., and Sommese, A., 1990, Numerical Continuation Methods for Solving Polynomial Systems Arising in Kinematics, *ASME Journal of Mechanical Design*, Vol. 112, pp. 59–68.

Wampler, C., Morgan, A., and Sommese, A., 1992, Complete Solution of the Nine-Point Path Synthesis Problem for Four-Bar Linkages, *Journal of Mechanical Design*, Vol. 114, pp. 153–159.

Wolfram, S., 1991, *Mathematica: A System for Doing Mathematics by Computer*, Addison-Wesley, 2nd edn.

The Design of Spherical 4R Linkages for Four Specified Orientations

D. Alan Ruth and J. Michael McCarthy

Robotics and Automation Laboratory, Department of Mechanical Engineering
University of California, Irvine, Irvine, California 92697-3975
jmmccart@uci.edu

Summary. This chapter describes a computer-aided design software system for spherical four-bar linkages that is based on Burmester's planar theory. The designer identifies a task in the form of four goal orientations for the floating link and the software generates an array of available linkages known as the Type Map. The equation of a spherical triangle provides a direct connection to the graphical techniques of Burmester. This paper further develops the linkage type characterization required for an assessment of the resulting designs.

Keywords. Burmester theory, spherical four-bar linkage, computer-aided design, rigid-body guidance, four specified orientations.

1. Introduction

In this chapter we describe the theoretical foundation of the design software SphinxPC which provides a convenient means to design spherical four-bar linkages to guide a body through four specified orientations (Fig. 1). These linkages can be viewed as spherical quadrilaterals with vertices that are hinged joints, denoted R for *revolute*. The axes of these hinges must intersect in a single point in order for the spherical $4R$ linkage to provide a one degree-of-freedom rotational, or *spherical*, movement.

The design theory underlying our formulation parallels the complex number formulation of Sandor and Erdman (1984), which is the basis of the design software LINCAGES. Murray and McCarthy (1994) showed that the equation of a spherical triangle yields equivalent results for spherical linkage design, while McCarthy (1995) showed that their "standard form" equation is easily transformed into the equation of a planar triangle. This formulation has been implemented in the design software SphinxPC presented here.

1.1 Burmester Theory

A four-bar linkage is formed when the floating link of a two-jointed open chain, called an RR dyad, is rigidly connected to the floating link of another one, to form a closed $4R$ chain. Burmester (1886) formulated this problem as locating the center-points for RR dyads that are compatible with a specified set of positions of the floating link. All the linkages that can be assembled by connecting the various dyads are solutions to the design problem. This approach is necessarily a two step process: First, the set of RR chains is identified, and then, the associated $4R$ linkages are assembled and evaluated. This is called the "design of a guiding linkage" by Hall (1961), "rigid-body guidance" by Suh and Radcliffe (1978), and "motion generation" by Sandor and Erdman (1984).

In the late 70's and early 80's, researchers transformed Burmester's geometric design procedures into computer algorithms with interactive graphics. The first was Kaufman's KINSYN (Kaufman 1978), followed shortly by Erdman's LINCAGES package (Erdman and Gustafson 1977), and later by Waldron's RECSYN (Waldron and Song 1981). These programs rapidly compute the planar RR dyads that reach a specified set of positions. Then, various strategies are used to evaluate and present the associated planar $4R$ linkages. The goal is to enable the inventor to quickly survey the dimensions and evaluate the movement of many different planar linkages that achieve the specified task (Erdman 1993). Our approach to spherical linkage design follows this same format.

1.2 The Design Methodology

The task for the spherical linkage is defined by four orientations. The computation of associated spherical RR dyads is based on Roth's version of Burmester's theorem which states, in effect: The central-axis of a spherical RR dyad "views" opposite sides of the complementary-axis quadrilateral in equal, or supplementary, angles (Bottema and Roth 1979). McCarthy (1995) shows that this theorem yields a construction that uses the *complementary-axis quadrilateral* as a "compatibility linkage" to generate central-axes. The result is a cubic cone of central-axes, called the *central-axis cone.*

Pairs of points on the central-axis cone define $4R$ linkages that reach the four specified orientations. The result is a two dimensional set of linkages that is displayed as a *Type Map.* Important characteristics of the overall movement of the linkage such as its type, and whether or not it passes through each position smoothly and in order are displayed by color coding. The designer interacts with the Type Map to select and examine various designs.

Our spherical linkage design process consists of three major phases: *i)* Task specification; *ii)* Type map generation; and *iii)* Design selection. See Figure 2.

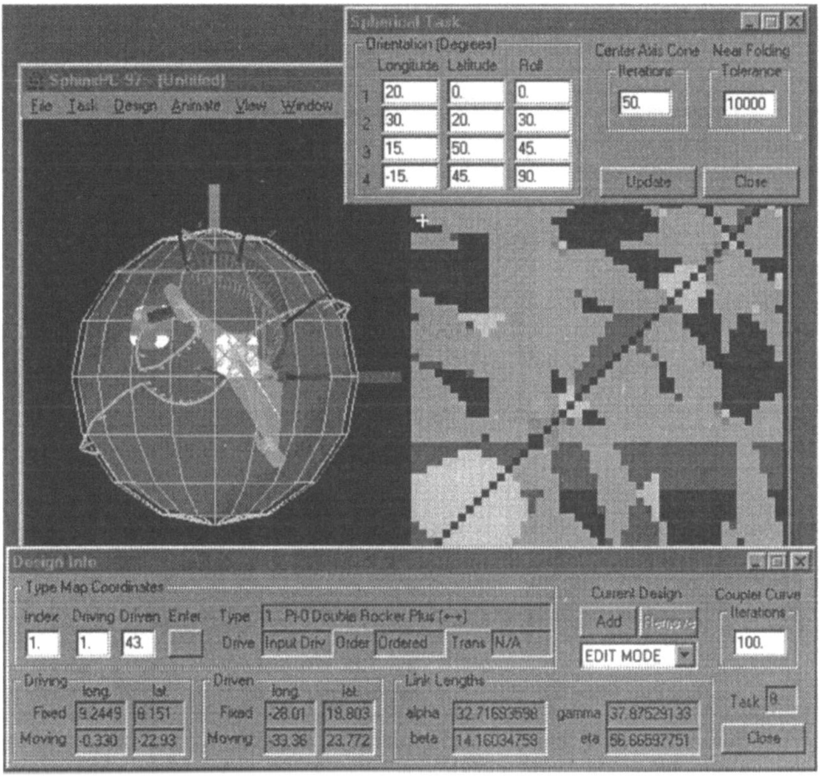

Fig. 1. The desk top of `SphinxPC` showing the workspace, type map, and the task and design dialog boxes.

2. Task Specification

The task for a spherical $4R$ linkage is defined in terms of a set of four orientations that the coupler link of the mechanism is to attain during its movement. Usually this consists of a start and a goal together with two intermediate orientations to guide the overall movement of the linkage. The task directly specifies, via the central-axis cone, all the spherical RR dyads that can be used to assemble linkages.

Figure 3 shows the dialog box which allows input of the desired orientations. We use "longitude" and "latitude" angles to position the Z-axis of the moving frame and a rotation angle about this axis, called "roll," to define each orientation. These coordinates allow the use of a globe as a convenient visualization tool.

The primary steps in this phase of the design process are: *i)* Construct the compatibility linkage, and *ii)* Generate the central-axis cone.

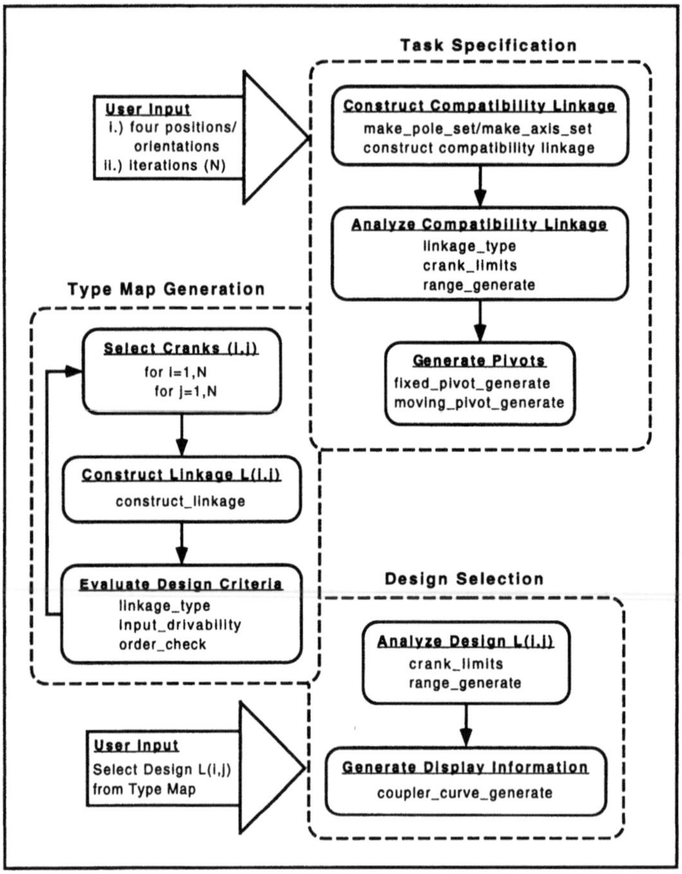

Fig. 2. Overview of spherical linkage design process implemented in SphinxPC.

2.1 Construct the Compatibility Linkage.

Given four orientations of the moving reference frame, we construct the relative rotations $[R_{12}]$, $[R_{23}]$, $[R_{34}]$ and $[R_{14}]$, and determine their rotation axes, $\mathbf{S}_{12}$, $\mathbf{S}_{23}$, $\mathbf{S}_{34}$, and $\mathbf{S}_{14}$. These axes define a spherical quadrilateral $\mathbf{S}_{12}\mathbf{S}_{23}\mathbf{S}_{34}\mathbf{S}_{14}$ known as a complementary-axis quadrilateral. We now consider this quadrilateral to be a spherical four-bar linkage (Figure 4), and use its configurations to determine the fixed axes compatible with the four orientations (McCarthy 1995). The angular lengths of the sides of this compatibility linkage are given by:

$$\alpha = \arccos(\mathbf{S}_{12}\cdot\mathbf{S}_{23}),\quad \beta = \arccos(\mathbf{S}_{14}\cdot\mathbf{S}_{34}),$$
$$\gamma = \arccos(\mathbf{S}_{12}\cdot\mathbf{S}_{14}),\quad \eta = \arccos(\mathbf{S}_{23}\cdot\mathbf{S}_{34}). \quad (1)$$

Its initial configuration is defined by the crank angle θ and coupler angle ϕ given by:

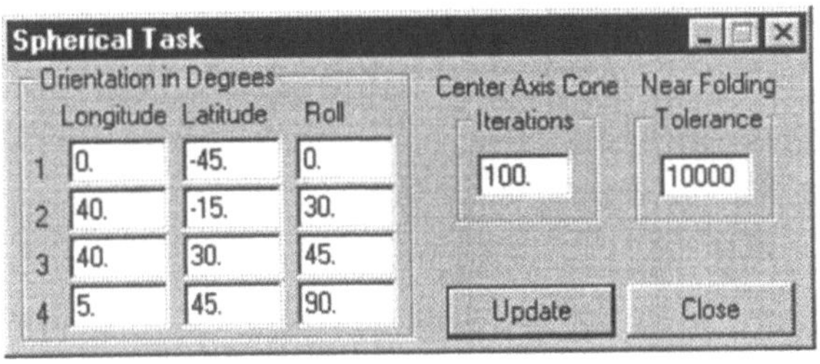

Fig. 3. The task of the linkage is defined by four orientations input through the Spherical Task dialog box.

$$\theta_0 = \arctan\frac{(\mathbf{S}_{12}\times\mathbf{S}_{14})\cdot\mathbf{S}_{23}}{(\mathbf{S}_{12}\times\mathbf{S}_{14})\cdot(\mathbf{S}_{12}\times\mathbf{S}_{23})},$$
$$\phi_0 = \arctan\frac{(\mathbf{S}_{12}\times\mathbf{S}_{23})\cdot\mathbf{S}_{34}}{(\mathbf{S}_{12}\times\mathbf{S}_{23})\cdot(\mathbf{S}_{23}\times\mathbf{S}_{34})}. \quad (2)$$

As the driving crank $\mathbf{S}_{12}\mathbf{S}_{23}$ rotates by the angle θ, the coupler $\mathbf{S}_{23}\mathbf{S}_{34}$ takes various orientations allowed by the dimensions of the linkage. The relative rotation axis of the coupler from its initial orientation to every other orientation is a central-axis compatible with the four specified orientations.

2.2 Generate the Central-axis Cone

In order for a central-axis $\mathbf{G}$ to view the sides $\mathbf{S}_{12}\mathbf{S}_{23}$ and $\mathbf{S}_{14}\mathbf{S}_{34}$ in the same angle κ, it must satisfy a pair of equations, each defining a spherical triangle. Consider the spherical triangle with vertex $\mathbf{G}$ and the arc joining $\mathbf{S}_{12}\mathbf{S}_{23}$ as its base. Let the interior angles at each end of the base be $\Delta\theta/2$ and $\Delta\phi/2$. Similarly, consider the spherical triangle formed by $\mathbf{G}$ and the arc $\mathbf{S}_{14}\mathbf{S}_{34}$, with interior angles $\Delta\psi/2$ and $\Delta\zeta/2$. Then, to be a central-axis the coordinates of $\mathbf{G}$ must satisfy the equations:

$$\begin{aligned}\sin\kappa\mathbf{G} &= \sin\frac{\Delta\theta}{2}\cos\frac{\Delta\phi}{2}\mathbf{S}_{12} + \sin\frac{\Delta\phi}{2}\cos\frac{\Delta\theta}{2}\mathbf{S}_{23}\\ &\quad + \sin\frac{\Delta\theta}{2}\sin\frac{\Delta\phi}{2}\mathbf{S}_{12}\times\mathbf{S}_{23},\\ \sin\kappa\mathbf{G} &= \sin\frac{\Delta\psi}{2}\cos\frac{\Delta\zeta}{2}\mathbf{S}_{14} + \sin\frac{\Delta\zeta}{2}\cos\frac{\Delta\psi}{2}\mathbf{S}_{34}\\ &\quad + \sin\frac{\Delta\psi}{2}\sin\frac{\Delta\zeta}{2}\mathbf{S}_{14}\times\mathbf{S}_{34}. \quad (3)\end{aligned}$$

These equations have a solution $\mathbf{G}$ when the angles θ, ϕ, ζ and ψ are the configuration angles of the compatibility linkage (McCarthy 1995).

The constraint equation of the complementary-axis quadrilateral considered as a spherical four-bar linkage determines the coupler angle ϕ as a function of θ, given by:

$$\phi(\theta) = \arctan(\frac{B}{A}) \pm \arccos(\frac{C}{\sqrt{A^2+B^2}}) \quad (4)$$

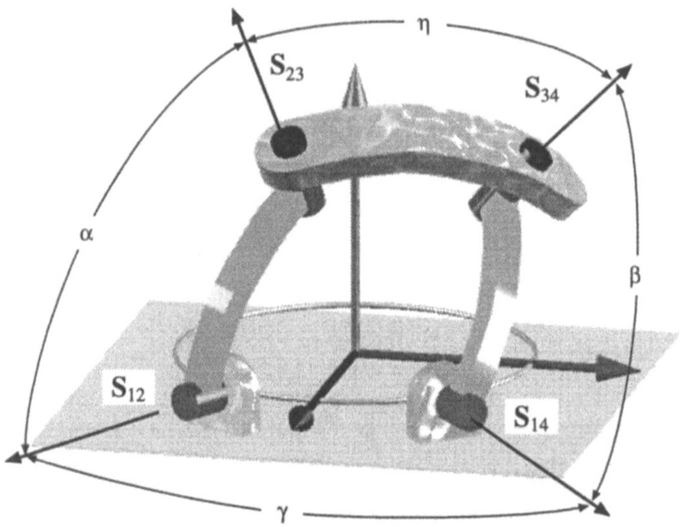

Fig. 4. The complementary-axis quadrilateral is used as a spherical 4R linkage to compute the central-axis cone.

where

$$\begin{aligned} A(\theta) &= \sin\eta(\sin\gamma\cos\alpha\cos\theta - \sin\alpha\cos\gamma), \\ B(\theta) &= -\sin\eta\sin\gamma\sin\theta, \\ C(\theta) &= \cos\beta - \cos\eta(\sin\alpha\sin\gamma\cos\theta + \cos\alpha\cos\gamma). \end{aligned} \tag{5}$$

The result is a parameterized equation of the central-axis cone:

$$\begin{aligned} \sin\frac{\beta}{2}\mathbf{G} &= \sin\frac{\Delta\theta}{2}\cos\frac{\Delta\phi(\theta)}{2}\mathbf{S}_{12} + \sin\frac{\Delta\phi(\theta)}{2}\cos\frac{\Delta\theta}{2}\mathbf{S}_{23} \\ &\quad + \sin\frac{\Delta\theta}{2}\sin\frac{\Delta\phi(\theta)}{2}\mathbf{S}_{12}\times\mathbf{S}_{23}, \end{aligned} \tag{6}$$

where $\Delta\theta = \theta - \theta_O$ and $\Delta\phi = \phi - \phi_O$.

2.2.1 Set the Bias for Each Axis. The fixed axes defined in Eq. (6) can be distinguished by whether or not they are directed toward the $Z > 0$ hemisphere. This provides a convenient way to keep track of the 16 equivalent forms that exist for a given spherical quadrilateral. By biasing all the axes toward the $Z > 0$ hemisphere, we can defined each of the various forms of the linkage by negating combinations of the four axes. Associated with each form is a different Type Map that can be selected by the designer.

To ensure that the central-axes are biased toward $Z > 0$, we identify the intersection of the central-axis cone with the $Z = 0$ plane, as the starting point for the cone generation routine. This intersection is obtained by solving the algebraic equation of the central-axis cone which becomes a cubic polynomial in the X-component of the central-axis:

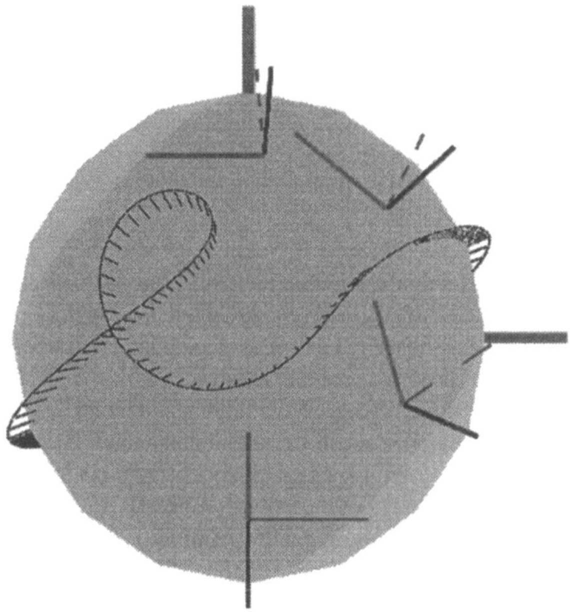

Fig. 5. The central-axis cone is the set of fixed axes of spherical RR dyads compatible with four specified orientations.

$$C_1x^3 + C_2x^2 + C_3x + C_4 = 0, \tag{7}$$

where

$$\begin{aligned}
C_1 &= -r_1q_2q_3q_4 + r_2q_1q_3q_4 + r_3q_1q_2q_4 - r_4q_1q_2q_3 + \\
&\quad q_1r_2r_3r_4 - q_2r_1r_3r_4 - q_3r_1r_2r_4 + q_4r_1r_2r_3, \\
C_2 &= r_1q_2q_3p_4 + r_1q_2p_3q_4 + r_1p_2q_3q_4 - q_1r_2q_3p_4 - \\
&\quad q_1r_2p_3q_4 - p_1r_2q_3q_4 - q_1q_2r_3p_4 - q_1p_2r_3q_4 - \\
&\quad p_1q_2r_3q_4 - r_1r_2r_3p_4 + q_1q_2p_3r_4 + q_1p_2q_3r_4 + \\
&\quad p_1q_2q_3r_4 + r_1r_2p_3r_4 + r_1p_2r_3r_4 - p_1r_2r_3r_4, \\
C_3 &= -r_1q_2p_3p_4 - r_1p_2q_3p_4 - r_1p_2p_3q_4 + q_1r_2p_3p_4 + \\
&\quad p_1r_2q_3p_4 + p_1r_2p_3q_4 + q_1p_2r_3p_4 + p_1q_2r_3q_4 + \\
&\quad p_1p_2r_3q_4 + r_1r_2r_3q_4 - q_1p_2p_3r_4 - p_1q_2p_3r_4 - \\
&\quad p_1p_2q_3r_4 - r_1r_2q_3r_4 - r_1q_2r_3r_4 + q_1r_2r_3r_4, \\
C_4 &= r_1p_2p_3p_4 - p_1r_2p_3p_4 - p_1p_2r_3p_4 - r_1r_2r_3p_4 + \\
&\quad p_1p_2p_3r_4 + r_1r_2p_3r_4 + r_1p_2r_3r_4 - p_1r_2r_3r_4;
\end{aligned} \tag{8}$$

and $\mathbf{S}_{12} = (p_1, q_1, r_1)^T$, $\mathbf{S}_{14} = (p_2, q_2, r_2)^T$, $\mathbf{S}_{23} = (p_3, q_3, r_3)^T$, and $\mathbf{S}_{34} = (p_4, q_4, r_4)^T$ are the coordinates of the vertices of the compatibility linkage.

Given that axes for the central-axis cone are biased toward the $Z > 0$ side of the sphere, the axes on the $Z < 0$ side are easily obtained by negating the original set. Figure 5 shows an example of a central-axis cone generated by `SphinxPC`.

3. Type Map Generation

The spherical linkages that solve a specific four-orientation task form a two-dimensional set that is color-coded based on linkage type, and form the *Type Map*. See Figure 6. The resolution of the the Type Map is defined by the number of points, N, computed on the central-axis cone.

A given spherical four-bar linkage with vertices **OABC** can be constructed with any combination of the four vectors negated. Thus, there are 16 forms for any given linkage. We bias the four vertices so they are directed toward the $Z > 0$ hemisphere, and denote this form as $(+,+,+,+)$. All other forms of the linkage are obtained by negating combinations of the axes, for example $(-,+,+,+)$ has $\mathbf{O}$ directed toward the $Z < 0$ hemisphere. Type Maps for all 16 forms of the linkage are generated by `SphinxPC`.

The steps required for this phase of the software are: *i)* Construct linkage $L(i,j)$ and *ii)* Evaluate linkage characteristics—specifically linkage type, input drivability, and order.

3.1 Construct the Linkage $L(i,j)$

Let $\mathbf{O}_i, i = 1,\ldots,N$ be the set of central-axes. For each of these axes, we determine the associated moving axis $\mathbf{A}_i$ using the constraint equations of the dyad. This is actually done in the moving frame, where for a given fixed axis we obtain a set of linear equations that define the moving axis. The resulting coordinate vector of the moving axis is transformed back to the fixed frame to yield the dyads $\mathbf{O}_i\mathbf{A}_i$. Two dyads $\mathbf{O}_i\mathbf{A}_i$ and $\mathbf{O}_j\mathbf{A}_j$ define the linkage $L(i,j)$. The other 16 forms of the linkage are generated by negating combinations of the fixed and moving axes.

The "X" coordinate of the Type Map is the driving crank and the "Y" coordinate is the driven crank. Reflection through the diagonal of the figure interchanges the driving and driven cranks. The angular dimensions of the four sides of the spherical linkage are given by:

$$\begin{aligned}\alpha &= \arccos(\mathbf{O}_i \cdot \mathbf{A}_i), \beta = \arccos(\mathbf{O}_j \cdot \mathbf{A}_j),\\ \gamma &= \arccos(\mathbf{O}_i \cdot \mathbf{O}_j), \eta = \arccos(\mathbf{A}_i \cdot \mathbf{A}_j).\end{aligned} \tag{9}$$

For each linkage, we determine the input angles θ_k and coupler angle ϕ_k when the linkage is assembled in each of the four design orientations, $k = 1, 2, 3, 4$. These are easily computed from the coordinates of the axes by the formulas:

$$\begin{aligned}
\theta_k &= \arctan\frac{(\mathbf{O}_i \times \mathbf{O}_j)\cdot \mathbf{A}_{ik}}{(\mathbf{O}_i \times \mathbf{O}_j)\cdot(\mathbf{O}\times \mathbf{A}_{ik})}, \\
\phi_k &= \arctan\frac{(\mathbf{O}_i \times \mathbf{A}_{ik})\cdot \mathbf{A}_{jk}}{(\mathbf{O}_i \times \mathbf{A}_{ik})\cdot(\mathbf{A}_{ik}\times \mathbf{A}_{jk})}, k=1,\ldots,4,
\end{aligned} \tag{10}$$

where $\mathbf{A}_{ik}$ is the ith moving axis located in the kth orientation.

3.2 Type Determination

The classification of the overall movement of a spherical $4R$ linkage can be described in the terms "crank" and "rocker," in a manner similar to planar linkages. In this context, a crank is a link coupled to the base that completely rotates relative to the base, and a rocker is a similar link that oscillates between limiting positions. Grashoff's criterion identifies the combination of four link lengths that ensure that at least one link can fully rotate. These linkages are said to be *Grashoff* and the remaining are *non-Grashoff*. It is the shortest link in a Grashoff linkage that can fully rotate; therefore, Grashoff linkages are further distinguished by which of the four links is the shortest: *i)* Crank-rocker (input); *ii)* Double-rocker (coupler); *iii)* Rocker-crank (output); and *iv)* Drag-link (ground). The non-Grashoff linkages are grouped together and called *triple rockers*; however, it is possible to further categorize these linkages (Murray 1996).

The presence of 16 variations of the same spherical linkage make the classification procedure a little more difficult. The usual approach is to change the dimensions of the linkage among these variations in order to obtain a "model linkage" for which an spherical version of Grashoff's criterion applies (Chiang 1988). Murray and McCarthy (1995) introduced an alternate approach which uses the actual dimensions of a given spherical linkage.

Consider a spherical linkage with vertices **OABC**. Let θ be the input angle of the driving link **OA** and ψ the output angle of the driven link **CB**. These angles are related by the condition that $\mathbf{B}\cdot\mathbf{A} = \cos\eta$ in all configurations of the linkage, which yields the constraint equation:

$$A(\theta)\cos\psi + B(\theta)\sin\psi = C(\theta), \tag{11}$$

where

$$\begin{aligned}
A(\theta) &= \sin\beta(\sin\alpha\cos\gamma\cos\theta - \sin\gamma\cos\alpha), \\
B(\theta) &= \sin\alpha\sin\beta\sin\theta, \\
C(\theta) &= \cos\eta - \cos\beta(\sin\alpha\sin\gamma\cos\theta + \cos\alpha\cos\gamma).
\end{aligned} \tag{12}$$

The solution is:

$$\psi(\theta) = \arctan(\frac{B}{A}) \pm \arccos(\frac{C}{\sqrt{A^2+B^2}}) \tag{13}$$

Notice that there are two solutions for each value of θ, one obtained for plus and one for minus in this equation. We call these the two *modes* of the solution.

3.2.1 Crank Rotation Limits. Notice that in order for a solution to exist the argument to arccos must be in the range $(-1, 1)$. This is equivalent to the condition:

$$A^2(\theta) + B^2(\theta) - C^2(\theta) \geq 0. \tag{14}$$

Set Eq. (14) equal to zero to obtain a quadratic polynomial in $\cos\theta$ with roots:

$$\begin{aligned} C_1 &= \frac{\cos(\eta - \beta) - \cos\alpha\cos\gamma}{\sin\alpha\sin\gamma}, \\ C_2 &= \frac{\cos(\eta + \beta) - \cos\alpha\cos\gamma}{\sin\alpha\sin\gamma}. \end{aligned} \tag{15}$$

These equations are simply the spherical cosine laws for the triangle formed when the coupler and output link are aligned. The root C_1 defines the smallest positive angle the input crank can reach, θ_{min}; the root C_2 determines the largest positive angle θ_{max}. Thus, we have:

$$\theta_{\text{min}} = \arccos C_1, \quad \text{and} \quad \theta_{\text{max}} = \arccos C_2. \tag{16}$$

However, it is possible that $|C_1| \geq 1$, in which case $\theta_{\text{min}} = 0$ and the crank can pass through zero. Similarly, if $|C_2| \geq 1$ then $\theta_{\text{max}} = \pi$ and the crank can pass through π.

A detailed study of Eq. (15) leads to the introduction of four parameters (Murray and McCarthy 1995):

$$\begin{aligned} T_1 &= \gamma - \alpha + \eta - \beta, T_2 = \gamma - \alpha - \eta + \beta, \\ T_3 &= \eta + \beta - \gamma - \alpha, \quad \text{and} \quad T_4 = 2\pi - (\eta + \beta + \gamma + \alpha). \end{aligned} \tag{17}$$

These parameters characterize the movement of the driving link as follows:

1. $T_1T_2 > 0$ and $T_3T_4 > 0$, the driving link fully rotates.
2. $T_1T_2 > 0$ and $T_3T_4 < 0$, the driving link rocks through $\theta = 0$.
3. $T_1T_2 < 0$ and $T_3T_4 > 0$, the driving link rocks through $\theta = \pi$.
4. $T_1T_2 < 0$ and $T_3T_4 < 0$, the driving link rocks over two ranges, neither of which includes 0 or π.

An analysis of the linkage in terms of the driven link angle ψ shows that the same parameters characterize the movement of the driven link:

1. $T_2T_4 < 0$ and $T_1T_3 < 0$, the driven link fully rotates.
2. $T_2T_4 < 0$ and $T_1T_3 > 0$, the driven link rocks through $\psi = 0$.
3. $T_2T_4 > 0$ and $T_1T_3 < 0$, the driven link rocks through $\psi = \pi$.
4. $T_2T_4 > 0$ and $T_1T_3 > 0$, the driven link rocks over two ranges, neither of which includes 0 or π.

Fig. 6. The set of linkages assembled from pairs of axes on the central-axis cone, color-coded for type, forms the Type Map.

3.2.2 The Parameters T_i. Let the function $\sigma(T_i)$ take on values $+1$, -1, or 0 depending on whether T_i is positive, negative or zero. This allows us to identify a signature $(\sigma(T_1), \sigma(T_2), \sigma(T_3), \sigma(T_4))$ for each spherical linkage, which we call its *type*.

It is easy to see that there are 81 spherical-linkage types, which can be separated into three general categories depending on the value of $\sigma(T_4)$. If $\sigma(T_4) = +1$, then the sum of the four angular link lengths is less than 2π. This means the linkage lies on one side of the sphere. There are 27 spherical-linkage types in this category which are equivalent to the well-known planar-linkage types. If $\sigma(T_4) = -1$, then the angular lengths of the four links add up to greater than 2π, and the linkage must wrap around the sphere. There are 27 linkage types in this category, which can be related to analogous planar-linkage types. Finally, there are 27 types for which $\sigma(T_4) = 0$. These linkages have a configuration in which the links can align to form a great circle on the sphere.

If any of the parameters T_i, for $i = 1, 2, 3, 4$, is zero, the linkage can collapse so that its axes become co-planar. These linkages are said to *fold*. Thus, there are 65 types of folding spherical linkages. The number of the parameters T_i that equal zero is the number of folding configurations. For

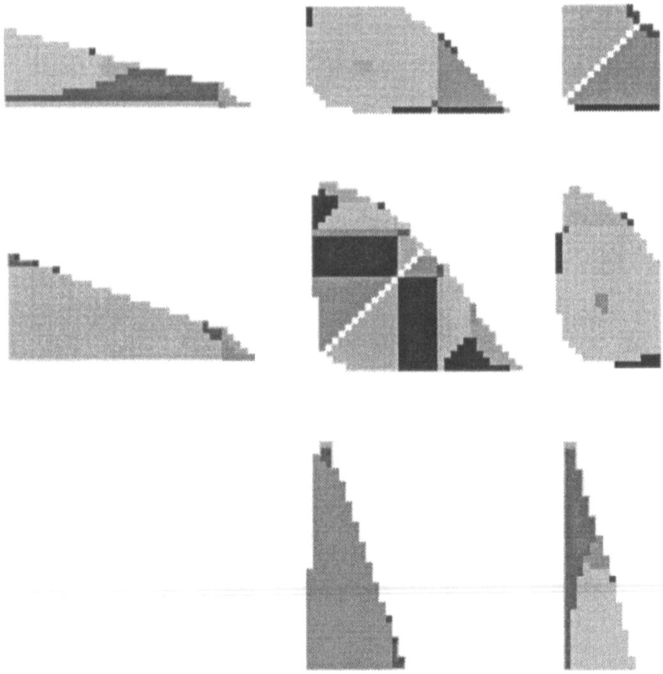

Fig. 7. The spherical linkage Type Map with the "input-drivable" and "order" filters applied.

example, the spherical rhombus with four sides of dimension $\frac{\pi}{2}$, the only linkage of type $(0, 0, 0, 0)$, has four folding configurations.

The 16 non-folding linkage types divide such that eight lie on one side of the sphere ($\sigma(T_4) = +1$) and eight wrap around the sphere ($\sigma(T_4) = -1$). The eight types in each category pair up to become the well-known four Grashoff linkages and the four triple rockers obtained using the classification scheme of Chiang (1988).

3.3 Evaluate Linkage Characteristics

For each linkage $L(i, j)$, we evaluate the parameters T_i and assign a color to represent the type. In addition, the designer can apply *filters* to eliminate linkages that have order and branching defects.

The color scheme for the Type Map uses shades of blue to represent non-Grashoff linkages, red to identify folding linkages, and other colors (yellow, orange, pink, and green) to represent the Grashoff linkages. We add black to obtain a darkened version of the same color scheme to identify linkages that wrap around the sphere.

The evaluation of branching defects consists of a series of checks. The first step is to determine that all of the goal positions are on the same solu-

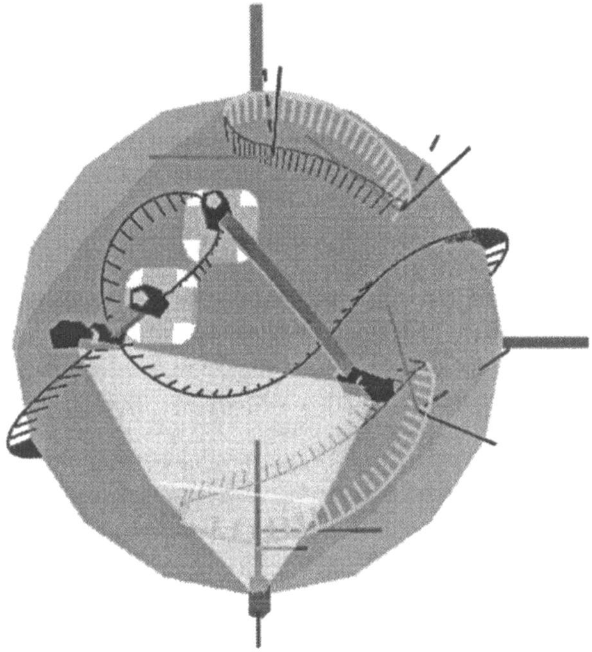

Fig. 8. The animation of a spherical linkage guiding its goal frame along a coupler curve is a useful evaluation tool.

tion of the linkage, non-Grashoff linkages that meet this criterion are called *input-drivable.* This means the driving crank will move the coupler smoothly through the four positions without jamming. For Grashoff linkages, the same linkage solution can be divided over two ranges of motion of the crank, so we must also check that the goal positions are reached from one range. Satisfying this check defines input-drivable Grashoff linkages. All linkages that are not input-drivable are flagged. The designer can eliminate them from consideration by selecting the input-drivable filter on the Type Map.

The order check determines if the linkage moves through the goal orientations in the specified order. Linkages that are not ordered are flagged and can be eliminated from the Type Map by applying the order filter. Figure 7 shows the type map with the input-drivable and order filters applied.

4. Design Selection

The designer interacts directly with the Type Map to select a linkage $L(i, j)$ and examine its characteristics. The detailed properties of this linkage are displayed in the Design Information dialog box. The primary tools for evaluation of a linkage are the display of its coupler curve, and the animation of its movement as it guides the moving frame along the coupler curve through the desired orientations. The operations of this phase of the software are: *i)* Analyze the linkage $L(i,j)$; and *ii)* Generate the display information.

Given the selection $L(i,j)$ on the Type Map, the software computes and displays the coupler curve, coded for the two solution modes. The arrow keys allow movement in the vertical and horizontal direction of the Type Map so a sequence of linkages and their evolving coupler curves can be viewed. This provides useful insight into characteristics of the task—see Figure 8, which shows a spherical linkage with its coupler curve.

An animation dialog box provides VCR style buttons to animate the linkage in forward and reverse directions with pause and fast forward. This provides a view of the linkage guiding the moving body through the desired positions. Various bias schemes for the joint axes provide different configurations for the movement of the supporting links.

There are also buttons to toggle between the two modes of solution as well as between the two assemblies available for Grashoff linkages. It is important to realize that while the separate circuits of a Grashoff linkage may correspond to separate modes of solution, it is also possible for the two solution modes to exist on each circuit (Fig. 8).

In the process of specifying a task and examining the resulting linkage designs, we have found that the designer develops a perception in the characteristics of the task that impact desirable features of these designs. This insight often directs a modification of the task that results in a larger set of acceptable linkages. While it seems to be easily done by the designer, we have yet to formulate a mathematical version of this strategy. Enhancing this capability for the designer may be fundamental to the success of spatial linkage design tools.

5. Conclusion

This chapter described the implementation of spherical four-orientation synthesis in the software `SphinxPC`. The software encodes a new formulation of classical Burmester theory based on the equation of a spherical triangle which yields a convenient parameterized equation for the central-axis cone. The linkage Type Map is used to provide a convenient means to display and evaluate the characteristics of the two-dimensional set of solutions to the four-orientation synthesis problem. The result is a design tool which, in the hands of students at several universities, has yielded spherical linkages that achieve a wide variety of spatial reorientation tasks.

Acknowledgment

The authors gratefully acknowledge the support of the National Science Foundation grant DMII9321936 and the contributions of Mohan Bodduluri, Pierre Larochelle and Andrew Murray to the evolution of the Sphinx linkage design software over the past decade.

References

Bottema, O., and Roth, B. (1979): *Theoretical Kinematics.* North-Holland, New York.

Burmester, L. (1886): *Lehrbuch der Kinematik.* Verlag Von Arthur Felix, Leipzig, Germany.

Chiang, C. H. (1988): *Kinematics of Spherical Mechanisms.* Cambridge Univ. Press.

Erdman, A., and Gustafson, J. (1977): LINCAGES: Linkage INteractive Computer Analysis and Graphically Enhanced Synthesis Packages, Technical Report 77-DET-5, American Society of Mechanical Engineers.

Erdman, A. G. (1993): *Modern Kinematics: Developments in the Last Forty Years.* John Wiley and Sons, New York.

Hall, A. S., Jr. (1961): *Kinematics and Linkage Design.* Prentice-Hall, Englewood Cliffs, New Jersey.

Kaufman, R. (1978): Mechanism Design by Computer, *Machine Design.* pp. 94-100.

McCarthy, J. M. (1993): The Opposite Pole Quadrilateral as a Compatibility Linkage for Parameterizing the Center Point Curve. *ASME Journal of Mechanical Design.* 115(2):332-336.

McCarthy, J. M., The Synthesis of Planar RR and Spatial CC Chains and the equation of a Triangle. *Transactions of the ASME - Special 50th Anniversary Design Issue.* 117:101-106.

Murray, A. P., and McCarthy, J. M. (1994): Five Position Synthesis of Spatial CC Dyads, *Proc. ASME Mechanisms Conference* Minneapolis, Sept.

Murray, A. P., and McCarthy, J. M. (1995): A Linkage Type Map for Spherical 4 Position Synthesis, *Proceedings of the 1995 ASME Design Technical Conferences.* Boston, MA, Sept.

Murray, A. P. (1996): *Global Properties of Constraint Manifolds in the Kinematic Synthesis of Closed Chains.* PhD Thesis, University of California, Irvine.

Roth, B. (1967): On the Screw Axes and Other Special Lines Associated with Spatial Displacements of a Rigid Body. *ASME Journal of Engineering for Industry.* February.

Sandor, G. N., and Erdman, A. G. (1984): *Advanced Mechanism Design: Analysis and Synthesis.* Vol. 2, Prentice-Hall, Englewood Cliffs, New Jersey.

Suh, C. H., and Radcliffe, C. W. (1978): *Kinematics and Mechanism Design.* John Wiley and Sons, New York.

Waldron, K. J., and Song, S. M. (1981): Theoretical and Numerical Improvements to an Interactive Linkage Design Program, RECSYN, *Proceedings of the Seventh Applied Mechanisms Conference*, Kansas City, Missouri, December.

On the Design of Efficient Parallel Mechanisms

Clément M. Gosselin

Département de Génie Mécanique, Université Laval, Québec, Québec, Canada, G1K 7P4, email: gosselin@gmc.ulaval.ca

Summary. The design of parallel mechanisms is addressed in this chapter using two main performance criteria, namely, the kinematic dexterity and the concept of static balancing. Static balancing is defined here as the set of conditions on mechanism dimensional and mass parameters which, when satisfied, ensure that the weight of the links does not produce any torque (or force) at the actuators for any configuration of the mechanism, under static conditions. Spherical three-degree-of-freedom parallel mechanisms are used as examples and applications to manipulators and motion simulation systems are studied.

Keywords: Parallel manipulators, parallel mechanisms, static balancing, dexterity, design, optimization, spherical mechanisms, spherical parallel manipulators, flight simulators

1. Introduction

Parallel mechanical architectures have first been introduced in tire-testing and motion simulation applications (Gough and Whitehall 1962; Stewart 1965), and later on in robotics (see for instance (Hunt 1983) and (Fichter 1986)). Parallel mechanisms are characterized by the fact that the end-effector is connected to the base via multiple kinematic chains and that all the actuators can be located on or close to the base. This leads to high stiffness and load-carrying capacity and to very good dynamic properties since the inertia of the moving parts is drastically reduced. However, although parallel mechanisms have attractive properties for several applications, the complexity of their kinematics makes it more difficult for a designer to determine a set of kinematic parameters which will efficiently produce prescribed performances. Indeed, the behavior of parallel mechanisms is far less intuitive than that of serial manipulators. Moreover, current parallel mechanisms typically require powerful actuators to overcome the weight of the links. In commercial flight simulators, for instance, maintaining the platform in static equilibrium requires considerable power because of the large forces induced in the actuators which lead to hydraulic fluid losses. Hence, the kinematic and kinetostatic design of parallel mechanisms is an important issue. Indeed, the optimization of the kinematic accuracy as well as the reduction or elimination of the static forces at the actuators due to the weight of the links would lead to designs requiring less powerful actuators and would also improve the efficiency.

In this chapter, the kinematic properties of parallel mechanisms are first reviewed and formulations are proposed for their optimization. Then, the static balancing of parallel mechanisms and manipulators is addressed. Static balancing is first defined and the formulation allowing the derivation of the conditions associated with balanced mechanisms is presented. Static balancing is a very important issue since it leads to a considerable reduction of the actuator torques (or forces) and hence to more efficient designs. Applications include flight simulators, positioning devices, manual manipulators, and high-performance robotic manipulators. Examples of spherical 3-dof parallel mechanisms are used here to illustrate the proposed methodology. Finally, two examples of applications — one in robotics and one in flight simulation — are given.

2. Kinematics of parallel mechanisms

As stated above, parallel mechanisms are defined as those in which the end-effector (or platform) is connected to the base by several kinematic chains and in which the actuators are located on or close to the base. The kinematics of parallel mechanisms has been studied by several authors in the last two decades (see (Merlet 1997) for an extensive literature review). The basic kinematic relationships associated with parallel mechanisms are now briefly recalled.

2.1 Inverse kinematic problem

Let vector $\mathbf{x}$ be defined as the vector of Cartesian coordinates of the platform of a parallel mechanism. This vector describes the *pose* — i.e. the position and orientation — of the platform. If the degree of freedom of the mechanism is equal to the dimension of the space in which the motion is embedded — e.g. 6-dof mechanisms for spatial systems — then vector $\mathbf{x}$ can be prescribed arbitrarily. The kinematic chains connecting the platform to the base can then be considered as serial manipulators, which leads to the solution of the inverse kinematic problem. Since this problem is well known and fully understood for serial manipulators, a solution to the problem associated with parallel mechanisms is readily available. Moreover, for most architectures of parallel mechanisms, the geometry of the kinematic chains connecting the base to the end-effector is simple, which leads to simple solutions for the inverse kinematics. If the vector of actuated joint coordinates is noted $\boldsymbol{\theta}$, then the solution of the inverse kinematic problem can be written as the following function

$$\boldsymbol{\theta} = \mathbf{f}(\mathbf{x}) \tag{2.1}$$

If the degree of freedom of the mechanism is smaller than the dimension of the space in which the motion is embedded — e.g. 4-dof spatial mechanisms — then vector $\mathbf{x}$ cannot be prescribed arbitrarily. Indeed, since the mechanism does not have full mobility, the motion of the platform must satisfy the constraints imposed by the kinematic chains. In general, a set of independent Cartesian coordinates can be defined by partitioning vector $\mathbf{x}$ as follows

$$\mathbf{x} = [\mathbf{x}_i^T \ \mathbf{x}_d^T]^T \tag{2.2}$$

where $\mathbf{x}_i$ is the vector of independent Cartesian coordinates and vector $\mathbf{x}_d$ is the vector of dependent Cartesian coordinates. Clearly, this partition is not unique and will be chosen according to the tasks to be performed by the mechanism and/or its architecture.

Once this partition is established, the kinematic constraints of the closed loops in the mechanism can be used to obtain a relationship between vectors $\mathbf{x}_d$ and $\mathbf{x}_i$ as

$$\mathbf{x}_d = \mathbf{g}(\mathbf{x}_i) \tag{2.3}$$

Once the above relationship is obtained, it can be used to compute the complete pose of the platform and the inverse kinematic problem can then be solved by considering the kinematic chains connecting the base to the platform as serial manipulators; one obtains

$$\boldsymbol{\theta} = \mathbf{f}(\mathbf{x}) = \mathbf{f}(\mathbf{x}_i, \mathbf{g}(\mathbf{x}_i)) \tag{2.4}$$

2.2 Velocity equations

For manipulators with complete mobility, differentiating eq.(2.1) with respect to time and rearranging (Gosselin and Angeles 1990) leads to

$$\mathbf{K}\dot{\boldsymbol{\theta}} = \mathbf{J}\mathbf{t} = \mathbf{J}\mathbf{R}\dot{\mathbf{x}} \tag{2.5}$$

where $\mathbf{K}$ and $\mathbf{J}$ are referred to as Jacobian matrices and vector $\mathbf{t}$ is the twist of the platform, defined as

$$\mathbf{t} = [\boldsymbol{\omega}^T \ \dot{\mathbf{p}}^T]^T \tag{2.6}$$

where $\boldsymbol{\omega}$ is the angular velocity of the platform and $\dot{\mathbf{p}}$ is the velocity of a reference point on the platform. Matrices $\mathbf{K}$ and $\mathbf{J}$ can respectively be used to determine the two different types of singularities of the mechanism (Gosselin and Angeles 1990) while matrix $\mathbf{R}$ characterizes the transformation between the time derivative of the pose parameters and the twist vector. The latter matrix can be used to assess the formulation singularities which may occur due to the choice of orientation parameters used in vector $\mathbf{x}$ (e.g. formulation singularities associated with Euler angles or other parameters).

For manipulators with incomplete mobility, the differentiation of eq.(2.1) and the use of eq.(2.2) leads to

$$\mathbf{K}\dot{\boldsymbol{\theta}} = \mathbf{J}_1\dot{\mathbf{x}}_i + \mathbf{J}_2\dot{\mathbf{x}}_d \tag{2.7}$$

and the differentiation of eq.(2.3) yields

$$\dot{\mathbf{x}}_d = \mathbf{L}\dot{\mathbf{x}}_i \tag{2.8}$$

The substitution of eq.(2.8) into eq.(2.7) then leads to

$$\mathbf{K}\dot{\boldsymbol{\theta}} = (\mathbf{J}_1 + \mathbf{J}_2\mathbf{L})\dot{\mathbf{x}}_i = \mathbf{J}'\dot{\mathbf{x}}_i = \mathbf{J}\mathbf{R}'\dot{\mathbf{x}}_i' \tag{2.9}$$

where matrices $\mathbf{K}$ and $\mathbf{J}$ are the effective Jacobian matrices of the manipulator and where vector $\dot{\mathbf{x}}_i'$ is a vector which may be required in order to clearly identify the formulation singularities. This will be necessary only if vector $\mathbf{x}_i$ contains a complete description of the orientation of the platform. When this occurs, matrix $\mathbf{R}'$ will be used to determine the architecture singularities.

2.3 Kinematic dexterity

Several measures have been proposed in the literature for the characterization of the kinematic accuracy of manipulators and mechanisms (see for instance (Salisbury and Craig 1982; Yoshikawa 1985; Klein and Blaho 1987)). One of the most widely accepted measure of kinematic accuracy or *dexterity*, is the one originally proposed in (Salisbury and Craig 1982), i.e., the condition number of the Jacobian matrix of the manipulator. In this work, the reciprocal of the condition number, noted ζ, will be used. This is written as

$$\zeta(\mathbf{x}) = \frac{1}{\kappa(\mathbf{K}^{-1}\mathbf{J})} \tag{2.10}$$

where $\kappa(\cdot)$ denotes the condition number of its matrix argument, which, in turn, can be defined as

$$\kappa(\mathbf{A}) = ||\mathbf{A}||\ ||\mathbf{A}^{-1}|| \tag{2.11}$$

where $||\cdot||$ denotes any norm of its matrix argument. The Euclidean norm is used in this work. Additionally, the notion of *characteristic length* introduced in (Tandirci et al. 1992) is also adopted here. This is necessary in cases where the Jacobian matrices involve quantities of different physical nature, i.e., when the Cartesian coordinates involve translations and rotations. The use of a characteristic length then ensures that the kinematic measures obtained are physically meaningful.

Since the dexterity measure introduced above is configuration dependent, a *global dexterity index* has also been introduced in (Gosselin and Angeles 1991) and can be defined as

$$\eta = \frac{1}{W}\int_{\mathcal{W}} \zeta(\mathbf{x})dW \tag{2.12}$$

where $\mathcal{W}$ denotes the workspace of the mechanism and W the volume of the workspace. This index characterizes the average dexterity of the mechanism over all its workspace.

3. Static balancing of mechanisms and manipulators

The balancing of mechanisms has been an important research topic for several decades (see for instance (Lowen et al. 1983) for a literature review). Mechanisms are said to be *force-balanced* when the total force applied by the mechanism on the fixed base is constant for any motion of the mechanism. In other words, a mechanism is force-balanced when its global center of mass remains fixed, for any arbitrary motion of the mechanism. This condition is very important in machinery since unbalanced forces on the base will lead to vibrations, wear and other undesirable side effects. For robotic manipulators or motion simulation devices, however, the forces on the base are usually not critical and designers are mostly concerned with the torques (or forces) which are required at the actuators to maintain the manipulator or mechanism in static equilibrium. Hence, in this context, manipulators or mechanisms are said to be *statically balanced*, when the weight of the links does not produce any torque (or force) at the actuators under static conditions, for any configuration of the manipulator or mechanism. This condition is also referred to as *gravity compensation*. Gravity-compensated serial manipulators have been designed in (Nathan 1985; Hervé 1986; Streit and Gilmore 1989; Ulrich and Kumar 1991; Walsh et al. 1991) using counterweights, springs and sometimes cams and/or pulleys. A hybrid direct-drive gravity-compensated manipulator has also been developed in (Kazerooni and Kim 1988). Moreover, a general approach for the equilibration of planar linkages using springs has been presented in (Streit and Shin 1990). When springs are used, static balancing can be defined as the set of conditions for which the total potential energy in the mechanism — including gravitational energy and the elastic energy stored in the springs — is constant for any configuration of the mechanism. When no springs — or other means of storing elastic energy — are used, then static balancing conditions imply that the center of mass of the mechanism does not move in the direction of the gravity vector, for any motion of the mechanism. As pointed out in (Streit and Gilmore 1989), the use of elastic elements (e.g. springs) is often preferred since it adds very little mass and inertia to the system. However, in some instances, balancing the mechanisms without introducing springs can also be of interest since the resulting mechanism is usually force-balanced, i.e., statically balanced for any orientation of its base with respect to the direction of gravity.

Let a general spatial n-degree-of-freedom mechanism be composed of n_b moving bodies and one fixed link. Moreover, let the position vector of the center of mass of each moving body with respect to a fixed reference frame be noted $\mathbf{c}_i$ and let the mass of the ith moving body be noted m_i. The position vector of the center of mass of the mechanism with respect to the fixed frame, noted $\mathbf{c}$, can be written as

$$\mathbf{c} = \frac{1}{M} \sum_{i=1}^{n_b} m_i \mathbf{c}_i \tag{3.1}$$

where M is the total mass of the moving links, i.e.,

$$M = \sum_{i=1}^{n_b} m_i \tag{3.2}$$

In general, vector $\mathbf{c}$ will be a function of the configuration of the mechanism, i.e.,

$$\mathbf{c} = \mathbf{c}(\boldsymbol{\theta}) \tag{3.3}$$

where $\boldsymbol{\theta}$ is the vector comprising all the joint coordinates of the mechanism.

Following this notation, the conditions for the force balancing of the mechanism can be written as

$$\mathbf{c} = \mathbf{c}_t \tag{3.4}$$

where $\mathbf{c}_t$ is an arbitrary constant vector, i.e., a vector which is independent from the joint coordinate vector, $\boldsymbol{\theta}$. Similarly, if no elastic elements are used, the condition for static balancing can be written as

$$\mathbf{e}_z^T \mathbf{c} = C_t \tag{3.5}$$

where C_t is an arbitrary constant and $\mathbf{e}_z$ is a unit vector oriented in the direction of gravity.

When elastic elements are used, the total potential energy in the mechanism, noted V, is defined as the sum of the gravitational and elastic potential energy and can be written as

$$V = g\mathbf{e}_z^T \sum_{i=1}^{n_b} m_i \mathbf{c}_i + \frac{1}{2} \sum_{j=1}^{n_s} k_j (s_j - s_j^o)^2 \tag{3.6}$$

where g is the magnitude of the gravitational acceleration, n_s is the number of linear elastic elements in the system, k_j is the stiffness of the jth elastic element, s_j is the length of the jth elastic element and s_j^o is its undeformed length. As mentioned above, when elastic elements are used, the condition for static balancing is that the total potential energy is constant, which can be written as

$$V = V_c \tag{3.7}$$

where V_c is an arbitrary constant.

In articulated mechanical systems, elastic components — e.g. springs — may provide an *exact* compensation of the gravitational forces for all configurations (Nathan 1985; Streit and Gilmore 1989). This compensation usually requires that the *effective* undeformed length of the springs be equal to zero. As shown in (Streit and Gilmore 1989), this assumption does not present any particular problem and is generally simple to achieve in a practical system. The exact compensation can be illustrated using the simple one-degree-of-freedom example shown in Fig. 3.1 and taken from (Streit and Gilmore 1989).

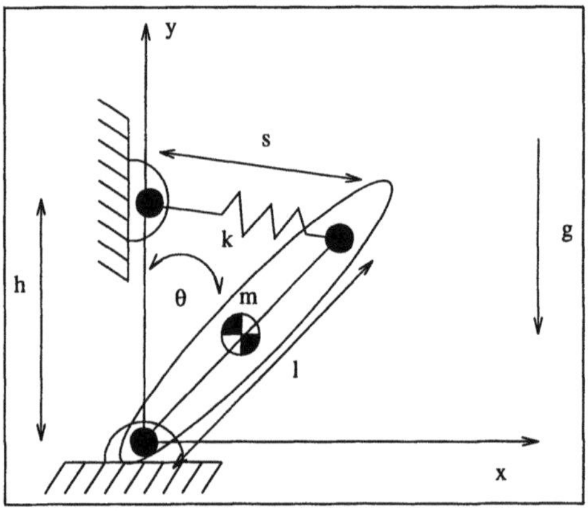

Fig. 3.1. Simple one-degree-of-freedom statically balanced system.

In this system, a single link of mass m is rotating in a vertical plane, as illustrated in the figure. The center of mass of the link is located at a distance c from the pivot and a spring is attached to the fixed link along the vertical axis, at a distance h from the pivot, as well as to the rotating link, at a distance l from the pivot. According to eq.(3.6), the total potential energy in the system can be written as

$$V = mgc\cos\theta + \frac{1}{2}k(s - s^o)^2 \tag{3.8}$$

where k is the stiffness of the spring while s and s^o are its length and its undeformed length, respectively. From the law of cosines, one can write

$$s^2 = h^2 + l^2 - 2hl\cos\theta \tag{3.9}$$

Moreover, if the undeformed length of the spring is equal to zero, i.e.,

$$s^o = 0 \tag{3.10}$$

then, eq.(3.9) can be substituted directly in eq.(3.8), which leads to

$$V = (mgc - khl)\cos\theta + \frac{1}{2}k(h^2 + l^2) \tag{3.11}$$

Hence, the total potential energy of the system will be constant if and only if one has

$$mgc = khl \tag{3.12}$$

or, in other words, if the stiffness of the spring is chosen such that

$$k = \frac{mgc}{hl} \tag{3.13}$$

The resulting statically balanced system will therefore be in static equilibrium for *any* configuration, as long as the orientation of the base of the mechanism with respect to the gravity acceleration vector is not changed. The latter observation is important: when springs are used, static balancing will be obtained for *one given orientation of the base* with respect to gravity. On the other hand, when static balancing is achieved using only a redistribution of the masses and the use of counterweights, it usually leads to force balancing and the resulting mechanisms are statically balanced for *any* orientation of the gravity. In the above example, a force-balanced mechanism would be obtained by removing the spring and adding a counterweight such that the center of mass of the link would be moved to the center of the pivot. Such a mechanism would be statically balanced for any orientation of the gravity.

4. Optimization of parallel mechanisms

Although parallel mechanisms have attractive properties for several applications, the complexity of their kinematics makes it more difficult for a designer to determine a set of kinematic parameters which will efficiently produce prescribed performances. Indeed, the behavior of parallel mechanisms is far less intuitive than that of serial manipulators. The geometric conditions associated with singularities, for instance, may be much more difficult to identify (Merlet 1989; Mayer St-Onge and Gosselin 1996). Therefore, more systematic analysis and optimization tools are required in order to make parallel mechanisms more accessible to designers (Gosselin et al. 1995). Two important categories of performance indices to be used in the optimization of parallel mechanisms are introduced in the following subsections.

4.1 Optimization of the kinematic properties

One very important kinematic property of manipulators or mechanisms is their workspace. For parallel mechanisms, this property is especially critical since the parallel architecture inherently tends to lead to smaller working volumes. In a context of design, a required or prescribed workspace is often specified from the outset, based on the tasks to be performed by the mechanism. The objective of a workspace optimization algorithm would then be to ensure that the actual workspace of the mechanism matches the prescribed one as closely as possible. As proposed in (Gosselin and Guillot 1991), an appropriate formulation of this optimization can be written as

$$\max_{\mathbf{h}} W(\mathcal{W} \cap \mathcal{W}_p) \qquad (4.1)$$

where $\mathbf{h}$ is the vector of kinematic parameters of the mechanism, $\mathcal{W}$ is the workspace of the mechanism, $\mathcal{W}_p$ is the prescribed workspace of the mechanism and $W(\cdot)$ denotes the volume of the region given as an argument. Therefore, the above index would tend to maximize the volume of the intersection between the prescribed and actual workspaces, which, in turn, would tend to make the actual workspace cover the prescribed one.

Another important property of mechanisms is the kinematic accuracy, which is also related to the singularities. For parallel mechanisms this quantity can be measured using the dexterity index defined in a previous section. This index is a function of the configuration of the mechanism — represented by vector $\boldsymbol{\theta}$ — and of the kinematic parameters assembled in vector $\mathbf{h}$. If the workspace of the mechanism to be designed is small, i.e., if only a small portion of the theoretical workspace is of interest, then the following formulation can be used for the optimization of the dexterity

$$\max_{\mathbf{h}} \left[\max_{\boldsymbol{\theta}} \zeta(\boldsymbol{\theta}, \mathbf{h}) \right] \tag{4.2}$$

The above formulation will tend to identify the best conditioned configuration and then try to maximize the dexterity associated to this configuration. If the small workspace can be located around this configuration, the manipulator obtained will evolve around a very accurate configuration and it is assumed that the neighboring configurations will also lead to a good accuracy. This approach is used, for instance, in (Tandirci et al. 1992).

Another approach for the optimization of the dexterity of parallel mechanisms is to use the global dexterity index defined above, i.e., to formulate the problem as follows

$$\max_{\mathbf{h}} \eta(\mathbf{h}) \tag{4.3}$$

Since the global dexterity index η characterizes the average dexterity over the workspace, the solutions obtained from the above optimization will tend to favor a good average dexterity rather than specific configurations of the workspace. This approach may be more appropriate when the workspace is larger. Neither of the two indices presented above can guarantee that singularities will be kept away from the workspace and hence, a singularity analysis still has to be performed on the mechanisms obtained after optimization.

4.2 Static balancing

As mentioned above, static balancing is an important design criterion for a parallel mechanism, especially if the weight of the moving links is large. In flight simulators, for instance, the mass of the platform is usually very large and powerful actuators are required to support the weight of the moving links. In such a system, static balancing can be used to eliminate the need for the actuators to support the weight of the links. This can be accomplished using eq.(3.5) when no elastic elements are used or using eq.(3.7) when elastic components are included. When a complete static balancing is not possible, the actuator forces or torques due to the weight of the links can be minimized by using the following formulation

$$\min_{\mathbf{h}} \left[\frac{1}{W} \int_{W} \boldsymbol{\tau}^T \boldsymbol{\tau} dW \right] \tag{4.4}$$

where $\boldsymbol{\tau}$ is the vector of forces or torques at the actuators when the mechanism is in static equilibrium in a given configuration of the workspace. Therefore, the above optimization will tend to minimize the magnitude of the gravity forces or torques at the actuators over the workspace.

The application of kinematic optimization and static balancing to spherical parallel manipulators will be presented in the next sections.

5. Application to the kinematic design of a spherical 3-dof 9R mechanism

This mechanism has been studied, for instance, in (Craver 1989) and (Gosselin and Angeles 1989). The solution of the inverse kinematic problem has been given in the latter references and a polynomial solution to the direct kinematic problem has been provided in (Gosselin et al. 1994). Moreover, the kinematic design of this mechanism has been addressed in (Gosselin and Lavoie 1993) and (Daniali et al. 1994) while its application to a high-performance camera-orienting device has been presented in (Gosselin and Hamel 1994) and (Gosselin and St-Pierre 1997). This architecture will now be used to illustrate the application of the kinematic optimization formulations introduced in the previous section.

The general architecture of the spherical 3-dof 9R parallel manipulator is represented schematically in Fig. 5.1.

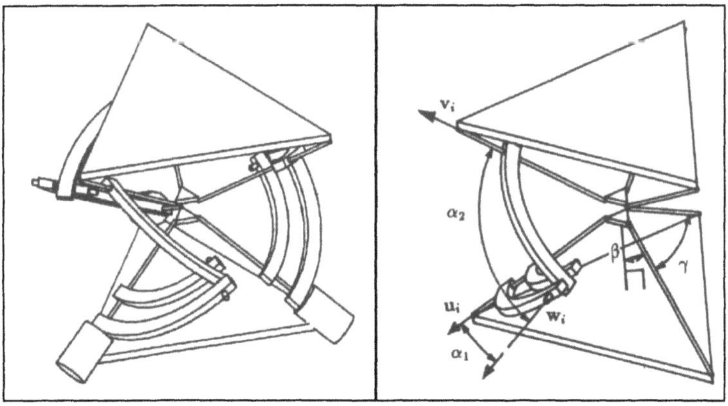

Fig. 5.1. General architecture of a spherical 3-dof 9R parallel manipulator and geometric parameters.

The moving platform — or end-effector in the case of a manipulator — is attached to the fixed link via three kinematic chains. Each of these chains is composed of three revolute joints and two intermediate links, which are termed the *proximal* and *distal* links, respectively. The geometry of the mechanism is such that the axes of all nine revolute joints intersect at one common point which is the *center of rotation* of the device. The three actuators are fixed to the base and are represented as cylinders on the figure.

The unit vectors directed along the axes of the actuators are referred to as vectors $\mathbf{u}_i$, $i = 1, 2, 3$ while the unit vectors directed along the axes of the revolute joints on the end-effector are referred to as vectors $\mathbf{v}_i$, $i = 1, 2, 3$. Moreover, the orientation of the end-effector with respect to the base is given by a rotation matrix $\mathbf{Q}$ and the actuator angles are noted θ_i, $i = 1, 2, 3$. The link angles of the manipulator are assumed to be the same on each of the chains connecting the end-effector to the base and are noted α_1 (for the proximal links) and α_2 (for the distal links). This is represented in Fig. 5.1 together with angles γ and β which define the geometry of the base and the end-effector links. It is clear, from the figure that angles γ and β are not independent. They are in fact related through the following geometric relationship

$$\sin\beta = \frac{2\sqrt{3}}{3}\sin(\frac{\gamma}{2}) \tag{5.1}$$

and hence, only one of these angles suffices to define the geometry of the base or end-effector link. Finally, unit vectors directed along the axis of the intermediate revolute joint of each of the kinematic chains are noted $\mathbf{w}_i$, $i = 1, 2, 3$.

A fixed reference frame $\mathcal{R}$ is attached to the base, with its origin at the center of rotation of the mechanism. This frame is oriented such that the z axis is pointing vertically upwards. Moreover, angle η_i, $i = 1, 2, 3$, is defined as the angle between the projection of vector $\mathbf{u}_i$ in the xy plane and the x axis of the fixed reference frame. Since the mechanism is assumed to be symmetric, one has

$$\eta_1 = 0, \quad \eta_2 = 2\pi/3, \quad \eta_3 = 4\pi/3 \tag{5.2}$$

and, from the geometry of the mechanism, vector $\mathbf{u}_i$ can be written as

$$\mathbf{u}_i = [\cos\eta_i \sin\beta \;\; \sin\eta_i \sin\beta \;\; -\cos\beta]^T, \quad i = 1, 2, 3 \tag{5.3}$$

The joint coordinate θ_i is measured around axis $\mathbf{u}_i$ in a counterclockwise direction, with respect to vector $\mathbf{t}_i$, defined as

$$\mathbf{t}_i = [-\sin\eta_i \;\; \cos\eta_i \;\; 0]^T, \quad i = 1, 2, 3 \tag{5.4}$$

Therefore, vector $\mathbf{w}_i$ can be written as, for $i = 1, 2, 3$,

$$\mathbf{w}_i = \begin{bmatrix} \cos\alpha_1 \sin\beta \cos\eta_i - \cos\theta_i \sin\alpha_1 \sin\eta_i + \sin\theta_i \sin\alpha_1 \cos\beta \cos\eta_i \\ \cos\alpha_1 \sin\beta \sin\eta_i + \cos\theta_i \sin\alpha_1 \cos\eta_i + \sin\theta_i \sin\alpha_1 \cos\beta \sin\eta_i \\ -\cos\alpha_1 \cos\beta + \sin\theta_i \sin\alpha_1 \sin\beta \end{bmatrix} \tag{5.5}$$

As mentioned above, the orientation of the upper platform of the mechanism is described by rotation matrix $\mathbf{Q}$, which is defined as

$$\mathbf{Q} = \begin{bmatrix} c\psi_1 c\psi_2 c\psi_3 - s\psi_1 s\psi_3 & -c\psi_1 c\psi_2 s\psi_3 - s\psi_1 c\psi_3 & c\psi_1 s\psi_2 \\ s\psi_1 c\psi_2 c\psi_3 + c\psi_1 s\psi_3 & -s\psi_1 c\psi_2 s\psi_3 + c\psi_1 c\psi_3 & s\psi_1 s\psi_2 \\ -s\psi_2 c\psi_3 & s\psi_2 s\psi_3 & c\psi_2 \end{bmatrix} \tag{5.6}$$

where $c\psi_i$ and $s\psi_i$ stand respectively for the cosine and the sine of angle ψ_i, for $i = 1, 2, 3$ and where ψ_1, ψ_2 and ψ_3 are three Euler angles used to define the orientation of the upper platform. Vector $\mathbf{v}_i$, which is attached to the platform, can therefore be computed as

$$\mathbf{v}_i = \mathbf{Q}\mathbf{v}_{i0}, \quad i = 1, 2, 3 \tag{5.7}$$

where $\mathbf{v}_{i0}$ represents the orientation of vector $\mathbf{v}_i$ associated with the reference orientation, i.e., the orientation obtained when $\mathbf{Q}$ is equal to the identity matrix. From symmetry and to prevent singularities (Gosselin and St-Pierre 1997), it is assumed that

$$\mathbf{v}_{10} = -\mathbf{u}_3, \quad \mathbf{v}_{20} = -\mathbf{u}_1, \quad \mathbf{v}_{30} = -\mathbf{u}_2 \tag{5.8}$$

which allows the determination of vectors $\mathbf{v}_i$, $i = 1, 2, 3$ as functions of the Euler angles and the geometric parameters of the mechanism.

Since symmetry has been assumed in the above description, the three kinematic parameters which are left free are α_1 and α_2, the angles defining the geometry of the proximal and distal links, as well as angle β — or alternatively γ —, the angle defining the geometry of the base and platform links. Therefore, the dexterity of the mechanism can be optimized over these three parameters in order to obtain the best possible designs.

In (Gosselin and Lavoie 1993) and (Daniali et al. 1994), it was shown that the optimization using the formulation of eq.(4.2) leads to infinitely many solutions, i.e., there exists a continuum of kinematic parameters which lead to isotropic manipulators, i.e. manipulators with an optimal local dexterity.

If the formulation defined in eq.(4.3) is used with the prescribed workspace defined in (Gosselin and Hamel 1994), then the mechanism presented in (Gosselin and Hamel 1994) is obtained. As an illustration of the dexterity of this mechanism, plots of the dexterity over the workspace are given in Fig. 5.2 for four values of one of the Cartesian coordinates covering the prescribed workspace. As can be seen from the plots, the dexterity is always good, throughout the workspace. Finally, it is interesting to point out that the mechanism presented in (Gosselin and Hamel 1994) is also an optimum according to the formulation of eq.(4.1) if the prescribed workspace is the entire space of rotations. Indeed, the geometric parameters of this design allow the mechanism to cover the entire space of rotations.

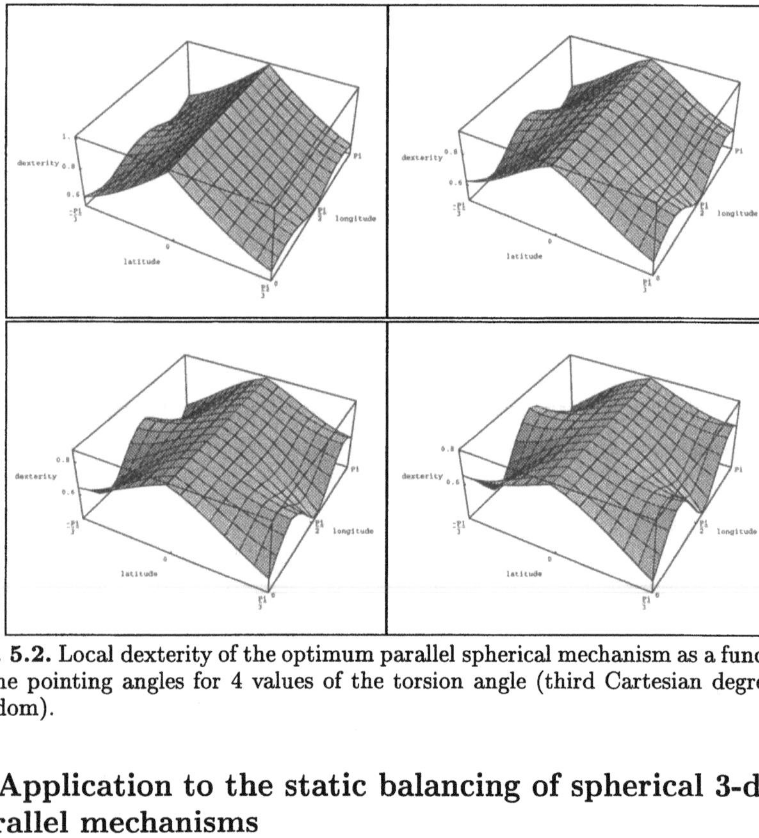

Fig. 5.2. Local dexterity of the optimum parallel spherical mechanism as a function of the pointing angles for 4 values of the torsion angle (third Cartesian degree of freedom).

6. Application to the static balancing of spherical 3-dof parallel mechanisms

Two different architectures of spherical parallel mechanisms as well as one simplified system with a single body will now be used to illustrate the application of static balancing to the design of efficient mechanisms. Expressions for the vertical coordinate of the center of mass of the mechanisms as functions of the dimensional and mass properties as well as the joint coordinates will be derived. Expressions for the potential energy stored in the elastic components will also be obtained when relevant. Eqs.(3.5) or (3.7) will then be applied by setting the coefficients of the latter expressions multiplying the functions of the joint or Cartesian coordinates to zero. This will lead to a constant potential energy and hence, to the conditions for static balancing.

6.1 Static balancing of a single-body spherical 3-dof mechanism with springs

In order to provide some insight into the balancing problem and more specifically into the use of springs, the conditions associated with a single body mounted on a spherical joint are first derived. The system is illustrated in Fig. 6.1.

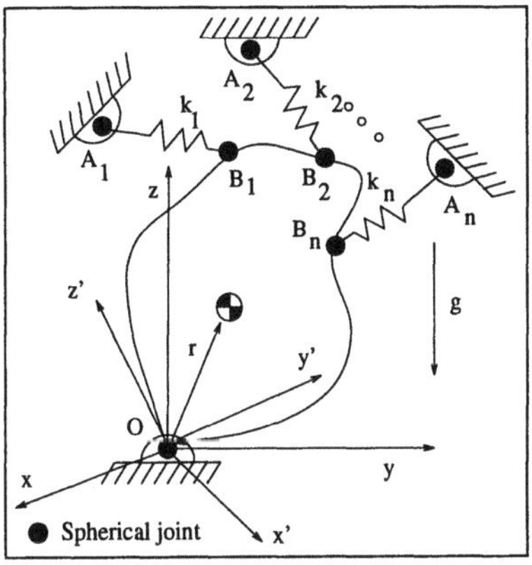

Fig. 6.1. Spherical 3-dof single-body mechanism with springs.

The rigid body is mounted on a 3-dof spherical joint located in O. The center of mass of the body is located at the tip of vector $\mathbf{r}$, as shown on the figure, and the mass of the body is noted m. A series of n springs are attached at one end to fixed spherical joints, noted A_i, and at the other end to spherical joints on the moving body, noted B_i. A fixed coordinate frame $\mathcal{R}(O - xyz)$ is attached to the base with its origin in O. The z axis of this reference frame is pointing in the opposite direction of the gravity vector. Moreover, a moving reference frame $\mathcal{R}'(O - x'y'z')$ is attached to the moving body and a reference configuration is defined for which both reference frames have the same orientation. The orientation of frame $\mathcal{R}'$ with respect to frame $\mathcal{R}$ is given by the rotation matrix $\mathbf{Q}$ defined in eq.(5.6) and in which ψ_1, ψ_2 and ψ_3 are three Euler angles used to define the orientation of the moving body. Additionally, vector $\mathbf{r}$ is a constant vector when expressed in frame $\mathcal{R}'$ and is defined as

$$[\mathbf{r}]_{\mathcal{R}'} = [r_x \; r_y \; r_z]^T \tag{6.1}$$

while vector $\mathbf{a}_i$ is defined as the vector from point O to point A_i and is written, in the fixed frame, as

$$[\mathbf{a}_i]_{\mathcal{R}} = [a_{ix} \; a_{iy} \; a_{iz}]^T, \quad i = 1, \ldots, n \tag{6.2}$$

Similarly, vector $\mathbf{b}_i$ is defined as the vector from point O to point B_i and is written, in the moving frame, as

$$[\mathbf{b}_i]_{\mathcal{R}'} = [b_{ix} \; b_{iy} \; b_{iz}]^T, \quad i = 1, \ldots, n \tag{6.3}$$

Vectors $\mathbf{r}$ and $\mathbf{b}_i$ can be written in the fixed frame using the above rotation matrix, i.e.,

$$[\mathbf{r}]_{\mathcal{R}} = \mathbf{Q}\,[\mathbf{r}]_{\mathcal{R}'}, \quad [\mathbf{b}_i]_{\mathcal{R}} = \mathbf{Q}\,[\mathbf{b}_i]_{\mathcal{R}'}, \quad i = 1, \ldots, n \tag{6.4}$$

The issue addressed with the present system is the balancing of the moving body using springs. The objective is to determine whether it is possible to choose attachment points and stiffness constants for the springs in order to obtain a statically balanced system for *any* orientation of the moving body. As mentioned in the preceding section, the conditions for balancing can be obtained by imposing that the total potential energy (gravitational and elastic) be constant. For the present mechanism, the total potential energy, noted V, can be written as

$$V = mg(\mathbf{r}^T\mathbf{e}_z) + \frac{1}{2}\sum_{i=1}^{n} k_i(\mathbf{a}_i^T\mathbf{a}_i + \mathbf{b}_i^T\mathbf{b}_i - 2\mathbf{a}_i^T\mathbf{b}_i) \tag{6.5}$$

where g is the gravitational acceleration, k_i is the stiffness constant of the ith spring, $\mathbf{e}_z$ is a unit vector in the direction of the z axis of the fixed frame and where all vectors are expressed in the fixed reference frame. Point O has been used as a datum for the gravitational potential energy. Moreover, as explained in a previous section, it has also been assumed, in the above equation, that the undeformed lengths of the springs are equal to zero, which leads to the second term of the equation. As shown in (Streit and Gilmore 1989), this assumption does not represent any particular practical problem since the physical springs can extend beyond the attachment points using guiding systems or pulleys and wires.

In order to obtain the conditions which will ensure that the potential energy is constant for any orientation of the body, eq.(6.5) is rewritten as

$$\begin{aligned} V = & \frac{1}{2}\sum_{i=1}^{n} k_i(\mathbf{a}_i^T\mathbf{a}_i + \mathbf{b}_i^T\mathbf{b}_i) + mg(r_x q_{31} + r_y q_{32} + r_z q_{33}) \\ & - \sum_{i=1}^{n} k_i a_{iz}(b_{ix} q_{31} + b_{iy} q_{32} + b_{iz} q_{33}) \\ & - \sum_{i=1}^{n} k_i a_{ix}(b_{ix} q_{11} + b_{iy} q_{12} + b_{iz} q_{13}) \\ & - \sum_{i=1}^{n} k_i a_{iy}(b_{ix} q_{21} + b_{iy} q_{22} + b_{iz} q_{23}) \end{aligned} \tag{6.6}$$

where q_{ij} stands for the ijth entry of the rotation matrix $\mathbf{Q}$, with $i, j = 1, 2, 3$. The first term of the right hand side of eq.(6.6) is constant. Moreover, since the different entries of matrix $\mathbf{Q}$ involve different expressions of the Euler angles and because of the orthogonality of the sine and cosine functions, the coefficients multiplying the entries of matrix $\mathbf{Q}$ must vanish for V to be a constant function. Therefore, collecting the terms in eq.(6.6), the conditions for the static balancing of the system are obtained as:

$$mgr_x - \sum_{i=1}^{n} k_i a_{iz} b_{ix} = 0 \tag{6.7}$$

$$mgr_y - \sum_{i=1}^{n} k_i a_{iz} b_{iy} = 0 \tag{6.8}$$

$$mgr_z - \sum_{i=1}^{n} k_i a_{iz} b_{iz} = 0 \tag{6.9}$$

$$\sum_{i=1}^{n} k_i a_{ix} b_{ix} = 0 \tag{6.10}$$

$$\sum_{i=1}^{n} k_i a_{ix} b_{iy} = 0 \tag{6.11}$$

$$\sum_{i=1}^{n} k_i a_{ix} b_{iz} = 0 \tag{6.12}$$

$$\sum_{i=1}^{n} k_i a_{iy} b_{ix} = 0 \tag{6.13}$$

$$\sum_{i=1}^{n} k_i a_{iy} b_{iy} = 0 \tag{6.14}$$

$$\sum_{i=1}^{n} k_i a_{iy} b_{iz} = 0 \tag{6.15}$$

Hence, the system of Fig. 6.1 will be balanced for any orientation if eqs.(6.7–6.15) are verified. The interpretation of these conditions will now be studied, for cases in which 1, 2 or more springs are used.

6.1.1 System with one spring ($n = 1$). If only one spring is used, then the summations in eqs.(6.7–6.15) have only one term. Hence eqs.(6.10–6.15) require that

$$a_{1x} = a_{1y} = 0 \tag{6.16}$$

since point B_1 cannot coincide with point O unless the center of mass of the body is located in O, which is a trivial case. Therefore, the fixed point of attachment of the spring, A_1, must be located on the vertical axis passing through point O. Since the summations in eqs.(6.7–6.9) also have only one term, vectors $\mathbf{r}$ and $\mathbf{b}_1$ must be proportional, i.e.,

$$\mathbf{r} = \frac{k_1 a_{1z}}{mg} \mathbf{b}_1 \tag{6.17}$$

which means that the point of attachment of the spring on the body must be on the line defined by points O and the center of mass. If the spring is attached to the fixed link *above* point O, then point B_1 is located on the same side of point O as the center of mass. Conversely, if the spring is attached to the fixed link *under* point O, then point B_1 will be located on the opposite side of point O from the center of mass. Hence, in a physical implementation, the spring can be attached to the base of the mechanism or to a *ceiling* rigidly attached to the base.

This result is important since it shows that a 3-dof body which can undergo arbitrary pure rotations about a fixed spherical joint can be balanced for all configurations using a single spring. The total potential energy in the system will remain constant for any orientation of the body and hence the system can be brought to static equilibrium without any external force or torque in any orientation. Two examples of balanced systems are illustrated in Fig. 6.2.

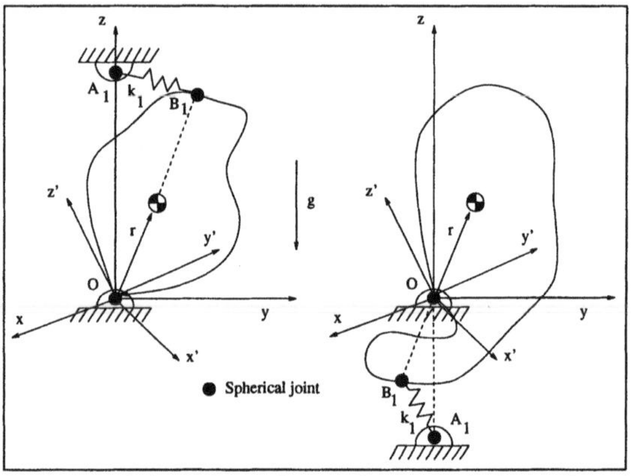

Fig. 6.2. Spherical 3-dof single-body mechanisms balanced with one spring.

6.1.2 System with two springs ($n = 2$). If two springs are used, then all the summations in eqs.(6.7–6.15) have two terms. Therefore, eqs.(6.10–6.15) can be rewritten as

$$\boldsymbol{\Lambda}\mathbf{h} = \mathbf{0}_6 \tag{6.18}$$

where $\boldsymbol{\Lambda}$ is a 6×6 matrix defined as

$$\boldsymbol{\Lambda} = \begin{bmatrix} k_1 a_{1x}\mathbf{1} & k_2 a_{2x}\mathbf{1} \\ k_1 a_{1y}\mathbf{1} & k_2 a_{2y}\mathbf{1} \end{bmatrix} \tag{6.19}$$

where $\mathbf{1}$ stands for the 3×3 identity matrix. Moreover, vector $\mathbf{h}$ is defined as

$$\mathbf{h} = [b_{1x}\ b_{1y}\ b_{1z}\ b_{2x}\ b_{2y}\ b_{2z}]^T \tag{6.20}$$

and $\mathbf{0}_6$ is a six-dimensional zero vector.

Equation (6.18) forms a homogeneous linear system of 6 equations in 6 unknowns. Therefore, it can be satisfied if vector $\mathbf{h}$ vanishes or if the determinant of matrix $\boldsymbol{\Lambda}$ is equal to zero. However, a vanishing vector $\mathbf{h}$ is not possible since this would imply, from eqs.(6.7–6.9), that the center of mass of the body is located in O, which is a trivial case. Therefore, the solutions are found by expanding the determinant of matrix $\boldsymbol{\Lambda}$, which leads to

$$\det(\boldsymbol{\Lambda}) = k_1 k_2 (a_{1x} a_{2y} - a_{2x} a_{1y}) = 0 \tag{6.21}$$

and hence, the mechanism can be balanced if and only if

$$a_{1x} a_{2y} = a_{2x} a_{1y} \tag{6.22}$$

which means that the fixed points of attachments no longer have to be on the vertical axis passing through point O. However, the line defined by points A_1 and A_2 must intersect the vertical axis passing through point O. Moreover, eqs.(6.10–6.15) can also be rewritten in vector form as

$$k_1 a_{1x} \mathbf{b}_1 + k_2 a_{2x} \mathbf{b}_2 = \mathbf{0}_3 \tag{6.23}$$

$$k_1 a_{1y} \mathbf{b}_1 + k_2 a_{2y} \mathbf{b}_2 = \mathbf{0}_3 \tag{6.24}$$

where $\mathbf{0}_3$ stands for the three-dimensional zero vector. Equations (6.23) and (6.24) clearly show that, if a_{1x}, a_{1y}, a_{2x} and a_{2y} are different from zero, then vectors $\mathbf{b}_1$ and $\mathbf{b}_2$ must be collinear, i.e.,

$$\mathbf{b}_2 = \frac{-k_1 a_{1x}}{k_2 a_{2x}} \mathbf{b}_1 = \frac{-k_1 a_{1y}}{k_2 a_{2y}} \mathbf{b}_1 \tag{6.25}$$

and moreover, considering eqs.(6.7–6.9), when $\mathbf{b}_1$ and $\mathbf{b}_2$ are aligned, they must also be aligned with the center of mass.

Therefore, when two springs are used, two possible cases arise. In the first case, the attachment points on the fixed link are located on a line intersecting the vertical axis passing through point O. In this case, the attachment points on the moving body must be located on the line defined by point O and the center of mass of the body. In the second case, the attachment points on the fixed link are located on the vertical axis passing through point O. In this case, the attachment points on the moving body must be located such that vectors $\mathbf{r}$, $\mathbf{b}_1$ and $\mathbf{b}_2$ are coplanar. Two examples of balanced systems with two springs corresponding to these two possible cases are illustrated in Fig. 6.3.

6.1.3 System with more than two springs ($n > 2$). When n is larger than 2, eq.(6.18) then becomes a linear system of 6 equations in $3n$ unknowns. Vector $\mathbf{h}$ is then defined as

$$\mathbf{h} = [b_{1x}\ b_{1y}\ b_{1z}\ b_{2x}\ b_{2y}\ b_{2z} \ldots b_{nx}\ b_{ny}\ b_{nz}]^T \tag{6.26}$$

and matrix $\boldsymbol{\Lambda}$ is a $6 \times 3n$ matrix which can be written as

$$\boldsymbol{\Lambda} = \begin{bmatrix} k_1 a_{1x} \mathbf{1} & k_2 a_{2x} \mathbf{1} & \ldots & k_n a_{nx} \mathbf{1} \\ k_1 a_{1y} \mathbf{1} & k_2 a_{2y} \mathbf{1} & \ldots & k_n a_{ny} \mathbf{1} \end{bmatrix} \tag{6.27}$$

Therefore, the system of eq.(6.18) is an underdetermined system of equations, which may have infinitely many solutions. Hence, the restrictions on the position of the fixed attachment points found in the first two cases are now replaced with linear relationships between the coordinates of these points.

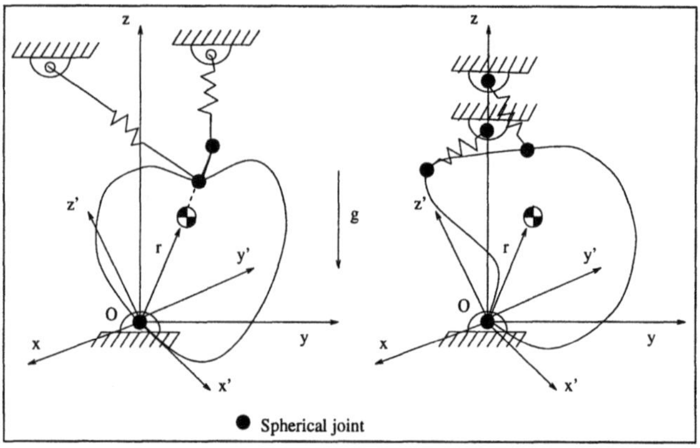

Fig. 6.3. Spherical 3-dof single-body mechanisms balanced with two springs.

Equations (6.7–6.15) then define the set of constraints on the position of the fixed and moving attachment points and the stiffness constants. Several solutions become possible. One particular case occurs when $n = 3$ and if a_{ix} and a_{iy} are different from zero for $i = 1, 2, 3$. In this case, eqs.(6.23) and (6.24) — written for $n = 3$ — will require that the three vectors $\mathbf{b}_1$, $\mathbf{b}_2$ and $\mathbf{b}_3$ be coplanar.

In the next sections, the balancing of parallel 3-dof spherical mechanisms will be addressed. Balancing conditions for systems with no springs will first be derived. Then, the results obtained in this section will be used in order to select the geometry of the spring systems and balancing conditions will be obtained for systems with springs.

6.2 Static balancing of a 3-dof 9R parallel spherical mechanism

This mechanism has been introduced above and used to illustrate the use of kinematic optimization formulations.

Before the balancing of the mechanism can be addressed, the position of the center of mass of the moving bodies must be defined. Since vectors $\mathbf{v}_i$, $i = 1, 2, 3$ are all attached to the platform and are not coplanar, they can be used as a basis in which the position of the center of mass of the platform, noted $\mathbf{r}_p$, can be defined. One writes,

$$\mathbf{r}_p = \delta_1 \mathbf{v}_1 + \delta_2 \mathbf{v}_2 + \delta_3 \mathbf{v}_3 \tag{6.28}$$

where δ_1, δ_2 and δ_3 are scalar quantities defining the position of the center of mass with respect to the center of rotation of the mechanism. Moreover, in what follows, it is assumed that the platform is a symmetric body and hence that

$$\delta_1 = \delta_2 = \delta_3 = \delta \tag{6.29}$$

and the results will be given for this specific case only, for the sake of conciseness. However, this assumption can easily be eliminated and more general results can be obtained.

Similarly, the position of the center of mass of the ith proximal link, noted $\mathbf{r}_{i1}$ can be written using vectors $\mathbf{u}_i$, $\mathbf{w}_i$ and their vector product as a basis. One writes

$$\mathbf{r}_{i1} = s_{11}\mathbf{u}_i + s_{12}\mathbf{w}_i + s_{13}(\mathbf{u}_i \times \mathbf{w}_i), \quad i = 1, 2, 3 \tag{6.30}$$

where s_{11}, s_{12} and s_{13} are scalar quantities defining the position of the center of mass, which are assumed to be the same for each of the proximal links, by symmetry.

Finally, the position of the center of mass of the ith distal link, noted $\mathbf{r}_{i2}$, is assumed to be located in the plane defined by vectors $\mathbf{w}_i$ and $\mathbf{v}_i$, i.e.,

$$\mathbf{r}_{i2} = s_{21}\mathbf{w}_i + s_{22}\mathbf{v}_i, \quad i = 1, 2, 3 \tag{6.31}$$

where s_{21} and s_{22} are scalar quantities defining the position of the center of mass, which are assumed to be the same for each of the distal links, by symmetry. The center of mass of all the moving bodies is defined with respect to the center of rotation of the mechanism.

6.2.1 Balancing without elastic elements. In this section, the conditions for the balancing of the spherical 3-dof 9R mechanism are obtained. Balancing is achieved here by redistribution of the masses only and no elastic elements are introduced in the system. The main advantage of this approach is that the conditions obtained lead to mechanisms whose center of mass is fixed, and hence the mechanisms are balanced for any orientation and magnitude of the gravity vector. This may be useful when the mechanism is to be installed in different orientations with respect to the ground. On the other hand however, balancing a mechanism by redistributing the masses usually leads to large counterweights and to a substantial increase of the inertia. Each of the approaches has its own merit.

Considering the spherical 3-dof 9R parallel mechanism described above and the definitions for the position of the center of mass of each of the links, one can write

$$M\mathbf{r} = m_p\mathbf{r}_p + \sum_{i=1}^{3}(m_1\mathbf{r}_{i1} + m_2\mathbf{r}_{i2}) \tag{6.32}$$

where M is the total mass of the moving links, $\mathbf{r}$ is the position vector of the global center of mass of the mechanism, m_p is the mass of the platform, m_1 is the mass of a proximal link and m_2 is the mass of a distal link.

Substituting eqs.(6.28), (6.30) and (6.31) into eq.(6.32), one obtains an expression for the global center of mass of the mechanism as a function of the Cartesian and joint variables. For instance, the z component, noted $[M\mathbf{r}]_z$ leads to

$$
\begin{aligned}
[M\mathbf{r}]_z = & -3m_1 s_{12}\cos\alpha_1\cos\beta - 3m_2 s_{21}\cos\alpha_1\cos\beta \\
& +3(\delta m_p + m_2 s_{22})\cos\beta\cos\psi_2 \\
& +m_1 s_{13}\sin\alpha_1\sin\beta(\cos\theta_1+\cos\theta_2+\cos\theta_3) \\
& +\sin\alpha_1\sin\beta(m_1 s_{12}+m_2 s_{21})(\sin\theta_1+\sin\theta_2+\sin\theta_3) \quad (6.33)
\end{aligned}
$$

and hence, the vertical component of the center of mass of the mechanism is constant if the following conditions are met:

$$\delta m_p + m_2 s_{22} = 0 \quad (6.34)$$

$$m_1 s_{12} + m_2 s_{21} = 0 \quad (6.35)$$

$$s_{13} = 0 \quad (6.36)$$

Moreover, although the expressions for the x and y components of the center of mass are slightly more complex, they lead to exactly the same balancing conditions. Therefore, the center of mass of the mechanisms satisfying eqs.(6.34–6.36) will remain fixed in all directions and these mechanisms will be force-balanced, i.e., they will be statically balanced for any orientation and magnitude of the gravity vector.

However, as can be inferred from theses equations, mechanisms which satisfy the balancing conditions will most likely involve large counterweights, especially if the mass of the platform is large. Indeed, from eq.(6.34), it is clear that if m_p is large and if δ is not equal to zero — i.e. the center of mass of the platform is not located at the center of rotation of the mechanism, which will be the case in most practical applications — then the product $m_2 s_{22}$ will be substantially large. The latter product represents the mass distribution of the distal links and therefore these links will have a large inertia. For instance, for a small flight simulation device, if it is assumed that m_p is equal to 1000 kg and that δ and s_{22} are respectively equal to 0.5 m and 2 m, then the mass of the distal link would be 250 kg. This means that the total mass of the distal links would be 750 kg, which is almost as much as the platform and is clearly out of proportion. Nevertheless, in applications involving smaller masses and for which the center of mass of the platform is close to the center of rotation, this approach may be useful. Moreover, as mentioned above, it will lead to a mechanism which is balanced for any orientation and magnitude of the gravity vector.

6.2.2 Balancing using elastic elements. The balancing of the spherical 3-dof 9R mechanism will now be addressed by considering the possibility of including springs in the mechanism. As mentioned in a previous section and in (Streit and Gilmore 1989), the main advantage of using springs is that they add very little inertia to the system since they eliminate the need for large counterweights. However, they will lead to a mechanism which is balanced for only one direction and magnitude of the gravity vector but this limitation is of no serious concern for most applications, including flight simulators.

In most applications where the 3-dof 9R parallel mechanism would be used, the platform of the mechanism would be the heaviest body in the system. Since the platform has three degrees of freedom and undergoes spherical motion around the center of rotation, it can be balanced using one or several springs, as shown in the previous section. Therefore, in what follows, springs are attached to a fixed pivot and to the moving platform. Three springs are used in order to preserve symmetry but fewer springs could be used, as shown in the previous section. The springs are attached at one end to a fixed pivot located on the vertical axis passing through the center of rotation of the mechanism and at the other end to the platform at points located along the direction of vectors $\mathbf{v}_1$, $\mathbf{v}_2$ and $\mathbf{v}_3$.

In order to balance the mechanism, an expression for the total potential energy must first be obtained. It is assumed that the point of attachment of the springs on the fixed body is located at a distance c_f from the center of rotation of the mechanism while the distance between the center of rotation and the points of attachment on the platform is noted c_p. Moreover, the vector connecting the center of rotation to the fixed point of attachment of the springs is noted $\mathbf{c}_f$ while the vectors connecting the center of rotation to the attachment points on the platform are noted $\mathbf{c}_{pi}$, where

$$\mathbf{c}_f = c_f \mathbf{e}_z, \qquad \mathbf{c}_{pi} = c_p \mathbf{v}_i, \ i = 1, 2, 3 \tag{6.37}$$

where $\mathbf{e}_z$ is a unit vector pointing in the direction of the z axis of the fixed coordinate frame, as defined previously. Therefore, the expression for the total potential energy can be written as

$$\begin{aligned} V &= [M\mathbf{r}]_z g + \frac{1}{2}k\sum_{i=1}^{3}(\mathbf{c}_f^T\mathbf{c}_f + \mathbf{c}_{pi}^T\mathbf{c}_{pi} - 2\mathbf{c}_f^T\mathbf{c}_{pi}) \\ &= [M\mathbf{r}]_z g + \frac{1}{2}k\sum_{i=1}^{3}(c_f^2 + c_p^2 - 2\mathbf{c}_f^T\mathbf{c}_{pi}) \end{aligned} \tag{6.38}$$

where $[M\mathbf{r}]_z$ is as given in eq.(6.33) and where all vectors are expressed in the fixed frame. Substituting eqs.(6.37) into the above equation leads to an expression of the total potential energy as a function of the Cartesian and joint coordinates. One obtains:

$$\begin{aligned} V = {} & \frac{3}{2}k(c_f^2 + c_p^2) - 3g\cos\alpha_1\cos\beta(m_1 s_{12} + m_2 s_{21}) \\ & + 3\cos\beta\cos\psi_2(g m_p \delta + g m_2 s_{22} - k c_f c_p) \\ & + g m_1 s_{13}\sin\alpha_1\sin\beta(\cos\theta_1 + \cos\theta_2 + \cos\theta_3) \\ & + g\sin\alpha_1\sin\beta(m_1 s_{12} + m_2 s_{21})(\sin\theta_1 + \sin\theta_2 + \sin\theta_3) \end{aligned} \tag{6.39}$$

For the mechanism to be statically balanced, the above quantity has to be independent from the mechanism's configuration and hence, the condition for static balancing can be written as

$$g(\delta m_p + m_2 s_2 2) - k c_f c_p = 0 \quad (6.40)$$
$$m_1 s_{12} + m_2 s_{21} = 0 \quad (6.41)$$
$$s_{13} = 0 \quad (6.42)$$

It is clear, when comparing eqs.(6.40–6.42) to eqs.(6.34–6.36), that the introduction of the springs allows the balancing of the platform (of mass m_p) without increasing the mass of the distal links. This is a significant improvement, especially if the mass of the platform is large. However, the constraint relating the mass of the proximal and distal links, as expressed by eq.(6.41) must still be satisfied, which may be a limitation for the design. In order to eliminate this constraint, additional springs can be mounted on the proximal links, which will introduce new terms in eq.(6.41) involving the stiffness of the springs and will hence allow more freedom in the design of the proximal and distal links.

6.3 Static balancing of a 3-dof 366 parallel spherical mechanism

A spherical 366 parallel mechanism is represented schematically in Fig. 6.4. This mechanism is referred to as a 366 mechanism since the platform is connected to the base via 3 kinematic chains: one 3-dof chain and two 6-dof chains. The 3-dof chain consists of three intersecting revolute joints and is therefore equivalent to a spherical joint. The axis of the first of the three revolute joints on this chain is vertical and this joint is actuated. This is not represented in the figure. The two other chains are as follows: the first joint attached to the base is a passive revolute joint with a vertical axis and the second revolute joint has a horizontal axis orthogonal to the first one. These two joints are equivalent to a Hooke joint and the second revolute joint is actuated. Then, the intermediate joint of the chain is a revolute joint whose axis is parallel to that of the actuated joint and finally, the chain is attached to the platform by a spherical joint. A fixed reference frame $\mathcal{R}$ is defined with its origin at the center of the equivalent spherical joint attached to the base and with its z axis vertical. Furthermore, as indicated on the figure, the plane defined by the z axis and the line connecting the origin of the fixed frame to the intersection of the axes of the first two joints of the ith 6-dof chain makes an angle η_i with plane xz of the fixed frame. Therefore, the position vector of point A_i in the fixed frame, noted $\mathbf{a}_i$, is written as

$$[\mathbf{a}_i]_{\mathcal{R}} = [l_0 \cos\eta_i \;\; l_0 \sin\eta_i \;\; -h_0]^T \quad i = 1, 2 \quad (6.43)$$

where l_0 and h_0 are dimensional parameters associated with the geometry of the base.

A moving reference frame, noted $\mathcal{R}'$ is attached to the platform with its origin at the center of rotation of the equivalent fixed spherical joint. The position vector of point B_i, noted $\mathbf{b}_i$, for $i = 1, 2$ and of the center of mass of the platform, noted $\mathbf{r}$ can be written in this frame as

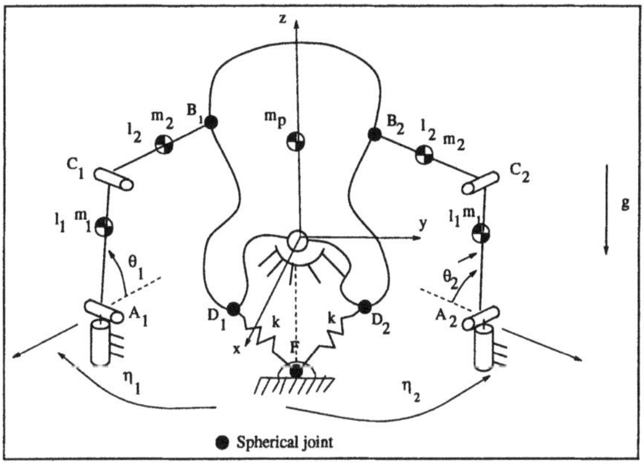

Fig. 6.4. Spherical 3-dof 366 parallel mechanism.

$$[\mathbf{b}_i]_{\mathcal{R}'} = [l_p \cos \eta_i \; l_p \sin \eta_i \; h_p]^T, \quad [\mathbf{r}]_{\mathcal{R}'} = [0 \; 0 \; \delta]^T \tag{6.44}$$

where l_p and h_p are dimensional parameters associated with the geometry of the platform.

When springs are used, they are assumed to be attached to point F on the fixed frame whose position vector $\mathbf{f}$ is written as

$$[\mathbf{f}]_{\mathcal{R}} = [0 \; 0 \; -f]^T \tag{6.45}$$

and to point D_i, on the platform, whose position vector $\mathbf{d}_i$ in the moving frame is written as

$$[\mathbf{d}_i]_{\mathcal{R}'} = [d \cos \eta_i \; d \sin \eta_i \; -v]^T, \quad i = 1, 2 \tag{6.46}$$

where d and v are dimensional parameters associated with the geometry of the platform. Additionally, the orientation of the platform with respect to the base is given by the rotation matrix defined in eq.(5.6).

The orientation of the proximal link of the ith 6-dof chain with respect to the base frame is represented by two angles: angle ϕ_i — associated with the first passive joint — is the angle between the plane formed by segments of length l_1 and l_2 and the xz plane of the fixed frame (this angle is not shown in the figure) while angle θ_i is the angle between the segment of length l_1 and the xy plane.

The center of mass of the ith proximal link — of length l_1 — is located along the line defined from A_i to C_i at a distance r_1 from point A_i. The vertical coordinate of this center of mass in the fixed frame, noted z_{pi}, can therefore be written as

$$z_{pi} = -h_0 + r_1 \sin \theta_i, \quad i = 1, 2 \tag{6.47}$$

Similarly, the center of mass of the ith distal link — of length l_2 — is located along the line defined from B_i to C_i and its position, noted $\mathbf{u}_i$ is given by parameter α, i.e.,

$$\mathbf{u}_i = \mathbf{c}_i + \alpha(\mathbf{b}_i - \mathbf{c}_i) = (1-\alpha)\mathbf{c}_i + \alpha\mathbf{b}_i, \quad i = 1,2 \tag{6.48}$$

where $\mathbf{b}_i$ and $\mathbf{c}_i$ are the position vectors of points B_i and C_i, respectively. Therefore, the vertical coordinate of the center of mass of the ith distal link in the fixed frame, noted z_{ui}, can be written as, for $i = 1, 2$,

$$\begin{aligned} z_{ui} &= (1-\alpha)(-h_0 + l_1 \sin\theta_i) + \alpha(-l_p \sin\psi_2 \cos\psi_3 \cos\eta_i \\ &\quad + l_p \sin\psi_2 \sin\psi_3 \sin\eta_i + h_p \cos\psi_2) \end{aligned} \tag{6.49}$$

Assembling the above equations, the vertical coordinate of the center of mass of the global mechanism, noted r_z, can be written as

$$\begin{aligned} r_z &= \frac{1}{M}[(m_p\delta + 2m_2\alpha h_p)\cos\psi_2 + m_1 \sum_{i=1}^{2}(-h_0 + r_1 \sin\theta_i) \\ &\quad + m_2(1-\alpha)\sum_{i=1}^{2}(-h_0 + l_1 \sin\theta_i) \\ &\quad + m_2\alpha \sum_{i=1}^{2}(-l_p \sin\psi_2 \cos\psi_3 \cos\eta_i + l_p \sin\psi_2 \sin\psi_3 \sin\eta_i)] \end{aligned} \tag{6.50}$$

where m_1 is the mass of a proximal link, m_2 is the mass of a distal link, m_p is the mass of the platform and M is the total mass of the moving links, defined as

$$M = m_p + 2m_1 + 2m_2 \tag{6.51}$$

6.3.1 Balancing without elastic elements. From eq.(6.50), it is clear that the vertical coordinate of the center of mass of the system will be constant if and only if the following conditions are satisfied

$$m_p\delta + 2m_2\alpha h_p = 0 \tag{6.52}$$

$$m_1 r_1 + m_2(1-\alpha)l_1 = 0 \tag{6.53}$$

$$\eta_1 - \eta_2 \pm \pi = 0 \tag{6.54}$$

The latter condition implies that the points of attachment of the 6-dof chains on the base and on the platform must be located at 180° from one another. Moreover, as in the case of the 9R mechanism, if the mass of the platform is large and if δ is not close to zero, the counterweights imposed on the distal links will be large, as prescribed by eq.(6.52).

6.3.2 Balancing with elastic elements. If elastic elements are used, the condition for static balancing is obtained using the total potential energy, which can then be written as

$$\begin{aligned} V &= Mgr_z + \frac{1}{2}k\sum_{i=1}^{2}[d^2 + v^2 + f^2 \\ &\quad + 2f(d \sin\psi_2 \cos\psi_3 \cos\eta_i - d \sin\psi_2 \sin\psi_3 \sin\eta_i - v\cos\psi_2)] \end{aligned} \tag{6.55}$$

where the expression for r_z is given in eq.(6.50) and where k is the stiffness of each of the springs. Eq.(6.55) then leads to new conditions for the static balancing which can be written as

$$g(m_p\delta + 2m_2\alpha h_p) - kfv = 0 \quad (6.56)$$

$$m_1r_1 + m_2(1-\alpha)l_1 = 0 \quad (6.57)$$

$$gm_2\alpha l_p - kfd = 0 \quad (6.58)$$

Similarly to what was observed in the case of the 9R mechanism, the springs now counterbalance the mass of the platform, which eliminates the need for large counterweights on the distal links. Moreover, angles η_1 and η_2 are no longer constrained. Again, the condition of eq.(6.57) can be modified by introducing some springs in the base of the 6-dof chains if required.

6.3.3 Example. An example of a motion simulation device based on a 3-dof 366 spherical parallel mechanism similar to the one described above is represented in Figs. 6.5 and 6.6. This motion simulator is being developed for a single-user facility. In this particular device, the center of mass of the platform is very close to the center of rotation of the system and therefore, no springs are used. Additionally, the kinematic architecture of the 6-dof chains is slightly different from the one described above. Indeed, the planar 2-dof 'serial' part of the 6-dof legs has been replaced with a 5-bar linkage. The balancing equations for the modified system are readily obtained and the resulting balancing conditions are similar. The use of spherical mechanisms in flight simulators is of particular interest since it has been shown recently (Pouliot et al. 1996) that they can produce motion simulation quality comparable to that obtained with conventional 6-dof flight simulators.

7. Conclusion

In this chapter, the design of efficient parallel mechanisms has been addressed using optimum kinematic design and static balancing. Optimum kinematic design has been defined using the concept of dexterity and a few optimization formulations have been proposed. Static balancing has been defined as the set of conditions on mechanism dimensional and mass parameters which, when satisfied, ensure that the weight of the links does not produce any torque (or force) at the actuators for any configuration of the mechanism, under static conditions. Examples of application of these principles to the design of spherical parallel manipulators have been provided. Among other applications, motion simulation systems, where the mass of the platform is large, are of particular interest. Future work also includes the design and fabrication of balanced prototypes.

Acknowledgements: This work has been supported by the Natural Sciences and Engineering Research Council of Canada (NSERC). The author would

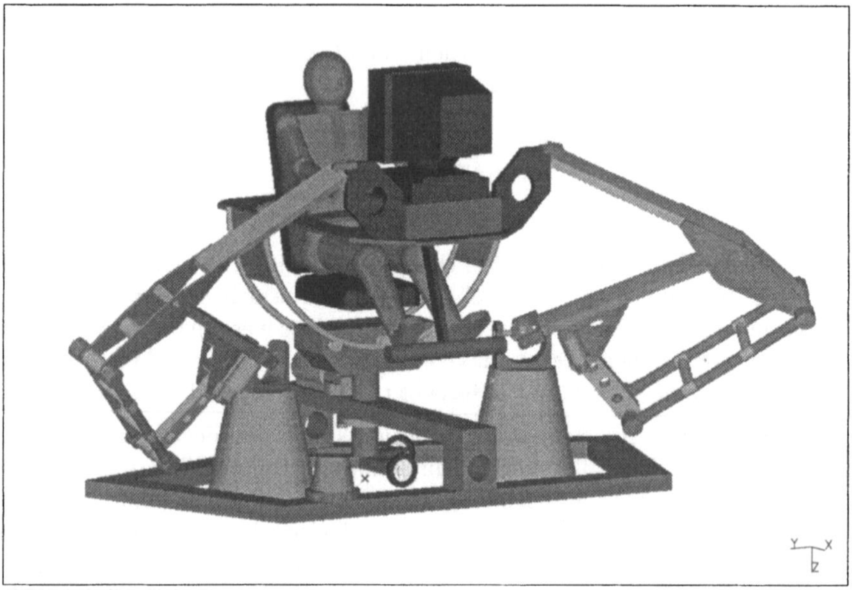

Fig. 6.5. A motion simulation device based on the 3-dof 366 spherical parallel mechanism.

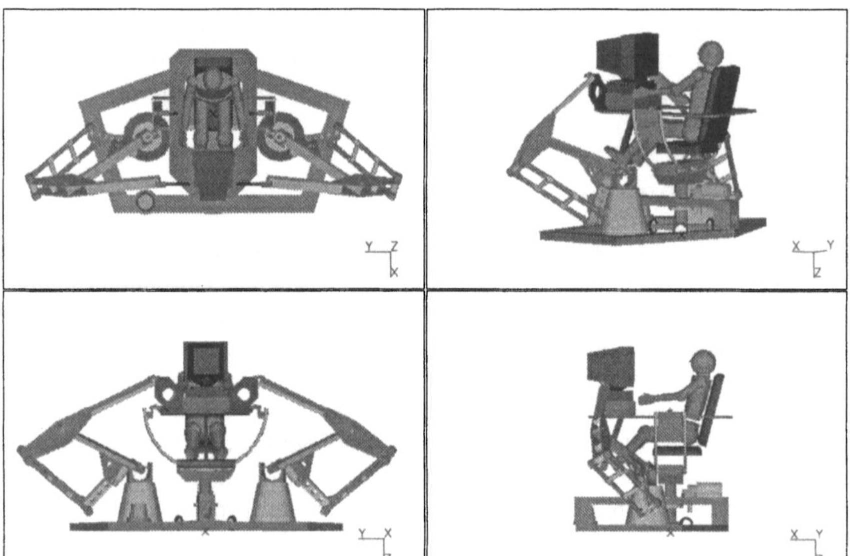

Fig. 6.6. A motion simulation device based on the 3-dof 366 spherical parallel mechanism (orthogonal views).

also like to thank Nicolas Pouliot and Éric Lavoie for their help with the preparation of the CAD models.

References

Craver, W.M., 1989, Structural analysis and design of a three-degree-of-freedom robotic shoulder module, Master's Thesis, The University of Texas at Austin, Department of Mechanical Engineering.

Daniali, H.R.M., Zsombor-Murray, P.J. and Angeles, J., 1994, Isotropic design of spherical parallel manipulators, Proceedings of the *ASME Design Automation Conference*, Minneapolis, Vol. 2.

Fichter, E.F., 1986, A Stewart platform-based manipulator: general theory and practical construction, *The International Journal of Robotics Research*, Vol. 5, No. 2, pp. 157–182.

Gosselin, C. and Angeles, J., 1989, The optimum kinematic design of a spherical three-degree-of-freedom parallel manipulator, *ASME Journal of Mechanisms, Transmissions, and Automation in Design*, Vol. 111, No. 2, pp. 202–207.

Gosselin, C. and Angeles, J., 1990, Singularity analysis of closed-loop kinematic chains, *IEEE Transactions on Robotics and Automation*, Vol. 6, No. 3, pp. 281–290.

Gosselin C. and Angeles, J., 1991, A global performance index for the kinematic optimization of robotic manipulators, *ASME Journal of Mechanical Design*, Vol. 113, No. 3, pp. 220–226.

Gosselin, C. and Guillot, M., 1991, The synthesis of manipulators with prescribed workspace, *ASME Journal of Mechanical Design*, Vol. 113, No. 4, pp. 451–455.

Gosselin, C. and Hamel, J.-F., 1994, The agile eye: a high-performance three-degree-of-freedom camera-orienting device, *Proceedings of the IEEE International Conference on Robotics and Automation*, San Diego, pp. 781–786.

Gosselin, C. and Lavoie, E., 1993, On the kinematic design of spherical three-degree-of-freedom parallel manipulators, *The International Journal of Robotics Research*, Vol. 12, No. 4, pp. 394–402.

Gosselin, C., Perreault, L. and Vaillancourt, C., 1995, Simulation and computer-aided kinematic design of three-degree-of-freedom spherical parallel manipulators, *Journal of Robotic Systems*, Vol. 12, No. 12, pp. 857–869.

Gosselin, C., Sefrioui, J., and Richard, M.J., 1994, On the direct kinematics of spherical three-degree-of-freedom parallel manipulators of general architecture, *ASME Journal of Mechanical Design*, Vol. 116, No. 2, pp. 594–598.

Gosselin, C. and St-Pierre, E., 1997, Development and experimentation of a fast three-degree-of-freedom camera-orienting device, to appear in *The International Journal of Robotics Research*.

Gough, V.E. and Whitehall, S.G., 1962, Universal tyre test machine, *Proceedings of the Ninth International Technical Congress F.I.S.I.T.A.*, May.

Hervé, J.M., 1986, Device for counter-balancing the forces due to gravity in a robot arm, United States Patent 4,620,829, May.

Hunt, K.H., 1983, Structural kinematics of in-parallel-actuated robot arms, *ASME Journal of Mechanisms, Transmissions, and Automation in Design*, Vol. 105, No. 4, pp. 705–712.

Kazerooni, H. and Kim, S., 1988, A new architecture for direct drive robots, *Proceedings of the IEEE Int. Conference on Robotics and Automation*, Philadelphia, pp. 442–445.

Klein, C.A. and Blaho, B.E., 1987, Dexterity measures for the design and control of kinematically redundant manipulators, *The International Journal of Robotics Research*, Vol. 6, No. 2, pp. 72–83.

Lowen, G.G., Tepper, F.R. and Berkof, R.S., 1983, Balancing of linkages – an update, *Mechanism and Machine Theory*, Vol. 18, No. 3, pp. 213–220.

Mayer St-Onge, B. and Gosselin, C., 1996, Singularity analysis and representation of spatial six-dof parallel manipulators, *Proceedings of Advances in Robot Kinematics*, Portorož-Bernardin, pp. 389–398.

Merlet, J.-P., 1989, Singular configurations of parallel manipulators and Grassmann geometry, *The International Journal of Robotics Research*, Vol. 8, No. 5, pp. 45–56.

Merlet, J.-P., 1997, *Les robots parallèles*, 2ème édition, Éditions Hermès, Paris.

Nathan, R.H., 1985, A constant force generation mechanism, *ASME Journal of Mechanisms, Transmissions, and Automation in Design*, Vol. 107, No. 4, pp. 508–512.

Pouliot, N., Nahon, M. and Gosselin, C., 1996, Analysis and comparison of the motion simulation capabilities of three-degree-of-freedom flight simulators, *Proceedings of the AIAA Flight Simulation Technologies Conference*, San Diego, pp. 37–47.

Salisbury, J. K., and Craig, J.J. 1982, Articulated hands: force control and kinematic issues, *International Journal of Robotics Research*, Vol. 1, No. 1, pp. 4–17.

Stewart, D., 1965, A platform with six degrees of freedom, *Proceedings of the Institution of Mechanical Engineers*, Vol. 180, No. 5, pp. 371–378.

Streit, D.A. and Gilmore, B.J., 1989, Perfect spring equilibrators for rotatable bodies, *ASME Journal of Mechanisms, Transmissions, and Automation in Design*, Vol. 111, No. 4, pp. 451–458.

Streit, D.A. and Shin, E., 1990, Equilibrators for planar linkages, Proceedings of the *ASME Mechanisms Conference*, Chicago, Vol. DE-25, pp. 21–28.

Tandirci, M., Angeles, J., and Ranjbaran, F., 1992, The characteristic point and the characteristic length of robotic manipulators, *Proceedings of the ASME Mechanisms Conference*, Scottsdale, DE-Vol. 45, pp. 203-208.

Ulrich, N. and Kumar, V., 1991, Passive mechanical gravity compensation for robot manipulators, *Proceedings of the IEEE International Conference on Robotics and Automation*, Sacramento, pp. 1536–1541.

Walsh, G.J., Streit, D.A. and Gilmore, B.J., 1991, Spatial spring equilibrator theory, *Mechanism and Machine Theory*, Vol. 26, No. 2, pp. 155–170.

Yoshikawa, T., 1985, Manipulability of robotic mechanisms, *The International Journal of Robotics Research*, Vol. 4, No. 2, pp. 3–9.

Optimization of Multi-DOF Mechanisms

Patrick Chedmail

IRCyN-UMR CNRS 6597, Ecole Centrale de Nantes
1, Rue de la Noë, F-44072 Nantes, France
Patrick.Chedmail@lan.ec-nantes.fr

Summary. The purpose of this chapter is to discuss recent results reported in the literature on mechanism synthesis using optimization techniques. We are interested in the optimum design of multi-DOF mechanisms taking into account their dimensions and their morphology. After introducing the general problem of optimum design in this context, we describe different formulations of objective functions used in order to represent the mechanical specifications of the mechanism at hand. A short survey of classical and nonclassical optimization algorithms is included with respect to their application to mechanism synthesis. Finally, several applications of mechanism optimization are discussed, as appearing in research and industrial environments. These applications comprise industrial manipulators and ergonomic layouts in encumbered environments.

Keywords. Mechanism synthesis, dimensional synthesis, topological synthesis, robot mechanism synthesis, ergonomic layout synthesis, local and global optimization, stochastic approaches.

1. Introduction

The use of optimization techniques in mechanical engineering is becoming more and more popular due to the sustained increase of computer power. This approach is neither general nor global because of the complexity of the design processes, which bring about many different considerations, namely,

- the function of a product;
- the mechanical behavior (geometric, kinematic, static or dynamic aspects) of a product;
- the manufacturability of the parts;
- the assemblability of the components;
- the maintainability of the product; and
- the cost.

Therefore, it is impossible to simultaneously take into account so many different constraints, variables, and objective functions. Human ability is necessary for managing (in project teams, for example) the different actors of the design process.

However, several domains of mechanical design can be handled with these techniques, as their objectives are clearly defined (uniquely if possible), and their number of variables and constraints are limited. This is the case, for instance, of structural optimization: it is now possible to optimize subsets of complex products such as the wings of an airplane. Optimization is thus one of the possible approaches to mechanism synthesis.

We are interested here in the optimum design of mechanisms taking into account their dimension and their morphology. We consider geometric ("reach a given zone of the environment without collisions with the obstacles"), kinematic ("follow a given trajectory with a point or a frame with a specified velocity"), static ("the force transmission ratio is specified"), or dynamic ("the accelerations are given during the motion of a part ") specifications.

We distinguish here the analysis phase ("given a set of design variables of a mechanism, which is its mechanical behavior?"), and the synthesis phase of a mechanism ("given an expected mechanical behavior of a mechanism, define its design variables"). These two opposite passages from the structural to the functional points of view (the analysis phase) and its inverse (the synthesis phase) have been pointed out, for instance, by Freudenstein and Maki (1979). The second problem is classical in mechanical engineering and pertains to the class of inverse problems. This is one of the current challenges of concurrent engineering.

One solution for solving the inverse problems mentioned above is the use of catalogs of various kinds, i.e.,

- the mental catalog of a designer who develops his or her knowledge from experience;
- the physical catalog, as described in a book; and
- the virtual catalog of a database of mechanisms in the computer memory.

The catalog permits, by scanning and testing the different items, to detect a solution of a problem. Along the lines of famous catalogs such as Leonardo da Vinci's, a well-known catalog is Artobolevski's (Artobolevski 1975). However, as noted by Sardain (1993), it seems today that the types of mechanisms that Artobolevski describes are very limited: no open-loop mechanisms such as those of manipulators; only one or two loops with only one dof; and mostly, planar mechanisms. Only 8% of the items are actual spatial mechanisms. Therefore, this catalog seems to be quite restrictive. The virtual catalog may be interesting if it results in an automatic generation of mechanisms with specified constraints. This solution was proposed by Mayourian and Freudenstein (1984) for general mechanisms, and by Delignières (1987) for manipulators. Finally, the human mental catalog is very particular because of the human ability to strictly couple one's database with his or her personal object-oriented approach.

Olson, Erdman and Riley (1985) propose to decompose the synthesis problem into the three stages:

- the specification of the problem: specification of the function of the mechanism and of its environment (obstacles, gravity effect, thermal effects, etc.);
- the topological synthesis of the mechanism; and
- the dimensional synthesis of the mechanism.

The first stage corresponds to the technical and functional specification phase of a product: it describes the task. As described by Jacquet, Petiot, Sommer and Koyama (1997), there exists an extremely rich literature on this stage, which we consider here out of the scope of the chapter.

The second stage is dedicated to the choice of the type of mechanism:

- is it composed of an open chain or not?;
- which is the nature of the joints (revolute, prismatic, etc.)?;
- how are these joints oriented with respect to each other (this last question can be considered also as pertaining to the dimensional synthesis stage of the mechanism)?

The last stage considers the instanciation of the variables which are associated with the chosen topology. It consists in choosing the variables and verifying the validity of the initial technical and functional specifications:

- lengths of the links;
- definition of the mechanical characteristics of the links (section, thickness, shape, material, etc.);
- distances and orientations between the joints;
- position and orientation of the mechanism with respect to the task;
- definition of the transmission ratio; and
- choice of the actuators.

Finally, the global process for designing mechanisms can be decomposed as proposed in Fig. 1.

In fact, in a general case, the design process is incompletely defined in Fig. 1, since the "Search for Solutions" phase is not described therein. We suppose henceforth that this phase is performed within an optimization approach; this will achieve the synthesis of the different constraints. The input of the process is the specification of the task (the "Model of the task"). The output is given with the description of the mechanism and eventually some undefined item of the task (as, for instance, the complete collision-free trajectories in the case of a manipulator working in an encumbered environment). Our purpose is to describe the possibilities of this approach.

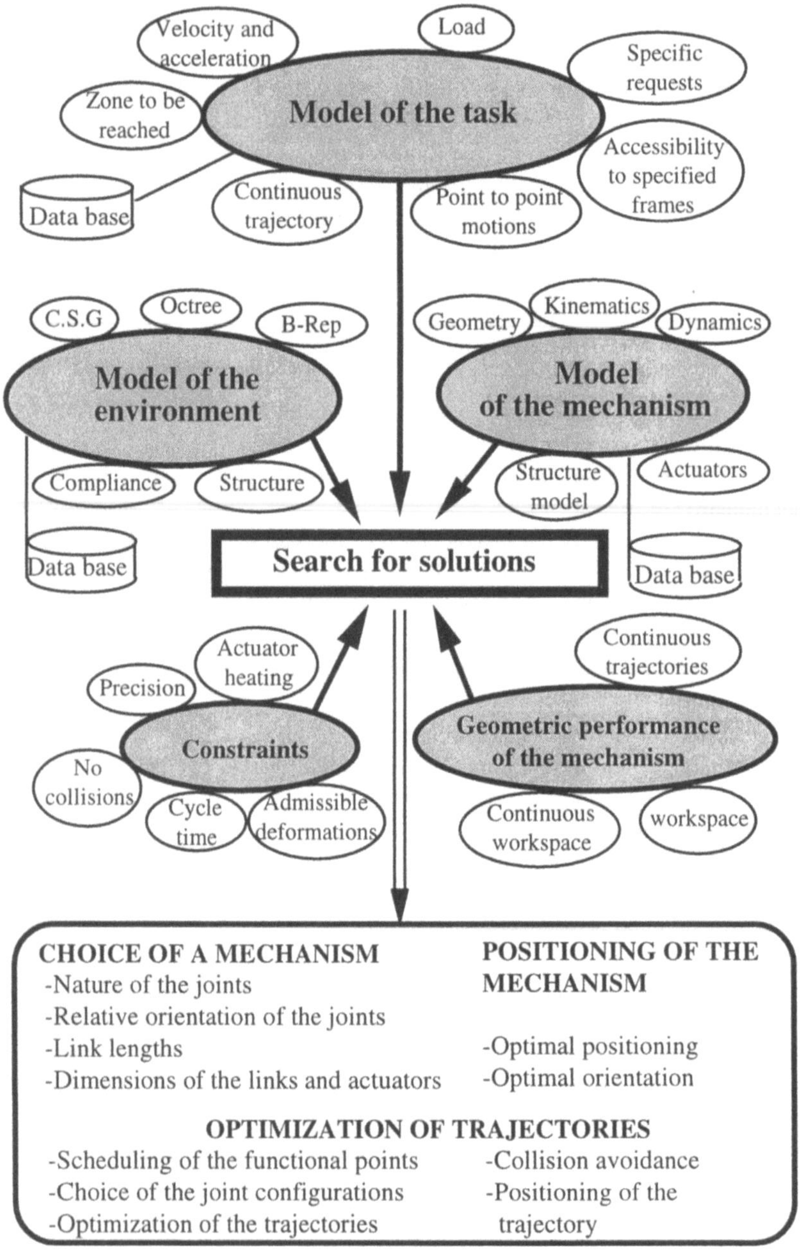

Fig. 1. Decomposition of the process of mechanism design

After a description of the state of the art on mechanism synthesis and its place in the field of research on mechanisms (§2), we describe several formulations associated with optimization problems (§3). We give a general description of the optimization variables (the design variables). We also propose an analysis of possible objective functions in order to represent the mechanical specifications of a mechanism: geometric, kinematic, dynamic constraints or performances. Section 4 is dedicated to the description of a few classical and nonclassical optimization techniques. We focus particularly on stochastic and nonstochastic methods for solving dimensional and topological synthesis problems (§4). In section 5, we include different applications drawn from the literature and from our own results in the field of robot manipulator and ergonomic analysis.

2. A Short Survey of Mechanism Synthesis

2.1 The Place of Mechanism Synthesis in Mechanism Research

The history of research and development in the field of mechanisms since 1945 has been described by Erdman and Sandor (1991). The authors consider different activities in this domain, which are related to graphical and CAD aspects, to the integration of finite elements in mechanism analysis, to the effects of high speed in mechanisms, to the optimization, and to the synthesis of mechanisms.

A first remark on this historical account is that it is strongly linked to that of the development of computer science. Therefore, the first software packages for analyzing mechanisms appear in universities and some companies in the early 1960s. The flexibility of the links in mechanisms is taken into account during the 1960s and the first commercial software package appears in the 1970s. Out of the graphical approaches, which are limited to simple mechanisms, a large number of solutions exist today for simulating mechanisms. There is dedicated software such as the well-known ADAMS (Automatic Dynamic Analysis of Mechanical Systems), MECANO (application of the SAMCEF package for rigid and flexible mechanisms), DADS (Dynamic Analysis and Design Systems), and SDS. One can find also in standard CAD packages several "kinematic" applications, e.g., in CATIA, AutoCAD, CADDS, EUCLID and I-DEAS. This coupling with CAD systems occurred in the mid 1980s, the late integration being mainly due to the differences between database formats originating from different software packages.

At the same time, high-speed mechanisms are studied but, due to the complexity of their associated nonlinear effects, industrial developments appear only in the late 1980s.

The synthesis of four-bar linkages is developed in the early 1960s, the first solutions being based on an algebraic approach. A few software packages exist today for solving this type of problems, such as AUCASYN (Hansen 1993), M.I.T.'s KINSIN (Rubel and Kaufman 1977) or LINCAGES-4 and -6, developed at the University of Minnesota (Erdman and Sandor 1991). However, the generalization to dimensional synthesis using optimization methods remained in the research domain, mainly due to the excessive CPU-time requirements.

Likewise, topological synthesis remained until recently the exclusive domain of enumeration methods. Its real development took place in the last ten years with the use of expert systems and new optimization techniques. Expert systems seemed highly suitable for grasping the knowledge of experts (mechanical engineers) during the design process. Unfortunately, this process is highly complex and many different considerations lead the designer to a compromise that he or she rarely analysis. New optimization techniques, such as stochastic methods, will permit the designer to take into account nonquantitative variables such as the nature of a joint (revolute or prismatic) or the topology of a mechanism. A problem in these techniques is the lack of convergence proofs. Moreover, there is no evidence of real industrial interest for the application of these methods.

2.2 A Survey of a Few Methods

We describe in this subsection some of the methods which have been proposed in the literature for solving the mechanism synthesis problem. Such survey has been also proposed by other authors, e.g., Sardain (1993) and Vasiliu (1997).

2.2.1 Catalogs and Manual Approaches

The first natural method consists of using a catalog of mechanisms and to interactively test each of its items and to compare the results with the specifications. The designer discriminates between good and wrong results using his or her own experience. Of course, the existence of automatic catalogs based on formalized criteria would be very useful for performing a first selection. Mayourian and Freudenstein (1984) proposed an algorithm in this context for creating morphologies as functions of the degree of freedom and of the number of links of the resulting mechanism. Delignières (1987) proposed something similar for the synthesis of manipulators.

2.2.2 The Contribution of Graphs

The use of the graph of a mechanism, whereby the nodes represent the links and the edges the joints, can be used during the design phase for qualifying the global kinematic properties of the mechanism. Mayourian and Freudenstein (1984) defined with the aid of this graph the partial and the global mobilities of a mechanism. Agrawal and Rao (1987) use graph theory for obtaining a systematic definition of the mobilities, Gosselin and Angeles (1987) defined for different subdomains the mobilities of planar and spherical mechanisms with revolute joints, while Liu and Yu (1995) studied the mobility of mechanisms with several degrees of freedom. These analyses can be used for an *a priori* selection of a morphology.

2.2.3 Expert Systems

Hoeltzel and Chieng (1990) reported the state of the art of the methods using knowledge-based systems for establishing a correspondence between the function and the structure of mechanisms. Because of the nonuniqueness of this relation, there are two families of solutions: the direct solution is based on a systematic enumeration of the design alternatives which satisfy the task constraints. The reverse solution relies on a database of cases and uses a pattern-matching approach. Sardain (1993) proposes to filter a database of mechanisms with logical criteria or with rules which are activated with the description of the task. The result of these approaches is a set of potential morphologies that the designer or the system has to validate in a next phase.

2.2.4 Algebraic Methods

These methods were proposed for instance by Freudenstein and Maki (1979). They are very popular and rely on an exact algebraic description of the following problem: we want to find the mechanism which permits one to reach a set of given points or frames along the trajectory. The unknown variables are the dimensional parameters of the mechanism. The method gives an exact solution when it exists. Of course, the number of points to be reached and the number of design variables is strictly related; this is the limitation of the method.

2.2.5 Virtual Reality Approach

The use of virtual reality may be very powerful for helping the designer in his or her choices. By observing the direct effects of the design purpose, the designer can immediately react and create a coupling between the virtual object and himself or herself. For instance, with the aid of USYCAMS package, developed by Gonzalez-Palacios and Angeles (1993), one finds this feature. This software package simulates the behavior of cam transmissions. All along the simulation, in an on-line process, the designer can modify the parameters of the transmission at will, the mechanism being continuously transformed. The designer can thus evaluate the design sensitivity and adapt his or her decisions accordingly. An interesting result is that for some singular values of the design variables, the mechanism changes its morphology. Using an analogy with the singularities in the configuration space of a robot, which separate different postures of the robot, we can interpret that feature by saying that the singularities in the design space correspond to the boundaries between different morphologies. Chedmail and

Damay (1996) have proposed to generalize this idea by coupling the virtual model of the mechanism, an optimization algorithm, and the action of the designer in a real-time process; the actions of the optimizer and of the designer are concurrent. The design variables may be shared or decoupled between the two actors. The performance of the global system for designing a mechanism depends not only on this variable distribution, but also on the sampling period between actions of the two actors.

2.2.6 Optimization Methods

The classical way for designing a mechanism is to use a sequential approach. This manual method consists in a sequence of simulations and modifications of the design variables. The validity of the solution is established by the designer, developments in computer performance increasing the interest of designers in such an approach, as the simulation phases become shorter and shorter. The virtual-reality approach is a final development of this solution (see §2.2.5). Optimization methods are used for automating this process, one of their important features being the automatic definition of the design variables during the modification phase (see Fig. 2).

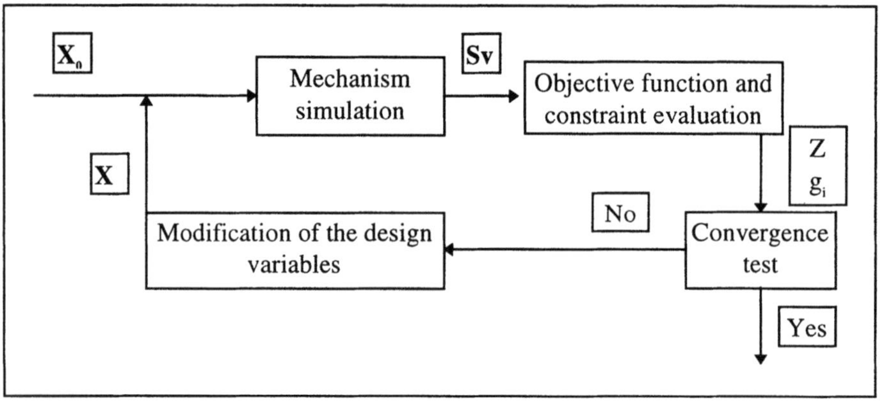

Fig. 2. The optimization scheme

The solution is obtained by minimizing an objective function Z(**X**) where vector **X** of the design variables is subjected to verify a set of constraints g_i, for i=1,...,m. The result of the mechanism simulation is a vector of "state variables" **Sv**.

2.2.6.1 The Continuation Method

This method consists of following a specified trajectory leading to the optimum solution. The idea (Angeles et al. 1988; Hansen 1993) is to improve the convergence process by introducing, between the initial design variable vector $\mathbf{X}_0$ and the desired optimum, intermediate objectives of the search.

2.2.6.2 The Energy Method

In the energy method developed by Avilés et al. (1994), the objective is to produce a trajectory. In this method we suppose that the mechanism follows the trajectory and the links are subjected to some virtual deformation. The method consists of minimizing the energy of deformation of the system in the space of the design variables. When the method succeeds, the energy is equal to zero, the mechanism is rigid, and the trajectory is followed.

2.2.6.3 Neural Networks

The use of neural networks for mechanism synthesis was recently proposed by Vasiliu (1997). The idea is to model the application between the function of a mechanism, i.e. its performance or the value of the objective function representing the task or the optimal length of the trajectory it can reach along the desired one, Z(**X**), and its structure (the set of the design variables **X**):

$$\mathbf{X} \rightarrow Z(\mathbf{X}). \tag{1}$$

In an example pertaining to the synthesis of planar mechanisms, Vasiliu uses a database of cases (about 30,000 different mechanisms) which have been simulated and estimated. Thereafter, he optimizes a neural network for matching this database. Finally he uses this model for obtaining a mechanism which will be adapted to a new given trajectory out of the database. This approach is limited to dimensional synthesis.

2.2.6.4 Genetic Algorithms

One of the main difficulties in mechanism design is topological synthesis. This arises with the presence of nonquantitative variables, such as the nature of a joint (we cannot compare a revolute and a prismatic joint, for there is no inequality relation between them, nor can two different morphologies of a mechanism be compared, for instance). Therefore, out of the above-mentioned enumeration and expert approaches, there was no solution to that problem. The idea of using stochastic algorithms, such as genetic algorithms, was proposed recently by different authors (Kim 1992; Chen and Burdick 1995; Chedmail and Ramstein 1996; and Farritor et al. 1996). This optimization method was originally proposed by John Holland and his coworkers, and developed by Goldberg (1989) and Michalewicz (1992). The idea is to optimize a set of solutions where the design variables can be the type of the joints or the morphology of the mechanism. We will come back to this question in §§4.2.2 and in §5.2.

2.2.6.5 Topological Structural Optimization

Quite recently, an original and promising method was proposed by Bendsoe (1996); Larsen *et al* (1996); and Frecker *et al.* (1997) for the topological synthesis of mechanisms. Contrary to previous methods, the mechanism is the result of a continuous process of optimization where the variables themselves are continuous. The mechanism can be considered as a singularity in the space of design variables. The authors use for that purpose the structural topology optimization approach which was proposed by Cheng during the 1970s (Rozvany *et al.* 1995 and Cheng 1997). This method is applied to structural optimization in continuum mechanics. The design variable is the density of voids in the material. An homogenization technique permits one to describe with equivalent parameters the associated virtual material. The result of this approach is the occurrence of singular structures as the ribs on a plate or bars or beams in a continuous structure.

Bendsoe, Larsen *et al.*, and Frecker *et al.* proposed to use the foregoing property for "finding" morphologies of mechanisms. The examples presented by Bendsoe or Frecker are related to the determination of micro-mechanisms for handling micro-parts. The specification of the mechanism is: "determine a transmission of effort from a piezo-electrical actuator to an effector". The use of the method leads to the occurrence of a structure with weakened zones corresponding to possible revolute joints of a mechanism.

3. A Global-Optimization Approach to Mechanism Synthesis

We describe in this section a global-optimization approach to mechanism synthesis. We study successively the set of the design variables and different objective functions or constraints. We end up by establishing a global-optimization problem of mechanism synthesis. This approach was used in robotic applications by Chedmail and Wenger (1989) and Chedmail (1990).

3.1 The Design Variables of a Mechanism

We use the formalism of the Denavit and Hartenberg (1955) joints parametrization. Moreover, we suppose that the mechanism under synthesis has an open-chain structure and derive the associated set of geometric relations between its joint and cartesian variables.

The vector $\mathbf{X}$ of design variables is composed of two different subvectors:

$$\mathbf{X} = [\mathbf{X}_{MD}{}^{T}, \mathbf{X}_{SD}{}^{T}]^{T}, \tag{2}$$

where $\mathbf{X}_{MD}$ is the design vector of the mechanism itself and $\mathbf{X}_{SD}$ is the design vector of the mechanism in its site.

Vector $\mathbf{X}_{MD}$ contains the parameters below:

- $[{}^n x_E, {}^n y_E, {}^n z_E]^T$, the location of the end-effector on link n;
- $[{}^n\psi_E, {}^n\theta_E, {}^n\phi_E]^T$, the orientation variables of the end-effector; the two foregoing vectors being grouped in vector ${}^n\mathbf{X}_E$;
- the nature of the joints; for joint i, $\sigma_i = 0$ for a revolute joint and 1 for a prismatic joint;
- the geometric parameters of the joints (distance and orientation between neighboring axis) will be represented with the vector of Denavit and Hartenberg parameters $\mathbf{d}$;
- the characteristics of the ith actuator (nominal torque, mass, transmission ratio, nominal velocity, etc.) will be represented with a vector $\mathbf{X}_{ai}$, all vectors $\mathbf{X}_{ai}$ being grouped in a vector $\mathbf{X}_a$ of dimension greater than each $\mathbf{X}_{ai}$;
- the structural characteristics of the jth link (mass, Young's modulus, moment of inertia of the cross section, etc.) will be defined with a vector $\mathbf{X}_{sj}$, all vectors $\mathbf{X}_{sj}$ being grouped in a vector $\mathbf{X}_s$ of dimension greater than each $\mathbf{X}_{sj}$;

In summary, we have

$$\mathbf{X}_{MD} = [\sigma^T, \mathbf{d}^T, {}^n\mathbf{X}_E^T, \mathbf{X}_s^T, \mathbf{X}_a^T]^T. \tag{3}$$

We can see on Fig. 3 an example of such a parametrization in the case of the design of a robot for airplane cleaning (Chedmail and Wenger 1995). The robot is supposed to clean 80% of the surface of the airplane and the problem is to define its morphology and its dimensions, vector $\mathbf{X}_{MD}$. This example will be detailed in §5.1 and in §§5.2.3.

The second set of parameters, grouped in vector $\mathbf{X}_{SD}$, describes the mechanism in its site. It is composed of the following variables:

- ${}^0\mathbf{X}_M$ represents the position and the orientation of the mechanism with respect to its task (the trajectory to follow, the frames to be reached, etc.);
- ${}^0\mathbf{X}_{env}$ represents the position and the orientation of the different components of the environment of the mechanism (the machines, structures, etc.) with respect to the task of the mechanism.

We thus have

$$\mathbf{X}_{SD} = [{}^0\mathbf{X}_M^T, {}^0\mathbf{X}_{env}^T]^T. \tag{4}$$

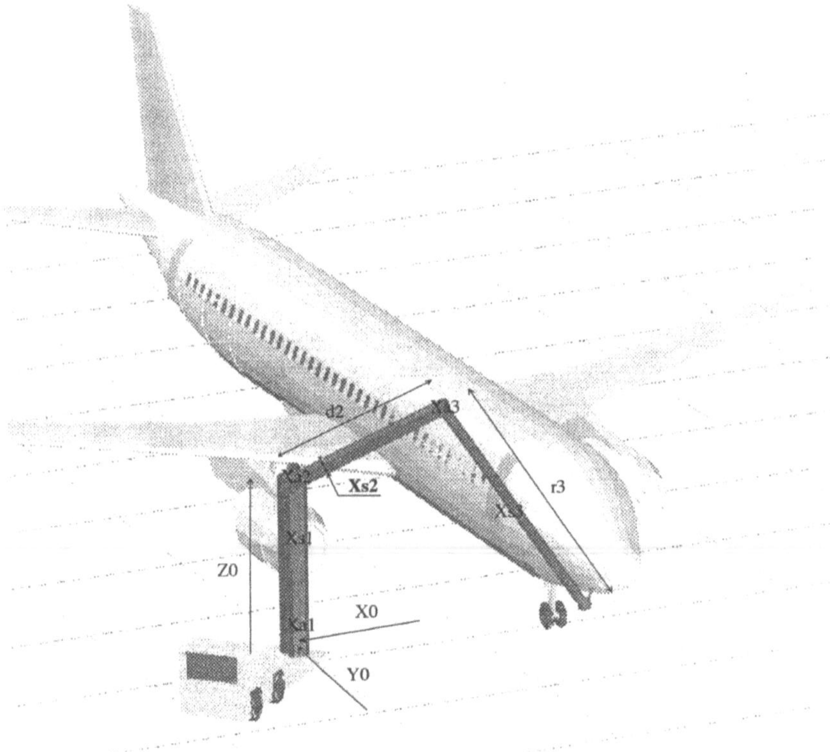

Fig. 3. $\mathbf{X}_{MD}$ and $\mathbf{X}_{SD}$ parameters of an airplane-cleaning robotic mechanism

3.2 Design Objective Functions and Constraints of a Mechanism

An optimization approach to mechanism synthesis implies to define the objective function and the constraints that describe the task. We propose below different solutions for solving this problem. Moreover, we do not always distinguish between objective functions and constraints. In fact, in the majority of the cases, the optimization technique (for solving a constrained problem) consists of transforming the constrained optimization problem into an unconstrained problem. This is the case of the popular penalty-function methods or of the augmented Lagrangian functions. Finally, we distinguish among geometric, kinematic, and mechanical constraints or objective functions.

3.2.1 Geometric Objective Functions and Constraints

We suppose that the mechanism is located in an environment with obstacles. Besides the task of reaching a specified zone, trajectory, frame, or point, it may be mandatory to avoid collisions during this task. We propose three different levels for defining the associated objective functions and constraints: the accessibility to a frame; the ability to continuously follow a trajectory; and the accessibility to a zone.

3.2.1.1 Accessibility to a Frame

We discuss here the problem of the verification of a human operator's ability to reach an object in an encumbered environment as depicted in Fig. 4.

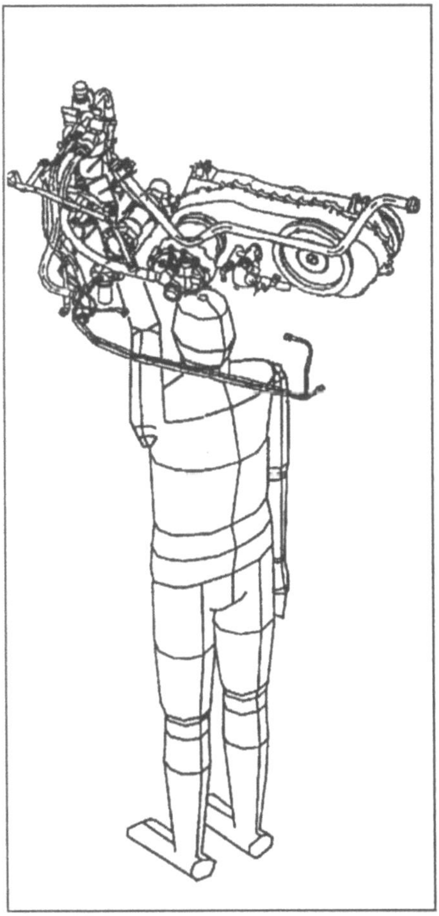

Fig. 4. Accessibility to a frame in an encumbered environment

We focus on the possibility of finding a posture and not of finding a path for reaching this posture. For example, does the human operator, in Fig. 4 reach a filter among the obstacles? Is there a position of the legs, of the trunk, of the arm which allows the performance of the task to do this without a collision? In this problem, the operator will be modeled with a mechanism whose topology and dimensions are completely defined. The design variables are the components of the $\mathbf{X}_{SD}$ vector. As proposed by Riffard and Chedmail (1996), the problem can be transformed in the following way: let us suppose that the task is performed, i.e. the hand of the operator is on the target. Now we suppose that the trunk has to be located in a given subspace P (a parallelepiped for instance) and the orientation is in a given subdomain C, a cone for instance (see Fig. 5). Moreover, we suppose that the degree of freedom, defined as the dimension of vector $\mathbf{q}_{glob}$, of the associated mechanism is that of the arm. Under these conditions, the problem consists of minimizing the distance d_1 of the trunk to P plus d_2, the "distance" of the orientation to the cone C without collision with the environment. Here, we write "distance" with quotation marks to emphasize that d_1 and d_2 are of quite different natures. The former is measured in meter, while the second, in radian. We thus need a weighting factor to treat d_1 and d_2 on an equal basis.

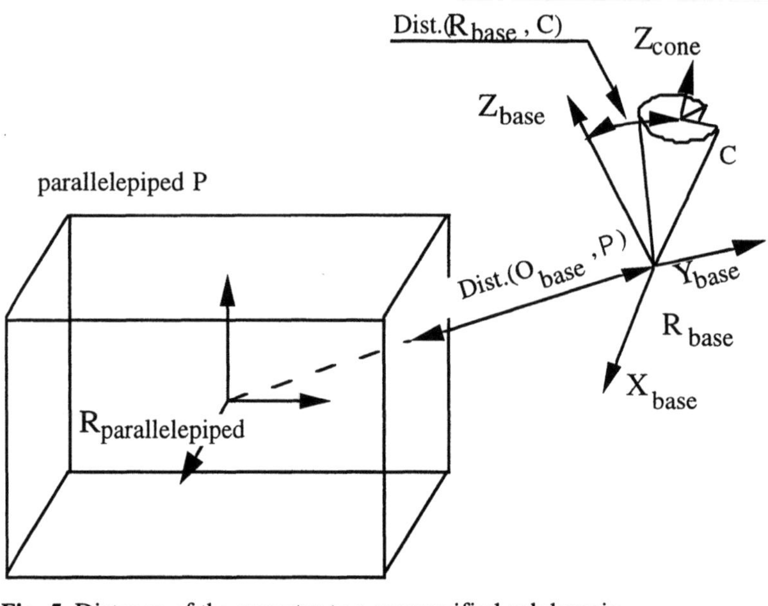

Fig. 5. Distance of the operator to a prespecified subdomain

Therefore, an objective function measuring the accessibility $Z(\mathbf{q}_{glob}, \mathbf{X})$ for this problem is

$$Z(\mathbf{q}_{glob}, \mathbf{X}) = \varepsilon(Z_1(\mathbf{q}_{glob}) + \sigma) + (1-\varepsilon)Z_2(\mathbf{q}_{glob}, \mathbf{X}), \tag{5}$$

where Z_1 represents the sum of the two distances d_1 and d_2; Z_2 represents a penalty term for the collisions of the operator with the environment (the surface of collision or the volume of the collision, for instance); $\varepsilon = 1$ if $d_1 \neq 0$ and $\varepsilon = 0$ if $d_1 = 0$. Moreover, σ is a parameter for regularizing the function $Z(\mathbf{q}_{glob}, \mathbf{X})$ and ensuring its local monotony when ε changes its value.

The results of the optimization will be discussed in §5.3.

3.2.1.2 Ability to follow a trajectory

The ability of a mechanism to follow a given trajectory is a well-known task in mechanism synthesis. We will not pursue this question since many authors, as Wenger and Chedmail (1991), have discussed it extensively.

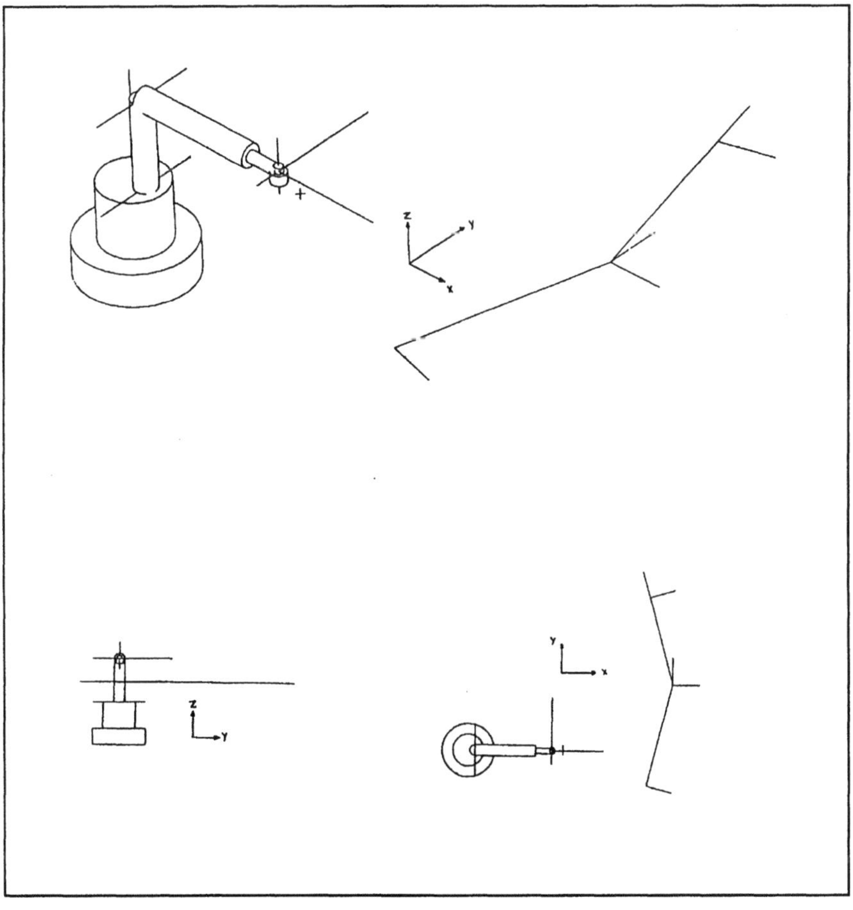

Fig. 6. Description of the robot and of the desired trajectory

We only want to point out two aspects:

- The measure of the distance between a desired trajectory and the trajectory obtained with a current mechanism (defined with its design vector **X**) is not as simple as it may appear. The first reason is that (if we consider the position *and* the orientation of a rigid solid) it is impossible to theoretically define such distance because SO3 (the group of the rotations in 3D space) does not have the structure of the Euclidean space. Moreover, the calculation of the distance between two curves in $\mathbf{R}^3$ (space points in 3D space), may lead to some difficulties due to the existence of artificial numerical singularities in the space of the design variables;
- It is possible to take into account simultaneously different objective functions/constraints such as the ability to follow a trajectory *and* to avoid collisions with the environment *and* minimize the cycle time along the trajectory. The question is how to determine a suitable weighting factor between these different objective functions or constraints.

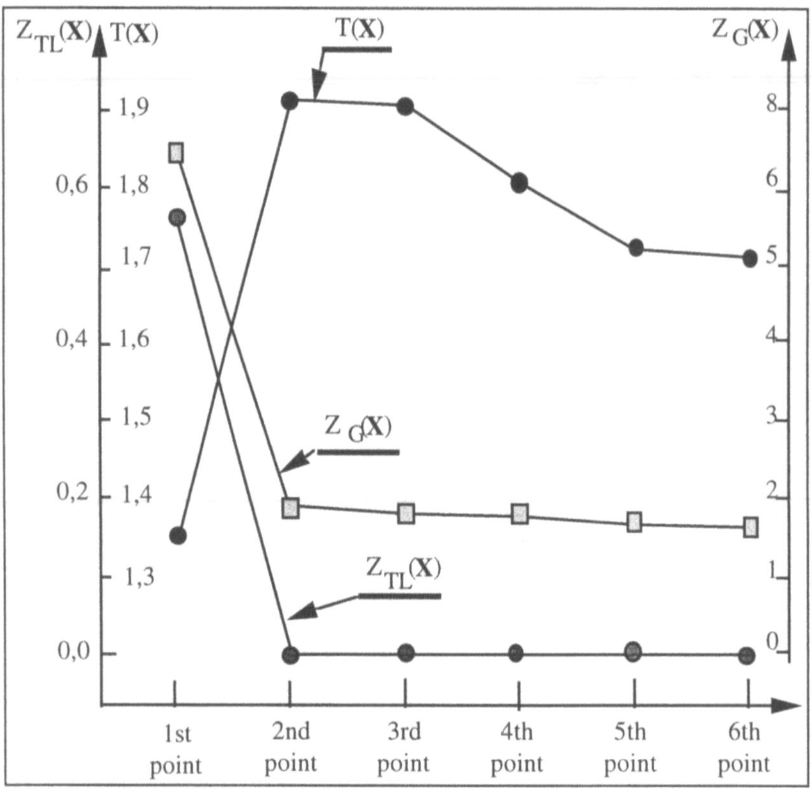

Fig. 7. Evaluation of $Z_G(X)$, $Z_{TL}(X)$, and T(X) for different positions of the robot

We give a simple example for illustrating the second remark without taking into account collisions. Let us consider the mechanism of the anthropomorphic manipulator of Fig. 6; its trajectory is composed of two segments; the design variable (**X**) is the position of its base.

Let us define the following objective function:

$$Z_G(\mathbf{X}) = \sigma^* Z_{TL}(\mathbf{X}) + T(\mathbf{X}), \tag{6}$$

where $T(\mathbf{X})$ is the cycle time for running along the trajectory; $Z_{TL}(\mathbf{X})$ is the length of the trajectory which can be followed; and σ is a penality term (10 in this example). One can see in Fig. 7 the values of $Z_G(\mathbf{X})$, $Z_{TL}(\mathbf{X})$ and $T(\mathbf{X})$ for different positions of the base of the robot. In fact, these positions correspond to different iterations of an optimization process; this function Z_G has been successfully used for optimizing the position of the base of the robot.

3.2.1.3 Accessibility and ability to move in a zone among obstacles

This last definition of an objective function was proposed by Chedmail and Wenger (1989). It is more dedicated to the design of manipulator mechanisms.

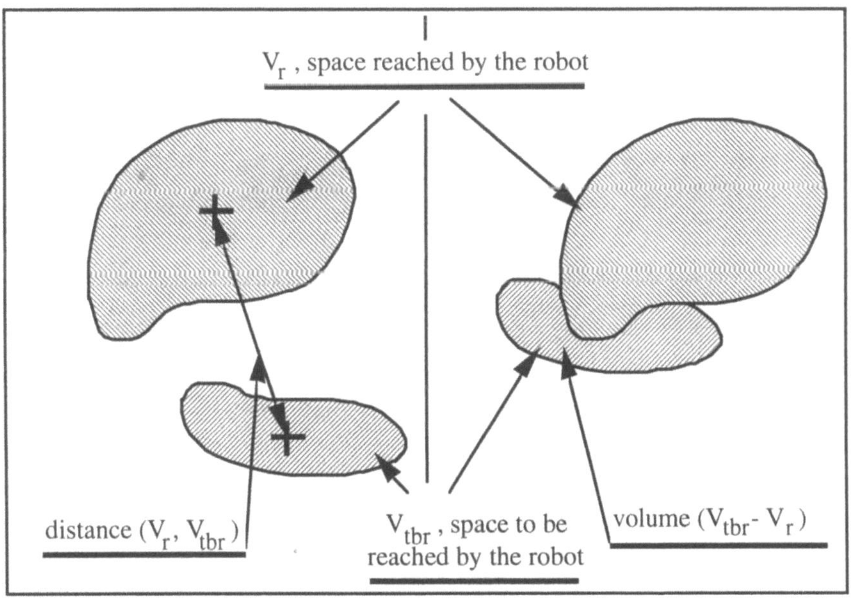

Fig. 8. Description of the objective function Z(**X**)

We consider an n-degrees-of-freedom robotic manipulator, its environment, which is encumbered with obstacles, and a zone of its task space to be reached. We want to know which is the best-suited set of design variables, grouped in vector **X**, for insuring the ability of the manipulator to continuously move anywhere in the given zone. This last question has been extensively studied by Wenger and Chedmail (1991).

So, let us consider Fig. 8, where V_{tbr} is the space to be reached and V_r is the space where the manipulator can continuously move without collisions with the environment. The objective function Z(**X**), which describes the intrusion of V_{tbr} by V_r is:

$$Z(\mathbf{X}) = \text{distance}\ (V_{tbr}, V_r) \text{ if } V_{tbr} - V_r = V_r \text{ (without intrusion);} \tag{7a}$$

$$Z(\mathbf{X}) = \text{volume}\ (V_{tbr} - V_r) \text{ if } V_{tbr} - V_r \neq V_r \text{ (with intrusion).} \tag{7b}$$

The objective function Z(**X**) depends on all the components of **X**. To illustrate this statement, let us consider the example depicted in Fig. 9: a first robot RO_1 has been dimensioned for supporting a 5-Kg load and a second robot RO_2 for a 25-Kg load. The cross sections of their links are different and the corresponding accessible spaces V_{r1} and V_{r2} are different as illustrated in Fig. 9. One can see the sensitivity of Z(**X**) with respect to variations of **X**.

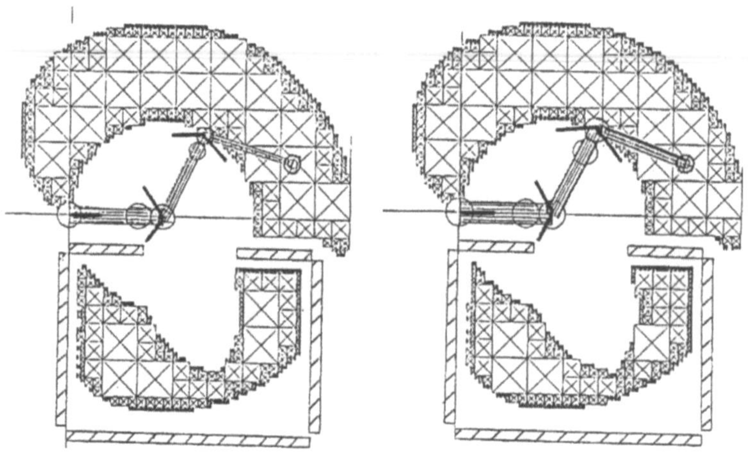

Robot RO_1 and V_{a1} Robot RO_2 and V_{a2}

Fig. 9. Two different accessible workspaces V_{a1} and V_{a2}

3.2.2 Other Mechanical Objective Functions and Constraints

One can find in the literature many different objective functions for optimizing a mechanism. The first one is the mass of the system, proposed for instance by Willmert et al. (1978). Another one is the accelerations by choosing the transmissions ratio (Seering and Pasch 1984). Lee and Tortorelli (1985) proposed to consider the position of the actuators and the dimensions of the cross sections of the links; the objective function is to obtain the isotropy of the instantaneous global inertial operator. Thomas *et al.* (1985) used as an objective function the maximum transported load or the maximum acceleration. Potkonjak (1986) introduced a thermal objective function on the actuators. Merlet (1996) proposed an objective function for improving the manipulability of a heavy load in the case of a Stewart platform. Angeles et al. (1992) used the Jacobian condition number as an objective function for the dimensioning of the DIESTRO manipulator. Otherwise, many authors have worked on the problem of static and dynamic syntheses of mechanisms: the problem here is to determine the best dimensions and the mass distribution for the balancing of the dynamic and the gravity effects. In the case of Martin and Gosselin (1996), for instance, the authors proposed a general methodology to balance the gravity effects with springs or masses.

Chedmail (1990) and Chedmail and Gautier (1990) proposed a general formulation for the optimum synthesis of robotic manipulators:

Determine **X**, the vector of the design variables of the manipulator, such that it minimizes

$$C = C(\mathbf{m}_a, \mathbf{m}_s, \mathbf{m}_t, \mathbf{m}_g) \tag{8}$$

subjected to the constraints:

$$g^{(1)}_i = \left\{ \max_{(\mathbf{q},\dot{\mathbf{q}},\ddot{\mathbf{q}}) \in C_{dyn.}} |\Gamma_{a,i}| \right\} - \Gamma_{imp,i} \leq 0 \tag{9}$$

$$g^{(2)}_i = \left\{ \max_{\mathbf{q}(t) \in TR_{réal.}} (T_{a,i}) \right\} - T_{na,i}{}^2 \leq 0 \tag{10}$$

$$g^{(3)} = \left\{ \max_{\mathbf{q} \in \mathbf{Q}} f(\mathbf{U}_E) \right\} - \mathbf{U}_{max.} \leq 0 \tag{11}$$

$$g^{(4)}_i = L_{ng,i} - \left\{ \min_{\mathbf{q}(t) \in TR_{réal.}} L_{g,i} \right\} \leq 0 \tag{12}$$

for i = 1, ..., n

$$g^{(5)} = \mathbf{ct}(\mathbf{X}) \leq 0 \tag{13}$$

where

- index i pertains to the ith link of the manipulator, which is supposed to have an open-loop structure with n joints and links,

- $C(\mathbf{m}_a, \mathbf{m}_s, \mathbf{m}_t, \mathbf{m}_g)$ is the cost function; it depends on the mass $\mathbf{m}_a$ of the actuators, the structure of the links ($\mathbf{m}_s$), the transmission and the driving components ($\mathbf{m}_t$ and $\mathbf{m}_g$),
- the constraints $g^{(1)}_i$, $g^{(2)}_i$, $g^{(3)}$, $g^{(4)}_i$, $g^{(5)}$ are related to the limited expected torque on each joint, the limited temperature in the actuators, the limited deformations of the links, the limited duration of the drives and several complementary technological constraints,
- C_{dyn} is the space of the dynamic admissible states of the robot,
- $T_{réal}$. is the space of the feasible trajectories, which are compatible with the trajectory generator of the robot,
- **Q** is the configuration space of the robot.

Different subproblems have been solved which have been extracted from this global one (Chedmail 1990; Chedmail and Gautier 1990).

One of the most challenging difficulties for solving the foregoing problem is the existence of the four internal optimization problems which are hidden in the constraints $g^{(1)}_i$, $g^{(2)}_i$, $g^{(3)}$, $g^{(4)}_i$. This question leads to the concept of *penalizing configuration or trajectory with respect to the considered constraint* (Chedmail and Gautier 1990).

4. Optimization Methods

The objective of this section is to discuss the choice of the most suitable optimization method for mechanism synthesis. One important feature of this type of problem is the existence of nonconvex objective functions and constraints. Therefore, in the majority of the cases, one cannot expect to reach the global minimum of the problem at hand. However, optimization methods will lead to a "better" solution than the initial guess, which may be acceptable in some industrial applications. However, when we look for the topology of a mechanism or its dimensions, it is unacceptable to obtain a trajectory that lies far from the desired one!

Another remark concerns the nonlinear nature of these problems and the existence of constraints. We suppose henceforth that the original constrained problems have been transformed into unconstrained ones by using a dual method, for instance, the penality-function or the augmented Lagrangian approaches. We thus have: minimize f(**X**) where **X** belongs to an open domain $\aleph$.

Finally, a number of unconstrained optimization methods are available, as described by Minoux (1983). A literature survey being outside of this chapter, we will outline some of the methods which seem to be most convenient for our purposes. We have chosen to distinguish between deterministic and stochastic methods. Deterministic methods use the knowledge of the topology of the objective function in the neighborhood of the current design vector **X**. On the contrary, stochastic methods do not use directly this information; instead, they use random processes for testing new points. In fact, it is possible to make a mixing of the two approaches, but we consider them separately for purposes of this discussion.

4.1 Deterministic Methods

Deterministic methods use the information on the local topology of the objective function. These methods may be performed directly, as in the case of descent methods, where the evolution of vector **X** is subjected to follow the direction of the local gradient:

$$\mathbf{X}^{(k+1)} = \mathbf{X}^{(k)} - \lambda^{(k)} \nabla f(\mathbf{X}^{(k)}) \tag{14}$$

In eq. (14), k is the index of the current iteration and $\nabla f(\mathbf{X}^{(k)})$ is the gradient of the objective function evaluated at $\mathbf{X}^{(k)}$.

In the case of Newton and quasi-Newton methods, this direction $\nabla f(\mathbf{X}^{(k)})$ is modified by multiplying it by the inverse of the Hessian, or its estimate, as follows:

$$\mathbf{X}^{(k+1)} = \mathbf{X}^{(k)} - (\nabla^2 f(\mathbf{X}^{(k)}))^{-1} \nabla f(\mathbf{X}^{(k)}) \text{ (Newton method)} \tag{15}$$

$$\mathbf{X}^{(k+1)} = \mathbf{X}^{(k)} - \lambda^{(k)} \mathbf{H}^{(k)} \nabla f(\mathbf{X}^{(k)}) \text{ (quasi-Newton methods)} \tag{16}$$

where $\nabla^2 f(\mathbf{X}^{(k)})$ is the Hessian of the objective function f evaluated at $\mathbf{X}^{(k)}$, and $\mathbf{H}^{(k)}$ is an approximation of the inverse of $\nabla^2 f(\mathbf{X}^{(k)})$ in quasi-Newton methods, as, for instance, in Powell's, Davidon-Fletcher-Powell's or Broyden-Fletcher-Goldfarb-Shanno's corrections.

The various deterministic methods (steepest descent, accelerated steepest descent, etc.) differ with the choice of $\lambda^{(k)}$.

Other methods, for instance, conjugate gradient methods, are related to the local properties of the Hessian $\nabla^2 f(\mathbf{X}^{(k)})$ of the objective function.

Finally, heuristic methods such as that of Hooke and Jeeves (1961) exploit also the local properties of the topology of the objective function .

The conclusion of this very short account is that *fundamentally* we cannot expect to obtain a global minimum of f with these methods. The good news is that these methods are quite fast, as compared with stochastic methods, from a CPU-time point of view.

4.2 Stochastic Methods

The two methods described in this section are not the only stochastic optimization approaches in the literature. There exist other ones such as TABOU search or, of course, purely random searches! The latter are only a reference when comparing the performances of the others.

4.2.1 Simulated Annealing

Simulated annealing was proposed by Kirkpatrick et al. (1983). Their purpose was to develop a method for finding the global minimum of a function f(**X**). Their algorithm is originally supported by the works of Metropolis *et al.* (1953) on the annealing of metals. The objective of Metropolis and his colleagues was to determine the equilibrium state of a material subjected to a temperature *T*. The material is supposed to be composed of N particles, and the authors try to find their optimum array leading to the equilibrium temperature *T*. The energy E of the material is supposed to be a function of the array **X**. The equilibrium state is obtained for the minimum value of E(**X**).
For solving this physical problem, Metropolis and his colleagues proposed the following algorithm:

- iteration 1: let $\mathbf{X}^{(0)}$ be the initial array of the particles;
- iteration k:
  * $\mathbf{X}^{(k-1)}$ is supposed to be known;
  * let $\mathbf{Y} = \mathbf{X}^{(k-1)} + \Delta\mathbf{X}^{(k-1)}$, where $\Delta\mathbf{X}^{(k-1)}$ is a random increment;
  * compute $\Delta E = E(\mathbf{Y}) - E(\mathbf{X}^{(k-1)})$;
  * if $\Delta E < 0$, then $\mathbf{X}^{(k)} = \mathbf{Y}$, and proceed to the next iteration k+1;
  * if $\Delta E > 0$, then determine a random value of the real δ such that $\delta \in [0, 1]$;
    - if $\delta < \exp\left(-\frac{\Delta E}{Tk_B}\right)$, then $\mathbf{X}^{(k)} = \mathbf{Y}$, and proceed to the next iteration k+1;
    - if $\delta > \exp\left(-\frac{\Delta E}{Tk_B}\right)$, then $\mathbf{X}^{(k)} = \mathbf{X}^{(k-1)}$, and proceed to the next iteration k+1.

In this algorithm, *T* represents the temperature and k_B the Boltzman constant. The last test is known as the Metropolis criterion, which means that it is possible to accept a worse solution than the initial one ($\Delta E > 0$) with a random test. The smaller ΔE, the more often we can expect such occurrence. Therefore, it becomes possible to accept a deterioration of the energy with the hope that a better result will appear later on.

The idea of the simulated-annealing algorithm is to perform a continuous decreasing of the temperature T or of the control term $c = k_B T$:

- external iteration 1: let $\mathbf{X}^{(0)}$ be the initial array of the particles, and $T^{(0)}$ the initial temperature (the sequence $T^{(k)}$ is available);
- external iteration k:
  * $\mathbf{X}^{(k-1)}$, $T^{(k-1)}$ and $\Delta T^{(k-1)}$ are supposed to be known;
  * execute the Metropolis algorithm; $\mathbf{X}^{(k)}$ is the result of this algorithm; proceed to the next external iteration k+1.

The simulated-annealing method demonstrated very quickly its powerful ability for determining the global minimum of a function. However, the difficulties below should be pointed out :

- the decreasing of T (or c) is difficult to determine: if it is too fast, the algorithm stops in a local minimum; if too slow, the time to convergence may become prohibitively long. The procedure needs an initial experimental approach;
- the internal iterations have to be stopped over a given value N_{iter}, which has to be initially defined; this needs also an experimental approach;
- the convergence of the internal and external iterations has to be tested. These tests have to be carefully chosen as, once again, the success of the procedure depends on their choice,
- the validation of the method has to be stochastic, with many tests. Fortunately, with the same set of data we can perform as many independent tests as we want if the random generator gives different results for two sequential tests.

In spite of these restrictions, the method has been successfully used in many different domains. We shall see an example of application in §5.3.

4.2.2 Genetic Algorithms

Genetic algorithms were introduced by John Holland and his colleagues, their use having been strongly encouraged by Goldberg (1989) and Michalewicz (1992).

The basic underlying idea corresponds to an analogy with the living world. Let us consider a population of individuals (a family of solutions or design vectors). Each individual is determined by its chromosome (the design variables). At each generation (each iteration), the population is transformed into a new one across three different transformations (three different operators):

- the *mutation transformation*: with a probability p_{mut}, some of the chromosomes of each individual can mutate (some of the components of each solution can be randomly modified);

- the *crossover transformation*: two individuals in the initial population, the "parents", can have "children" who will share the chromosomes of their "parents". This transformation is obtained with a probability p_{cross} (two design vectors can exchange some of their components);
- the *selection*: from one generation to the following, the individuals having the best performances with respect to a given objective function have a bigger chance to be selected in the next generation. However, this procedure is not completely deterministic and some of the individuals who have bad performances can remain in the next generation with a given probability.

Finally, after several generations (iterations), the population can converge toward a given composition where there exists a dominant type of solution.

This algorithm gained popularity in a short time. This is probably because of (a) the high ease of programming it and (b) its ability to take into account any type of variables (quantitative or not). However, we must notice the following points:

- because of the high number of solutions N to be evaluated at each generation, the convergence may be quite slow: the exploration of the space of design variables requires high amounts of CPU time;
- it is necessary to predetermine the p_{mut} and the p_{cross} parameters and the number of individuals N in the family. This will be performed experimentally;
- there is no proof of convergence;
- as compared with a pure random approach, the gain is rarely greater than a factor of 5.

We include in §5.2 several results which have been obtained with this algorithm in mechanism synthesis.

5. Applications

We present below a few applications of the methods that we have described in the foregoing sections, namely,

- an application of a classical deterministic method to the optimal design of a robot for reaching a prespecified zone;
- three examples of mechanism synthesis using genetic algorithms;
- an application of simulated annealing for solving a problem of optimal positioning of a human operator for reaching a target.

5.1 Design of an Airplane Cleaner

This example was described in §3 (Chedmail and Wenger 1995). The manipulator under synthesis is shown in Fig. 3. The objective here is to define the morphology of a robot, its dimensions, and the positioning of its base for cleaning more than 80% of the external surface of the airplane.

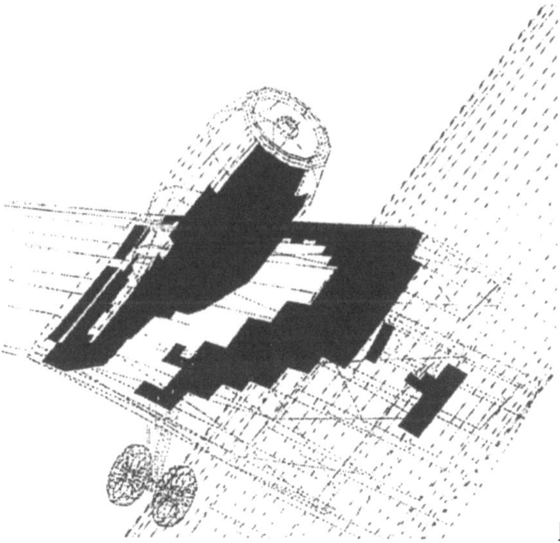

Fig. 10. Unreachable zones

In this example, we have chosen *a priori* an anthropomorphic and a SCARA robot, which have been optimized separately. The objective function is the area of the surface of the airplane, which is reached without collision, eq.(7b). The optimization method is that due to Hooke and Jeeves (1961). The result is the robot of Fig. 3 whose link lengths, their cross sections, its actuators, and the position of the base of the robot (there are six different zones to be cleaned and six different positions) were found with the aid of Hooke and Jeeves' algorithm. The CPU-time was about 3 hours on an old VAX 8700 minicomputer. The unreachable zones are located under the airplane (see Fig. 10), especially under the engine which is at 0.6 m from the ground. These zones represent less than 10% of the surface of the airplane.

5.2 Topological Synthesis of a Mechanism

5.2.1 Design of a Walking Robot

Farritor, Dubowski, Rutman, and Cole (1996) proposed an original approach to the synthesis of a walking robot. The problem is to determine the best suitable mechanism aimed at crossing the keel of an ancient ship for inspection purposes. The objective function is the cycle time for crossing the ship. The authors chose a genetic algorithm in order to introduce the topological variables of the mechanism in the optimization procedure: the number of legs, the number and the nature of the joints (revolute or prismatic). The method converges in 10 000 iterations.

5.2.2 Design of a Simple Mechanism

In this example (Chedmail and Ramstein 1996), the objective is to determine the best morphology, RR or RP, for reaching a pre-specified trajectory among obstacles.

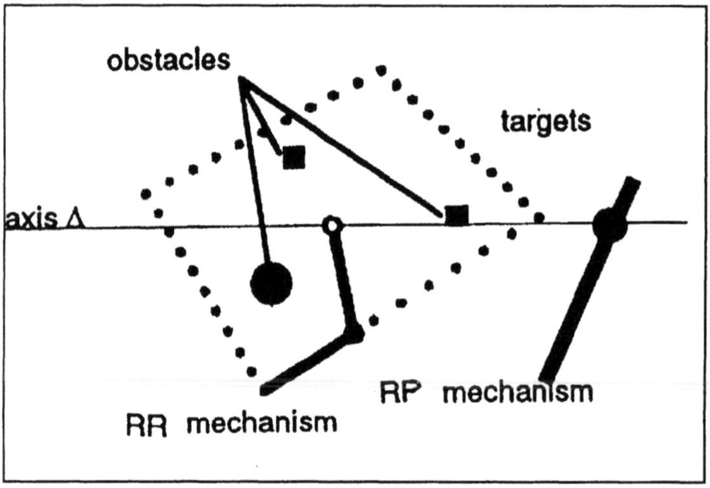

Fig. 11. Description of the site and the two morphologies

The other variable is the location of the mechanism along the Δ axis (Fig. 11).

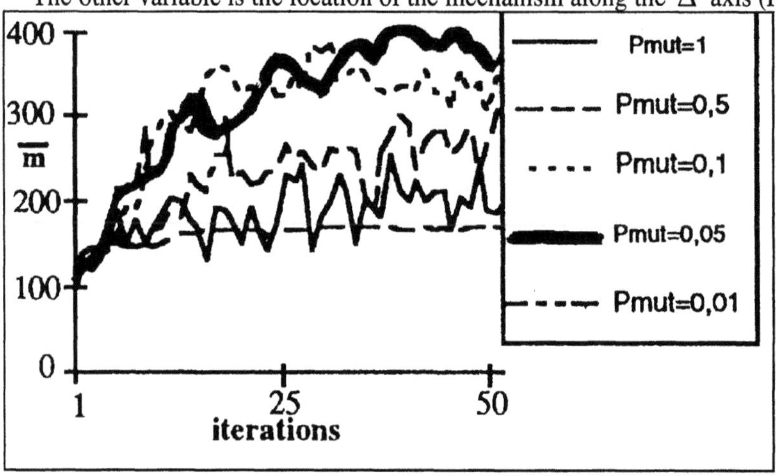

Fig. 12. Evolution of the mean performance of the family with respect to the number of iterations and the value of $p_{mut.}$ ($p_{cross} = 0.5$)

We use a genetic algorithm for this purpose. The population is composed of 21 individuals. Each of them has two components, the type of morphology and the position along the Δ axis. One can see in Fig. 12 the convergence of the algorithm for different values of p_{mut} and $p_{cross} = 0.5$. The value of $\overline{m}$ represents the mean performance of the population. Of course, when $p_{mut} = 1$, there is a mutation at each iteration, and the algorithm becomes a purely random method. When $p_{mut} = 0.01$, at each iteration the occurrence of mutations is less frequent, and the process is damped. The optimal value of p_{mut} was found to be 0.05.

5.2.3 Design of an Airplane Cleaner

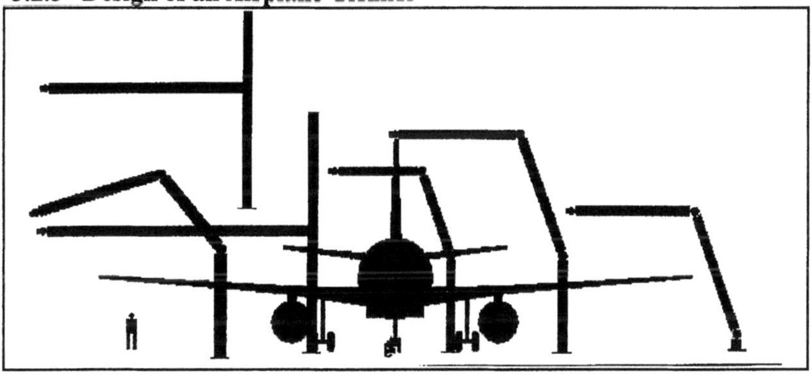

Fig. 13. Profile of the airplane and the different morphologies of robots

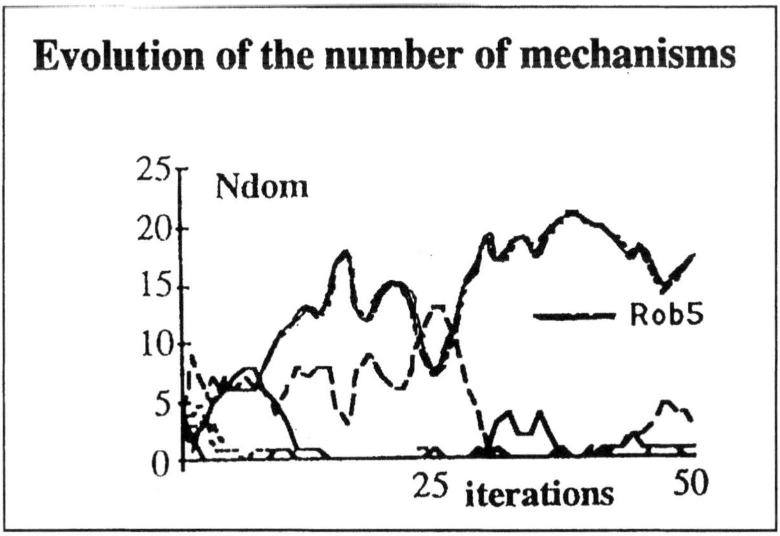

Fig. 14. Convergence of the genetic algorithm

After the foregoing simple example, we come back to the airplane cleaner (see §5.1), while taking into account the morphology of the robot as a variable of the problem.

Figure 13 describes the two types of morphology (SCARA and anthropomorphic), the other variables being the link lengths. The objective function is still the surface of the airplane which can be reached without collision.

One can observe in Fig. 14 the convergence of the algorithm. N_{dom} represents the number of individuals of the dominant type. The solution is obtained after 45 iterations (i.e. 950 tests of the objective function) with the anthropomorphic morphology on the left side of Fig. 13.

5.3 Optimum Positioning of a Human Operator in an Encumbered Environment

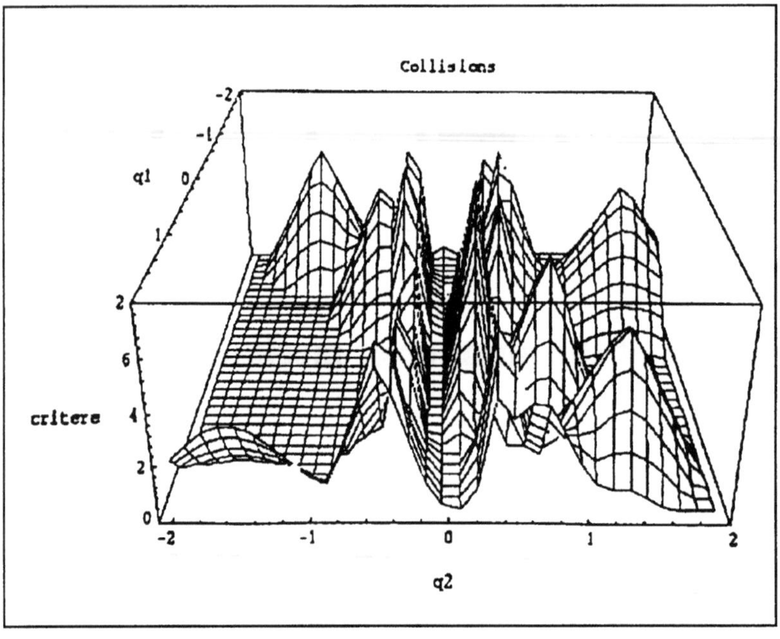

Fig. 15. Projection of the collision penalty function in the subspace (q_1, q_2)

This example is related to the problem described in §3.2.1.1., Fig. 4. The solution was developed by Riffard and Chedmail (1996). The objective function is the distance between the position and the orientation of the base of the trunk and a prespecified subdomain. This function is penalized with a function representing the collisions with the environment. The hand of the operator is located on the target, the optimization variables being the joint coordinates of the arm of the operator. We use here the simulated-annealing algorithm. The existence of many local minima (see Fig. 15) of the collision penalty function justifies this choice. The environment in this example (see Fig. 16) is composed of two vertical poles. The operator has to reach a target which is located behind the holes. The convergence of the algorithm was obtained in 94% of the tests (see Fig. 17). Only 6 tests over 100 led to a failure of the method. However, two consecutive failures were never reported. Therefore, we can consider that the probability to have 2 consecutive failures is negligibly low.

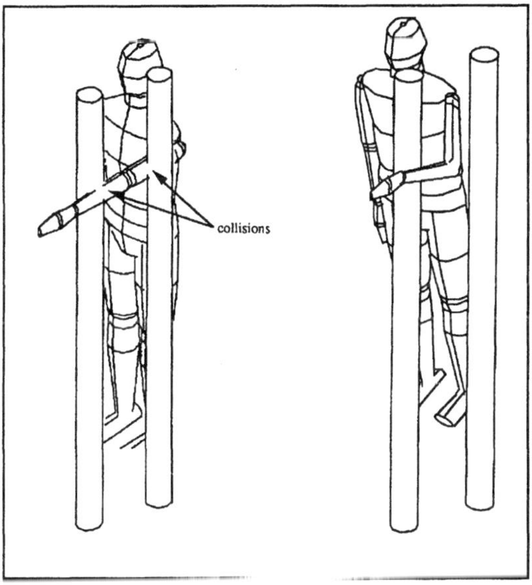

Fig. 16. Description of the site

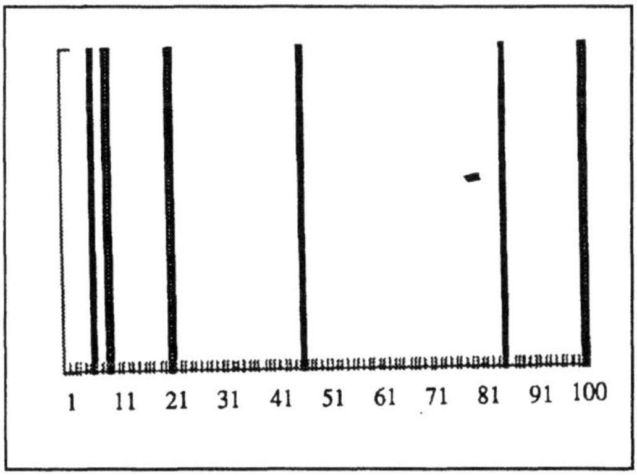

Fig. 17. Occurrence of failure

6. Conclusions

An overview of the optimization of multi-dof mechanisms was given in this chapter, with regard to formulation and solution techniques. Both classical, or deterministic, and stochastic techniques were reviewed. The main efforts of researchers have been concentrated on the dimensional synthesis of the mechanisms. The topology synthesis remained until recently a challenge. With regard to stochastic techniques, two methods were outlined, namely, simulated annealing and genetic algorithms. These methods are more generally applicable than deterministic approacheses in that they can accommodate nonquantitative variables, e.g., the type of kinematic pair and the type of mechanism (its topology). Recent papers have demonstrated their applicability in industrial environment. However, their generalization is still limited by their high demands on CPU-time to reach convergence.

Acknowledgments

NATO SEAD's financial support is gratefully acknowledged. A part of the work reported here is under the support of Dassault Aviation S.A. and CBC S.A.

References

Agrawal V., and Rao J. (1987): The mobility properties of kinematic chains. Mechanism and Machine Theory, **22** 5, 497-504.

Angeles J., Alivizatos A., and Akhras R. (1988): An unconstrained non-linear least-square method of optimization of RRRR planar path generators. Mechanism and Machine Theory, **23** 5, 343-353.

Artobolevski I. (1975): Les Mécanismes dans la technique moderne. Mir Publishing Co., Moscow.

Avilés R., Ajuria N.B., and Garcia De Jalon J. (1994): A fairly general method for optimum synthesis of planar mechanisms. Mechanism and Machine Theory, **22** 5, 321-328.

Bendsoe M.P. (1996): Structural Optimization of and with Advanced Materials. In Tatsumi T., Watanabe E., and Kambe T. (Editors): Theoritical and Applied Mechanics, 19th ICTAM, Elsevier, 269-284.

Chedmail P., and Wenger P. (1989): Design and positioning of a robot in an environment with obstacles using optimal search. Proc. IEEE Int. Conf. on Robotics and Automation, Scottsdale, May, 1069-1074.

Chedmail P. (1990): Synthèse de robots et de sites robotisés - Modélisation des robots souples. Thèse de Doctorat ès Sciences, University of Nantes and E.C.N., January.

Chedmail P., and Gautier M. (1990): Optimum choice of robot actuators. Trans. ASME, Int. Journal of Engineering in Industry, **112** 4, 361-367.

Chedmail P., and Wenger P. (1995): Conception optimale d'un site robotisé de nettoyage d'avion. Proceedings of the International Conference on Industrial Engineering, Montréal, October, 367-376.

Chedmail P., Damay E., and Rouchon C. (1996): Integrated design and artificial reality: accessibility and path planning. Proc. ASME Engineering System Design and Analysis, ESDA'96, Vol. 8, Montpellier, 149-156.

Chedmail P., and Ramstein E. (1996): Robot mechanism synthesis and genetic algorithms. Proc. IEEE Int. Conf. on Robotics and Automation, Minneapolis, April, 3466-3471.

Chen I. and Burdick J. (1995): Determining task optimal robot assembly configurations. Proc. IEEE Int. Conf. on Robotics and Automation, Nagoya, Aichi, May, 132-137.

Cheng G. (1996): Some development of structural topology optimization. In Tatsumi T., Watanabe E., and Kambe T. (Editors): Theoretical and Applied Mechanics, 19th ICTAM, Elsevier, 379-394.

Delignières S. (1987): Choix de morphologies de robots. Doctoral Thesis University of Nantes, ENSM, November 1987.

Erdman A.G., and Sandor G.N. (1991): Mechanism Design, Analysis and Synthesis, Prentice Hall, Englewood Cliffs, N. J.

Farritor S., Dubowski S., Rutman N., and Cole J. (1996): A system-level modular design approach to field robotics. Proc. IEEE Int. Conf. on Robotics and Automation, Minneapolis, April, 2890-2895.

Frecker M.I., Ananthasuresh G.K., Nishiwaki S., Kikushi N., and Kota S. (1997): Topological synthesis of compliant mechanisms using multi-criteria optimization. Journal of Mechanical Design, Trans. ASME, **119**, 238-245.

Freudenstein F., and Maki E.R. (1979): The creation of mechanisms according to kinematic structure and function. Environment and Planning B, **6**, 375-391.

Goldberg D. (1989): Genetic algorithms in search optimization and machine learning. Addison-Wesley, Reading, MA.

Gonzalez-Palacios M.A., and Angeles J. (1993): Cam Synthesis, Kluwer Academic Publishers, Dordrecht.

Gosselin C., and Angeles J. (1987): Représentation graphique de la région de mobilité des mécanismes plans et sphériques à barres articulées. Mechanism and Machine Theory, **22** 6, 557-562.

Hansen J.M. (1993): Synthesis of spatial mechanisms using optimization and continuation methods, in Seabra Pereira, M.F.O. and Ambrosio, J. (Editors): Computer Aided Analysis of Rigid and Flexible Mechanical Systems, NATO ASI Series E, Vol. **268**, Kluwer Academic Publishers, Dordrecht, 441-453.

Hoeltzel D., and Chieng W.H. (1990): Knowledge-based approaches for the creative synthesis of mechanisms. Computer-Aided Design, **22** 1, 57-67.

Hooke R., and Jeeves T. (1961): Direct search solution of numerical and statistical problems. J.A.C.M., **8**, 212-229.

Jacquet L., Petiot J.F., Sommer J.L., and Koyama Y. (1997): Analysis of the setting up of a concurrent engineering scenario. In Chedmail P., Bocquet J.C., and Dornfeld D. (editors): Integrated Design and Manufacturing in Mechanical Engineering. Kluwer Academic Publishers. Dordrecht, 3-13.

Kim J.O. (1992): Task based kinematic design of robot manipulators. Ph.D Thesis, Robotics Institute, Carnegie-Mellon University, Pittsburg.

Kirkpatrick S., Gellatt J.R., and Vecchi M.P. (1985): Optimization by simulated annealing. Science, **220**, 671.

Larsen U.D., Sigmund O., and Bouwstra S. (1996): Design and fabrication of compliant micromechanisms and structures with negative Poisson's ratio, Proc. of the 9th IEEE Workshop on Micro Electro Mechanical Systems, February, San Diego, California, 365-371.

Lee K., and Tortorelli D.A. (1985): A CAD system for designing robotic manipulators. Proc. IEEE Int. Conf. on Robotics and Automation, St Louis, March, 376-380.

Liu T., and Yu C.H. (1995): Identification and classification of multi-degree-of-freedom and multi-loop mechanisms. Mechanism and Machine Theory, **117** 3, 104-111.

Martin J., and Gosselin C. (1996): Static balancing of planar parallel manipulators. Proceedings of the IEEE International Conference on Robotics and Automation, Minneapolis, April, 3732-3737.

Mayourian M., and Freudenstein F. (1984): The development of an atlas of the kinematic structures of mechanisms. ASME J. of Mechanisms, Transmissions, and Automatic in Design. **106** 12, 458-461.

Merlet J.P. (1996): Workspace oriented methodology for designing a parallel manipulator. Proceedings of the IEEE International Conference on Robotics and Automation, Minneapolis, April, 3726-3731.

Metropolis N., Rosenbluth A., Rosenbluth N., Teller A., and Teller E. (1953): Equations of state- Calculations by fast computing machines. Journal of Chemical Physics, **21**, 1087.

Michalewicz Z. (1992): Genetic Algorithms + Data Structures = Evolution Programs. Springer-Verlag, Berlin, 3rd edn. 1996.

Minoux M. (1983): Programmation mathématique, Théorie et algorithmes, Tome 1, Dunod - Bordas Editors and CNET-ENST, Paris.

Olson D.G., Erdman A.G., and Riley D.R. (1985): A systematic procedure for type synthesis of mechanisms with literature review. Mechanism and Machine Theory, **20** 4, 285-295.

Potkonjak V. (1986): Thermal criterion for the selection of d.c. drives for industrial robots. Proceedings ISIR'86, Brussels, Sept.-Oct., 129-140.

Riffard V., and P. Chedmail (1996): Optimal posture of a human operator and CAD in robotics. IEEE International Conference on Robotics and Automation, 22-28 April, Minneapolis, 1199-1205.

Rozvany G.I.N., Bendsoe M.P., and Kirsch U. (1995): Layout optimization of structures. Applied Mechanics Review, **48** 2, 41-119.

Rubel A.J., and Kaufman R.E. (1977): KINSYN 3: A new human engineered system for interactive computer aided design of planar mechanisms, Trans. ASME, J. Engineering in Industry, **99**, 440-448.

Sardain P. (1993): Contribution à la synthèse de mécanismes - Un environnement CAO pour la synthèse topologique et dimensionnelle de générateurs de mouvement. Doctoral Thesis, University of Poitiers, June.

Seering W.P., and Pasch K.A. (1984): Methods for choosing actuators and transmission components for high performance manipulators. In: M. Brady, L.A. Gerhardt, M.F.

Davidson (eds.), Robotics and Artificial Intelligence, NATO ASI, Series F, Vol. **11**, Springer-Verlag, Berlin.

Thomas M., Yuan Chou H.C, and Tesar D. (1985): Optimal actuator sizing for robotic manipulators based on local dynamic criteria. Trans. ASME, J. Mechanisms, Transmissions, and Auto. in Design, **107**, June, 163-169.

Vasiliu A.A. (1997): Une approche CAO pour la préconception des mécanismes plans générateurs de trajectoire: Realisme. Doctoral Thesis, Ecole Centrale de Paris, Châtenay-Malabry, February.

Wenger P., and Chedmail P. (1991): Ability of a robot to travel through its free workspace. The Int. J. of Robotics Research, Vol. **10** 3, 214-227.

Willmert K.D., Thornton W.A., and Khan M. R. (1978): A hierarchy of methods for analysis of elastic mechanisms with design applications. Proc. ASME. Design Eng. Tech. Conf., Minneapolis, September, 24-27.

Computational Issues in the Kinematic Design of Tactile Sensing Fixtures

Walter Willem Nederbragt[1] and Bahram Ravani[2]

Department of Mechanical and Aeronautical Engineering
University of California at Davis, Davis, California 95616

[1] Now at the National Institute of Standards and Technology
Building 220, Room A127, Gaithersburg, Maryland 20899
email: wwneder@cme.nist.gov

[2] email: bravani@ucdavis.edu

Summary. This paper discusses computational issues in kinematic design of tactile sensing fixtures used in robotics applications. It deals with mechanical fixtures built or modeled by feature surfaces consisting of planes, spheres, and cylinders. It develops the governing equations for locating each of these geometric objects using tactile sensing probes. It shows that although four points are needed to locate a sphere, in many applications sensing three points is sufficient for referencing. In the case of a cylinder it is shown that in general six points are necessary and that in many applications five points are sufficient for locating the cylinder. The paper reduces the governing equations for a cylinder to a set of polynomial equations consisting of a second-degree and a third-degree equation. The solutions of this set are found using symbolic computations. The results are applied to the kinematic design and analysis of a mechanical fixture consisting of a sphere and a cylinder as its feature surfaces.

Keywords. Fixture, tactile sensing, robotics, calibration, part-referencing.

1. Introduction

Tactile sensing fixtures are used in robotics and manufacturing – see, for example, Duffie et al. (1984), McCallion and Pham (1984), Mooring and Pack (1987), Ravani and Ge (1991), and Slocum (1988) – for part referencing and calibration. This is the process of determining the relative location of a part with respect to a tool (such as a machine tool, a robot, or a material-handling system) or with respect to a world coordinate system. In robot calibration, the position of the end-effector is usually measured at a set of predetermined locations using some form of a sensing system. This data is then combined with joint-encoder readings from the same set of locations to update the kinematic parameters of the robot in its programming system – see, for example, Roth, Mooring and Ravani (1987) or Hollerbach (1988) – to improve its positioning accuracy. Since both part-referencing and calibration require measurements of relative locations between two objects, tactile-sensing mechanical fixtures are generally used to simplify the sensing function and to improve repeatability.

In order for a tactile sensing mechanical fixtures to be useful, a reference frame must be created out of the geometric contacts made to the surface of the fixture. Therefore, the geometric feature surfaces and contact conditions are critical to the design and operation of a practical fixture. This chapter deals with computational aspects of localization of feature surfaces using the contact conditions. In particular, we develop methods for dealing with feature surfaces consisting of planes, spheres, and cylinders. The results for planes and spheres (in the general case) are known in some form or another, but the results for cylinder are new as well as some special cases for spheres.

The work presented in this paper addresses the computational aspect of some of our recent work (Nederbragt and Ravani, 1996 and 1997) in developing a scientific foundation for the kinematic design of mechanical fixtures.

The organization of the chapter is as follows: First, we discuss part-referencing based on touch-sensing. We then give methods for finding a plane, a sphere, and a cylinder in three-dimensional Euclidean space using a finite number of points on the surface of these geometric elements. Finally, learning from our computational analysis methods, we apply these methods to create a reference frame for a simple cylinder-sphere reference fixture.

2. Part-Referencing Based on Touch-Sensing

Part-referencing using tactile sensing involves bringing a sensing element into contact with a surface of the part, activating the touch sensor, and measuring the location of the touch point in the sensor coordinate system. This was the case, for example, in the system described by Duffie et al. (1984) where a touch sensor was attached to the end of a robot and it was moved until it contacted the spherical surface of a fixture. In such a system the location of the touch is only known in the robot manipulator frame. The shape of the touch surface on the fixture is, however, completely known. (In this case it is a sphere.)

If several touches are made to the surface, then enough information may be obtained to determine the relative location of the two frames. Duffie et al. (1984) used a fixture consisting of three separate spheres of known radii. They found that four separate touches to each of the spheres made it possible to determine the location of the fixture with respect to the robot. McCallion and Pham (1984) used three non-collinear touches to a plane to determine its location in space, and, using three perpendicular planes of a cube, the relative location of the robot to the fixture was found. Nederbragt and Ravani (1996, 1997) developed a scientific framework for design of such fixtures. In their work they created several new fixture geometries and also enumerated the geometric contact combinations necessary for the elimination of all continuous motions.

With the introduction of many new possible fixture geometries, the next step is to find methods for localizing the feature geometries in space based on the contact conditions. This may require the solution of several non-linear equations using computational methods. In the next sections, methods for determining the location of a plane, a sphere, and a cylinder are given. In the case of the plane, the methods for finding their location are simple and well known (McCallion and Pham, 1984), but for completeness a common method is given. The sphere case is also well known. However, we have developed a method that can reduce the number of points necessary for the determination of the location of the sphere in certain instances. A method for locating a cylinder is also given. The cylinder case is significantly more complicated than the sphere or the plane and, therefore, is the main focus of this paper. The method for finding its location has applications in fixture design, coordinate-measuring machines (CMM) and the Burmester theory for the design of linkages.

The three feature surfaces of planes, spheres and cylinders are the building blocks for a large number of practical fixture geometries.

3. Determination of the Location of a Plane Using Three Points

It is well known that the location of a plane can be found in three-dimensional Euclidean space if the location of three points on the plane are known (McCallion and Pham, 1984) (see Figure 1). However, for completeness, a common method is given.

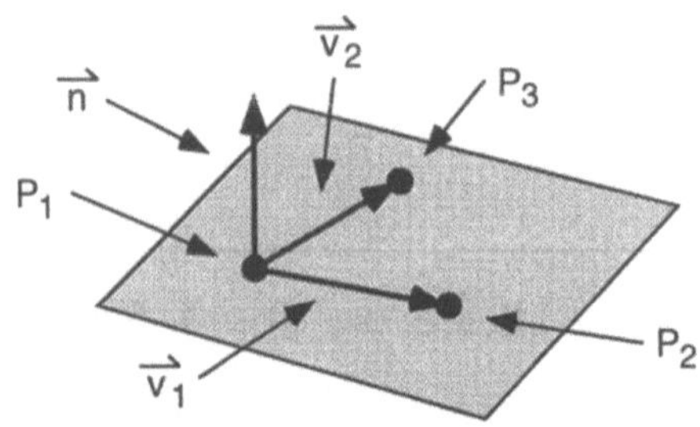

Fig. 1. Three points on a plane

Let points p_1, p_2, and p_3 be on a plane where $p_i = (x_i, y_i, z_i)$. We now define two vectors as $\mathbf{v}_1 = p_2 - p_1$ and $\mathbf{v}_2 = p_3 - p_1$. Let $\mathbf{n} = \mathbf{v}_1 \times \mathbf{v}_2 = (a, b, c)$. Since $\mathbf{v}_1$ and $\mathbf{v}_2$ are parallel to the plane, $\mathbf{n}$ must be normal to the plane. From algebraic geometry (1939), the equation of the plane is:

$$a(x - x_1) + b(y - y_1) + c(z - z_1) = 0 \tag{1}$$

or, in general form

$$ax + by + cz + d = 0 \tag{2}$$

where $d = -ax_1 - by_1 - cz_1$.

4. Determination of the Location of a Sphere Using Three Points

In general, it takes four points on the surface of a sphere to determine the radius and the center of that sphere (Duffie et al., 1984). This is well known. However, if the radius of the sphere is already known (this is always the case for referencing fixtures), then three points on the surface of the sphere will give two possible solutions (see Figure 2). If a touch probe is being used to make the point contacts to the sphere, then the orientation of the probe during contact can be used to possibly eliminate the incorrect sphere out of the two possible solutions. This makes the use of spheres for referencing more appealing because a three-sphere fixture may require significantly less than the twelve contacts used by Duffie et al. (1984) to make a complete reference measurement.

An algebraic method is given in the next section that will determine the location of a sphere of known radius. An example is also given that uses this algebraic method.

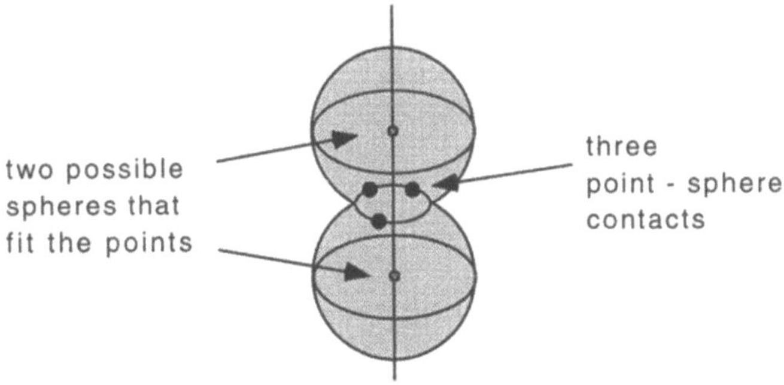

Fig. 2. Three points on a sphere gives two possible solutions

4.1 Algebraic Method

The general equation for a sphere in space is

$$(x-a)^2+(y-b)^2+(z-c)^2-r^2=0, \tag{3}$$

where (a,b,c) is the center of the sphere and r is the radius of the sphere.

Given the points $p_i=(x_i,y_i,z_i)$ for $i=1$ to 3 on the surface of a sphere of radius r, the following three equations must be satisfied:

$$(x_1-a)^2+(y_1-b)^2+(z_1-c)^2=r^2, \tag{4}$$

$$(x_2-a)^2+(y_2-b)^2+(z_2-c)^2=r^2, \tag{5}$$

$$(x_3-a)^2+(y_3-b)^2+(z_3-c)^2=r^2. \tag{6}$$

If we subtract eq. 5 from eq. 4 we obtain

$$\begin{aligned}(x_2-x_1)a+(y_2-y_1)b+(z_2-z_1)c+\\ 0.5({x_1}^2-{x_2}^2+{y_1}^2-{y_2}^2+{z_1}^2-{z_2}^2)=0.\end{aligned} \tag{7}$$

If we subtract eq. 6 from eq. 4 we obtain

$$\begin{aligned}(x_3-x_1)a+(y_3-y_1)b+(z_3-z_1)c+\\ 0.5({x_1}^2-{x_3}^2+{y_1}^2-{y_3}^2+{z_1}^2-{z_3}^2)=0.\end{aligned} \tag{8}$$

Using eq. 7 and eq. 8, we can solve for a and b in terms of c, namely,

$$\begin{aligned}b=[2(-x_2z_1+x_3z_1+x_1z_2-x_3z_2-x_1z_3+x_2z_3)c+{x_1}^2x_2\\ -x_1{x_2}^2-{x_1}^2x_3+{x_2}^2x_3+x_1{x_3}^2-x_2{x_3}^2+x_2{y_1}^2-x_3{y_1}^2\\ -x_1{y_2}^2+x_3{y_2}^2+x_1{y_3}^2-x_2{y_3}^2+x_2{z_1}^2-x_3{z_1}^2-x_1{z_2}^2\\ +x_3{z_2}^2+x_1{z_3}^2-x_2{z_3}^2]M\end{aligned} \tag{9}$$

and

$$\begin{aligned}a=[2(y_2z_1-y_3z_1-y_1z_2+y_3z_2+y_1z_3-y_2z_3)c+{x_2}^2y_1\\ -{x_3}^2y_1-{x_1}^2y_2+{x_3}^2y_2-{y_1}^2y_2+y_1{y_2}^2+{x_1}^2y_3-{x_2}^2y_3\\ +{y_1}^2y_3-{y_2}^2y_3-y_1{y_3}^2+y_2{y_3}^2-y_2{z_1}^2+y_3{z_1}^2+y_1{z_2}^2\\ -y_3{z_2}^2-y_1{z_3}^2+y_2{z_3}^2]M\end{aligned} \tag{10}$$

where $M = 1/[2(x_2 y_1 - x_3 y_1 - x_1 y_2 + x_3 y_2 + x_1 y_3 - x_2 y_3)]$.

Equations 9 and 10 can be substituted into eq. 4 to leave a quadratic equation in terms of the variable c. This equation can be solved for the two values of c. These values can be substituted back into eqs. 9 and 10 to obtain two center points for the sphere. Let the center points be $Center_1 = (a_1, b_1, c_1)$ and $Center_2 = (a_2, b_2, c_2)$.

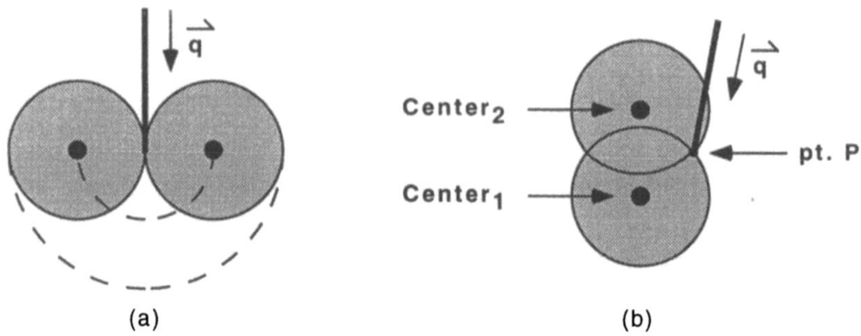

Fig. 3. Elimination of one of two sphere centers

If a touch sensor is being used to make contact with the sphere, then the knowledge of the orientation of the sensor relative to the two possible spheres may possibly eliminate one of the two mathematical solutions, hence, leaving the correct solution. Let $\mathbf{q}$ be the vector pointing in the direction of the touch sensor. Let p be the position of the touch to the surface of the sphere. From Figure 3a, it can be seen that the center of the sphere must be located "lower" than point p. From Figure 3b, one of the two spheres cannot be possible because the touch sensor must intersect that sphere to reach point p. Therefore, we can eliminate the sphere that does not satisfy the following equation:

$$(p - Center_i) \cdot \mathbf{q} < 0. \tag{11}$$

If both sphere centers pass this test for all three points, then another point will be necessary to eliminate one of the sphere centers.

4.2 An Example

If we let $r = 2$, $p_1 = (1, 0, 0)$, $p_2 = (0, 1, 0)$, $p_3 = (-1, 0, 0)$, and $\mathbf{q} = (-4, 1, 1)$ for point p_1, then eqs. 4, 5, and 6 become

$$(1 - a)^2 + (b)^2 + (c)^2 = 4, \tag{12}$$

$$(a)^2 + (1 - b)^2 + (c)^2 = 4, \tag{13}$$

$$(-1 - a)^2 + (b)^2 + (c)^2 = 4. \tag{14}$$

From eqs. 7 and 8 we find that $a = b = 0$. Plugging these results into eq. 12, we find $c = \pm\sqrt{3}$. Therefore, our possible center points are $Center_1 = (0, 0, +\sqrt{3})$ or $Center_2 = (0, 0, -\sqrt{3})$. Using $\mathbf{q}$, $Center_1$ cannot be possible because

$$\begin{aligned} [P_1 - Center_1] \cdot \mathbf{q} &= [(1, 0, 0) - (0, 0, \sqrt{3})] \cdot (-4, 1, 1) \\ (1, 0, -\sqrt{3}) \cdot (1, 1, -4) &= 1 + 4\sqrt{3} > 0. \end{aligned} \tag{15}$$

Hence, the sphere center must be $Center_2 = (0, 0, -\sqrt{3})$.

5. Determination of the Location of a Cylinder Using Five Points

The Location of a cylinder of known radius, in general, can be found in three-dimensional Euclidean space if five points on the surface of the cylinder are known. There are several different ways of representing a cylinder. However, a method used by Schaal (1985) develops algebraic equations using a minimal set of variables. This method has been extended for application here. In the next section, the algebraic equations for finding a cylinder are derived. An example using the derived method is also given.

5.1 Equation Formulation

In order to find the location of a cylinder in space using a finite number of points on its surface, a general equation for a cylinder needs to be formulated. We use the equation developed by Schaal (1985) since it describes a cylinder in a form suitable for our subsequent developments. Here we give a derivation of this equation.

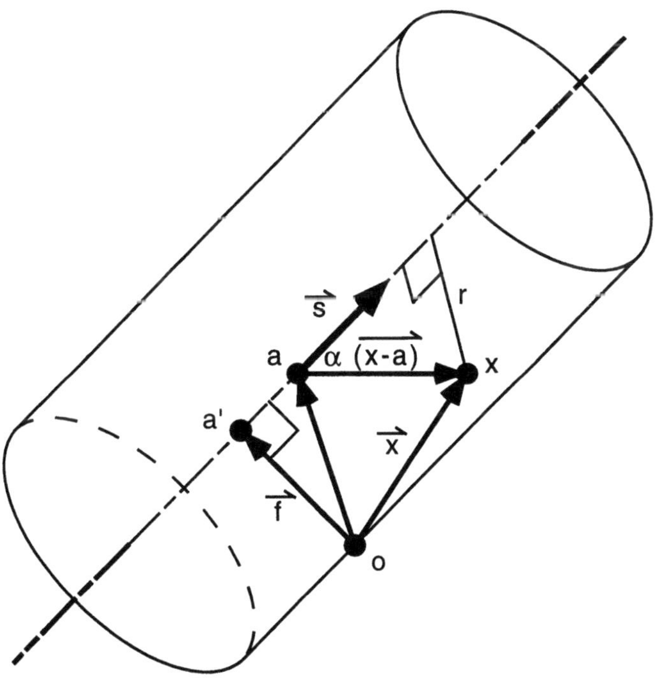

Fig. 4. Analytical model of a cylinder

In Figure 4, a cylinder is given with the following properties: r is the radius of the cylinder, x and o are points on the surface of the cylinder, a and a' are points on the axis of the cylinder, **s** is a vector in the direction of the axis of the cylinder, α is the angle between **s** and (x-a), **x** is a vector from point o to point x, and **f** is a vector from point o to the axis of the cylinder where $\mathbf{f}$ is perpendicular to s. From Figure 4 and vector analysis it is obvious that

$$r = |(x-a)| \sin\alpha. \tag{16}$$

Also from vector analysis we know that the cross product between two vectors is equal in magnitude to the product of the magnitudes of the vectors times the sine of the angle between the two vectors. Using this property and eq. 16 we obtain,

$$|(x-a) \times \mathbf{s}| = |(x-a)||\mathbf{s}| \sin\alpha = (|(x-a)| \sin\alpha)|\mathbf{s}| = r|\mathbf{s}|. \tag{17}$$

Letting $||\mathbf{v}||^2 = \mathbf{v} \cdot \mathbf{v}$, eq. 17 can be written as,

$$||(x-a) \times \mathbf{s}||^2 = r^2||\mathbf{s}||^2 \quad or \quad ||(x-a) \times \mathbf{s}||^2 - r^2||\mathbf{s}||^2 = 0. \tag{18}$$

Equation 18 is a general equation for a cylinder. We will now proceed to change this equation to only leave the variable s.

The left side of eq. 18, substituting $(x-a) = \mathbf{x} - \mathbf{f}$, can be written as

$$||(x-a) \times \mathbf{s}||^2 = [(\mathbf{x} - \mathbf{f}) \times \mathbf{s}] \cdot [(\mathbf{x} - \mathbf{f}) \times \mathbf{s}]. \tag{19}$$

Using the vector equation

$$(\mathbf{a} + \mathbf{b}) \times \mathbf{c} = (\mathbf{a} \times \mathbf{c}) + (\mathbf{b} \times \mathbf{c}), \tag{20}$$

$(\mathbf{x} - \mathbf{f}) \times \mathbf{s}$ can be written as

$$(\mathbf{x} - \mathbf{f}) \times \mathbf{s} = (\mathbf{x} \times \mathbf{s}) - (\mathbf{f} \times \mathbf{s}). \tag{21}$$

Substituting $\boldsymbol{\delta} = \mathbf{f} \times \mathbf{s}$ into the right side of eq. 21, we obtain

$$(\mathbf{x} \times \mathbf{s}) - (\mathbf{f} \times \mathbf{s}) = (\mathbf{x} \times \mathbf{s}) - \boldsymbol{\delta}. \tag{22}$$

Substituting eq. 22 into eq. 19, we obtain

$$||(x-a) \times \mathbf{s}||^2 = [(\mathbf{x} \times \mathbf{s}) - \boldsymbol{\delta}] \cdot [(\mathbf{x} \times \mathbf{s}) - \boldsymbol{\delta}]. \tag{23}$$

Using the vector equation

$$(\mathbf{a} + \mathbf{b}) \cdot (\mathbf{c} + \mathbf{d}) = \mathbf{a} \cdot \mathbf{c} + \mathbf{a} \cdot \mathbf{d} + \mathbf{b} \cdot \mathbf{c} + \mathbf{b} \cdot \mathbf{d}, \tag{24}$$

equation 23 can be written as

$$||(x-a) \times \mathbf{s}||^2 = ||\mathbf{x} \times \mathbf{s}||^2 + ||\boldsymbol{\delta}||^2 - 2[(\mathbf{x} \times \mathbf{s}) \cdot \boldsymbol{\delta}]. \tag{25}$$

We already know that

$$(\mathbf{x} \times \mathbf{s}) \cdot \boldsymbol{\delta} = (\mathbf{x} \times \mathbf{s}) \cdot (\mathbf{f} \times \mathbf{s}) \tag{26}$$

from our definition of δ. Using the vector equation

$$(\mathbf{a} \times \mathbf{b}) \cdot (\mathbf{c} \times \mathbf{d}) = (\mathbf{a} \cdot \mathbf{c})(\mathbf{b} \cdot \mathbf{d}) - (\mathbf{a} \cdot \mathbf{d})(\mathbf{b} \cdot \mathbf{c}), \tag{27}$$

equation 26 can be written as

$$(\mathbf{x} \times \mathbf{s}) \cdot (\mathbf{f} \times \mathbf{s}) = (\mathbf{x} \cdot \mathbf{f})(\mathbf{s} \cdot \mathbf{s}) - (\mathbf{x} \cdot \mathbf{s})(\mathbf{s} \cdot \mathbf{f}). \tag{28}$$

If we let s be normal to $\mathbf{f}$ then $(\mathbf{s} \cdot \mathbf{f}) = 0$. This changes eq. 28 to

$$(\mathbf{x} \times \mathbf{s}) \cdot (\mathbf{f} \times \mathbf{s}) = (\mathbf{x} \cdot \mathbf{f})(\mathbf{s} \cdot \mathbf{s}) = (\mathbf{x} \cdot \mathbf{f})||\mathbf{s}||^2. \tag{29}$$

Substituting eq. 29 into eq. 25, we obtain

$$||(x-a)\times \mathbf{s}||^2 = ||\mathbf{x}\times \mathbf{s}||^2 + ||\boldsymbol{\delta}||^2 - 2(\mathbf{x}\cdot \mathbf{f})||\mathbf{s}||^2. \tag{30}$$

Substituting eq. 30 into eq. 18, we obtain

$$||\mathbf{x}\times \mathbf{s}||^2 + ||\boldsymbol{\delta}||^2 - 2(\mathbf{x}\cdot \mathbf{f})||\mathbf{s}||^2 - r^2||\mathbf{s}||^2 = 0. \tag{31}$$

Since we let $\mathbf{f}$ be normal to $\mathbf{s}$, the magnitude of $\mathbf{f}$ must be equal to the radius of the cylinder. Therefore, $|\boldsymbol{\delta}| = |(\mathbf{f}\times \mathbf{s})| = |\mathbf{f}||\mathbf{s}|$. Hence,

$$||\boldsymbol{\delta}||^2 - r^2||\mathbf{s}||^2 = 0. \tag{32}$$

Using eq. 32, eq. 31 becomes

$$||\mathbf{x}\times \mathbf{s}||^2 - 2(\mathbf{x}\cdot \mathbf{f})||\mathbf{s}||^2 = 0. \tag{33}$$

Let x_1, x_2, x_3, x_4, and x_5 be the points on the surface of the cylinder. Let x_5 be the point o, and let $\mathbf{p}_i = x_i - x_5$ for $i = 1$ *to* 4. With these changes, eq. 33 can be written as

$$(\mathbf{p}_i\cdot \mathbf{f}) - \frac{1}{2||\mathbf{s}||^2}||\mathbf{p}_i\times \mathbf{s}||^2 = 0 \quad for\ i = 1\ to\ 4. \tag{34}$$

Expanding eq. 34 into a component form using $\mathbf{p}_1$, $\mathbf{p}_2$, and $\mathbf{p}_3$, we obtain

$$\begin{bmatrix} p_{1_x} & p_{1_y} & p_{1_z} \\ p_{2_x} & p_{2_y} & p_{2_z} \\ p_{3_x} & p_{3_y} & p_{3_z} \end{bmatrix} \begin{bmatrix} f_x \\ f_y \\ f_z \end{bmatrix} - \frac{1}{2||\mathbf{s}||^2} \begin{bmatrix} ||\mathbf{p}_1\times \mathbf{s}||^2 \\ ||\mathbf{p}_2\times \mathbf{s}||^2 \\ ||\mathbf{p}_3\times \mathbf{s}||^2 \end{bmatrix} = 0. \tag{35}$$

If we let

$$M = \begin{bmatrix} p_{1_x} & p_{1_y} & p_{1_z} \\ p_{2_x} & p_{2_y} & p_{2_z} \\ p_{3_x} & p_{3_y} & p_{3_z} \end{bmatrix}, \tag{36}$$

then

$$M^{-1} = \frac{1}{det[M]} \begin{bmatrix} (\mathbf{p}_2\times \mathbf{p}_3)_x & (\mathbf{p}_3\times \mathbf{p}_1)_x & (\mathbf{p}_1\times \mathbf{p}_2)_x \\ (\mathbf{p}_2\times \mathbf{p}_3)_y & (\mathbf{p}_3\times \mathbf{p}_1)_y & (\mathbf{p}_1\times \mathbf{p}_2)_y \\ (\mathbf{p}_2\times \mathbf{p}_3)_z & (\mathbf{p}_3\times \mathbf{p}_1)_z & (\mathbf{p}_1\times \mathbf{p}_2)_z \end{bmatrix}. \tag{37}$$

Multiplying both sides of eq. 35 by M^{-1} gives

$$\begin{bmatrix} f_x \\ f_y \\ f_z \end{bmatrix} = \frac{W}{2||\mathbf{s}||^2\, det[M]} \begin{bmatrix} ||\mathbf{p}_1\times \mathbf{s}||^2 \\ ||\mathbf{p}_2\times \mathbf{s}||^2 \\ ||\mathbf{p}_3\times \mathbf{s}||^2 \end{bmatrix} \tag{38}$$

where

$$W = \begin{bmatrix} (\mathbf{p}_2\times \mathbf{p}_3)_x & (\mathbf{p}_3\times \mathbf{p}_1)_x & (\mathbf{p}_1\times \mathbf{p}_2)_x \\ (\mathbf{p}_2\times \mathbf{p}_3)_y & (\mathbf{p}_3\times \mathbf{p}_1)_y & (\mathbf{p}_1\times \mathbf{p}_2)_y \\ (\mathbf{p}_2\times \mathbf{p}_3)_z & (\mathbf{p}_3\times \mathbf{p}_1)_z & (\mathbf{p}_1\times \mathbf{p}_2)_z \end{bmatrix}. \tag{39}$$

Let $\mathbf{n}_1 = \mathbf{p}_2\times \mathbf{p}_3$, $\mathbf{n}_2 = \mathbf{p}_3\times \mathbf{p}_1$, and $\mathbf{n}_3 = \mathbf{p}_1\times \mathbf{p}_2$. Substituting $\mathbf{n}_1$, $\mathbf{n}_2$, and $\mathbf{n}_3$ into eq. 38 and multiplying both sides of the equation by $\mathbf{s}$ we obtain

$$[\mathbf{s}] \begin{bmatrix} f_x \\ f_y \\ f_z \end{bmatrix} = \frac{\mathbf{s}}{2||\mathbf{s}||^2\, det[M]} \begin{bmatrix} n_{1_x} & n_{2_x} & n_{3_x} \\ n_{1_y} & n_{2_y} & n_{3_y} \\ n_{1_z} & n_{2_z} & n_{3_z} \end{bmatrix} \begin{bmatrix} ||\mathbf{p}_1\times \mathbf{s}||^2 \\ ||\mathbf{p}_2\times \mathbf{s}||^2 \\ ||\mathbf{p}_3\times \mathbf{s}||^2 \end{bmatrix}. \tag{40}$$

Since $\mathbf{s} \cdot \mathbf{f} = 0$, the two sides of eq. 40 must be equal to zero. The right side of the equation now becomes

$$\begin{bmatrix} s_x & s_y & s_z \end{bmatrix} \begin{bmatrix} n_{1_x} & n_{2_x} & n_{3_x} \\ n_{1_y} & n_{2_y} & n_{3_y} \\ n_{1_z} & n_{2_z} & n_{3_z} \end{bmatrix} \begin{bmatrix} \|\mathbf{p}_1 \times \mathbf{s}\|^2 \\ \|\mathbf{p}_2 \times \mathbf{s}\|^2 \\ \|\mathbf{p}_3 \times \mathbf{s}\|^2 \end{bmatrix} = 0. \tag{41}$$

Equation 41 is the equation Schaal (1985) derived for a cylinder. This equation describes a cylinder in three-dimensional Euclidean space using vector $\mathbf{s}$ and four points on the surface of the cylinder.

Vector $\mathbf{s}$ describes the direction of the center line of the cylinder. It can be denoted as (s_x, s_y, s_z). The actual magnitude of the vector is not important for our case; therefore, we can set one of the vector components equal to one. Let $s_z = 1$. For this case, vector $\mathbf{s}$ must not be parallel to the x-y plane; if it is, then s_x or s_y will tend to infinity during a calculation of $\mathbf{s}$. If this happens, then either s_x or s_y should be set to one instead of s_z. Note that it is unlikely that a vector will have any directional components equal to zero using an actual robot end-effector frame; hence, any real calculations should work with $s_z = 1$.

Using $\mathbf{s} = (s_x, s_y, 1)$ means that there are two unknowns, s_x and s_y, yet we only have one equation, eq. 41. Therefore, another equation is necessary for the calculation of $\mathbf{s}$. If we add another point on the surface, x_4 where $\mathbf{p}_4 = x_4 - x_5$, to eq. 35, then we obtain

$$\begin{bmatrix} p_{1_x} & p_{1_y} & p_{1_z} \\ p_{2_x} & p_{2_y} & p_{2_z} \\ p_{3_x} & p_{3_y} & p_{3_z} \\ p_{4_x} & p_{4_y} & p_{4_z} \end{bmatrix} \begin{bmatrix} f_x \\ f_y \\ f_z \end{bmatrix} - \frac{1}{2\|\mathbf{s}\|^2} \begin{bmatrix} \|\mathbf{p}_1 \times \mathbf{s}\|^2 \\ \|\mathbf{p}_2 \times \mathbf{s}\|^2 \\ \|\mathbf{p}_3 \times \mathbf{s}\|^2 \\ \|\mathbf{p}_4 \times \mathbf{s}\|^2 \end{bmatrix} = 0. \tag{42}$$

From linear algebra (Roberts, 1985), Eq. 42 is only valid if

$$\det \begin{bmatrix} p_{1_x} & p_{1_y} & p_{1_z} & \|\mathbf{p}_1 \times \mathbf{s}\|^2 \\ p_{2_x} & p_{2_y} & p_{2_z} & \|\mathbf{p}_2 \times \mathbf{s}\|^2 \\ p_{3_x} & p_{3_y} & p_{3_z} & \|\mathbf{p}_3 \times \mathbf{s}\|^2 \\ p_{4_x} & p_{4_y} & p_{4_z} & \|\mathbf{p}_4 \times \mathbf{s}\|^2 \end{bmatrix} = 0. \tag{43}$$

Equation 43 is a second-degree equation. Using eqs. 41 and 43, vector $\mathbf{s}$ is, in general, solvable using $s_z = 1$. The two equations can be written in the general form

$$a{s_x}^3 + b{s_y}^3 + c{s_x}^2 s_y + d s_x {s_y}^2 + e{s_x}^2 + f{s_y}^2 + g s_x s_y + h s_x + i s_y + j = 0 \tag{44}$$

and

$$k{s_x}^2 + l{s_y}^2 + m s_x s_y + n s_x + p s_y + q = 0. \tag{45}$$

Equations 44 and 45 can be combined to form one equation of degree six in only one variable using Sylvester's Method (Boehm and Prautsch, 1994). This equation can be solved numerically to obtain six possible solutions for one of the variables, s_x or s_y. Using eq. 45, the other variable can be found. Therefore, we will have six possible solutions for the cylinder axis direction $\mathbf{s}$. We still need to find a point on the center line of the cylinder to define the location of the cylinder in space. We will do this by using three of the points on the surface to create a circle on a plane perpendicular to the cylinder and find the center of this circle, as shown in Figure 5.

When we find the center of the circle, we can also calculate the radius of the circle, which is also the radius of the cylinder. Since we know the radius of the

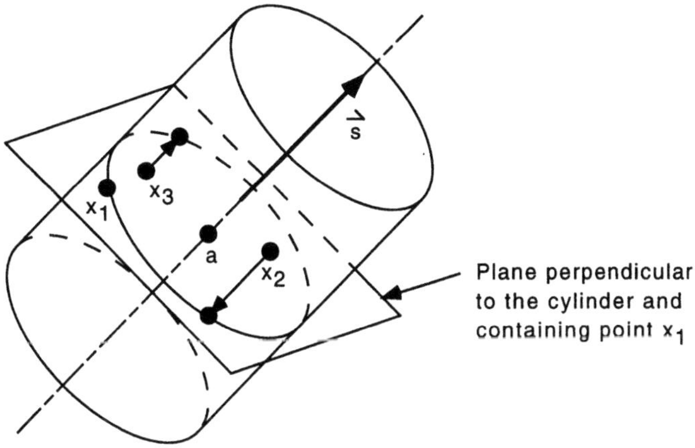

Fig. 5. Finding a point on the center line of a cylinder

cylinder already, we can eliminate the cases where the calculated radius does not match the actual radius. This should leave only the correct cylinder result. In the next section, an example is given to illustrate the cylinder finding process.

5.2 An Example

Let x_1 = (2.927025, 2.000000, 1.625000), x_2 = (1.775000, 2.800000, 1.443785), x_3 = (1.372975, 2.600000, 3.116620), x_4 = (2.075000, 3.40000, 2.185405), and x_5 = (3.527025, 3.200000, 3.108240) be five points on the surface of a cylinder of radius r = 1.000000. These points correspond to a cylinder of radius one and a center line (defined using a vector and point) of vector $\mathbf{s}$ = (0.300000, 0.600000, 0.741620) and point O = (2, 2, 2). Since the answer is known, the results of the example can be verified. Using equation $\mathbf{p}_i = x_i - x_5$ for i = 1 to 4, we find $\mathbf{p}_1$ = (-0.600000, -1.200000, -1.483240), $\mathbf{p}_2$ = (-1.752025, -0.400000, - 1.664455), $\mathbf{p}_3$ = (-2.154050, -0.600000, 0.008380), and $\mathbf{p}_4$ = (-1.452025, 0.200000, -0.922835).

Using the values for $\mathbf{p}_i$, eq. 41 written in the form of eq. 44 becomes

$$\begin{aligned}-1.785361 - 5.257678s_x + 3.124352s_x^2 - 0.504446s_x^3 \\ +0.489010s_y + 16.119020s_x s_y + 0.854990s_x^2 s_y - 3.701228s_y^2 \\ +4.371015s_x s_y^2 - 2.048000s_y^3 = 0.\end{aligned} \tag{46}$$

Using the values for $\mathbf{p}_i$, eq. 43 written in the form of eq. 45 becomes

$$\begin{aligned}0.724860 - 9.049719s_x + 0.875140s_x^2 + 0.316760s_y \\ -5.049719s_x s_y + 6.400000s_y^2 = 0.\end{aligned} \tag{47}$$

Using an equation-solver, eqs. 46 and 47 are solved for $\mathbf{s}$. Six solutions exist for this case, two of which are real, namely, $\mathbf{s}$ = (-257.042261,-71.59786,1.00000) and $\mathbf{s}$ = (0.40452,0.80904,1.00000).

Using the values of $\mathbf{s}$, we calculated the values for the radius of the cylinder and a point on the axis of the cylinder. Table 1 shows the results. We know the radius for this particular case is exactly one. Hence, the case that has a radius that is very close to one should be the correct cylinder. The case with $\mathbf{s}$ =

Table 1. List of possible cylinder locations

s	Radius of cyl.	Point on the axis
$(-257, -71.6, 1)$	$r = 0.9165$	$(2.751, 2.640, 2.257)$
$(0.405, 0.809, 1)$	$r = 1.00000$	$(2, 2, 2)$

(0.404520,0.809040,1.000000) has a mathematically determined radius of 1.000000. Therefore, it is the correct cylinder. This matches the expected result (note: **s** needs to be normalized to match the expected result).

If the actual radius of the cylinder was not known ahead of time, then an additional point on the cylinder would be needed to eliminate the incorrect solutions. However, this would not happen when working with reference fixtures because the user designs the fixture.

6. Fixture Analysis Example

We now have the computational tools to find a plane, a sphere, and a cylinder in space. To emphasize the use of these methods, we will give an example of a fixture with feature surfaces consisting of a sphere and a cylinder (a cylinder-sphere fixture). Figure 6 shows a possible design of a cylinder-sphere fixture. The limited touch area on the cylinder is needed to determine the direction of **Y**. Vector **Y** must be parallel to the axis of the cylinder; however, this leaves two possible directions. By limiting the touch sensitive surface of the cylinder to be above or below the sphere, we can always pick the correct direction.

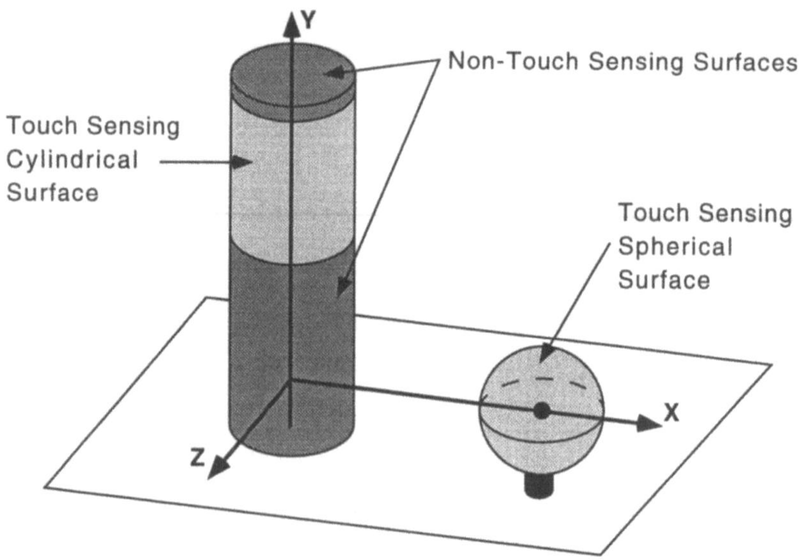

Fig. 6. A cylinder-sphere fixture and its frame

The creation of a reference frame from the geometric elements is quite simple as can be seen in Figure 6. Vector **Y** points along the axis of the cylinder, vector **X** is parallel to the shortest line between the axis of the cylinder and the center of the sphere, and vector **Z** is the cross product $\mathbf{X} \times \mathbf{Y}$. The origin of the frame is the intersection between the axis of the cylinder and the line created by finding the shortest distance between the center of the sphere and the axis of the cylinder.

Based on the frame construction just described, the information needed to create the frame is the location of the axis of the cylinder and the center of the sphere. Therefore, given five point-contact locations on the cylinder and three point-contact locations on the sphere, we should be able to create a reference frame using the methods described earlier. For this case, if the shortest distance between the sphere and the cylinder is known, then one less point on the sphere is necessary. In the area of fixture design, the fixture is designed; therefore, the distance between all of its geometric elements are known. This is shown in the example calculations.

For this example we will use the cylinder values from the cylinder analysis in the previous section. Hence, we know that $\mathbf{s}$ = (0.300000, 0.600000, 0.741620) and O = (2, 2, 2). Let x_6 = (5.456074, 4.600000, 2.543645) and x_7 = (5.981074, 4.400000, 3.841479). The points x_6 and x_7 are two points on the surface of the sphere. The analysis model for this cylinder-sphere example is shown in Figure 7. Let the shortest distance d between the sphere and the axis of the cylinder be 3.000000. All of these values were chosen to correspond to a sphere of radius one with a center located at (5.681074, 3.800000, 3.099860). In our example we will let the center of the sphere be (c_x, c_y, c_z), and, using the given values, try to find the correct center value.

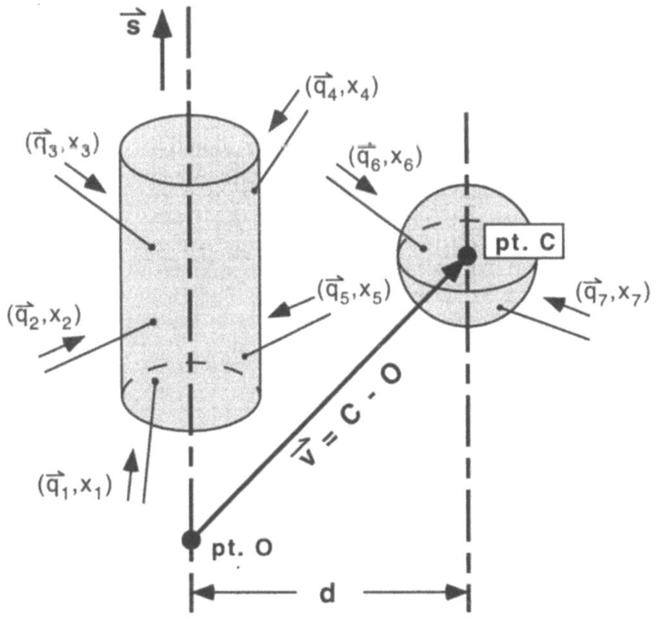

Fig. 7. Analysis of a cylinder-sphere fixture

We know that the two points on the sphere must satisfy the general equation for a sphere of unit radius. Therefore, the following equations must hold:

$$(5.456074 - c_x)^2 + (4.600000 - c_y)^2 + (2.543645 - c_z)^2 = 1. \tag{48}$$

$$(5.981074 - c_x)^2 + (4.400000 - c_y)^2 + (3.841479 - c_z)^2 = 1. \tag{49}$$

Also the shortest distance between the axis of the cylinder and the center of the sphere must be 3.00000. From Figure 7 and vector analysis, we know

$$d = \frac{|\mathbf{s} \times \mathbf{v}|}{|\mathbf{s}|}. \tag{50}$$

Vector $\mathbf{s}$ is given as a unit vector for our calculations. Therefore, the denominator for eq. 50 is one. Equation 50 can be rewritten as

$$d^2 = (\mathbf{s} \times \mathbf{v}) \cdot (\mathbf{s} \times \mathbf{v}). \tag{51}$$

Equations 48, 49, and 51 can be solved to find the possible values of the sphere center. However, if eq. 48 is subtracted from eq. 49, then a linear equation is produced. If this resultant equation is substituted for one of the quadratic sphere equations, then there will be only four solutions instead of eight. Therefore, this is a good computational simplification.

Solving these equations we get four solutions, two of which are real. They are (5.600545, 5.180286, 3.345141) and (5.681074, 3.80000, 3.09986). Obviously, the second solution is the one we are looking for. If we have the probe orientation vectors for each contact $\mathbf{q}_i$, then we may be able to eliminate the incorrect solution as we did in the previous sphere example. With the correct sphere center and the axis of the cylinder known, the complete reference frame can be created.

7. Conclusion

In this paper, we presented computational methods encountered in kinematic design of tactile sensing fixtures. The methods given make it possible to localize spheres, planes, and cylinders in space using relatively simple algebraic-geometric principles. A detailed example of a cylinder-sphere fixture was given to illustrate these ideas. The results, in addition to their applications in fixture design and analysis, can be applied in other areas of kinematics, including in three-dimensional Burmester theory.

References

Boehm W. and Prautzsch H. (1994), Geometric Concepts for Geometric Design, A K Peters, Wellesley, MA.

Duffie N., Bollinger J., Van Aken L., et al. (1984), A Sensor Based Technique for Automated Robot Programming, *Journal of Manufacturing Systems*, vol. 3, pp. 13–26.

Hollerback J. M. (1988), A survey of Kinematic Calibration, *In Robotics Review*, pp. 208–242.

McCallion H. and Pham D. T. (1984), On Machine Perception of the Relative Position of Two Objects Using Bilaterial Tactile Sensing Systems, *Proceedings of the Institution of Mechanical Engineers*, vol. 198B, pp. 179–186.

Mooring B. W. and Pack T. J. (1987), Aspects of Robot Repeatability, *Robotica*, pp. 223–239.

Nederbragt W. W. and Ravani B. (1996), Type Synthesis of Contact Sensing Elements for Robotic Fixturing, *Eleventh CISM-IFToMM Symposium on Theory and Practice of Robots and Manipulators.*

Nederbragt W. W. and Ravani B. (1997), Design of Tactile Fixtures for Robotics and Manufacturing, *ASME Journal of Mechanical Design*, June.

Ravani B. and Ge Q. J. (1991), Kinematic Localization for World Model Calibration in Off-Line Robot Programming Using Clifford Algebra, *Proc. IEEE International Conference on Robotics and and Automation, Sacramento, California*, pp. 584–589.

Roberts A. W. (1985), *Elementary Linear Algebra*, Benjamin-Cummings Publishing, Reading, MA.

Roth Z. S., Mooring B. W. and Ravani B. (1987), An Overview of Robot Calibration, *IEEE Journal of Robotics and Automation*, vol. Ra-3, pp. 377–385.

Schaal H. (1985), *Ein Geometrisches Problem der Metrischen Getriebesyntheses*, Sitzungsberichte der Osterreichischen Akademie der Wissenschaften, Wien.

Slocum A. H. (1988), Kinematic Couplings For Precision Fixturing – Part 1, *Precision Engineering*, vol. 10, pp. 85–91.

Spencer W. A. (1939), *Basic Principles of Analytic Geometry*, The Orthovis Company – educational publishers, Chicago.

Numerical Methods for Mechanism and Manipulator Workspace Analysis

Edward J. Haug*, Frederick A. Adkins**, and Chi-Mei Luh**

*Department of Mechanical Engineering and **Program in Applied Mathematics and Computational Sciences, The University of Iowa, Iowa City, Iowa 52242, USA, email: haug@ccad.uiowa.edu

Summary. Numerical algorithms for mapping boundaries of mechanism and manipulator workspaces are developed and illustrated. Analytical criteria derived are applicable for both manipulators having the same number of input and output coordinates and redundantly controlled manipulators with a larger number of inputs than outputs. A numerical method is presented for finding an initial point on the boundary, from which a continuation method that accounts for simple and multiple bifurcation of one-dimensional solution curves is used. Planar and spatial manipulators are analyzed, determining both the exterior boundaries of workspaces and interior curves that represent local impediments to motion control.

Keywords. Workspace analysis, manipulator workspace, singular configuration analysis.

1. Introduction

Criteria for boundaries of workspaces of mechanisms and manipulators, using conditions associated with singularity of constraint Jacobian or velocity transformation matrices have been developed by a number of authors (Litvin 1980). Analytical conditions associated with the geometry of specific manipulators have been used by a number of authors to obtain explicit criteria for boundaries of workspaces (Tsai and Soni 1981; Yang and Lee 1983; Freudenstein and Primrose 1984; Spanos and Kohli 1985). Singularity of the velocity transformation between input and output coordinates has likewise been used to characterize singular surfaces of manipulators (Waldron, Wang, and Bolin 1985; Kumar and Waldron 1981).

Numerical approaches for mapping boundaries of workspaces of mechanisms and manipulators have been presented by Jo and Haug (1989a, 1989b), Haug, Wang, and Wu (1992), and Wang and Wu (1993). Studies by Tsai and Waldron (1987), Nielsen, deWit, and Hagander (1991), and Haug, Luh, Adkins, and Wang (1996) have shown that such interior curves and surfaces are often associated with real restriction on motion control.

The purpose of this chapter is to present criteria for boundary curves and a method for continuation calculation, in order to deal with broad classes of manipulators. Generalized coordinates $\mathbf{q} = \left[q_1, \ldots, q_{n_q}\right]^T \in R^{n_q}$ are defined that characterize the position and orientation of each body in the mechanism, subject to independent kinematic constraint equations of the form

$$\mathbf{\Phi}(\mathbf{q}) = \mathbf{0} \tag{1}$$

where $\mathbf{\Phi}: R^{n_q} \rightarrow R^m$ is a smooth function. The constraint Jacobian (or Jacobian) of the function $\mathbf{\Phi}(\mathbf{q})$ of eq.(1), at an assembled configuration $\mathbf{q}^0$, is the $m \times n_q$ matrix

$$\mathbf{\Phi}_\mathbf{q}(\mathbf{q}^0) \equiv \left[\frac{\partial \Phi_i}{\partial q_j}(\mathbf{q}^0) \right] \tag{2}$$

Its row rank (or rank) is denoted Rank $\mathbf{\Phi}_\mathbf{q}(\mathbf{q}^0)$. At a regular point $\mathbf{q}^0$ where the Jacobian has full row rank, there is a neighborhood U of $\mathbf{q}^0$, i.e., an open set of R^{n_q} that contains $\mathbf{q}^0$, throughout which the Jacobian has full row rank. The number of degrees of freedom of the mechanism in this neighborhood U is $\mathrm{DOF} = n_q - \mathrm{Rank}\ \mathbf{\Phi}_\mathbf{q}(\mathbf{q}^0)$.

The motion of the mechanism is to be controlled by specifying the values of selected generalized coordinates, called input coordinates, $\mathbf{v} = [v_1, \ldots, v_{n_v}]^T$. Output coordinates are a subset of mechanism generalized coordinates that define the useful functionality of the mechanism, $\mathbf{u} = [u_1, \ldots, u_{n_u}]^T$. Generalized coordinates that are neither input coordinates nor output coordinates are called intermediate coordinates, $\mathbf{w} = [w_1, w_2, \ldots, w_{n_w}]^T$, where $n_w = n_q - n_v - n_u$. The choice of input coordinates $\mathbf{v}$ and output coordinates $\mathbf{u}$ for a mechanism defines a mechanical system that has an intended function, i.e., a manipulator.

Inequality constraints are often imposed on generalized coordinates in real mechanisms, of the form $q_i^{min} \le q_i \le q_i^{max}$. Such constraints can be accounted for, without introducing slack variables (Jo and Haug 1989b), by defining new generalized coordinates q_i^* through the transformation $q_i = a_i + b_i \sin q_i^*$, where $a_i = (q_i^{max} + q_i^{min})/2$ and $b_i = (q_i^{max} - q_i^{min})/2$ are the mid point and half range of the inequality constraint. Replacing q_i with q_i^*, using this transformation, the inequality constraints are identically satisfied.

2. Criteria for Workspace Boundaries

2.1 Accessible Output Sets

The accessible output set of a manipulator is the collection of all achievable output coordinates $\mathbf{u}$ of a point on the working body of the manipulator. To characterize the accessible output set for a manipulator, consider the partitioning of generalized coordinates $\mathbf{q} = \left[\mathbf{u}^T, \mathbf{v}^T, \mathbf{w}^T\right]^T \equiv \left[\mathbf{u}^T, \mathbf{z}^T\right]^T$, where the input and intermediate coordinates are consolidated in the vector $\mathbf{z} = \left[\mathbf{v}^T, \mathbf{w}^T\right]^T$. In terms of this partitioning of generalized coordinates, the constraint equations for the manipulator of Eq. 1 may be written in the form $\mathbf{\Phi}(\mathbf{u}, \mathbf{z}) = \mathbf{0}$, where $\mathbf{\Phi}(\mathbf{u}, \mathbf{z})$ has at least three continuous derivatives. The accessible output set A is thus

$$A \equiv \left\{\mathbf{u} \in R^{n_u} : \mathbf{\Phi}(\mathbf{u}, \mathbf{z}) = \mathbf{0}, \text{ for some } \mathbf{z}\right\} \tag{3}$$

Interest centers here on manipulators whose accessible output sets contain open subsets U of R^{n_u}. This requires that eq.(1) can be satisfied by some function $\mathbf{z}(\mathbf{u})$, for all $\mathbf{u}$ in U. Thus, the number n_z of coordinates in $\mathbf{z}$ must be greater than or equal to the number m of independent equations in eq.(1); where it is assumed that the Jacobian of eq.(2) has full row rank for $\mathbf{q} = \left[\mathbf{u}^T, \mathbf{z}(\mathbf{u})^T\right]^T$, with $\mathbf{u} \in U$.

At a point $\mathbf{u}^*$ on the boundary of the accessible output set of eq.(3), there exists no maximal-dimension square submatrix of $\mathbf{\Phi}_z$ that is nonsingular. Otherwise, for fixed values of the input and intermediate coordinates that are not associated with this submatrix, and for any $\mathbf{u}$ in a neighborhood of $\mathbf{u}^*$, there exists a unique solution for the remaining independent and intermediate coordinates; i.e., $\mathbf{u}^*$ is not on the boundary of the accessible output set. Thus, the boundary of the accessible output set for a manipulator is a subset of the accessible output set at which the sub-jacobian $\mathbf{\Phi}_z$ of the kinematic constraint functions of eq.(1) is row rank deficient.

To reduce this condition to analytical form, note that a matrix is row-rank deficient if and only if its transpose are linearly dependent. Thus, a necessary condition for the boundary of the accessible output set for a manipulator is

$$\partial A \subset S \equiv \left\{\mathbf{u} \in A : \mathbf{\Phi}_z^T(\mathbf{u}, \mathbf{z})\boldsymbol{\xi} = \mathbf{0}, \boldsymbol{\xi}^T\boldsymbol{\xi} = 1, \mathbf{\Phi}(\mathbf{u}, \mathbf{z}) = \mathbf{0}, \text{ for some } \mathbf{z} \text{ and } \boldsymbol{\xi} \in R^m\right\} \tag{4}$$

2.2 Operational Envelopes

The set of all points that are occupied by a point in the working body of a mechanism or manipulator, for some kinematically admissible configuration, is defined as its operational envelope. The formulation must thus involve kinematic equations that determine the set of admissible positions and orientations of a body-fixed reference frame on the working body and its shape representation relative to this reference frame. Attention here is limited to working bodies with boundaries that are smooth and strictly convex, to avoid technical difficulties associated with nondifferentiability at corners and difficulties associated with nonconvex boundary segments.

A point P in the working body is located in space by the vector $\mathbf{u} = [x, y, z]^T$. Point P is located relative to the $x' - y' - z'$ body-fixed reference frame by the vector $\mathbf{s}'^{P}$. Defining the vector $\mathbf{r}(\mathbf{q}^{k})$ from the origin of the global x-y-z reference frame to the origin of the x'-y'-z' frame and the orientation transformation matrix $\mathbf{A}(\mathbf{q}^{k})$ from the body-fixed reference frame to the global frame, both of which are functions of the mechanism generalized coordinates $\mathbf{q}^{k}$, the vector $\mathbf{u}$ from the origin of the global reference frame to point P is $\mathbf{u}=\mathbf{r}(\mathbf{q})+\mathbf{s}^{P}=\mathbf{r}(\mathbf{q})+\mathbf{A}(\mathbf{q})\mathbf{s}'^{P}$. A typical point P in the working body can be defined as a function of a parameter vector $\boldsymbol{\mu}$, which is contained in $\mathbf{q}=\left[\mathbf{q}^{kT},\boldsymbol{\mu}^{T}\right]^{T}$, in the form $\mathbf{s}'^{P}=\mathbf{f}(\boldsymbol{\mu})$. To avoid technical difficulties, it is presumed that the function $\mathbf{f}(\boldsymbol{\mu})$ is at least twice continuously differentiable.

The analytical condition for the location of points in the working body, as it moves under the influence of kinematic constraints on the underlying mechanism, is thus

$$\boldsymbol{\Psi}(\mathbf{u},\mathbf{q})\equiv\begin{bmatrix}\mathbf{u}-\mathbf{r}(\mathbf{q}^{k})-\mathbf{A}(\mathbf{q}^{k})\mathbf{f}(\boldsymbol{\mu})\\ \boldsymbol{\Phi}(\mathbf{q}^{k})\end{bmatrix}=\mathbf{0} \tag{5}$$

The operational envelope of the working body is thus the set of all points $\mathbf{u}$ that satisfy eq.(5); i.e.,

$$\mathrm{OE}=\left\{\mathbf{u}\in R^{\mathbf{n_u}}:\boldsymbol{\Psi}^{P}(\mathbf{u},\mathbf{q})=\mathbf{0},\text{ for some }\mathbf{q}\right\} \tag{6}$$

The boundary of OE determines the nature of the operational envelope of the working body, which may not be either convex or simply connected. As in the case of the accessible output set, necessary conditions for the boundary of the operational envelope are that the matrix

$$\boldsymbol{\Psi}_{\mathbf{q}}(\mathbf{q})=\begin{bmatrix}\left\{-\mathbf{r}(\mathbf{q}^{k})-\mathbf{A}(\mathbf{q}^{k})\mathbf{f}(\boldsymbol{\mu})\right\}_{\mathbf{q}^{k}} & -\mathbf{A}(\mathbf{q}^{k})\mathbf{f}_{\boldsymbol{\mu}}(\boldsymbol{\mu})\\ \boldsymbol{\Phi}_{\mathbf{q}^{k}}(\mathbf{q}^{k}) & \mathbf{0}\end{bmatrix} \tag{7}$$

must be row-rank deficient. Analytically, the associated boundary is characterized by

$$\partial\mathrm{OE}\subset\mathrm{S}=\left\{\mathbf{u}\in R^{n_u}:\boldsymbol{\Psi}_{\mathbf{q}}^{T}(\mathbf{q})\boldsymbol{\xi}=\mathbf{0},\ \boldsymbol{\xi}^{T}\boldsymbol{\xi}=1,\boldsymbol{\Psi}(\mathbf{u},\mathbf{q})=\mathbf{0},\text{ for some }\mathbf{q}\text{ and }\boldsymbol{\xi}\right\} \tag{8}$$

2.3 Domains of Interference

The sets of all points in pairs of working bodies of mechanisms and manipulators that can coincide, for some kinematically admissible configuration of the underlying mechanism are called domains of interference. An extensive literature has developed addressing the problem of contact between pairs of bodies in mechanisms and manipulators (Lee and Preparata 1984; Cameron 1985; Gilbert, Johnson, and Keerthi 1988). Most of this work has focused on monitoring the distance between nearest points of approach on boundaries of bodies that may interfere as the mechanism follows a trajectory in configuration space. Interference is reported when the distance is zero. The focus here is design-oriented, i.e., eliminating the possibility of interference, rather than on control algorithms that would assure the avoidance of interference that could occur in kinematically admissible configurations of the underlying mechanism.

Since interference between a pair of working bodies depends on both the mechanism generalized coordinates and the shapes of the domains of the working bodies, the respective domains must be parameterized. Points P in the domains of bodies 1 and 2 are characterized in terms of parameter vectors $\boldsymbol{\mu}_1$ and $\boldsymbol{\mu}_2$, in their respective body-fixed reference frames, as $\mathbf{s}_1'^P = \mathbf{f}_1(\boldsymbol{\mu}_1)$ and $\mathbf{s}_2'^P = \mathbf{f}_2(\boldsymbol{\mu}_2)$. In case inequalities involving parameter vectors $\boldsymbol{\mu}_i$ are required to characterize the bounded domain of the associated working body; e.g., using slack variable constraints (Jo and Haug 1989b), the associated equations are included in kinematic equations that describe admissible configurations of the mechanism. Combining the set of geometric parameter vectors $\boldsymbol{\mu} = \left[\boldsymbol{\mu}_1^T, \boldsymbol{\mu}_2^T\right]^T$ with kinematic coordinates, the generalized coordinate vector is $\mathbf{q} = \left[\mathbf{q}^{kT}, \boldsymbol{\mu}^T\right]^T$. It is presumed here that the boundaries of working bodies are smooth, the bodies are strictly convex, and the functions $\mathbf{f}_1(\boldsymbol{\mu}_1)$ and $\mathbf{f}_2(\boldsymbol{\mu}_2)$ are at least twice continuously differentiable. While these assumptions may be relaxed, technical challenges are anticipated.

The condition that some point P is occupied by a material point in each of bodies 1 and 2 is $\mathbf{r}_2(\mathbf{q}^k) + \mathbf{A}_2(\mathbf{q}^k)\mathbf{f}_2(\boldsymbol{\mu}_2) - \mathbf{r}_1(\mathbf{q}^k) - \mathbf{A}_1(\mathbf{q}^k)\mathbf{f}_1(\boldsymbol{\mu}_1) = \mathbf{0}$. The domain of interference is defined here as the set of points $I_1' D_2$ in body 1 that interfere with points in the domain of body 2, for some kinematically admissible configuration of the mechanism.

Let $\mathbf{u}_1'$ be a point in body 1, characterized in its body-fixed reference frame, i.e., $\mathbf{u}_1' = \mathbf{f}_1(\boldsymbol{\mu}_1)$, that is in $I_1' D_2$. Then, there exist $\mathbf{q}^k$ and $\boldsymbol{\mu}$ that satisfy

$$\boldsymbol{\Psi}\left(\mathbf{u}_1', \mathbf{q}^k, \boldsymbol{\mu}\right) \equiv \begin{bmatrix} \mathbf{u}_1' - \mathbf{f}_1(\boldsymbol{\mu}_1) \\ \mathbf{r}_2(\mathbf{q}) + \mathbf{A}_2(\mathbf{q}^k)\mathbf{f}_2(\boldsymbol{\mu}_2) - \mathbf{r}_1(\mathbf{q}) - \mathbf{A}_1(\mathbf{q}^k)\mathbf{f}_1(\boldsymbol{\mu}_1) \\ \boldsymbol{\Phi}(\mathbf{q}^k, \boldsymbol{\mu}) \end{bmatrix} = \mathbf{0} \tag{9}$$

Thus, the domain of interference $I_1' D_2$ is

$$I_1' D_2 = \left\{\mathbf{u}_1' : \boldsymbol{\Psi}_1\left(\mathbf{u}_1', \mathbf{q}^k, \boldsymbol{\mu}\right) = \mathbf{0}, \text{for some } \mathbf{q}^k \text{ and } \boldsymbol{\mu}\right\} \tag{10}$$

If a point $\mathbf{u}_1'$ in $I_1' D_2$ is on its boundary, denoted $\partial I_1' D_2$, then it is necessary that the sub-jacobian matrix

$$\boldsymbol{\Psi}_{q}(\mathbf{u}_1',\mathbf{q})=$$

$$\begin{bmatrix} \mathbf{0} & -f_{1_{\mu_1}}(\mu_1) & \mathbf{0} \\ \left\{\mathbf{r}_2(\mathbf{q}^k)+\mathbf{A}_2(\mathbf{q}^k)f_2(\mu_2)-\mathbf{r}_1(\mathbf{q}^k)-\mathbf{A}_1(\mathbf{q}^k)f_1(\mu_1)\right\}_{\mathbf{q}^k} & -\mathbf{A}_1(\mathbf{q}^k)f_{1_{\mu_1}}(\mu_1) & \mathbf{A}_2(\mathbf{q}^k)f_{2_{\mu_2}}(\mu_2) \\ \boldsymbol{\Phi}_{\mathbf{q}^k}(\mathbf{q}^k,\mu) & \boldsymbol{\Phi}_{\mu_1}(\mathbf{q}^k,\mu) & \boldsymbol{\Phi}_{\mu_2}(\mathbf{q}^k,\mu) \end{bmatrix} \tag{11}$$

is row-rank deficient, for some **q** that, together with **u**, satisfies eq.(9); i.e.,

$$\partial I_1'D_2 \subset S \equiv$$

$$\left\{\mathbf{u}_1' \in I_1'D : \boldsymbol{\Psi}_{1q,\mu}^{T}(\mathbf{u}_1',\mathbf{q}^k,\mu)\xi=\mathbf{0},\ \xi^T\xi=1,\ \boldsymbol{\Psi}_1(\mathbf{u}_1',\ \mathbf{q}^k,\ \mu)=\mathbf{0},\ \text{for some } \mathbf{q} \text{ and } \xi\right\} \tag{12}$$

3. Mapping The Set S

3.1 Conditions for Mapping S

In the inflated space of $\mathbf{x}=\left[\mathbf{u}^T,\mathbf{q}^T,\xi^T\right]^T \in R^n$, $n=m+n_q$, points on sets S defined in Section 2 are projections of solutions of the following equation onto the **u**-space:

$$\mathbf{G}(\mathbf{x}) \equiv \begin{bmatrix} \boldsymbol{\Psi}(\mathbf{u},\mathbf{z}) \\ \boldsymbol{\Psi}_{\mathbf{z}}^T(\mathbf{u},\mathbf{z})\xi \\ \xi^T\xi-1 \end{bmatrix} = \mathbf{0} \tag{13}$$

where $\boldsymbol{\Psi} \equiv \boldsymbol{\Phi}$ and $\mathbf{q} \equiv \mathbf{z}$ in the case of the accessible output set. This is a system of $m+nz+1$ equations in $nu+nz+m$ variables. Thus, the dimension of the solution set of eq.(13), hence the dimension of the S, is $\dim S = n_u - 1$. If **u** is two-dimensional, then S is one-dimensional. If **u** is three-dimensional, then S is two-dimensional, which is the case for manipulators in spatial applications. If the output coordinate **u** represents both position and orientation capability, or a broader generalization such that $\dim \mathbf{u} > 3$, then S must be viewed in a higher-dimensional setting.

A practical approach to mapping S for $\dim \mathbf{u} \geq 3$ is to define a plane (or hyperplane if $\dim \mathbf{u} > 3$) that intersects S in one dimensional sets, i.e., define linear relations among the output coordinates **u** of the form $\mathbf{Lu}=\mathbf{b}$, where matrix **L** is of dimension $(n_u-2)\times n_u$. Adding this condition to the constraint equations of eq.(13) yields a solution set that is one-dimensional, for selected values of **b**; i.e., an (n_u-2)-dimensional family of curves on S as **b** varies. For $n_u=3$, this is a one-dimensional family of generators for S.

3.2 Finding an Initial Point on S

For realistic applications, it is a nontrivial task to find an initial point on S. An algorithm for doing so involves starting from an assembled configuration of the manipulator $\mathbf{q}^0$, with $\mathbf{u}^0$ interior to S. A unit vector $\mathbf{c}$ in the output space is selected, and the ray emanating from $\mathbf{u}^0$ along vector $\mathbf{c}$ is traced until a point on S is encountered. Discrete movement along this ray is defined by $\mathbf{u}^i = \mathbf{u}^{i-1} + h\mathbf{c}$, where $i = 1,2,\ldots$ and h is the step size. In taking a step along this ray, with $\mathbf{u}^i$ specified, associated values $\mathbf{z}^i$ must be determined. To satisfy eq.(1) with $\mathbf{u}^i$ fixed, a generalized Newton-Raphson algorithm is employed, i.e.,

$$\begin{aligned} &\mathbf{\Phi}_{\mathbf{z}}\left(\mathbf{u}^i, \mathbf{z}^{i(j)}\right)\Delta\mathbf{z}^{(j)} = -\mathbf{\Phi}\left(\mathbf{u}^i, \mathbf{z}^{i(j)}\right) \\ &\mathbf{z}^{i(j+1)} = \mathbf{z}^{i(j)} + \Delta\mathbf{z}^{(j)} \end{aligned} \tag{14}$$

where $j = 0,1,\ldots$ is the iteration counter and $\mathbf{z}^{i(0)} = \mathbf{z}^{i-1}$.

In the general case, the sub-jacobian matrix $\mathbf{\Phi}_{\mathbf{z}}$ has more columns than rows, so eq.(14) has multiple solutions. The goal in moving along the ray defined from $\mathbf{u}^0$ along vector $\mathbf{c}$ is to define assembled configurations, until this is no longer possible beyond S. A strategy for obtaining a solution of eq.(11) is to find the solution $\Delta\mathbf{z}^{i(j)}$ with minimum norm, i.e., minimize $(1/2)\Delta\mathbf{z}^T\Delta\mathbf{z}$, subject to the condition $\mathbf{\Phi}_{\mathbf{z}}\Delta\mathbf{z} = -\mathbf{\Phi}$. As discussed by Haug, Luh, Adkins, and Wang (1996), this procedure leads to use of the Moore-Penrose inverse.

3.3 Mapping One-Dimensional Solution Curves in S

The desired one-dimensional curve of S is a solution set of eq.(13). A tangent function $\mathbf{h}(\mathbf{x})$ to the one-dimensional set defined by eq.(13) is assured by the following theorem (Rheinboldt 1986; Allgower and Georg 1990):

Let $\mathbf{G}: S \subset R^n \to R^{n-1}$, $n \geq 2$, be of class $C^r, r \geq 1$, on the open set $S \subset R^n$. Then, if $\mathbf{G}_{\mathbf{x}}(\mathbf{x})$ has a full row-rank, a unit tangent vector $\mathbf{h}(\mathbf{x}) \in R^n$ is uniquely defined by

$$\begin{aligned} &\mathbf{G}_{\mathbf{x}}(\mathbf{x})\mathbf{h}(\mathbf{x}) = \mathbf{0} \\ &\mathbf{h}(\mathbf{x})^T\mathbf{h}(\mathbf{x}) = 1 \\ &\mathrm{Det}\begin{bmatrix} \mathbf{G}_{\mathbf{x}}(\mathbf{x}) \\ \mathbf{h}(\mathbf{x})^T \end{bmatrix} > 0 \end{aligned} \tag{15}$$

and the map $\mathbf{x} \to \mathbf{h}(\mathbf{x})$ is of class C^{r-1}.

As long as $\mathbf{G}_{\mathbf{x}}(\mathbf{x})$ has full row-rank, the tangent vector $\mathbf{h}(\mathbf{x})$ given by eq.(15) can be computed at each point along the solution of eq.(13) as the basis for proceeding stepwise along the solution curve in a continuation method (Jo and Haug 1989a; Rheinboldt 1986; Allgower and Georg 1990). A well-established computer code, PITCON (Rheinboldt 1986), for implementing this continuation method is called to map solutions of eq.(13) until a closed trajectory is found. Then, **b** is perturbed and an adjacent one-dimensional solution set of eq.(13) is mapped to obtain another closed curve. Proceeding in this fashion, a family of one-dimensional generators is numerically constructed that includes S.

In many applications, multiple solution curves cross at bifurcation points of eq.(13). At such points, $\mathbf{G_x}$ is row rank deficient and eq.(15) fails to determine a unique tangent. A continuation algorithm at such points that involves a second-order Taylor expansion of a reduced set of equations from eq.(13) is presented by Haug, Luh, Adkins, and Wang (1996), yielding both simple and multiple bifurcation points.

An experimental computer code that automatically traces one dimensional solution sets in S has been constructed using the software packages PITCON, ADOL-C, and LINPACK. PITCON (Rheinboldt 1986) computes points along a one-dimensional solution manifold of eq.(13). The capability to detect and locate bifurcation points has been added to version 5.0 of PITCON. The automatic differentiation code ADOL-C (Griewank et al. 1994) evaluates the Jacobian of the left side eq.(13) and the Hessian of the second-order Taylor expansion. LINPACK (Dongarra et al. 1979) is used to solve the linear systems involved.

4. Numerical Examples

4.1 Planar Stewart Platform

The planar Stewart platform shown in Fig. 1 consists of a moving upper platform and three actuators that control the three degrees of freedom of the platform. The working point is chosen at the center of the top platform, so $\mathbf{s}'^{P} = \mathbf{0}$. The actuator lengths ℓ_1, ℓ_2, and ℓ_3 are input coordinates that control the position $\mathbf{r} = [x, y]^T$ of point P and the orientation ϕ of the top platform. Actuator length unilateral constraints are $0 < \ell_i^{min} \le \ell_i \le \ell_i^{max}$, $i = 1, 2, 3$, where $\ell_1^{min} = \ell_2^{min} = \sqrt{2}$, $\ell_3^{min} = 1$, $\ell_1^{max} = \ell_2^{max} = 2$, and $\ell_3^{max} = \sqrt{3}$. Introducing new input coordinates v_1, v_2, and v_3 so that $\ell_i = a_i + b_i \sin v_i$, $i = 1, 2, 3$, where $a_1 = a_2 = \left(\sqrt{2}+2\right)/2$, $a_3 = \left(1+\sqrt{3}\right)/2$, $b_1 = b_2 = \left(2-\sqrt{2}\right)/2$, and $b_3 = \left(\sqrt{3}-1\right)/2$, the actuator unilateral constraints are automatically satisfied. The generalized coordinate vector is thus $\mathbf{q} = [x, y, v_1, v_2, v_3, \phi]^T$ and the constraint equations for this manipulator are

$$\mathbf{\Phi}(\mathbf{q}) \equiv \begin{bmatrix} \ell_1^2 - (x - \cos\phi + 1)^2 - (y - \sin\phi)^2 \\ \ell_2^2 - (1 - x + \cos\phi)^2 - (y - \sin\phi)^2 \\ \ell_3^2 - (x + \cos\phi - 2)^2 - (y + \sin\phi)^2 \end{bmatrix} = \mathbf{0} \tag{16}$$

where ℓ_i is a function of v_i. Since $\mathbf{s}'^{P} = \mathbf{0}$, $\mathbf{u} = [x, y]^T \equiv \mathbf{g}(\mathbf{q})$.

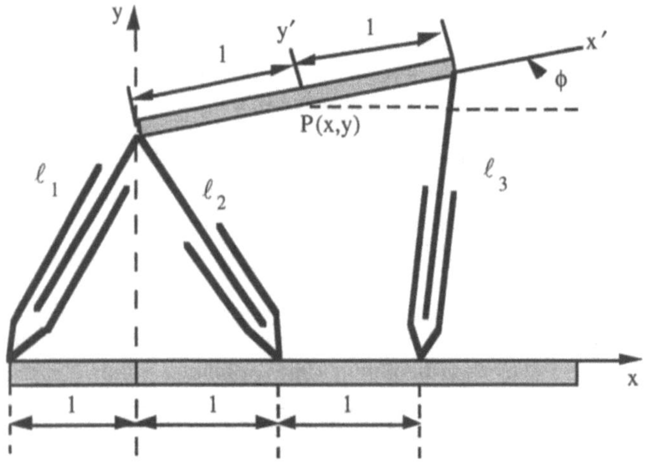

Fig. 1. Planar stewart platform

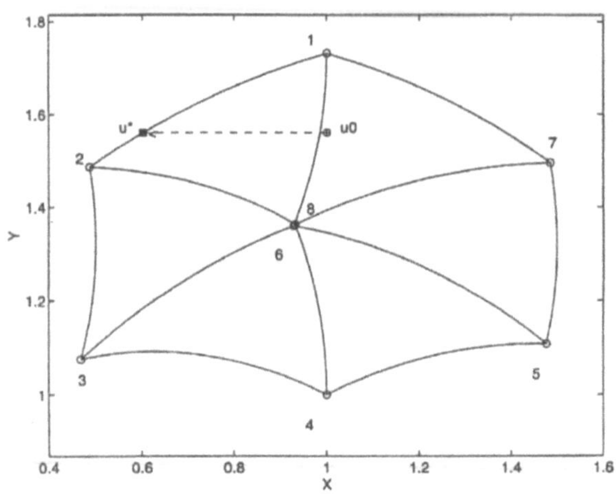

Fig. 2. Accessible output set

To obtain an initial point on the boundary of the accessible output set to begin numerical calculations, the platform is raised to a mid-point in which the upper platform is horizontal; i.e., $\phi = 0$. The initial values for the generalized coordinates for this point interior to A are thus $\mathbf{q}^0 = [1,\ 1.5607,\ 0.5236,\ 0.5236,\ 0.5236,\ 0.0]^T$. Selecting a search direction parallel to the x-asis; i.e., $\mathbf{c} = [-1, 0]^T$, the search process of Subsection 3.2 for an initial point on ∂A is carried out, yielding the point $\mathbf{u}^*$ on ∂A shown in Fig.

2. The final configuration on ∂A obtained in this process is $\mathbf{q}^* = [0.6018, 1.5607, -0.5209, 1.5694, 1.5658, 0.1235]^T$.

With this starting point, candidate curves on ∂A are mapped using the continuation method outlined in Subsection 3.3. The curves obtained are shown in Fig. 2, where numbered points are bifurcation points. Points 6 and 8, while appearing to be close together, are actually the projections onto the **u**-space of two distinctly different bifurcation points. The generalized coordinate vector for point 6 is $\mathbf{q} = [x, y, f]^T = [0.928571, 1.36090, -0.38075]^T$ and that for point 8 is $\mathbf{q} = [0.931663, 1.366332, 0.37183]^T$.

Consider next the operational envelope of the projection dome of Fig. 3 that is attached to the moving body of the planar Stewart platform of Fig. 1 to form a driving simulator. The generalized coordinate vector is $\mathbf{q} = [x, y, v_1, v_2, v_3, \phi]^T$ and the constraint equations for the manipulator are given in eq.(16). The equation for the boundary of the dome in Fig. 3 is $\mu_1^2 + \left(2 + 1.3(\mu_2 - 0.5)^3\right)^2 \left((\mu_2 - 0.5)^2 - 1\right) = 0$. The domain of the dome is thus characterized by the inquality $\mu_1^2 + \left(2 + 1.3(\mu_2 - 0.5)^3\right)^2 \left((\mu_2 - 0.5)^2 - 1\right) \leq 0$. To represent the domain of the dome in equation form, a slack variable μ_3 is introduced and the equation for the domain is

$$B(\mu_1) = \mu_{11}{}^2 + (2 + 1.3(\mu_{12} - 0.5)^3)^2((\mu_{12} - 0.5)^2 - 1) + \mu_{13}{}^2 = 0$$

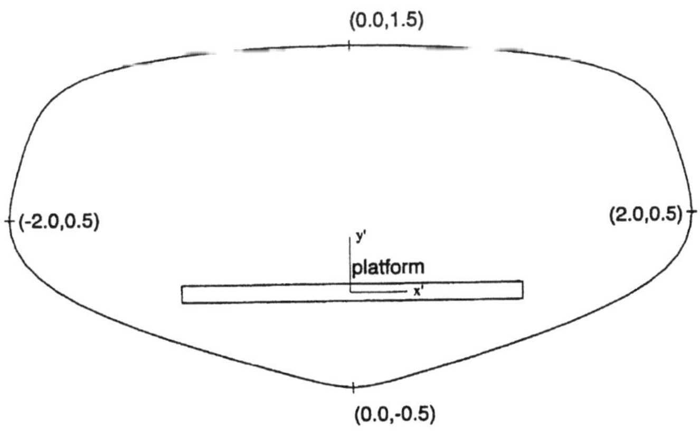

Fig. 3. Projection dome

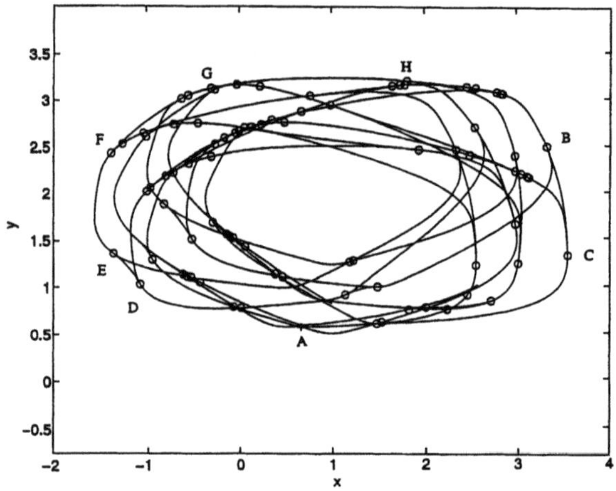

Fig. 4. Singular boundary curves

A typical point in the domain of the dome is thus given, in the platform body-fixed $x' - y'$ frame, as $\mathbf{s}_2'^P = [\mu_1, \mu_2]^T \equiv \mathbf{f}(\boldsymbol{\mu})$. The equation defining the operational envelope of the domain of the dome of the planar Stewart platform is thus

$$\boldsymbol{\Psi}(\mathrm{u},\mathrm{q},\mu) \equiv \begin{bmatrix} \mu_1 - x - \mu_1 \cos\phi + \mu_2 \sin\phi \\ \mu_2 - y - \mu_1 \sin\phi - \mu_2 \cos\phi \\ {\mu_1}^2 + (2 + 1.3(\mu_2 - 0.5)^3)^2((\mu_2 - 0.5)^2 - 1) + {\mu_3}^2 \\ \ell_1^2 - (\mathrm{x} - \cos\phi + 1)^2 - (\mathrm{y} - \sin\phi)^2 \\ \ell_2^2 - (1 - \mathrm{x} + \cos\phi)^2 - (\mathrm{y} - \sin\phi)^2 \\ \ell_3^2 - (\mathrm{x} + \cos\phi - 2)^2 - (\mathrm{y} + \sin\phi)^2 \end{bmatrix} = \mathbf{0}$$

To obtain a point interior to OE, the mechanism legs were set to their maximum length, which determines the values of x, y, and ϕ. The values $\mu_1 = 0.0$, $\mu_2 = 0.5$, and $\mu_3 = 2.0$ were chosen, and the search direction was selected as $\mathbf{c} = [0,1]^T$. Searching in this direction, using the method of Subsection 3.2 from the configuration $\left[\mathbf{u}^{0^T}, \mathbf{q}^{0^T}, \boldsymbol{\mu}^{0^T}\right]^T = \left[1.0, \sqrt{3}, 1.0, \sqrt{3}, \frac{\pi}{2}, \frac{\pi}{2}, \frac{\pi}{2}, 0, 0, 0.5, 2\right]^T$, yielded the following point on $\partial\mathrm{OE}: \left[\mathbf{u}^{*^T}, \mathbf{q}^{*^T}, \boldsymbol{\mu}^{*^T}\right]^T = [1.0, 3.232, 1.0, 1.732, 1.571, 1.571, 1.571, 0, 0, 0.5, 2]^T$.

Starting from $\mathbf{x}^*$, the boundary is traced using the method outlined in Subsection 3.3. The resulting set of boundary segments is shown in Fig. 4. The outer curves define the boundary of the operational envelope. In Fig. 4, several points of interest are marked. There are points at which the projections of boundary curves cross, but the curves computed do not intersect in the higher-dimensional space. An example of such points is point A. Points such as these pose a problem when attempting to calculate the boundary of the operational envelope without also calculating the interior singular curves. Many segments on the boundary of the operational envelope arise from tracing the shape of the dome with the platform in a fixed configuration. Examples of this behavior are segments EF and GH. There are, however, boundary segments that result from moving the platform and the dome, sweeping out a curve from one configuration of the platform to another. Examples of such segments are BC and DE. On segment BC, the mechanism is moving by changing the third leg from its maximum to its minimum length; hence, v_3 changes from $\pi/2$ to $-\pi/2$. On segment DE, the mechanism is moving by changing the second leg from its minimum to its maximum length; hence, v_2 changes from $-\pi/2$ to $\pi/2$.

Consider next the planar Stewart platform and dome operating in a room that is 3.0 units high and 5.4 units wide. The global reference frame is used as the local frame for the room and the center of the room is at (0.7, 1.5). The room is selected so that the dome will interfere with the ceiling and the right wall, but not with the floor and the left wall. The coordinates of points outside of the room are $\mathbf{s}_2^{\prime P} = [\mu_{21}, \; \mu_{22}]^T$. Equations are thus needed to represent the right wall and the ceiling. Points to the right of the right wall can be given as $\mu_{21} = 3.4 + \mu_{23}^2$, and points above the ceiling can be given as $\mu_{22} = 3.0 + \mu_{24}^2$, where μ_{23} and μ_{24} are slack variables. Interference equations are $\mu_{21} - x - \mu_{11}\cos(\phi) + \mu_{12}\sin(\phi) = 0$ and $\mu_{22} - y - \mu_{11}\sin(\phi) - \mu_{12}\cos(\phi) = 0$. The governing equations for $I_1^{\prime}D_2$ relative to the volume to the right of the right wall are

$$\Psi_1(\mathbf{u},\mathbf{q},\boldsymbol{\mu}) \equiv \begin{bmatrix} u_1 - \mu_{11} \\ u_2 - \mu_{12} \\ \mu_{21} - x - \mu_{11}\cos\phi + \mu_{12}\sin\phi \\ \mu_{22} - y - \mu_{11}\sin\phi - \mu_{12}\cos\phi \\ \ell_1^2 - (x - \cos\phi + 1)^2 - (y - \sin\phi)^2 \\ \ell_2^2 - (1 - x + \cos\phi)^2 - (y - \sin\phi)^2 \\ \ell_3^2 - (x + \cos\phi - 2)^2 - (y + \sin\phi)^2 \\ \mu_{11}{}^2 + (2 + 1.3(\mu_{12} - 0.5)^3)^2((\mu_{12} - 0.5)^2 - 1) + \mu_{13}{}^2 \\ \mu_{21} - 3.4 - \mu_{23}^2 \end{bmatrix} = \mathbf{0}$$

The governing equations for $I_1^{\prime}D_2$ above the ceiling are obtained by replacing the last equation above with $\mu_{22} - 3.0 - \mu_{24}^2 = 0$. Initial points on the boundaries of these sets were determined by geometry, and the method of

Subsection 3.3 was used to map boundary segments. Results for $I_1'D_2$ are displayed in Fig. 5.

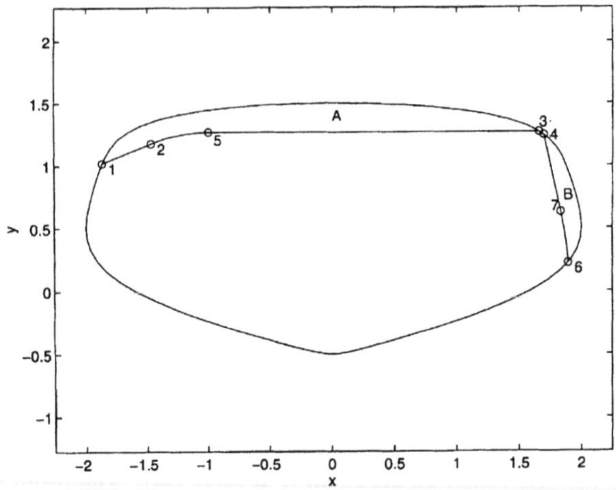

Fig. 5. $I_1'D_2$ for Stewart platform with dome and room

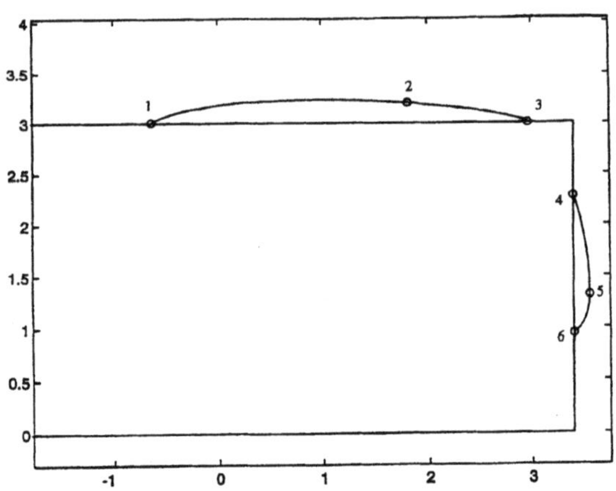

Fig. 6. $I_2'D_1$ for Stewart platform with dome and room

Points in region A of Fig. 5 result from interference of the dome with the ceiling and points in region B result from interference with the right wall. The boundary segment from point 3 to point 5 in Fig. 5, is due to the edge of the ceiling penetrating the dome, with all legs of the mechanism at full extension.

The boundary segment from point 2 to point 5 is the result of the mechanism moving by reducing the length of the third actuator, with the other two legs held at their maximum lengths. The boundary segment from point 6 to point 7 is a result of varying ℓ_3 to its minimum length, with ℓ_1 at maximum and ℓ_2 at minimum length. The segment from point 7 to point 4 is the result of the edge of the right wall being traced with ℓ_1 at maximum and both ℓ_2 and ℓ_3 at minimum length.

Results for $I_2'D_1$ are shown in Fig. 6. The boundary segment from point 1 to point 2 in Fig. 6 is due to the edge of the dome penetrating into the ceiling, with all legs of the mechanism at full extension. The boundary segment from point 2 to point 3 is the result of the mechanism moving by reducing the length of the second actuator, with the other two legs held at their maximum lengths. The boundary segment from point 4 to point 5 is a result of varying ℓ_3 to its minimum length, with ℓ_1 at maximum and ℓ_2 at minimum length. The segment from point 5 to point 6 is the result of the edge of the dome being traced with ℓ_1 at maximum and both ℓ_2 and ℓ_3 at minimum length.

4.2 Spatial Stewart Platform

The spatial Stewart platform shown schematically in Fig. 7 consists of a moving upper platform attached to ground by six extendable actuators. The lengths of the six actuators serve as inputs, with upper and lower bounds imposed on the lengths of each, to control the motion of the six-degree-of-freedom platform. The Iowa Driving Simulator (IDS) shown in Fig. 8 is a high fidelity ground vehicle simulator that incorporates such a Stewart platform as the motion base for six-degree-of-freedom motion control. Details of the kinematic formulation are given by Adkins and Haug (1997).

Attached to the motion base is a projection dome for wide-field visual display. This projection dome is modelled as a body of revolution about the vertical axis of the cross section shape presented in Fig. 3.

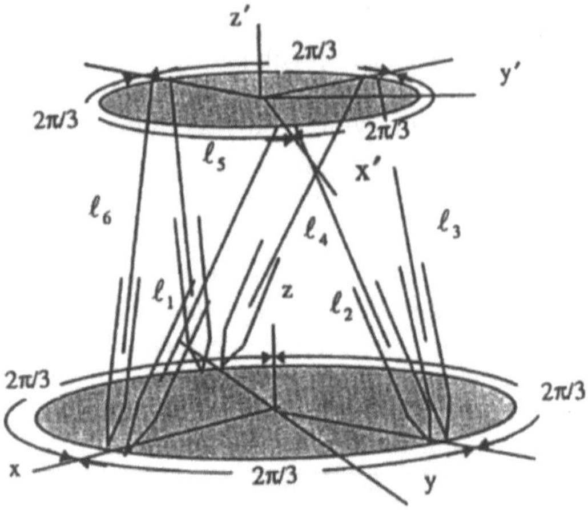

Fig. 7. A spatial Stewart platform

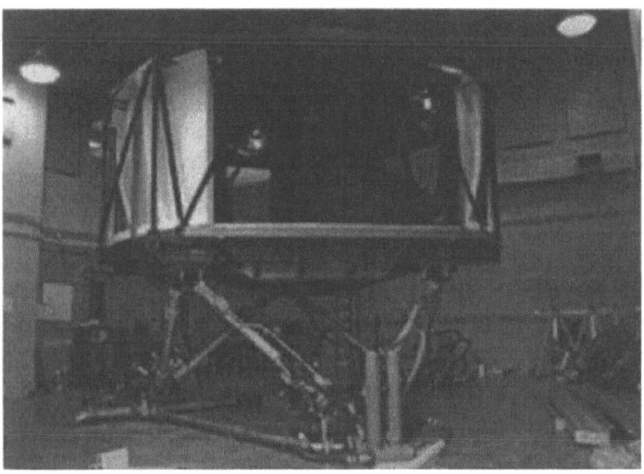

Fig. 8. Iowa Driving Simulator

A brief discussion of the kinematics formulation and the associated accessible output set is given here. Limitations on the lengths of the six actuators are specified as unilateral constrains of the form $0 < \ell_i^{min} \le \ell_i \le \ell_i^{max}, i = 1, \ldots 6$. They are accounted for by introducing auxiliary input variables v_i through the relations $\ell_i = \ell_i^0 + \ell_i^r \sin v_i$, where $\ell_i^0 = \ell_i^{max} + \ell_i^{min}/2, \ell_i^r = \ell_i^{max} + \ell_i^{min}/2$, $i = 1, \ldots 6$. Kinematic constraints for the Stewart platform manipulator are composed of six distance constraints associated with the six actuators and one normalization constraint on the unit vector $\boldsymbol{\beta}$ parallel to the axis of rotation (Haug 1989). The distance constraint equation for the *i*-th actuator is

$$\Phi_i = \left(\mathbf{r} + \mathbf{A}(\alpha,\boldsymbol{\beta})\mathbf{a}_i' - \mathbf{t}_i\right)^T\left(\mathbf{r} + \mathbf{A}(\alpha,\boldsymbol{\beta})\mathbf{a}_i' - \mathbf{t}_i\right) - \left(\ell_i^0 + \ell_i^r \sin v_i\right)^2 = 0$$

where the vectors from the origin of the x - y - z frame to the origin of the $x' - y' - z'$ frame in Fig. 7 and $\mathbf{a}_i'$ and $\mathbf{t}_i$ are the constant local coordinate vectors to the upper and lower actuator attachment locations. Actual data from the Iowa Driving Simulator's Stewart Platform motion base are used in the computations described below. The matrix (Haug 1989)

$$\mathrm{A}(\alpha,\boldsymbol{\beta}) = \cos\alpha\,\mathbf{I} + (1-\cos\alpha)\boldsymbol{\beta}\boldsymbol{\beta}^T + \sin\alpha\tilde{\boldsymbol{\beta}}$$

is the orientation transformation matrix for the upper platform, where α is the angle of rotation of the upper platform about a rotation axis that is specified by the unit vector $\boldsymbol{\beta} = \left[\beta_x, \beta_y, \beta_z\right]^T$, and the skew symmetric matrix $\tilde{\boldsymbol{\beta}}$ is

$$\tilde{\boldsymbol{\beta}} = \begin{bmatrix} 0 & -\beta_z & \beta_y \\ \beta_z & 0 & -\beta_x \\ -\beta_y & \beta_x & 0 \end{bmatrix}$$

The normalization constraint on $\boldsymbol{\beta}$ is $\boldsymbol{\Phi}_7 = \boldsymbol{\beta}^T\boldsymbol{\beta} - 1 = 0$. Thus, the kinematic constraint equations are $\boldsymbol{\Phi}(\mathbf{q}) = [\Phi_1, \Phi_2, \ldots, \Phi_7]^T = \mathbf{0}$, where the vector of generalized coordinates is $\mathbf{q} = \left[\mathbf{r}^T, \alpha, \boldsymbol{\beta}^T, \mathbf{v}^T\right]^T \in R^{13}$.

A typical point P in the dome, which is the body of revolution with shape shown in Fig. 3, is prescribed as a function of a vector $\boldsymbol{\mu} = \left[\mu_1, \mu_2, \mu_3, \mu_4\right]^T$ in the form

$$\mathbf{s}'^P = \mathbf{f}(\boldsymbol{\mu}) \equiv \left[\mu_1, \mu_2, \mu_3\right]^T$$

where $\boldsymbol{\mu}$ is subject to an equation that defines the domain of the projection dome,

$$\mathrm{B}(\boldsymbol{\mu}) = \mu_1^2 + \mu_2^2 + \left(2 + 1.3(\mu_3 - 0.5)^3\right)^2\left((\mu_3 - 0.5)^2 - 1\right) + \mu_2^4$$

Here, μ_1, μ_2, and μ_3 are spatial coordinates in the dome and μ_4 is a slack variable. The vector $\mathbf{u} = \left[u_1, u_2, u_3\right]^T$ of output coordinates from the origin of the global reference frame to point P in the projection dome is

$$\mathbf{u} = \mathbf{r}(\mathbf{q}) + \mathrm{A}(\alpha,\boldsymbol{\beta})\mathbf{s}'^P$$

The analytical condition for points in the projection dome as it moves under the influence of the underlying Stewart platform is thus

$$\Psi(\mathbf{u},\mathbf{q},\boldsymbol{\mu}) \equiv \begin{bmatrix} \mathbf{u} - \mathrm{r}(\mathbf{q}) - \mathbf{A}(\alpha,\boldsymbol{\beta})\mathrm{f}(\boldsymbol{\mu}) \\ \mathrm{B}(\boldsymbol{\mu}) \\ \boldsymbol{\Phi}(\mathbf{q}) \end{bmatrix} = \mathbf{0}$$

Analytical criteria from Section 3 are used to define one dimensional generators on the boundary ∂OE of the operational envelope. This involves forming the continuation equation

$$\mathbf{G}(\mathbf{x}) = \begin{bmatrix} \boldsymbol{\Psi} \\ \boldsymbol{\Psi}_{q,\mu}^{T}\xi \\ \xi^{T}\xi - 1 \\ u_3 - b \end{bmatrix} = \mathbf{0}$$

where $\mathbf{x} = \left[\mathbf{u}^{T}, \mathbf{q}^{T}, \xi^{T}\right]^{T}$. Since ∂OE is a surface in R^3, horizontal cutting planes $u_3 - b = 0$ are added to define a family of one dimensional generators for ∂OE. Due to the necessary, but not sufficient, conditions used in the analytical criteria for determining ∂OE, the computed candidate curves for ∂OE contain some interior segments. These sets are shown in Figs. 9 and 10 for two different cutting planes. The points marked by circles are bifurcation points where curves branch in the **x**-space.

The outer boundary was extracted from each slice and collected to form the operational envelope boundary surface shown in Fig. 11.

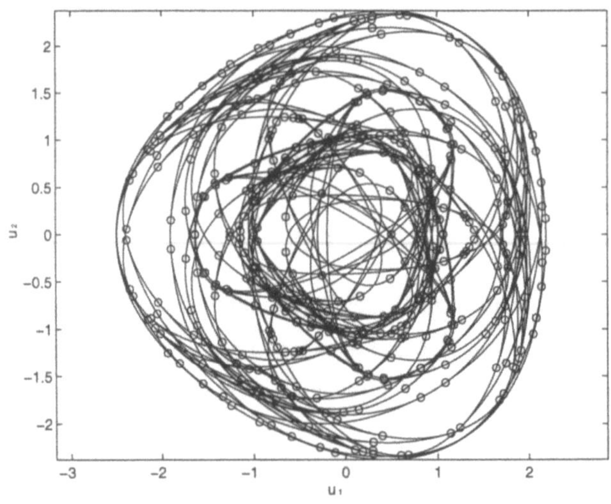

Fig. 9. Singular curves at μ_3=4.4 meters

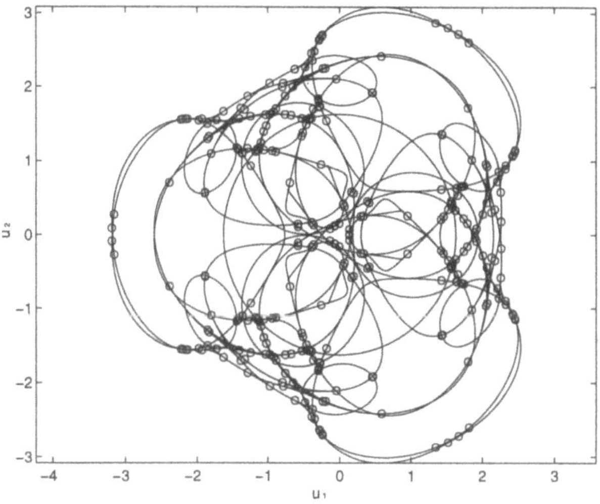

Fig. 10. Singular curves at μ_3=1.7 meters

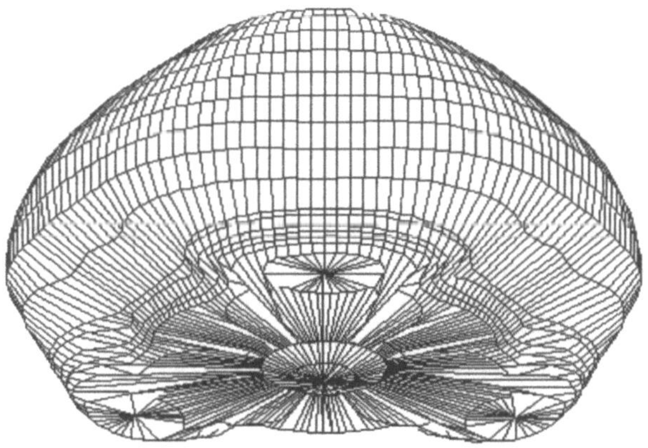

Fig. 11. ∂OE for IDS Stewart platform and dome

5. Conclusions

Necessary conditions for boundaries of three fundamentally different forms of workspaces: accessible output sets; operational envelopes; and domains of interference, are shown to be the same form of row-rank deficiency associated with a Jacobian of the expression defining admissible points in the workspace. Continuation methods that account for bifurcation of one dimensional generators of the boundary are shown to be directly applicable for each of the three workspace types treated.

Numerical results presented for planar and spatial Stewart platforms demonstrate the breadth of applicability of both analytical criteria and numerical methods for mapping workspace boundaries. Numerous curves satisfying necessary conditions are shown to exist within the interior of each of the workspaces studied. They are shown in the literature cited to represent barriers to motion control in critical configurations of the underlying mechanism. The outer boundary of the workspace is the envelope of all such singular surfaces.

Acknowledgement

This research was funded by the US Army Tank Automotive Command Center (TACOM) through the Automotive Research Center (Department of Defense contract number DAAE07-94-C-R094).

References

Adkins, F.A. and Haug, E.J. (1997): Operational Envelope of a Spatial Stewart Platform. Journal of Mechanical Design, **119**: 330-332

Allgower, E.L. and Georg, K. (1990): Numerical Continuation Methods, Springer-Verlag, Berlin

Cameron, S.A. (1985): A Study of the Clash Detection Problem in Robotics. Proc. IEEE International Conference on Robotics and Automation: 488-493

Dongarra, J.J., Moler, C.B., Bunch, J.R., and Stewart, G.W. (1979): LINPACK Users' Guide, SIAM, Philadelphia

Freudenstein, F. and Primrose, E.J. (1984): On the Analysis and Synthesis of the Workspace of a Three-Link, Turning-Pair Connected Robot Arm. Journal of Mechanisms, Transmissions, and Automation in Design, **106**: 365-370

Gilbert, E.G., Johnson, D.W., and Keerthi, S.S. (1988): A Fast Procedure for Computing the Distance Between Comples Objects in Three Dimensional Space. IEEE Journal of Robotics and Automation, **4(2)**: 193-203

Griewank, A., Juedes, D., Yoder, D., and Utke, J. (1994): ADOL-C: A Package for the Automatic Differentiation of Algorithms Written in C/C++, Version 1.5. ACM-TOMS

Haug, E.J. (1989): Computer Aided Kinematics and Dynamics. Prentice-Hall, Englewood Cliffs, New Jersey

Haug, E.J., Luh, C.-M., Adkins, F.A., and Wang, J.Y. (1996): Numerical Algorithms for Mapping Boundaries of Manipulator Workspaces. Journal of Mechanical Design, **118**: 228-234

Haug, E.J., Wang, J.Y., and Wu, J.K. (1992): Dextrous Workspaces of Manipulators; I, Analytical Criteria. Mechanics of Structures and Machines, **20(3)**: 321-361

Jo, D.Y. and Haug, E.J. (1989a): Workspace Analysis of Multibody Mechanical Systems Using Continuation Methods. Journal of Mechanisms, Transmissions, and Automation in Design, **3(4)**: 581-589

Jo, D.Y. and Haug, E.J. (1989b): Workspace Analysis of Closed-Loop Mechanisms with Unilateral Constraints. Advances in Design Automation, ASME DE, **19(3)**: 53-60

Kumar, A. and Waldron, K.J. (1981): The Workspaces of a Mechanical Manipulator. Journal of Mechanical Design, **103**: 665-672

Lee, D.T. and Preparata, F.P. (1984): Computational Geometry - A Survey. IEEE Trans. Computation, C-33, 1072-1101

Litvin, F.L. (1980): Application of Theorem of Implicit Function System Existence for Analysis and Synthesis of Linkages. Mechanism and Machine Theory, **15**: 115-125

Nielsen, L., de Wit, C.C., and Hagander, P. (1991): Controllability Issues of Robots in Singular Configurations. Proceedings of the 1991 IEEE International Conference on Robotics and Automation: 2210-2215

Rheinboldt, W.C. (1986): Numerical Analysis of Parameterized Nonlinear Equations. John Wiley & Sons, New York

Spanos, J. and Kohli, D. (1985): Workspace Analysis of Regional Structures of Manipulators. Journal of Mechanisms, Transmissions, and Automation in Design, **107**: 216-222

Tsai, M.J. and Waldron, K.J. (1987): Operating Barriers within the Workspaces of Manipulators. Society of Manufacturing Engineers, Robots II/17th ISIR, Chicago: 27-30

Tsai, Y.C. and Soni, A.H. (1981): Accessible Region and Synthesis of Robot Arms. Journal of Mechanical Design, **103**: 803-811

Waldron, K.J., Wang, S., and Bolin, S.J. (1985): A Study of the Jacobian Matrix of Serial Manipulators. Journal of Mechanisms, Transmissions, and Automation in Design, **107**: 230-238

Wang, J.Y. and Wu, J.K. (1993): Dextrous Workspaces of Manipulators, Part 2: Computational Methods. Mechanics of Structures and Machines, **21(4)**: 471-506

Yang, D.C.H. and Lee, T.W. (1983): On the Workspace of Mechanical Manipulators. Journal of Mechanisms, Transmissions, and Automation in Design, **105**: 62-69

PART II
Dynamics and Control of Rigid-Body Systems

Equations of Motion in Tensor Variables and Their Application to Multibody Systems

Ljubomir Lilov[1], Nikolina Vassileva[2]

[1]Faculty of Math. and Inf., Sofia University, 5 J. Bourchier St., BG-1126 Sofia, Bulgaria
lkl@ucc.uni-sofia.bg
[2]Department of Math., University of Mining and Geology, BG-1156 Sofia, Bulgaria
nmv@staff.mgu.bg

Summary. Two basic methods for deriving motion equations are used in analytical mechanics: the Newton-Euler approach and the Lagrangian approach. A well-known advantage of the first method is the description of the system's behavior by means of vector and tensor quantities with clear geometrical and mechanical sense such as mass center, inertia tensor, angular velocity, momentum, moment of momentum, etc. A disadvantage of this method is the necessary introduction of the constraint forces in the motion equations and then their elimination which very often turns out to be a complicated problem. On the other hand the second approach for deriving the motion equations – the Lagrangian formalism – uses as generalized parameters only scalar quantities which usually have a formal character and do not reflect directly the geometry and the dynamics of the system considered. In this chapter the Lagrangian formalism is generalized in such a way that not only scalar but also vector and tensor quantities can be chosen directly as generalized parameters. The derived tensor Lagrangian equations unify the advantages of both the basic methods and open new opportunities for investigations especially in the field of multibody systems. They have clear geometrical and mechanical meaning and build an effective base for analytical and numerical analysis of the system's behavior.

Keywords. Newton-Euler equations, Lagrangian equations, tensor generalized parameters, quasi-coordinates, holonomic and nonholonomic systems, multibody systems.

1. Introduction

Two basic methods are used in analytical mechanics for deriving motion equations: the Newton-Euler approach and the Lagrangian approach. A well-known advantage of the first method is the description of the system's behaviour by means of vector and tensor quantities with clear mechanical sense such as mass center, inertia tensor, angular velocity, momentum, moment of momentum, etc. At the same time additional work has to be done to first introduce the reaction forces and later to eliminate them from the motion equations and, what is worse, this process can not be automated.

The most important advantage of the second type of methods is the applicability of different computing techniques in the very beginning of the investigation - the generating of motion equations. According to d'Alembert's principle, for instance, the motion equations of an arbitrary mechanical system under ideal constraints can be derived from the relation

$$\int(\ddot{\mathbf{r}}dm - d\mathbf{f})\cdot\delta\mathbf{r} = 0\,, \tag{1}$$

where dm is the mass of a typical system's particle P, $d\mathbf{f}$ is the total force applied on P, $\delta\mathbf{r}$ is the virtual displacement of P, $\ddot{\mathbf{r}}$ is the acceleration of P and $\int(...)$ is summation over all of the particles of the considered discrete and/or continuous system. The position of P is given by $\overline{OP} = \mathbf{r} = \mathbf{r}(t, q_i)$, where O is a fixed point in the inertial space, $q_1, q_2, \ldots, q_n$ are some generalized parameters and the virtual displacement $\delta\mathbf{r}$ of P is defined as

$$\delta\mathbf{r} = \sum_{i=1}^{n}\frac{\partial\mathbf{r}}{\partial q_i}\delta q_i\,. \tag{2}$$

The expression (1) can be rewritten with (2) as

$$\sum_{i=1}^{n}\left(\int\frac{\partial\mathbf{r}}{\partial q_i}\cdot\ddot{\mathbf{r}}\,dm\right)\delta q_i = \sum_{i=1}^{n}\left(\int\frac{\partial\mathbf{r}}{\partial q_i}\cdot d\mathbf{f}\right)\delta q_i \tag{3}$$

and further represented in the Lagrangian form

$$\sum_{i=1}^{n}E_i\delta q_i = \sum_{i=1}^{n}Q_i\delta q_i\,, \tag{4}$$

where $E_i = \frac{d}{dt}\frac{\partial T}{\partial \dot{q}_i} - \frac{\partial T}{\partial q_i}$ is the Euler/Lagrange operator, $T = \frac{1}{2}\int\dot{\mathbf{r}}^2 dm$ is the kinetic energy of the system, $Q_i = \int\frac{\partial\mathbf{r}}{\partial q_i}\cdot d\mathbf{f}$ is the generalized force corresponding to the virtual displacement δq_i.

The main problem in the derivation of motion equations is how to construct independent equations, starting from the identity (4). It could be said that the way of splitting the equality (4) into separate equations, forming with the constraint equation a minimal or redundant set of final equations, is the division point between the different forms of motion equations of the Lagrangian type.

In the case of independent generalized parameters (generalized coordinates) $q_1, q_2, \ldots, q_n$ the motion equations follow from Eq. (4) in the form of Lagrange's equations $E_i = Q_i$ ($i=1,\ldots,n$) (Lagrange 1788). If the generalized parameters are not independent or if the mechanical system is nonholonomic, then the derivatives of the generalized parameters $q_1, q_2, \ldots, q_n$ satisfy m linear equations ($0 \le m < n$)

$$\sum_{j=1}^{n} a_{ij}\dot{q}_j + a_i = 0 \qquad (i=1,...,m) \tag{5}$$

and the variations δq_i of q_i (i=1,2,...,n) satisfy the homogenous system

$$\sum_{j=1}^{n} a_{ij}\delta q_j = 0 \qquad (i=1,...,m). \tag{6}$$

The form of the solution of equations (6) and the way of excluding the virtual displacements δq_i from the identity (4) result in different classical forms of motion equations such as Maggi's equations (Maggi 1901), Lagrangian equations with multipliers, Boltzmann-Hamel's equations (Boltzmann 1902, Hammel 1904) and so on.

The Lagrangian approach allows one to avoid, if desired, the determination of the constraints' reactions and with the help of a minimal number of parameters to directly write down the motion equations. The form of these equations is invariant and does not depend on the chosen generalized parameters. Its main disadvantage is the use of only scalar quantities as generalized parameters which usually have a formal character and do not reflect directly the geometry and the dynamics of the system considered. This leads to a complex form of the differential equations and creates difficulties not only in the analytical but even in the numerical integration of the equations. Usually, the scalar variables are the coordinates of certain geometrical objects relative to appropriate frames of references. The possibility of singularities appearing through these representations and the large amount of variables complicates the mathematical side of the problem. At the same time the equations loose their direct physical meaning. A simple illustration is the classical case of a rigid body with a fixed point O, where the use of any set of independent parameters such as Euler angles, Bryant angles, etc., always requires the consideration of singular points — a well-known topological fact. The corresponding Lagrangian equations in these generalized coordinates have a very unsymmetric form which is inconvenient for applications. The natural representation of the position of the body in this case is through the rotation tensor $\mathbf{Z}(t)$ determining the orientation of a body-fixed orthonormal basis $\mathbf{e}_1, \mathbf{e}_2, \mathbf{e}_3$ with respect to a Cartesian frame of reference $O\,\mathbf{e}_1^{(0)}, \mathbf{e}_2^{(0)}, \mathbf{e}_3^{(0)}$, i.e.

$$\mathbf{Z} = \mathbf{e}_1\mathbf{e}_1^{(0)} + \mathbf{e}_2\mathbf{e}_2^{(0)} + \mathbf{e}_3\mathbf{e}_3^{(0)} \tag{7}$$

Then, for the position vector $\mathbf{r}$ of an arbitrary point of the body we have $\mathbf{r} = \mathbf{Z}(t)\cdot\mathbf{u}$, where $\mathbf{u}$ is a constant vector. The Euler approach in this case (the theorem of moment of momentum) leads to the equation of motion in the simple vector form $\mathbf{J}_O\cdot\dot{\boldsymbol{\omega}} + \boldsymbol{\omega}\times\mathbf{J}_O\cdot\boldsymbol{\omega} = \mathbf{M}_O$, where $\mathbf{J}_O$ and $\mathbf{M}_O$ are the inertia tensor of the body and the resultant external torque with respect to the fixed point O, respectively. As a matter of fact this equation can be obtained also through the

Lagrangian approach in the form of Boltzmann-Hamel equations introducing quasi-coordinates which have time derivatives which are the components ω_1, ω_2, ω_3 of the angular velocity $\boldsymbol{\omega} = \omega_1\mathbf{e}_1 + \omega_2\mathbf{e}_2 + \omega_3\mathbf{e}_3$. However, the quasi-coordinates have an artificial nature and are far from the clear geometrical sense of the tensor **Z**.

Contemporary problems in analytical mechanics require first and foremost the investigation of those properties inherent to the mechanical system itself and not those which are connected with the mathematical model due to the choice of the generalized parameters. It is worth noticing that sometimes authors using different approaches receive results which differ only in the form of representation from already known solutions. To establish this fact is not always an easy task as shown in (Gorr et al. 1978). In this way a conclusion can be drawn that the language of description of mechanical systems plays a major role in determining the generality of the systems and the transparency of the solutions. The use of natural geometrical characteristics which have not only scalar but also vector or tensor form is an effective way to reach greater generality and derive the equations of the dynamics in more compact form. From the large variety of illustrative investigations just a few concrete examples will be mentioned. The vector form of the motion equations received in (Wittenburg 1977) for a tree-like multibody system with ball-and-socket joints gives a simple mechanical interpretation of the dynamical interaction between the system's bodies. The determination of a large class of steady-state motions for a single loop multibody system of *n* Lagrangian gyrostats in (Kharlamov 1972) and (Lilov et al. 1984) is a result of the algebraic structure of the equations given in vector form. Vector quantities are used not only in deriving the equations but also in the investigation of the stability of steady-state motions as, for instance, in (Wittenburg et al. 1975), where the expression for the reduced potential energy and its variation are written in compact form. This approach - the use of natural vector parameters like unit vectors and position vectors of points attached to the bodies is suggested also in some problems of numerical integration, for instance in (García de Jalón et al. 1986 and Lebedev 1986). Other examples where the use of vector quantities lead to more convenient forms of motion equations are presented in the works (Everett et al. 1986, Quinn 1990, Lilov 1993, Papastavridis 1994, Barhorst et al. 1995 etc).

The given examples show that the use of a vector-and-tensor apparatus has unquestionable advantages for deriving equations and qualitative investigation of the motion They give greater compactness and simplicity of the equations derived as well as independence from the choice of the coordinate system. On the other hand the best developed and most often applied approach for deriving the motion equations — the Lagrangian formalism — uses as generalized parameters only scalar quantities. From this point of view in order to extend the application of vector and tensor quantities for investigation of the motion it is necessary to generalise the Lagrangian formalism in such a way that these quantities can be considered also as generalized parameters.

In the following exposition the necessary mathematical apparatus is developed and the corresponding equations in terms of these generalized parameters are derived. These equations can be named the tensor equations of motion. The operations and expressions in the equations have clear geometrical and mechanical meaning and build an effective base for analytical and numerical investigation of the system's behaviour.

2. General Background

Some necessary notations and formulae will be introduced before obtaining the motion equations.

Consider a three-dimensional Euclidean space $\mathfrak{R}$ with notation **a.b** for the scalar product and **ab** for the tensor (dyadic) product of two vectors **a**, $\mathbf{b}\in\mathfrak{R}$. If $\underline{\mathbf{e}}=(\mathbf{e}_1,\mathbf{e}_2,\mathbf{e}_3)^{\mathrm{T}}$ is an orthonormal basis of $\mathfrak{R}$, then the tensors $\mathbf{e}_{i_1}...\mathbf{e}_{i_m}$ of order m, (here $i_1,...,i_m$ is an arbitrary arrangement of the numbers 1, 2 and 3) generate a basis of the tensor space $\mathfrak{R}_m$ of order m, i.e. for an arbitrary tensor $\mathbf{A}\in\mathfrak{R}_m$ it can be written $\mathbf{A}=\alpha_{i_1...i_m}\mathbf{e}_{i_1}...\mathbf{e}_{i_m}$ (summation convention for repeated indices).

The following notations will be used:

2.1. Let $\mathbf{A}\in\mathfrak{R}_m$, $\mathbf{B}\in\mathfrak{R}_n$ and k, $1\le k\le\min(m,n)$, is an integer. The notation $\mathbf{A}\underset{k}{*}\mathbf{B}$ will be used for a tensor of $\mathfrak{R}_{m+n-2k}$, derived by k-series summation by the indices: the last index of **A** with the first index of **B**, the before last index of **A** with the second index of **B** and so on. The tensor $\mathbf{A}\underset{k}{*}\mathbf{B}$ will be named k-series general product (general k-product). It may be put $\mathbf{A}\underset{0}{*}\mathbf{B}=\mathbf{AB}$, $\mathbf{A}\underset{1}{*}\mathbf{B}=\mathbf{A}\cdot\mathbf{B}$.

For the tensor $\mathbf{E}_{2m}=\mathbf{e}_{i_1}...\mathbf{e}_{i_m}\mathbf{e}_{i_m}...\mathbf{e}_{i_1}$ the identities

$$\mathbf{E}_{2m}\underset{m}{*}\mathbf{A}=\mathbf{A}\underset{m}{*}\mathbf{E}_{2m}=\mathbf{A},\quad \mathbf{A}\in\mathfrak{R}_{2m},\qquad \mathbf{E}_{2m}\underset{m}{*}\mathbf{A}=\mathbf{A}\underset{m}{*}\mathbf{E}_{2m}=\mathbf{A},\quad \mathbf{A}\in\mathfrak{R}_m,$$

hold, i.e. the tensor $\mathbf{E}_{2m}$ can be considered as the unit tensor of $\mathfrak{R}_{2m}$ and the identity mapping of $\mathfrak{R}_m$.

The norm of a tensor $\mathbf{A}\in\mathfrak{R}_m$ is defined as

$$\|\mathbf{A}\|=\sqrt{\sum\alpha^2_{i_1..i_m}}=\sqrt{\mathbf{E}_{2m}\underset{2m}{*}\mathbf{A}\mathbf{A}^{\mathrm{T}}}.$$

2.2. ${}^{[k]}\mathbf{A}$, $\mathbf{A}^{[k]}$ and $\mathbf{A}^{\mathrm{T}}$ are the tensors derived from the tensor $\mathbf{A}\in\mathfrak{R}_m$ by the relations ${}^{[k]}\alpha_{i_1...i_m}=\alpha_{i_{k+1}...i_m i_1...i_k}$, $\alpha^{[k]}_{i_1...i_m}=\alpha_{i_{m-k+1}...i_m i_1...i_{m-k}}$ $(0<k<m)$,

$\alpha^{\mathrm{T}}_{i_1\ldots i_{m-1}i_m} = \alpha_{i_m i_{m-1}\ldots i_1}$, where $\alpha_{i_1\ldots i_m}$, ${}^{[k]}\alpha_{i_1\ldots i_m}$, $\alpha^{[k]}_{i_1\ldots i_m}$ and $\alpha^{\mathrm{T}}_{i_1\ldots i_m}$ are the coordinates of the tensors $\mathbf{A}$, ${}^{[k]}\mathbf{A}$, $\mathbf{A}^{[k]}$ and $\mathbf{A}^{\mathrm{T}}$ with respect to one and the same basis. If k=0 or k=m, then ${}^{[0]}\mathbf{A} = \mathbf{A}^{[0]} = {}^{[m]}\mathbf{A} = \mathbf{A}^{[m]} = \mathbf{A}$ by definition.

The following two properties of the general k-product are true (Vasileva et al.1988):

1. If $\mathbf{A} \in \Re_m$, $\mathbf{B} \in \Re_n$, then

$$\mathbf{A} \underset{k}{*} \mathbf{B} = ({}^{[k]}\mathbf{B} \underset{k}{*} \mathbf{A}^{[k]})^{[m-k]} = {}^{[n-k]}({}^{[k]}\mathbf{B} \underset{k}{*} \mathbf{A}^{[k]}); \tag{8}$$

2. If $\mathbf{B} \in \Re_n$ and $n \geq l + k$, then

$$\mathbf{A} \underset{l}{*} (\mathbf{B} \underset{k}{*} \mathbf{C}) = (\mathbf{A} \underset{l}{*} \mathbf{B}) \underset{k}{*} \mathbf{C}. \tag{9}$$

2.3. In the three-dimensional space an important role is played by the skew-symmetrical tensor $\mathbf{\Theta} \in \Re_3$. Its coordinates with respect to an arbitrary bases are

$$\theta_{ijk} = \begin{cases} (-1)^{[i,j,k]} & \text{if the indeses } i,j,k \text{ are permutations of the numbers 1,2,3} \\ 0 & \text{in the other cases} \end{cases}.$$

Here $[i,j,k]$ is the number of the inversions in the permutation (i,j,k).

For the tensor $-\mathbf{\Theta} \cdot \mathbf{A} \in \Re_{m+1}$ ($\mathbf{A} \in \Re_m$) we introduce the notation $\widetilde{\mathbf{A}}$. For example, if $\mathbf{a} \neq \mathbf{0}$, then $\widetilde{\mathbf{a}} \cdot \mathbf{b} = \mathbf{a} \times \mathbf{b}$ and the solution of the equation $\mathbf{a} \cdot \mathbf{x} = 0$ is presented in the form $\mathbf{x} = \widetilde{\mathbf{a}} \cdot \mathbf{p}$, where $\mathbf{p}$ is an arbitrary vector.

We should remark that from the relationship $\mathbf{\Theta} \underset{2}{*} \mathbf{\Theta} = -2\mathbf{E}_2$, which may be checked directly, the following expression for an arbitrary vector $\mathbf{a}$

$$\mathbf{a} = -\frac{1}{2}(\mathbf{\Theta} \underset{2}{*} \mathbf{\Theta}) \cdot \mathbf{a} = \frac{1}{2}\mathbf{\Theta} \underset{2}{*} (-\mathbf{\Theta} \cdot \mathbf{a}) = \frac{1}{2}\mathbf{\Theta} \underset{2}{*} \widetilde{\mathbf{a}} \tag{10}$$

holds.

2.4. Let, in the open nonvoid set $U \in \Re_s$ a continuous mapping f:$U \to \Re_n$ be given. Considering the space $\Re_{s+n}$ as the space of the continuous linear mappings of $\Re_s$ into $\Re_n$, we define the derivative $\dfrac{df}{d\mathbf{T}}$ of the function f at the point $\mathbf{T} \in U$ as a tensor $\mathbf{A} \in \Re_{s+n}$, satisfying the equality $f(\mathbf{T}+\mathbf{H}) = f(\mathbf{T}) + \mathbf{A} \underset{s}{*} \mathbf{H} + o(\mathbf{H})$, where $\lim_{\|\mathbf{H}\|\to 0} \frac{\|o(\mathbf{H})\|}{\|\mathbf{H}\|} = 0$, for every **H**, **T**+**H**$\in U$. The differential $df \in \mathfrak{R}_n$ of the function $f(\mathbf{T})$ is given by the expression $df = \frac{df}{d\mathbf{T}} \underset{s}{*} d\mathbf{T}$ and is a function of the variables $\mathbf{T}\in U$ and $d\mathbf{T}\in \mathfrak{R}_s$.

Some of the rules of differentiation are (Vasileva et al. 1988):

1. $\frac{d\mathbf{T}}{d\mathbf{T}} = \mathbf{E}_{2s}$

This property follows from the equality $\mathbf{T}+\mathbf{H} = \mathbf{T}+\mathbf{E}_{2s} \underset{s}{*} \mathbf{H}$ which is satisfied by an arbitrary tensor $\mathbf{H}\in \mathfrak{R}_s$.

2. $\frac{d}{d\mathbf{T}}(f \underset{k}{*} g) = {}^{[s]}\left[\frac{df}{d\mathbf{T}}^{[s]} \underset{k}{*} g\right] + f \underset{k}{*} \frac{dg}{d\mathbf{T}}$ or

$$\frac{d}{d\mathbf{T}}(f \underset{k}{*} g) = \left[{}^{[k]}g \underset{k}{*} {}^{[m-k]}\frac{df}{d\mathbf{T}}\right]^{[m-k]} + f \underset{k}{*} \frac{dg}{d\mathbf{T}}, \tag{11}$$

suppose that the functions f: $\mathfrak{R}_s \to \mathfrak{R}_m$ and g: $\mathfrak{R}_s \to \mathfrak{R}_n$ are differentiable at the point $\mathbf{T}\in U$, $U\subset \mathfrak{R}_s$, and $0 \le k \le \min(m,n)$ is an integer. In particular, in the case when the tensor **A** is independent of the variable $\mathbf{T}\in \mathfrak{R}_s$, then

$$\frac{d}{d\mathbf{T}}(\mathbf{A} \underset{k}{*} \mathbf{T}) = \mathbf{A} \underset{k}{*} \mathbf{E}_{2s}.$$

3. The derivative $\dot{\mathbf{Y}}$ and the differential $d\mathbf{Y}$ of the function $\mathbf{Y}(t) = \mathbf{Y}(t, \mathbf{X}_1(t), \ldots, \mathbf{X}_n(t))$, where t is a real variable and $\mathbf{X}_i(t) \in \mathfrak{R}_{\mathrm{v}_i}$, can be written as

$$\dot{\mathbf{Y}} = \frac{d\mathbf{Y}}{dt} = \frac{\partial \mathbf{Y}}{\partial t} + \sum_{i=1}^{n} \frac{\partial \mathbf{Y}}{\partial \mathbf{X}_i} \underset{\mathrm{v}_i}{*} \dot{\mathbf{X}}_i, \tag{12}$$

$$d\mathbf{Y} = \frac{\partial \mathbf{Y}}{\partial t} dt + \sum_{i=1}^{n} \frac{\partial \mathbf{Y}}{\partial \mathbf{X}_i} \underset{\mathrm{v}_i}{*} d\mathbf{X}_i, \tag{13}$$

where $\frac{\partial \mathbf{Y}}{\partial t}$ and $\frac{\partial \mathbf{Y}}{\partial \mathbf{X}_i}$ are the partial derivatives of $\mathbf{Y}$.

4. The differentiation by $\dot{\mathbf{X}}_i$ of the Eq. (12) yields

$$\frac{\partial \dot{\mathbf{Y}}}{\partial \dot{\mathbf{X}}_i} = \frac{\partial \mathbf{Y}}{\partial \mathbf{X}_i} \underset{\nu_i}{*} \mathbf{E}_{2\nu_i} = \frac{\partial \mathbf{Y}}{\partial \mathbf{X}_i}. \tag{14}$$

2.5. Let the rotation tensor (7), determining the orientation of a given orthonormal basis $\underline{\mathbf{e}} = (\mathbf{e}_1, \mathbf{e}_2, \mathbf{e}_3)^{\mathrm{T}}$ in the inertial space is a function of the time t. The following well-known relation between the derivative $\dot{\mathbf{Z}}$ and the angular velocity $\boldsymbol{\omega}$ of the basis takes place

$$\tilde{\boldsymbol{\omega}} = \dot{\mathbf{Z}} \cdot \mathbf{Z}^{\mathrm{T}} = -\mathbf{Z} \cdot \dot{\mathbf{Z}}^{\mathrm{T}} = \dot{\mathbf{e}}_1 \mathbf{e}_1 + \dot{\mathbf{e}}_2 \mathbf{e}_2 + \dot{\mathbf{e}}_3 \mathbf{e}_3.$$

From the Eq. (10) the expression

$$\boldsymbol{\omega} = \frac{1}{2} \boldsymbol{\Theta} \underset{2}{*} (\dot{\mathbf{Z}} \cdot \mathbf{Z}^{\mathrm{T}}) = \frac{1}{2} (\mathbf{e}_1 \times \dot{\mathbf{e}}_1 + \mathbf{e}_2 \times \dot{\mathbf{e}}_2 + \mathbf{e}_3 \times \dot{\mathbf{e}}_3)$$

can be obtained, by means of which the velocity $\boldsymbol{\omega}$ is represented in the form

$$\boldsymbol{\omega} = \mathbf{e}_1 \times \dot{\mathbf{e}}_1 - (\dot{\mathbf{e}}_3 \cdot \mathbf{e}_2) \mathbf{e}_1 = \mathbf{e}_2 \times \dot{\mathbf{e}}_2 - (\dot{\mathbf{e}}_1 \cdot \mathbf{e}_3) \mathbf{e}_2 = \mathbf{e}_3 \times \dot{\mathbf{e}}_3 - (\dot{\mathbf{e}}_2 \cdot \mathbf{e}_1) \mathbf{e}_3 .$$

Another connection between $\mathbf{Z}$, $\dot{\mathbf{Z}}$, and $\boldsymbol{\omega}$ is the relationship (Vasileva et al. 1988):

$$\tilde{\mathbf{Z}} \underset{2}{*} {}^{[1]}\!\left(\frac{\partial \boldsymbol{\omega}}{\partial \dot{\mathbf{Z}}} \right) = \mathbf{E}_2 . \tag{15}$$

2.6. Let $\mathbf{a}_1 = \mathbf{a}_1(t)$, $\mathbf{a}_2 = \mathbf{a}_2(t)$, $\mathbf{a}_3 = \mathbf{a}_3(t)$ be linearly independent vectors. Introducing the reciprocal vectors

$$\hat{\mathbf{a}}_1 = \frac{\mathbf{a}_2 \times \mathbf{a}_3}{\mathbf{a}_1 \cdot (\mathbf{a}_2 \times \mathbf{a}_3)}, \quad \hat{\mathbf{a}}_2 = \frac{\mathbf{a}_3 \times \mathbf{a}_1}{\mathbf{a}_1 \cdot (\mathbf{a}_2 \times \mathbf{a}_3)}, \quad \hat{\mathbf{a}}_2 = \frac{\mathbf{a}_1 \times \mathbf{a}_2}{\mathbf{a}_1 \cdot (\mathbf{a}_2 \times \mathbf{a}_3)} \tag{16}$$

and taking into account the relations

$$\mathbf{a}_i \cdot \hat{\mathbf{a}}_j = \begin{cases} 1 & \text{if } i = j \\ 0 & \text{if } i \neq j \end{cases}, \quad \text{i.e.} \quad \underline{\mathbf{a}} \cdot \underline{\hat{\mathbf{a}}}^{\mathrm{T}} = \begin{pmatrix} 1 & 0 & 0 \\ 0 & 1 & 0 \\ 0 & 0 & 1 \end{pmatrix} = \underline{E}_2, \tag{17}$$

we can express the coefficients x_1, x_2, x_3 in the representation

$$\mathbf{x} = x_1 \mathbf{a}_1 + x_2 \mathbf{a}_2 + x_3 \mathbf{a}_3 \tag{18}$$

for an arbitrary vector $\mathbf{x}$ in the form $x_i = \mathbf{x} \cdot \hat{\mathbf{a}}_i$. Therefore,

$$\mathbf{x} = \mathbf{a}_1 (\hat{\mathbf{a}}_1 \cdot \mathbf{x}) + \mathbf{a}_2 (\hat{\mathbf{a}}_2 \cdot \mathbf{x}) + \mathbf{a}_3 (\hat{\mathbf{a}}_3 \cdot \mathbf{x}) = (\mathbf{a}_1 \hat{\mathbf{a}}_1 + \mathbf{a}_2 \hat{\mathbf{a}}_2 + \mathbf{a}_3 \hat{\mathbf{a}}_3) \cdot \mathbf{x} .$$

Consider the tensors

$$\mathbf{A} = \mathbf{a}_1\mathbf{e}_1^{(0)} + \mathbf{a}_2\mathbf{e}_2^{(0)} + \mathbf{a}_3\mathbf{e}_3^{(0)} \quad \text{and} \quad \hat{\mathbf{A}} = \mathbf{e}_1^{(0)}\hat{\mathbf{a}}_1 + \mathbf{e}_2^{(0)}\hat{\mathbf{a}}_2 + \mathbf{e}_3^{(0)}\hat{\mathbf{a}}_3.$$

It follows from $\mathbf{A}\cdot\mathbf{e}_i^{(0)} = \mathbf{a}_i$ that the tensor $\mathbf{A}$ defines a linear mapping transforming the vectors $\mathbf{e}_i^{(0)}$ to the vectors $\mathbf{a}_i$, hence, determining the position of the basis $\underline{\mathbf{a}}$. It is easy to check the relationships

$$\hat{\mathbf{A}}\cdot\mathbf{A} = \mathbf{E}_2,\ \mathbf{A}\cdot\hat{\mathbf{A}} = \mathbf{a}_1\hat{\mathbf{a}}_1 + \mathbf{a}_2\hat{\mathbf{a}}_2 + \mathbf{a}_3\hat{\mathbf{a}}_3.$$

In this way it can be written $\mathbf{x} = (\mathbf{A}\cdot\hat{\mathbf{A}})\cdot\mathbf{x} = \mathbf{A}\cdot(\hat{\mathbf{A}}\cdot\mathbf{x}) = \mathbf{A}\cdot\mathbf{u}$, where

$$\mathbf{u} = \hat{\mathbf{A}}\cdot\mathbf{x} = \mathbf{e}_1^{(0)}(\hat{\mathbf{a}}_1\cdot\mathbf{x}) + \mathbf{e}_2^{(0)}(\hat{\mathbf{a}}_2\cdot\mathbf{x}) + \mathbf{e}_3^{(0)}(\hat{\mathbf{a}}_3\cdot\mathbf{x}) = \mathbf{e}_1^{(0)}x_1 + \mathbf{e}_2^{(0)}x_2 + \mathbf{e}_3^{(0)}x_3.$$

Having in mind Eq. (18), we notice that the coordinates of $\mathbf{x}$ with respect to the basis $\underline{\mathbf{a}}$ coincide with the coordinates of $\mathbf{u}$ with respect to the inertial orthonormal basis $\underline{\mathbf{e}}^{(0)}$. For the derivative $\dot{\mathbf{x}}$ we have

$$\dot{\mathbf{x}} = \dot{\mathbf{A}}\cdot\mathbf{u} + \mathbf{A}\cdot\dot{\mathbf{u}} = \dot{\mathbf{A}}\cdot(\hat{\mathbf{A}}\cdot\mathbf{A})\cdot\mathbf{u} + \mathbf{A}\cdot\dot{\mathbf{u}} = (\dot{\mathbf{A}}\cdot\hat{\mathbf{A}})\cdot(\mathbf{A}\cdot\mathbf{u}) + \mathbf{A}\cdot\dot{\mathbf{u}} = \mathbf{W}\cdot\mathbf{x} + \mathbf{A}\cdot\dot{\mathbf{u}},$$

where

$$\dot{\mathbf{A}}\cdot\hat{\mathbf{A}} = \dot{\mathbf{a}}_1\hat{\mathbf{a}}_1 + \dot{\mathbf{a}}_2\hat{\mathbf{a}}_2 + \dot{\mathbf{a}}_3\hat{\mathbf{a}}_3 = \mathbf{W}. \tag{19}$$

As

$$\mathbf{A}.\dot{\mathbf{u}} = (\mathbf{a}_1\mathbf{e}_1^{(0)} + \mathbf{a}_2\mathbf{e}_2^{(0)} + \mathbf{a}_3\mathbf{e}_3^{(0)})\cdot\left[\dot{x}_1\mathbf{e}_1^{(0)} + \dot{x}_2\mathbf{e}_2^{(0)} + \dot{x}_3\mathbf{e}_3^{(0)}\right] = \dot{x}_1\mathbf{a}_1 + \dot{x}_2\mathbf{a}_2 + \dot{x}_3\mathbf{a}_3,$$

it is natural to call the vector $\overset{\circ}{\mathbf{x}} = \mathbf{A}\cdot\dot{\mathbf{u}}$ a derivative of $\mathbf{x}$ with respect to the basis $\mathbf{a}_1, \mathbf{a}_2, \mathbf{a}_3$. In this way

$$\dot{\mathbf{x}} = \mathbf{W}\cdot\mathbf{x} + \overset{\circ}{\mathbf{x}}. \tag{20}$$

If the basis $\mathbf{a}_1, \mathbf{a}_2, \mathbf{a}_3$ is orthonormal, i.e. **A=Z,** and $\boldsymbol{\omega}$ is its angular velocity, then $\mathbf{W} = \dot{\mathbf{a}}_1\mathbf{a}_1 + \dot{\mathbf{a}}_2\mathbf{a}_2 + \dot{\mathbf{a}}_3\mathbf{a}_3 = \tilde{\boldsymbol{\omega}}$, $\mathbf{W}\cdot\mathbf{x} = (\dot{\mathbf{a}}_1\mathbf{a}_1 + \dot{\mathbf{a}}_2\mathbf{a}_2 + \dot{\mathbf{a}}_3\mathbf{a}_3)\cdot\mathbf{x} = \tilde{\boldsymbol{\omega}}\cdot\mathbf{x} = \boldsymbol{\omega}\times\mathbf{x}$. In a similar way, every tensor $\mathbf{X}\in\mathfrak{R}_2$ can be represented in the form $\mathbf{X} = \mathbf{A}\cdot\mathbf{U}\cdot\mathbf{A}^{\mathrm{T}}$, where $\mathbf{U} = \hat{\mathbf{A}}\cdot\mathbf{X}\cdot\hat{\mathbf{A}}^{\mathrm{T}}$ is a tensor, such that its coordinate matrix relative to the basis $\underline{\mathbf{e}}^{(0)}$ is the same as the coordinate matrix $\underline{X}$ of the tensor $\mathbf{X}$ relative to the basis $\underline{\mathbf{a}}$. Differentiating we obtain

$$\dot{\mathbf{X}} = \dot{\mathbf{A}}\cdot\mathbf{U}\cdot\mathbf{A}^{\mathrm{T}} + \mathbf{A}\cdot\mathbf{U}\cdot\dot{\mathbf{A}}^{\mathrm{T}} + \mathbf{A}\cdot\dot{\mathbf{U}}\cdot\mathbf{A}^{\mathrm{T}} =$$

$$= \dot{\mathbf{A}}\cdot(\hat{\mathbf{A}}\cdot\mathbf{A})\cdot\mathbf{U}\cdot\mathbf{A}^{\mathrm{T}} + \mathbf{A}\cdot\mathbf{U}\cdot(\hat{\mathbf{A}}\cdot\mathbf{A})^{\mathrm{T}}\cdot\dot{\mathbf{A}}^{\mathrm{T}} + \mathbf{A}\cdot\dot{\mathbf{U}}\cdot\mathbf{A}^{\mathrm{T}} =$$

$$= \dot{\mathbf{A}}\cdot\hat{\mathbf{A}}\cdot\mathbf{X} + \mathbf{X}\cdot(\dot{\mathbf{A}}\cdot\hat{\mathbf{A}})^{\mathrm{T}} + \mathbf{A}\cdot\dot{\mathbf{U}}\cdot\mathbf{A}^{\mathrm{T}}.$$

The tensor $\overset{\circ}{\mathbf{X}} = \mathbf{A}\cdot\dot{\mathbf{U}}\cdot\mathbf{A}^{\mathrm{T}} = \underline{\mathbf{a}}\,\underline{\mathbf{e}}^{(0)^{\mathrm{T}}}\cdot(\underline{\mathbf{e}}^{(0)}\,\underline{\dot{X}\mathbf{e}}^{(0)^{\mathrm{T}}})\cdot\underline{\mathbf{e}}^{(0)}\,\underline{\mathbf{a}}^{\mathrm{T}} = \underline{\mathbf{a}}\,\underline{\dot{X}}\,\underline{\mathbf{a}}^{\mathrm{T}}$ may be called the time-derivative of the tensor **X** with respect to $\underline{\mathbf{a}}$. Hence, we can write $\dot{\mathbf{X}} = \mathbf{W}\cdot\mathbf{X} + \mathbf{X}\cdot\mathbf{W}^{\mathrm{T}} + \overset{\circ}{\mathbf{X}}$. The last formula together with Eq. (20) are generalizations of the relationships

$$\dot{\mathbf{X}} = \tilde{\boldsymbol{\omega}}\cdot\mathbf{X} - \mathbf{X}\cdot\tilde{\boldsymbol{\omega}} + \overset{\circ}{\mathbf{X}}, \qquad \dot{\mathbf{x}} = \tilde{\boldsymbol{\omega}}\cdot\mathbf{x} + \overset{\circ}{\mathbf{x}} \tag{21}$$

between the time derivatives of a tensor and a vector relative to two orthonormal bases.

2.7. Let the tensor $\mathbf{A} = \mathbf{a}_1\mathbf{e}_1^{(0)} + \mathbf{a}_2\mathbf{e}_2^{(0)} + \mathbf{a}_3\mathbf{e}_3^{(0)}$ be an explicit and implicit (through the tensors $\mathbf{X}_1 = \mathbf{X}_1(t), \ldots, \mathbf{X}_n = \mathbf{X}_n(t)$) function of the time t, i.e. $\mathbf{A}(t) = \mathbf{A}(t, \mathbf{X}_1(t), \ldots, \mathbf{X}_n(t))$. If the vector **x** is invariant with respect to the basis $\underline{\mathbf{a}}$, i.e. $\overset{\circ}{\mathbf{x}} = \mathbf{0}$, then $\dot{\mathbf{x}} = \mathbf{W}\cdot\mathbf{x}$ and its partial derivatives, according to Eqs. (14), (11) and (8), are

$$\frac{\partial\mathbf{x}}{\partial\mathbf{X}_k} = \frac{\partial\dot{\mathbf{x}}}{\partial\dot{\mathbf{X}}_k} = \left(\mathbf{x}\cdot{}^{[1]}\frac{\partial\mathbf{W}}{\partial\dot{\mathbf{X}}_k}\right)^{[1]}, \qquad {}^{[1]}\frac{\partial\dot{\mathbf{x}}}{\partial\dot{\mathbf{X}}_k} = \mathbf{x}\cdot{}^{[1]}\frac{\partial\mathbf{W}}{\partial\dot{\mathbf{X}}_k} = {}^{[2]}\frac{\partial\mathbf{W}}{\partial\dot{\mathbf{X}}_k}\cdot\mathbf{x} \tag{22}$$

If the tensor **A**=**Z** is a rotation tensor and **ω** is its corresponding angular velocity the relationships

$$\dot{\mathbf{x}} = \tilde{\boldsymbol{\omega}}.\mathbf{x} = -\tilde{\mathbf{x}}.\boldsymbol{\omega},$$

$${}^{[1]}\frac{\partial\mathbf{x}}{\partial\mathbf{X}_k} = {}^{[1]}\frac{\partial\dot{\mathbf{x}}}{\partial\dot{\mathbf{X}}_k} = {}^{[2]}\frac{\partial\tilde{\boldsymbol{\omega}}}{\partial\dot{\mathbf{X}}_k}\cdot\mathbf{x} = -{}^{[2]}\left(\boldsymbol{\Theta}\cdot\frac{\partial\boldsymbol{\omega}}{\partial\dot{\mathbf{X}}_k}\right)\cdot\mathbf{x} = {}^{[1]}\frac{\partial\boldsymbol{\omega}}{\partial\dot{\mathbf{X}}_k}\cdot\tilde{\mathbf{x}} \tag{23}$$

hold for an arbitrary vector **x** with constant coordinates in the orthonormal basis $\underline{\mathbf{a}}$.

2.8. We define the variation of the rotation tensor **Z**, when it is a function $\mathbf{Z} = \mathbf{Z}(t, \mathbf{X}_1, \ldots, \mathbf{X}_n)$ of the variables $t, \mathbf{X}_1, \ldots, \mathbf{X}_n$ as the differential $\delta\mathbf{Z} = \sum_{i=1}^{n}\frac{\partial\mathbf{Z}}{\partial\mathbf{X}_i}\underset{\nu_i}{*}\,\delta\mathbf{X}_i$ obtained from (13) by fixing time t. It follows from $\mathbf{Z}\cdot\mathbf{Z}^{\mathrm{T}} = \mathbf{Z}^{\mathrm{T}}\cdot\mathbf{Z} = \mathbf{E}_2$ that **Z** and its variation $\delta\mathbf{Z}$ are related by the equation $\delta\mathbf{Z}\cdot\mathbf{Z}^{\mathrm{T}} + \mathbf{Z}\cdot\delta\mathbf{Z}^{\mathrm{T}} = \mathbf{0}_2$. Since $\delta\mathbf{Z}^{\mathrm{T}} = (\delta\mathbf{Z})^{\mathrm{T}}$, then $\delta\mathbf{Z}\cdot\mathbf{Z}^{\mathrm{T}} = -\mathbf{Z}\cdot(\delta\mathbf{Z})^{\mathrm{T}} = -(\delta\mathbf{Z}\cdot\mathbf{Z}^{\mathrm{T}})^{\mathrm{T}}$, i.e. $\delta\mathbf{Z}\cdot\mathbf{Z}^{\mathrm{T}}$ is a skew-symmetric tensor, thus,

$$\delta\mathbf{Z}\cdot\mathbf{Z}^{\mathrm{T}} = \delta\widetilde{\pi}, \quad \delta\mathbf{Z} = \delta\widetilde{\pi}\cdot\mathbf{Z} = \delta\pi\cdot\widetilde{\mathbf{Z}}, \tag{24}$$

where $\delta\pi$ is a vector.

Let $\mathbf{x} = \mathbf{Z}(t)\cdot\mathbf{u}\,(t)$. Then $\delta\mathbf{x} = \delta\mathbf{Z}\cdot\mathbf{u} + \mathbf{Z}\cdot\delta\mathbf{u} = \delta\widetilde{\pi}\cdot\mathbf{x} + \mathbf{Z}\cdot\delta\mathbf{u}$ and

$$\mathbf{u} = (\mathbf{x}\cdot\mathbf{e}_1)\mathbf{e}_1^{(0)} + (\mathbf{x}\cdot\mathbf{e}_2)\mathbf{e}_2^{(0)} + (\mathbf{x}\cdot\mathbf{e}_3)\mathbf{e}_3^{(0)},$$

$$\delta\mathbf{u} = \delta(\mathbf{x}\cdot\mathbf{e}_1)\mathbf{e}_1^{(0)} + \delta(\mathbf{x}\cdot\mathbf{e}_2)\mathbf{e}_2^{(0)} + \delta(\mathbf{x}\cdot\mathbf{e}_3)\mathbf{e}_3^{(0)},$$

$$\mathbf{Z}\cdot\delta\mathbf{u} = \delta(\mathbf{x}\cdot\mathbf{e}_1)\mathbf{e}_1 + \delta(\mathbf{x}\cdot\mathbf{e}_2)\mathbf{e}_2 + \delta(\mathbf{x}\cdot\mathbf{e}_3)\mathbf{e}_3,$$

i.e. the coordinates of $\mathbf{Z}\cdot\delta\mathbf{u}$ relative to the basis $\underline{\mathbf{e}}$ are equal to the variation of the coordinates of $\mathbf{x}$ relative to $\underline{\mathbf{e}}$. In this way the vector $\overset{\circ}{\delta}\mathbf{x} = \mathbf{Z}\cdot\delta\mathbf{u}$ may be called the variation of the vector $\mathbf{x}$ relative to the basis $\underline{\mathbf{e}}$ and (Wittenburg et al. 1975, Everett et al. 1986, Lilov 1993, Barhorst et al. 1995) $\delta\mathbf{x} = \delta\widetilde{\pi}\cdot\mathbf{x} + \overset{\circ}{\delta}\mathbf{x}$, where $\delta\pi$ is given by the Eq. (24).

It can be shown that if $\mathbf{Z}_2 = \mathbf{\Phi}\cdot\mathbf{Z}_1$, where $\mathbf{Z}_2$, $\mathbf{\Phi}$ and $\mathbf{Z}_1$ are rotation tensors, and $\delta\mathbf{Z}_2 = \delta\widetilde{\pi}_2\cdot\mathbf{Z}_2$, $\delta\mathbf{\Phi} = \delta\widetilde{\kappa}\cdot\mathbf{\Phi}$, $\delta\mathbf{Z}_1 = \delta\widetilde{\pi}_1\cdot\mathbf{Z}_1$ are their variations, then $\delta\widetilde{\pi}_2 = \delta\widetilde{\kappa} + \delta\widetilde{\pi}_1$.

3. Equations of Motion

The motion of a mechanical system is entirely determined in the inertial frame $O\mathbf{e}_1^{(0)}, \mathbf{e}_2^{(0)}, \mathbf{e}_3^{(0)}$ if the position vector of an arbitrary point M of the system is given as a function of the time. In many cases it is convenient to introduce an additional frame of reference $O'\mathbf{a}_1,\mathbf{a}_2,\mathbf{a}_3$, relative to which the motion of the point M is known. If $\mathbf{r}_{O'} = \overline{OO'}$ and $\mathbf{A} = \mathbf{a}_1\mathbf{e}_1^{(0)} + \mathbf{a}_2\mathbf{e}_2^{(0)} + \mathbf{a}_3\mathbf{e}_3^{(0)}$ is the tensor of the angular position of this basis, then $\mathbf{r} = \mathbf{r}_{O'} + \boldsymbol{\rho}$, $\dot{\mathbf{r}} = \dot{\mathbf{r}}_{O'} + \dot{\boldsymbol{\rho}} = \dot{\mathbf{r}}_{O'} + \overset{\circ}{\boldsymbol{\rho}} + \mathbf{W}\cdot\boldsymbol{\rho}$ (see Eq.(20)), where $\mathbf{W}$ is defined by Eq. (19).

Let us assume that the points of the system are grouped in N sets $S_1,\ldots,S_N$ (discrete or continuous) whose motions are determined with respect to N coordinate systems with origins denoted by $P_1,\ldots,P_N$. The tensors $\mathbf{A}_1,\ldots,\mathbf{A}_N$ describing the angular orientations of the coordinate systems can be represented in the form $\mathbf{A}_i = \mathbf{a}_1^{(i)}\mathbf{e}_1^{(0)} + \mathbf{a}_2^{(i)}\mathbf{e}_2^{(0)} + \mathbf{a}_3^{(i)}\mathbf{e}_3^{(0)}$ (i=1,...,N). For simplicity we suppose that the point sets $S_1,\ldots,S_N$ are fixed in the corresponding frames. In this case every set of tensor quantities determining the positions of the reference frames in the inertial space forms a set of parameters determining the system. The constraints (holonomic or linear nonholonomic) are supposed to be written in a differential form. Often, there are simple relationships between the vector and tensor parameters such as $\mathbf{e}\cdot\mathbf{e}=1$ for an unit vector parameter, or $\mathbf{a}\cdot\mathbf{b}=0$ for mutually perpendicular vectors, or $\mathbf{Z}\cdot\mathbf{Z}^{\mathrm{T}}=\mathbf{Z}^{\mathrm{T}}\cdot\mathbf{Z}=\mathbf{E}_2$ for a rotation tensor. The differential form of these relations are

$$\mathbf{e}\cdot\mathbf{e}=1 \quad\Rightarrow\quad \dot{\mathbf{e}}\cdot\mathbf{e}=0,$$

$$\mathbf{a}\cdot\mathbf{b}=0 \quad\Rightarrow\quad \dot{\mathbf{a}}\cdot\mathbf{b}+\mathbf{a}\cdot\dot{\mathbf{b}}=0,$$

$$\mathbf{Z}\cdot\mathbf{Z}^{\mathrm{T}}=\mathbf{E}_2 \quad\Rightarrow\quad \dot{\mathbf{Z}}\cdot\mathbf{Z}^{\mathrm{T}}+\mathbf{Z}\cdot\dot{\mathbf{Z}}^{\mathrm{T}}=\mathbf{0}_2 .$$

For the mechanical system Σ let a set $\mathbf{X}_k\in\Re_{\nu_k}$ (k=1,...,n) of tensor generalized parameters be given and let the constraints on the system be represented analogously to (5) in the form

$$\sum_{k=1}^{n}\mathbf{F}_{ik}\underset{\alpha_{ik}}{*}\dot{\mathbf{X}}_k+\mathbf{F}_i=\mathbf{0}_{\beta_i},\quad \mathbf{F}_i\in\Re_{\beta_i},\quad \mathbf{F}_{ik}\in\Re_{2\alpha_{ik}+\beta_i-\nu_k},\ (i=1,\dots,m). \tag{25}$$

where $\mathbf{F}_i$ and $\mathbf{F}_{ik}$ are given functions of the time and the generalized parameters $\mathbf{X}_k$. Some of the equations (or all of them) may be a differential form of the holonomic constraints of the system. Represented in suitable frames of reference, for instance in the inertial frame, Eqs. (25) generate a system of $\sum_{i=1}^{m}3^{\beta_i}$ scalar equations for $\sum_{k=1}^{n}3^{\nu_k}$ scalar unknowns — the coordinates of the tensors $\mathbf{X}_k$ with respect to the chosen frames of reference. The coordinates of the functions $\mathbf{F}_{ik}$ form a $\sum_{i=1}^{m}3^{\beta_i}\times\sum_{k=1}^{n}3^{\nu_k}$ scalar matrix. Its rank r will be named a rank of the tensor matrix as well.

Together with Eqs. (25) the homogenous system

$$\sum_{k=1}^{n}\mathbf{F}_{ik}\underset{\alpha_{ik}}{*}\delta\mathbf{X}_k=\mathbf{0}_{\beta_i}\qquad (i=1,\dots,m) \tag{26}$$

is considered. The solutions $\delta\mathbf{X}_k$, called variations of the tensor generalized parameters $\mathbf{X}_k$, are the differential increments of $\mathbf{X}_k$ compatible with the "time-frozen" constraints (25). As

$$rank(\mathbf{F}_{ik})=r,$$

the variations $\delta\mathbf{X}_k$ can be expressed by

$$l = \sum_{k=1}^{n} 3^{\nu_k} - r$$

independent scalars parameters. For a holonomic system l is equal to the number of the system's degrees-of-freedom.

The position of a typical system particle M is given by $\overline{OM} = \mathbf{r} = \mathbf{r}(t, \mathbf{X}_1, ..., \mathbf{X}_n)$ and the virtual displacement $\delta\mathbf{r}$ of M is defined as the differential of $\mathbf{r}$ by "frozen" time and in accordance with Eq. (13) is $\delta\mathbf{r} = \sum_{k=1}^{n} \frac{\partial \mathbf{r}}{\partial \mathbf{X}_k} \underset{\nu_k}{*} \delta\mathbf{X}_k$. By means of this expression and using Eqs. (8)-(9) we obtain for the Eq. (1)

$$\sum_{k=1}^{n} \int\limits_{\Sigma}^{[1]} \frac{\partial \mathbf{r}}{\partial \mathbf{X}_k} \cdot \ddot{\mathbf{r}}\, dm \underset{\nu_k}{*} \delta\mathbf{X}_k = \sum_{k=1}^{n} \int\limits_{\Sigma}^{[1]} \frac{\partial \mathbf{r}}{\partial \mathbf{X}_k} \cdot d\mathbf{f} \underset{\nu_k}{*} \delta\mathbf{X}_k . \tag{27}$$

This identity can be written in different forms:

1. After applying the relationship (14) it may be shown, that the equations of motion can be represented in the following Lagrangian form

$$\sum_{k=1}^{n} \left(\frac{d}{dt} \frac{\partial T}{\partial \dot{\mathbf{X}}_k} - \frac{\partial T}{\partial \mathbf{X}_k} \right) \underset{\nu_k}{*} \delta\mathbf{X}_k = \sum_{k=1}^{n} \mathbf{Q}_k \underset{\nu_k}{*} \delta\mathbf{X}_k , \tag{28}$$

where T is the kinetic energy of the system. The quantity $\mathbf{Q}_k = \int\limits_{\Sigma}^{[1]} \frac{\partial \mathbf{r}}{\partial \mathbf{X}_k} \cdot d\mathbf{f} \in \Re_{\nu_k}$ (k=1,...,n) may be named a generalized force corresponding to the tensor generalized parameters $\mathbf{X}_k$. The order of $\mathbf{Q}_k$ is equal to the order of the tensor generalized parameters $\mathbf{X}_k$.

2. Consider now the integrals in Eq. (27). As each particle of the system belongs to one of the sets $S_1, ..., S_N$ one can obtain

$$\int\limits_{\Sigma}^{[1]} \frac{\partial \mathbf{r}}{\partial \mathbf{X}_k} \cdot \ddot{\mathbf{r}}\, dm = \sum_{i=1}^{N} \int\limits_{\Sigma}^{[1]} \frac{\partial (\mathbf{r}_{P_i} + \boldsymbol{\rho})}{\partial \mathbf{X}_k} \cdot (\ddot{\mathbf{r}}_{P_i} + \ddot{\boldsymbol{\rho}}) dm =$$

$$= \sum_{i=1}^{N} \left(\overset{[1]}{\frac{\partial \mathbf{r}_{P_i}}{\partial \mathbf{X}_k}} \cdot m_i (\ddot{\mathbf{r}}_{P_i} + \ddot{\boldsymbol{\rho}}_{C_i}) + \int_{S_i} \overset{[1]}{\frac{\partial \boldsymbol{\rho}}{\partial \mathbf{X}_k}} dm \cdot \ddot{\mathbf{r}}_{P_i} + \int_{S_i} \overset{[1]}{\frac{\partial \boldsymbol{\rho}}{\partial \mathbf{X}_k}} \cdot \ddot{\boldsymbol{\rho}} \, dm \right) =$$

$$= \sum_{i=1}^{N} \left[\overset{[1]}{\frac{\partial \mathbf{r}_{P_i}}{\partial \mathbf{X}_k}} \cdot m_i \ddot{\mathbf{r}}_i + \overset{[2]}{\frac{\partial \mathbf{W}}{\partial \dot{\mathbf{X}}_k}} \underset{2}{*} \left(m_i \boldsymbol{\rho}_{C_i} \ddot{\mathbf{r}}_{P_i} + \int_{S_i} \boldsymbol{\rho} \ddot{\boldsymbol{\rho}} \, dm \right) \right] \quad (k=1,\ldots,n), \tag{29}$$

where m_i and C_i are the mass and the mass centre of S_i, $\boldsymbol{\rho}_{C_i} = \overline{P_i C_i}$, $\mathbf{r}_i = \mathbf{r}_{P_i} + \boldsymbol{\rho}_{C_i}$ is the position vector of the mass centre C_i and $\mathbf{W}_i = \dot{\mathbf{A}}_i \cdot \hat{\mathbf{A}}_i$ (see Eqs.(19) and (22)).

Consider the tensor $\int_{S_i} \boldsymbol{\rho} \ddot{\boldsymbol{\rho}} \, dm$. In the special case, when

$$\boldsymbol{\rho} = x_1 \mathbf{a}_1 + x_2 \mathbf{a}_2 + x_3 \mathbf{a}_3, \quad x_i = const,$$

we have

$$\boldsymbol{\rho}\ddot{\boldsymbol{\rho}} = \left(\mathbf{a}_1^{(i)}, \mathbf{a}_2^{(i)}, \mathbf{a}_3^{(i)} \right) \begin{pmatrix} x_1^2 & x_1 x_2 & x_1 x_3 \\ x_1 x_2 & x_2^2 & x_2 x_3 \\ x_1 x_3 & x_2 x_3 & x_3^2 \end{pmatrix} \begin{pmatrix} \ddot{\mathbf{a}}_1^{(i)} \\ \ddot{\mathbf{a}}_2^{(i)} \\ \ddot{\mathbf{a}}_3^{(i)} \end{pmatrix}.$$

In this way

$$\int_{S_i} \boldsymbol{\rho} \ddot{\boldsymbol{\rho}} \, dm = \underline{\mathbf{a}}^{(i)\mathrm{T}} \underline{I} \underline{\ddot{\mathbf{a}}}^{(i)}, \tag{30}$$

where $\underline{I}$ is the coordinate matrix of the tensor $\mathbf{I} = \int_{S_i} \boldsymbol{\rho}\boldsymbol{\rho} \, dm$ relative to the basis $\underline{\mathbf{a}}$.

The generalized force $\mathbf{Q}_k$ can be represented in a similar to Eq. (29) form

$$\mathbf{Q}_k = \sum_{i=1}^{N} \int_{S_i} \frac{\partial (\mathbf{r}_{P_i} + \boldsymbol{\rho})}{\partial \mathbf{X}_k} \cdot d\mathbf{f} = \sum_{i=1}^{N} \left(\overset{[1]}{\frac{\partial \mathbf{r}_{P_i}}{\partial \mathbf{X}_k}} \cdot \mathbf{f}_i + \overset{[2]}{\frac{\partial \mathbf{W}_i}{\partial \dot{\mathbf{X}}_k}} \underset{2}{*} \mathbf{M}_{P_i} \right) \quad (k=1,\ldots,n),$$

where

$$\mathbf{f}_i = \int_{S_i} d\mathbf{f}, \qquad \mathbf{M}_{P_i} = \int_{S_i} \boldsymbol{\rho} \, d\mathbf{f}, \quad (i = 1, \ldots, N) \tag{31}$$

are determined by the forces acting on the points of the set S_i. In addition to these forces, the second-ordered tensor $\mathbf{M}_{P_i}$ depends on the origin P_i. Obviously, if the position vector $\mathbf{r}_{P_i}$ is independent of the tensor generalized parameter $\mathbf{X}_k$, then $\mathbf{Q}_k$ does not depend on the external forces acting on the points of the set S_i and if the tensor $\mathbf{W}_i$ is independent of the derivative $\dot{\mathbf{X}}_k$, then $\mathbf{Q}_k$ does not depend on the tensor $\mathbf{M}_{P_i}$. In other cases the influence of $\mathbf{f}_i$ and $\mathbf{M}_{P_i}$ on the generalized force $\mathbf{Q}_k$ is filtered by the linear operators $\overset{[1]}{\dfrac{\partial \mathbf{r}_{P_i}}{\partial \mathbf{X}_k}}$ and $\overset{[2]}{\dfrac{\partial \mathbf{W}_i}{\partial \dot{\mathbf{X}}_k}}$.

The equations of the motion can be obtained from the identity

$$\sum_{i=1}^{N}\sum_{k=1}^{n}\left[\overset{[2]}{\frac{\partial \mathbf{W}_i}{\partial \dot{\mathbf{X}}_k}} \underset{2}{*} \left(\int_{S_i} \rho\ddot{\rho}dm + m_i \rho_{P_i} \ddot{\mathbf{r}}_{P_i} - \mathbf{M}_{P_i}\right) + \overset{[1]}{\frac{\partial \mathbf{r}_{P_i}}{\partial \mathbf{X}_k}}(m_i \ddot{\mathbf{r}}_i - \mathbf{f}_i)\right] \underset{\nu_k}{*} \delta\mathbf{X}_k = 0, \tag{32}$$

where the variations $\delta\mathbf{X}_k$ satisfy the Eqs. (26). It is clear that the resultant $\mathbf{f}_i$ of the forces acting on S_i influences the translation motion of S_i, while the tensor $\mathbf{M}_{P_i}$ — the angular orientation of S_i.

The identities (28) and (32) are written in an invariant tensor form. The derivatives have to be evaluated taking into account the equations of constraints (25). For further development, in the usual case the equations must be represented in a scalar form and numerical methods ought to be applied. We shall consider some special cases when the solutions of Eqs. (26) can be expressed in an invariant form.

3. Let us assume now that the variations of the tensor generalized parameters can be written in the form

$$\delta\mathbf{X}_k = \sum_{s=1}^{d} \delta\boldsymbol{\pi}_s \underset{\mu_s}{*} \mathbf{A}_{ks}, \quad \delta\boldsymbol{\pi}_s \in \Re_{\mu_s}, \quad \mathbf{A}_{ks} \in \Re_{\mu_s+\nu_k} \quad (k=1,\dots,n), \tag{33}$$

where the variables $\delta\boldsymbol{\pi}_s$ (s=1,...,d) are independent and $rank\,(\mathbf{A}_{ks}) = l$. Having in mind Eqs. (8)-(9) we obtain

$$\delta\mathbf{r} = \sum_{k=1}^{n} \frac{\partial \mathbf{r}}{\partial \mathbf{X}_k} \underset{\nu_k}{*} \delta\mathbf{X}_k = \sum_{k=1}^{n}\sum_{s=1}^{d} \delta\boldsymbol{\pi}_s \underset{\mu_s}{*} \left(\mathbf{A}_{ks} \underset{\nu_k}{*} \overset{[1]}{\frac{\partial \mathbf{r}}{\partial \mathbf{X}_k}}\right).$$

Substitution of this expression and Eq. (29) into Eq.(1) and keeping in mind that the variables $\delta\boldsymbol{\pi}_s$ are independent, reduces the equations of motion to

$$\sum_{i=1}^{N}\sum_{k=1}^{n}\mathbf{A}_{ks}\underset{\nu_k}{*}\left[\overset{[2]}{\frac{\partial\mathbf{W}_i}{\partial\dot{\mathbf{X}}_k}}\underset{2}{*}\left(\int_{S_i}\boldsymbol{\rho}\ddot{\boldsymbol{\rho}}\,dm+m_i\boldsymbol{\rho}_{C_i}\ddot{\mathbf{r}}_{P_i}-\mathbf{M}_{P_i}\right)+\overset{[1]}{\frac{\partial\mathbf{r}}{\partial\mathbf{X}_k}}.(m_i\ddot{\mathbf{r}}_{P_i}-\mathbf{f}_i)\right]=\mathbf{0}_{\mu_s}$$

(*s*=1,...,*d*) (34)

or

$$\sum_{k=1}^{n}\mathbf{A}_{ks}\underset{\nu_k}{*}\left(\frac{d}{dt}\frac{\partial T}{\partial\dot{\mathbf{X}}_k}-\frac{\partial T}{\partial\mathbf{X}_k}-\mathbf{Q}_k\right)=\mathbf{0}_{\mu_s}\quad (s=1,...,d) \tag{35}$$

with the kinetic energy $T=\dfrac{1}{2}\int\dot{\mathbf{r}}^2dm$ of the system.

Substituting $\overset{[1]}{\dfrac{\partial\mathbf{r}_{P_i}}{\partial\mathbf{X}_k}}=\overset{[1]}{\dfrac{\partial\dot{\mathbf{r}}_{P_i}}{\partial\dot{\mathbf{X}}_k}}$ in Eqs. (34) we obtain another form of motion equations. This form is a tensor analogue of the equations given, for instance, in (Barhorst et al. 1995) and derived from the Gibbs-Appell equations using generalized speeds .

The equations (34) and (35) take the form of Maggi's equations for the considered system when the generalized coordinates are represented in a scalar form, i.e. every vector is expressed by three scalar quantities and every tensor — by 9 scalar quantities. The number of the scalar unknowns is equal to $\sum_{k=1}^{n}3^{\nu_k}$ .The scalar equations together with the constraint equations form a system whose rank is

$$rank(\mathbf{F}_{ik})+rank(\mathbf{A}_{ks})=r+l=r+\sum_{k=1}^{n}3^{\nu_k}-r=\sum_{k=1}^{n}3^{\nu_k}\,.$$

The simplest relationships between the tensor generalized parameters which lead to the expressions similar to Eq. (33) are:

1. When $\mathbf{X}_k=\mathbf{e}$ is a unit vector, i.e. $\mathbf{e}\cdot\mathbf{e}=1$, then $\delta\mathbf{e}\cdot\mathbf{e}=0$ and

$$\delta\mathbf{e}=\delta\boldsymbol{\pi}\cdot\tilde{\mathbf{e}}\ \Rightarrow\ \mathbf{A}_{k1}=\tilde{\mathbf{e}}\,. \tag{36}$$

2. When $\mathbf{X}_k=\mathbf{c}\in\Re$ and $\mathbf{a}\cdot\mathbf{c}=const$, $\mathbf{a}=\boldsymbol{const}$, then

$$\delta\mathbf{c}=\delta\boldsymbol{\pi}\cdot\tilde{\mathbf{a}}\ \Rightarrow\ \mathbf{A}_{k1}=\tilde{\mathbf{a}}\,.$$

3. When $\mathbf{X}_k=\mathbf{Z}$ is a rotation tensor, then the variation of $\mathbf{Z}$ satisfies the Eq. (24), thus,

$$\mathbf{A}_{k1}=\tilde{\mathbf{Z}}\,. \tag{37}$$

We have to notice that if $\mathbf{W}_i = \widetilde{\boldsymbol{\omega}}_i$, i.e. if the set S_i moves like a rigid body and $\boldsymbol{\omega}_i$ is its angular velocity, then

$$\frac{{}^{[2]}\partial \mathbf{W}_i}{\partial \dot{\mathbf{X}}_k} \underset{2}{*} \left(\int_{S_i} \boldsymbol{\rho}\ddot{\boldsymbol{\rho}}\,dm + m_i \boldsymbol{\rho}_{C_i} \ddot{\mathbf{r}}_{P_i} - \mathbf{M}_{P_i} \right) = \frac{{}^{[2]}\partial \widetilde{\boldsymbol{\omega}}_i}{\partial \dot{\mathbf{X}}_k} \underset{2}{*} \left(\int_{S_i} \boldsymbol{\rho}\ddot{\boldsymbol{\rho}}\,dm + m_i \boldsymbol{\rho}_{C_i} \ddot{\mathbf{r}}_{P_i} - \mathbf{M}_{P_i} \right) =$$

$$= \frac{{}^{[1]}\partial \boldsymbol{\omega}_i}{\partial \dot{\mathbf{X}}_k} \cdot \left(\int_{S_i} \widetilde{\boldsymbol{\rho}} \cdot \ddot{\boldsymbol{\rho}}\,dm + m_i \widetilde{\boldsymbol{\rho}}_{C_i} \cdot \ddot{\mathbf{r}}_{P_i} + \boldsymbol{\Theta} \underset{2}{*} \mathbf{M}_{P_i} \right) = \frac{{}^{[1]}\partial \boldsymbol{\omega}_i}{\partial \dot{\mathbf{X}}_k} \cdot \left((\mathbf{J}_{P_i} \cdot \boldsymbol{\omega})^{\cdot} + m_i \boldsymbol{\rho}_{C_i} \times \ddot{\mathbf{r}}_{P_i} - \mathbf{m}_{P_i} \right)$$

where $\mathbf{J}_{P_i} = \int_{S_i} (\boldsymbol{\rho}^2 \mathbf{E}_2 - \boldsymbol{\rho}\boldsymbol{\rho})\,dm$ and $\mathbf{m}_{P_i} = \int_{S_i} \boldsymbol{\rho} \times d\mathbf{f}$ are the inertia tensor of the set S_i and the resultant torque acting on the S_i with respect to the point P_i, respectively.

4. Equations of Motion of a Rigid Body System

The equations of motion for a system of N rigid bodies when the sets $S_1, \ldots, S_N$ are the bodies of the system, $\mathbf{r}_{P_i} = \overline{OP_i}$ are the position vectors of arbitrary points fixed in the bodies and $\boldsymbol{\omega}_i$ are the absolute angular velocities of S_i, respectively, take the form

$$\sum_{i=1}^{N} \sum_{k=1}^{n} \mathbf{A}_{ks} \underset{\nu_k}{*} \left[\frac{{}^{[1]}\partial \boldsymbol{\omega}_i}{\partial \dot{\mathbf{X}}_k} . \left((\mathbf{J}_{P_i} . \boldsymbol{\omega}_i)^{\cdot} + m_i\ \boldsymbol{\rho}_{C_i} \times \ddot{\mathbf{r}}_{P_i} - \mathbf{m}_{P_i} \right) + \frac{{}^{[1]}\partial \mathbf{r}_{P_i}}{\partial \mathbf{X}_k} . (m_i \ddot{\mathbf{r}}_i - \mathbf{f}_i) \right] = \mathbf{0}_{\mu_s}$$

$$(s=1,\ldots,d). \qquad (38)$$

If the mass centre C_i of the body S_i is chosen for the point P_i and $\mathbf{J}_{P_i} = \mathbf{J}_{C_i} = \mathbf{J}_i$, $\mathbf{m}_{P_i} = \mathbf{m}_{C_i} = \mathbf{m}_i$ are the principal inertia tensor and the resultant torque of the external and doing work internal forces acting on body S_i, the equations are reduced to

$$\sum_{i=1}^{N} \sum_{k=1}^{n} \mathbf{A}_{ks} \underset{\nu_k}{*} \left[\frac{{}^{[1]}\partial \boldsymbol{\omega}_i}{\partial \dot{\mathbf{X}}_k} \cdot \left((\mathbf{J}_i \cdot \boldsymbol{\omega}_i)^{\cdot} - \mathbf{m}_i \right) + \frac{{}^{[1]}\partial \mathbf{r}_i}{\partial \mathbf{X}_k} \cdot (m_i \ddot{\mathbf{r}}_i - \mathbf{f}_i) \right] = \mathbf{0}_{\mu_s} \quad (s=1,\ldots,d). \quad (39)$$

Bearing in mind that the kinetic energy of a system of N rigid bodies is represented by the expressions (Wittenburg 1977)

$$T=\frac{1}{2}\sum_{i=1}^{N}m_i\dot{\mathbf{r}}_i^2+\frac{1}{2}\sum_{i=1}^{N}\boldsymbol{\omega}_i\cdot\mathbf{J}_i\cdot\boldsymbol{\omega}_i,$$

$$T=\frac{1}{2}\sum_{i=1}^{N}m_i\dot{\mathbf{r}}_{P_i}^2+\sum_{i=1}^{N}m_i\dot{\mathbf{r}}_{P_i}\cdot(\boldsymbol{\omega}_i\times\boldsymbol{\rho}_{C_i})+\frac{1}{2}\sum_{i=1}^{N}\boldsymbol{\omega}_i\cdot\mathbf{J}_{P_i}\cdot\boldsymbol{\omega}_i$$

and that for the vector product the equality

$$m_i\boldsymbol{\rho}_{C_i}\times\ddot{\mathbf{r}}_{P_i}=\frac{d}{dt}(m_i\boldsymbol{\rho}_{C_i}\times\dot{\mathbf{r}}_{P_i})-m_i\dot{\boldsymbol{\rho}}_{C_i}\times\dot{\mathbf{r}}_{P_i}=\frac{d}{dt}(m_i\boldsymbol{\rho}_{C_i}\times\dot{\mathbf{r}}_{P_i})+m_i\dot{\mathbf{r}}_{P_i}\times\dot{\mathbf{r}}_{C_i},$$

takes place, we come to the relationships

$$\frac{\partial T}{\partial\boldsymbol{\omega}_i}=\mathbf{J}_{P_i}\cdot\boldsymbol{\omega}_i+m_i\boldsymbol{\rho}_{C_i}\times\dot{\mathbf{r}}_{P_i},\quad\frac{\partial T}{\partial\dot{\mathbf{r}}_{P_i}}=m_i\dot{\mathbf{r}}_{C_i},$$

$$(\mathbf{J}_{P_i}\cdot\boldsymbol{\omega})^{\cdot}+m_i\boldsymbol{\rho}_{C_i}\times\ddot{\mathbf{r}}_{P_i}=\frac{d}{dt}(\mathbf{J}_{P_i}\cdot\boldsymbol{\omega}+m_i\boldsymbol{\rho}_{C_i}\times\dot{\mathbf{r}}_{P_i})+m_i\dot{\mathbf{r}}_{P_i}\times\dot{\mathbf{r}}_{C_i}.$$

By means of these formulae we can easily come to the following form of the equations of motion

$$\sum_{i=1}^{N}\sum_{k=1}^{n}\mathbf{A}_{ks}\underset{\nu_k}{*}\left[\frac{\partial\boldsymbol{\omega}_i}{\partial\dot{\mathbf{X}}_k}\overset{[1]}{\cdot}\left(\frac{d}{dt}\frac{\partial T}{\partial\boldsymbol{\omega}_i}+\dot{\mathbf{r}}_{P_i}\times\frac{\partial T}{\partial\dot{\mathbf{r}}_{P_i}}-\mathbf{m}_i\right)+\frac{\partial\mathbf{r}_{P_i}}{\partial\mathbf{X}_k}\overset{[1]}{\cdot}\left(\frac{\partial T}{\partial\dot{\mathbf{r}}_{P_i}}-\mathbf{f}_i\right)\right]=\mathbf{0}_{\mu_s}$$

$(s=1,\ldots,d)$.

which are derived in the case of one rigid body, for instance, in (Milne-Thomson 1938).

Consider a holonomic system of rigid bodies described by a set of n mutually independent tensor generalized parameters $\mathbf{X}_1,\ldots,\mathbf{X}_n$, which can be scalars, vectors and second-ordered tensors. Let us suppose that a part of the generalized parameters are unit vectors and rotation tensors, i.e. they are subject to conditions in the forms $\mathbf{c}\cdot\mathbf{c}=1,\ \mathbf{Z}\cdot\mathbf{Z}^{\mathrm{T}}=\mathbf{Z}^{\mathrm{T}}\cdot\mathbf{Z}=\mathbf{E}_2$.

Each of the vectorial generalized parameters of the system is determined by means of three (two if it is a unit vector) and each of the tensor coordinates - by means of nine (three if it is rotation tensor) independent scalar parameters. The total number of all scalar parameters identifying the system is the number of the system's degrees-of-freedom. In this spacial case, taking into account Eq. (36) and (37), the relations (33) can be represented as

$$\delta\mathbf{X}_k=\delta\boldsymbol{\pi}_k\underset{\mu_k}{*}\mathbf{A}_k,\quad\delta\boldsymbol{\pi}_k\in\mathfrak{R}_{\mu_k},$$

$$\mathbf{A}_k = \begin{cases} \widetilde{\mathbf{X}}_k \in \Re_{\nu_k+1} & \text{if } \mathbf{X}_k \in \Re_{\nu_k} \text{ is an unit vector or rotation tensor} \\ \mathbf{E}_{2\nu_k} & \text{otherwise} \end{cases},$$

where $\delta\pi_k$ is an arbitrary scalar for a scalar generalized coordinate (μ_k=0), an arbitrary vector if $\mathbf{X}_k$ is a vector or a rotation tensor (μ_k=1) and, finally, an arbitrary tensor variable if $\mathbf{X}_k$ is a common second-ordered tensor (μ_k=2). From Eq. (38) it follows ($\mathbf{A}_{ks} = \mathbf{A}_{k1} = \mathbf{A}_k$), that the equations of motion are

$$\mathbf{A}_k \underset{\nu_k}{*} \sum_{i=1}^{N} \left[\frac{\overset{[1]}{\partial} \boldsymbol{\omega}_i}{\partial \dot{\mathbf{X}}_k} \cdot \left((\mathbf{J}_{P_i} \cdot \boldsymbol{\omega}_i)^{\cdot} + m_i \boldsymbol{\rho}_{C_i} \times \ddot{\mathbf{r}}_{P_i} - \mathbf{m}_{P_i} \right) + \frac{\overset{[1]}{\partial} \mathbf{r}_{P_i}}{\partial \mathbf{X}_k} \cdot (m_i \ddot{\mathbf{r}}_i - \mathbf{f}_i) \right] = \mathbf{0}_{\mu_k}$$

$$(k=1,\ldots,n) \qquad (40)$$

Each of the equations (40) corresponding to an unit-vector generalized parameter **e** is equivalent to two independent scalar equations since it expresses a condition for vectors to lie on a plane whose normal vector is **e.** The number of related scalar equations is equal to three when the corresponding generalized parameter in (40) is a rotation tensor.

It can be noticed that the form of Eqs. (39)-(40) is close to the well known in literature equations

$$\sum_{i=1}^{N} \left[\frac{\overset{[1]}{\partial} \boldsymbol{\omega}_i}{\partial \dot{\pi}_k} \cdot \left((\mathbf{J}_i . \boldsymbol{\omega}_i)^{\cdot} - \mathbf{m}_i \right) + \frac{\overset{[1]}{\partial} \dot{\mathbf{r}}_i}{\partial \dot{\pi}_k} \cdot (m_i \ddot{\mathbf{r}}_i - \mathbf{f}_i) \right] = 0 \quad (k=1,\ldots,n),$$

dealing with quasi-coordinates $\pi_1,\ldots,\pi_n$ of the system under consideration. However, the derivateves of $\boldsymbol{\omega}_i$ and $\dot{\mathbf{r}}_i$ in Eqs (39)-(40) are evalueted according to the rules of differentation (11),while those in the latter expretion are notations of a special chain rule (Papastavridis 1994, 1991).

Further transformations are possible if the tensor generalized parameters are separated into two disjoint sets. The first one consists of the parameters, describing the angular orientations of the point sets $S_1,\ldots,S_N$. The set of their indices is denoted by U. The second one is composed of the rest of the generalized parameters (describing the translations of the bodies) and the set of their indices is denoted by V. Then, using the Eq. (14) and supposing that $\mathbf{X}_1,\ldots,\mathbf{X}_n$ are mutually independent, we have

$$\mathbf{A}_p \underset{\nu_p}{*} \sum_{i=1}^{N} \frac{\overset{[1]}{\partial} \boldsymbol{\omega}_i}{\partial \dot{\mathbf{X}}_p} \cdot \left[\left((\mathbf{J}_i \cdot \boldsymbol{\omega}_i)^{\cdot} - \mathbf{m}_i \right) + \sum_{j=1}^{N} \frac{\overset{[1]}{\partial} \dot{\mathbf{r}}_j}{\partial \boldsymbol{\omega}_i} \cdot (m_j \ddot{\mathbf{r}}_j - \mathbf{f}_j) \right] = \mathbf{0}_{\mu_p} \quad (p \in U)$$

$$\mathbf{A}_q \underset{\nu_q}{*} \sum_{i=1}^{N} \frac{\overset{[1]}{\partial} \mathbf{r}_i}{\partial \mathbf{X}_q} \cdot (m_i \ddot{\mathbf{r}}_i - \mathbf{f}_i) = \mathbf{0}_{\mu_q} \quad (q \in V).$$

This representation is useful because of the simple relationship (21) between the absolute derivative of a vector and the angular velocity of an orthonormal basis. Under the same conditions, if the Lagrangian form (35) is preferred, the equations of motion can be reduced to a more convenient form, too. The derivative $\dot{\mathbf{Z}}_i$ of the tensor $\mathbf{Z}_i = \mathbf{Z}_i(t, \mathbf{X}_p)$ ($p \in U$, $i=1,\ldots,N$) determining the angular orientation of the body S_i depends on the time, $\mathbf{X}_p$ and $\dot{\mathbf{X}}_p$ ($p \in U$) but by the relationship $\dot{\mathbf{Z}}_i = \tilde{\boldsymbol{\omega}}_i \cdot \mathbf{Z}_i$ it can be represented as a function of the time, $\mathbf{X}_p$ ($p \in U$) and the absolute angular velocity $\boldsymbol{\omega}_i$. In this way the kinetic energy can be expressed as a function $T = \breve{T}(t, \mathbf{X}_p, \mathbf{X}_q, \dot{\mathbf{X}}_q, \boldsymbol{\omega}_i)$. Then as the equations of motion one can obtain (Vasileva et al. 1988)

$$\mathbf{A}_p \underset{\nu_p}{*} \left[\sum_{i=1}^{N} {}^{[1]} \frac{\partial \boldsymbol{\omega}_i}{\partial \dot{\mathbf{X}}_p} \cdot \left(\frac{d}{dt} \frac{\partial \breve{T}}{\partial \boldsymbol{\omega}_i} - \boldsymbol{\omega}_i \times \frac{\partial \breve{T}}{\partial \boldsymbol{\omega}_i} \right) - \frac{\partial \breve{T}}{\partial \mathbf{X}_p} - \mathbf{Q}_p \right] = \mathbf{0}_{\mu_p}, \quad (p \in U)$$

$$\mathbf{A}_q \underset{\nu_q}{*} \left(\frac{d}{dt} \frac{\partial \breve{T}}{\partial \dot{\mathbf{X}}_q} - \frac{\partial \breve{T}}{\partial \mathbf{X}_q} - \mathbf{Q}_q \right) = \mathbf{0}_{\mu_q}, \quad (q \in V) \tag{41}$$

Comparing Eqs. (35), (41) with Eqs. (34), (40) we have to remark that the application of the Lagrangian equations requires more analytical work because of the double differentiating of the quadratic form T. Equations (41) are close to the well-known Bolzmann-Hamel's equations of motion which are applied when a set of quasi-coordinates is used (Quinn 1990 and Papastavridis 1991).

5. Examples

5.1. Rigid body. The equations of motion (40) will be applied to a rigid body with mass m, mass-centre C and body-fixed basis $P\mathbf{e}_1, \mathbf{e}_2, \mathbf{e}_3$. Three cases will be considered with one, two and three degrees of freedom for the rotational motion of the body, respectively. It will be shown that in each of these cases the rotation tensor $\mathbf{Z}$ determining the orientation of the body fixed triple $\mathbf{e}_1, \mathbf{e}_2, \mathbf{e}_3$ can be expressed as a function of only one tensor coordinate $\mathbf{X}$, i.e. $\mathbf{Z}=\mathbf{Z}(\mathbf{X})$, $\boldsymbol{\omega} = \boldsymbol{\omega}(\mathbf{X}, \dot{\mathbf{X}})$. This coordinate is a scalar in the first case, a unit vector in the second case and the tensor $\mathbf{Z}$ itself in the third case.

5.1.1. Rigid body with one degree of freedom. Consider a rigid body with a fixed axis. Let, for example, the rotation axis of the body be the axis $O\mathbf{e}_3^{(0)}$ of the inertial frame of reference $O\mathbf{e}_1^{(0)}, \mathbf{e}_2^{(0)}, \mathbf{e}_3^{(0)}$ and $P=O$, $\mathbf{e}_3 = \mathbf{e}_3^{(0)}$. Then $\boldsymbol{\omega} = \dot{\varphi}\, \mathbf{e}_3^{(0)}$, where $\varphi = \angle(\mathbf{e}_1^{(0)}, \mathbf{e}_1)$ and $\dfrac{\partial \boldsymbol{\omega}}{\partial \dot{\varphi}} = \mathbf{e}_3^{(0)}$. It follows from Eq. (40), where now $\mathbf{A}_1 = 1$, $P_1 = O$, $\mathbf{r}_O = \mathbf{0}$, that the equation of motion is

$$\mathbf{e}_3^{(0)} \cdot [(\mathbf{J}_O \cdot \boldsymbol{\omega})^{\cdot} - \mathbf{m}_O] = 0 .$$

Here $\mathbf{J}_O$ and $\mathbf{m}_O$ are the inertia tensor and the resultant torque of the external forces acting on the body with respect to the point O.

5.1.2. Rigid body with two degrees of freedom. Let the motion of a rigid body with two degrees of freedom be realised by a universal joint with perpendicular axes. Let $O=P$ be the centre of the universal joint and $O\mathbf{e}_1^{(0)}, \mathbf{e}_2^{(0)}, \mathbf{e}_3^{(0)}$ the inertial frame of reference where, for instance, the first basis vector $\mathbf{e}_1^{(0)}$ is parallel to the axis of the joint fixed in the inertial space and $\mathbf{e}_3$ is the unit vector collinear to the second axis of the joint. Since the axes of the joint are mutually perpendicular, the vector $\mathbf{e}_1^{(0)}$ is for all the time in the coordinate plane $(\mathbf{e}_1, \mathbf{e}_2)$. First we shall see that the rotation tensor of the basis $\mathbf{e}_1, \mathbf{e}_2, \mathbf{e}_3$ can be represented as a function of one of the vectors $\mathbf{e}_1$ and $\mathbf{e}_2$.

In fact, the unit vectors $\mathbf{e}_1, \mathbf{e}_2$ and $\mathbf{e}_3$ satisfy the relationships

$$\mathbf{e}_3 = \frac{\mathbf{e}_1 \times \mathbf{e}_1^{(0)}}{\mathbf{e}_1^{(0)} \cdot \mathbf{e}_2}, \qquad \mathbf{e}_2 = \mathbf{e}_3 \times \mathbf{e}_1,$$

if the vectors $\mathbf{e}_1^{(0)}$ and $\mathbf{e}_1$ are not collinear ($\mathbf{e}_1^{(0)} \times \mathbf{e}_1 \neq \mathbf{0}$). This yields

$$\mathbf{e}_3 = \lambda \frac{\mathbf{e}_1 \times \mathbf{e}_1^{(0)}}{\left|\mathbf{e}_1 \times \mathbf{e}_1^{(0)}\right|}, \quad \mathbf{e}_2 = \mathbf{e}_3 \times \mathbf{e}_1,$$

where the parameter $\lambda=1$ when the trihedral $\mathbf{e}_1$, $\mathbf{e}_1^{(0)}$ and $\mathbf{e}_3$ is right-handed, otherwise $\lambda= -1$. It can be shown (Vasileva et al. 1988) that the vector $\boldsymbol{\omega}$ can be represented in the form

$$\boldsymbol{\omega} = -(\dot{\mathbf{e}}_3 \cdot \mathbf{e}_2)\mathbf{e}_1 + \mathbf{e}_1 \times \dot{\mathbf{e}}_1 = \left[-\frac{\mathbf{e}_1^{(0)} \cdot \mathbf{e}_1}{\mathbf{e}_1^{(0)} \cdot \mathbf{e}_2} \mathbf{e}_1\mathbf{e}_3 + \widetilde{\mathbf{e}}_1 \right] \cdot \dot{\mathbf{e}}_1 . \tag{42}$$

In this way the Eqs. (7) and (42) express the tensor $\mathbf{Z}$ and the angular velocity $\boldsymbol{\omega}$ as functions of the vector $\mathbf{e}_1$ and its derivative $\dot{\mathbf{e}}_1$. From the Eq. (42) we have $\dfrac{\partial \boldsymbol{\omega}}{\partial \dot{\mathbf{e}}_1} = -\dfrac{\mathbf{e}_1^{(0)} \cdot \mathbf{e}_1}{\mathbf{e}_1^{(0)} \cdot \mathbf{e}_2} \mathbf{e}_1\mathbf{e}_3 + \widetilde{\mathbf{e}}_1$, and if $\mathbf{z}$ is an arbitrary vector, then

$$\widetilde{\mathbf{e}}_1 \cdot \overset{[1]}{\frac{\partial \boldsymbol{\omega}}{\partial \dot{\mathbf{e}}_1}} \cdot \mathbf{z} = \frac{\mathbf{e}_1^{(0)} \cdot \mathbf{z}}{\mathbf{e}_1^{(0)} \cdot \mathbf{e}_2} \mathbf{e}_2 + (\mathbf{e}_3 \cdot \mathbf{z})\mathbf{e}_3, \quad (\mathbf{e}_1^{(0)} \times \mathbf{e}_1 \neq \mathbf{0})$$

In the same way if $\mathbf{e}_1^{(0)} \times \mathbf{e}_2 \neq \mathbf{0}$, then $\mathbf{e}_3 = \lambda \dfrac{\mathbf{e}_1^{(0)} \times \mathbf{e}_2}{|\mathbf{e}_1^{(0)} \times \mathbf{e}_2|}$, $\mathbf{e}_1 = \mathbf{e}_2 \times \mathbf{e}_3$ and $\boldsymbol{\omega}$ is represented in the form

$$\boldsymbol{\omega} = -(\dot{\mathbf{e}}_3 \cdot \mathbf{e}_1)\mathbf{e}_2 + \mathbf{e}_2 \times \dot{\mathbf{e}}_2 = \left[\frac{\mathbf{e}_1^{(0)} \cdot \mathbf{e}_2}{\mathbf{e}_1^{(0)} \cdot \mathbf{e}_1} \mathbf{e}_2 \mathbf{e}_3 + \widetilde{\mathbf{e}}_2 \right] \cdot \dot{\mathbf{e}}_2 .$$

Then $\dfrac{\partial \boldsymbol{\omega}}{\partial \dot{\mathbf{e}}_2} = \dfrac{\mathbf{e}_1^{(0)} \cdot \mathbf{e}_2}{\mathbf{e}_1^{(0)} \cdot \mathbf{e}_1} \mathbf{e}_2 \mathbf{e}_3 + \widetilde{\mathbf{e}}_2$ and

$$\widetilde{\mathbf{e}}_2 \cdot \overset{[1]}{\frac{\partial \boldsymbol{\omega}}{\partial \dot{\mathbf{e}}_2}} \cdot \mathbf{z} = \frac{\mathbf{e}_1^{(0)} \cdot \mathbf{z}}{\mathbf{e}_1^{(0)} \cdot \mathbf{e}_1} \mathbf{e}_1 + (\mathbf{e}_3 \cdot \mathbf{z})\mathbf{e}_3, \quad (\mathbf{e}_1^{(0)} \times \mathbf{e}_2 \neq \mathbf{0}) .$$

As a result each of the equations $\widetilde{\mathbf{e}}_1 \cdot \overset{[1]}{\dfrac{\partial \boldsymbol{\omega}}{\partial \dot{\mathbf{e}}_1}} \cdot \mathbf{z} = 0$ and $\widetilde{\mathbf{e}}_2 \cdot \overset{[1]}{\dfrac{\partial \boldsymbol{\omega}}{\partial \dot{\mathbf{e}}_2}} \cdot \mathbf{z} = 0$ is equivalent to the scalar equations $\mathbf{e}_1^{(0)} \cdot \mathbf{z} = 0$ and $\mathbf{e}_3 \cdot \mathbf{z} = 0$ independently of which of the cases $\mathbf{e}_1^{(0)} \times \mathbf{e}_2 \neq \mathbf{0}$ or $\mathbf{e}_1^{(0)} \times \mathbf{e}_1 \neq \mathbf{0}$ has been realised. This means that each of the parametrizations of the tensor $\mathbf{Z}$ — by the vector $\mathbf{e}_1$ or by the vector $\mathbf{e}_2$ — leads to the same equations of motion

$$\mathbf{e}_1^{(0)} \cdot [(\mathbf{J}_O \cdot \boldsymbol{\omega})^{\cdot} - \mathbf{m}_O] = 0, \qquad \mathbf{e}_3 \cdot [(\mathbf{J}_O \cdot \boldsymbol{\omega})^{\cdot} - \mathbf{m}_O] = 0.$$

5.1.3. Free rigid body. We accept as tensor generalized parameters the position vector $\mathbf{r}_P$ of a point P fixed in the body and the rotation tensor $\mathbf{Z}$ determining the angular orientation of the body. The equation of motion can be written now in the classical form

$$m\ddot{\mathbf{r}}_C = \mathbf{f}, \qquad (\mathbf{J}_P \cdot \boldsymbol{\omega})^{\cdot} + m\boldsymbol{\rho}_C \times \ddot{\mathbf{r}}_P = \mathbf{m}_P,$$

since in this case (see Eq. (15))

$$\frac{\partial \mathbf{r}_P}{\partial \mathbf{r}_P} = \mathbf{E}_2, \quad \frac{\partial \boldsymbol{\omega}}{\partial \dot{\mathbf{r}}_P} = \mathbf{0}_2, \quad \frac{\partial \mathbf{r}_P}{\partial \mathbf{Z}} = \mathbf{0}_3, \quad \widetilde{\mathbf{Z}}\underset{2}{*} \overset{[1]}{\frac{\partial \boldsymbol{\omega}}{\partial \dot{\mathbf{Z}}}} = \mathbf{E}_2 .$$

5.2. Nonholonomic system. We shall derive and solve the equations of motion for a homogenous heavy sphere with mass m and radius r, which rolls without slipping on a rough horizontal plane (Fig. 1). The latter one spins about a fixed

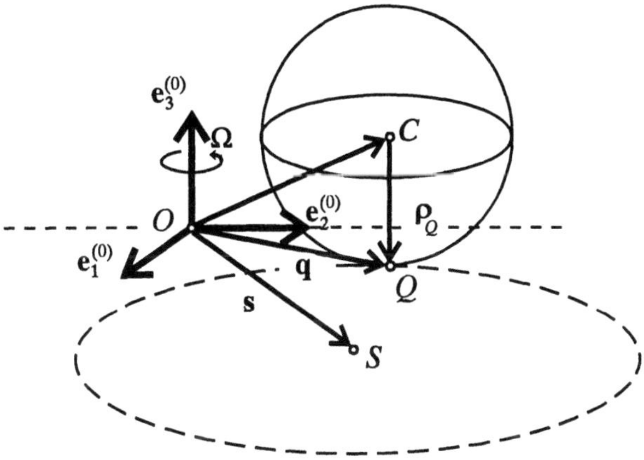

Fig. 1

vertical axis with an angular velocity $\Omega=\Omega(t)$. Many authors consider this example applying different approaches, for instance (Neimark et al. 1967, Lévy-Leblond 1986, Gersten et al. 1991, Papastavridis 1991, Essén 1994, Lewis et al. 1994). The solution given below is close to the vector form, represented in (Lévy-Leblond 1986).

Let the following Cartesian frames of reference be introduced: the inertial frame $O\mathbf{e}_1^{(0)}, \mathbf{e}_2^{(0)}, \mathbf{e}_3^{(0)}$, where $\mathbf{e}_3^{(0)}$ is a unit vector directed opposite to the force of gravity, the plane-fixed frame $O\,\mathbf{e}_1', \mathbf{e}_2', \mathbf{e}_3'$, $\mathbf{e}_3' = \mathbf{e}_3^{(0)}$, and the body-fixed frame $C\mathbf{e}_1, \mathbf{e}_2, \mathbf{e}_3$, where C is the mass-centre of the sphere. The rotation tensor and the absolute angular velocity of the basis are denoted with $\mathbf{Z}$ and $\boldsymbol{\omega}$, respectively. Let Q be the contact point of the sphere with the plane and $\mathbf{q} = \overline{OQ}$, $\boldsymbol{\rho}_Q = -r\mathbf{e}_3^{(0)}$.

The tensor $\mathbf{X}_1$=$\mathbf{Z}$ and the vector $\mathbf{X}_2$=$\mathbf{q}$ form a set of dependent generalized parameters. As the sphere rolls without slipping, the absolute velocity $\mathbf{v}_Q$ of the point Q, as a point of the plane, is equal to $\Omega\,\mathbf{e}_3^{(0)}\times\mathbf{q}$. On the other hand, $\mathbf{v}_Q$= $\frac{d}{dt}\overline{OC}+\boldsymbol{\omega}\times\boldsymbol{\rho}_Q = \dot{\mathbf{q}}+\boldsymbol{\omega}\times\boldsymbol{\rho}_Q$ as the velocity of a point of the sphere. In this way, $\dot{\mathbf{q}} = \mathbf{e}_3^{(0)}\times(\Omega\mathbf{q}-r\boldsymbol{\omega})$ which is the constraint equation. From $\widetilde{\mathbf{e}}_3^{(0)}\cdot\boldsymbol{\omega} =$ $= \mathbf{e}_3^{(0)}\cdot\widetilde{\boldsymbol{\omega}} = -\mathbf{e}_3^{(0)}\cdot\mathbf{Z}\cdot\dot{\mathbf{Z}}^{\mathrm{T}}$ we obtain $-r\mathbf{e}_3^{(0)}\cdot\mathbf{Z}\cdot\dot{\mathbf{Z}}^{\mathrm{T}}+\dot{\mathbf{q}}-\Omega\mathbf{e}_3^{(0)}\times\mathbf{q} = \mathbf{0}$, i.e., in

Eq. (25) we have to replace $\mathbf{F}_{11} = -r\mathbf{e}_3^{(0)} \cdot \mathbf{Z} \in \Re$, $\mathbf{F}_{12} = \mathbf{E}_2 \in \Re_2$, $\mathbf{F}_1 = \Omega \mathbf{e}_3^{(0)} \times \mathbf{q}$. Thus, the homogenous constraint equation is

$$-r\mathbf{e}_3^{(0)} \cdot \mathbf{Z} \cdot \delta\mathbf{Z}^{\mathrm{T}} + \delta\mathbf{q} = \mathbf{0}. \tag{43}$$

According to the Eq. (24), the variation of the variable $\mathbf{X}_1$=$\mathbf{Z}$ is $\delta\mathbf{X}_1 = \delta\mathbf{Z} = = \delta\tilde{\boldsymbol{\pi}} \cdot \mathbf{Z} = \delta\boldsymbol{\pi} \cdot \tilde{\mathbf{Z}}$, where $\delta\boldsymbol{\pi}$ is an arbitrary vector. Since $\mathbf{Z} \cdot \delta\mathbf{Z}^{\mathrm{T}} = -\delta\mathbf{Z} \cdot \mathbf{Z}^{\mathrm{T}} = = -\delta\tilde{\boldsymbol{\pi}} \cdot \mathbf{Z} \cdot \mathbf{Z}^{\mathrm{T}} = -\delta\tilde{\boldsymbol{\pi}}$ from Eq. (43) it follows that the variation of the variable $\mathbf{X}_2$=$\mathbf{q}$ is $\delta\mathbf{X}_2 = \delta\mathbf{q} = r\,\delta\boldsymbol{\pi} \cdot \tilde{\mathbf{e}}_3^{(0)}$.

Through the last formulae we determine the tensors $\mathbf{A}_{11} = \tilde{\mathbf{Z}}$, $\mathbf{A}_{21} = r\tilde{\mathbf{e}}_3^{(0)}$ for the motion equation (39). Further, we have $\mathbf{r}_C = \mathbf{r}(\mathbf{X}_2) = \mathbf{q} + r\mathbf{e}_3^{(0)}$, **m=0**, $\mathbf{f} = -mg\mathbf{e}_3^{(0)}$, $\dfrac{\partial\boldsymbol{\omega}}{\partial\dot{\mathbf{X}}_1} = \dfrac{\partial\boldsymbol{\omega}}{\partial\dot{\mathbf{Z}}}$, $\dfrac{\partial\boldsymbol{\omega}}{\partial\dot{\mathbf{X}}_2} = \mathbf{0}_2$, $\dfrac{\partial\mathbf{r}_C}{\partial\mathbf{X}_1} = \mathbf{0}_3$, $\dfrac{\partial\mathbf{r}_C}{\partial\mathbf{X}_2} = \dfrac{\partial\mathbf{r}_C}{\partial\mathbf{q}} = \mathbf{E}_2$ and the motion equation takes the form

$$\tilde{\mathbf{Z}} \underset{2}{*} \overset{[1]}{\frac{\partial\boldsymbol{\omega}}{\partial\dot{\mathbf{Z}}}} \cdot (\mathbf{J} \cdot \boldsymbol{\omega})^{\cdot} + r\mathbf{e}_3^{(0)} \times (m\ddot{\mathbf{q}} + mg\mathbf{e}_3^{(0)}) = \mathbf{0}.$$

Finally, having in mind that $\tilde{\mathbf{Z}} \underset{2}{*} \overset{[1]}{\dfrac{\partial\boldsymbol{\omega}}{\partial\dot{\mathbf{Z}}}} = \mathbf{E}_2$, $\mathbf{J} \cdot \boldsymbol{\omega} = \dfrac{2}{5}mr^2\boldsymbol{\omega}$, we obtain the following system of vector equations

$$\dot{\mathbf{q}} = \mathbf{e}_3^{(0)} \times (\Omega\mathbf{q} - r\boldsymbol{\omega}), \quad \frac{2}{5}mr^2\dot{\boldsymbol{\omega}} + mr\mathbf{e}_3^{(0)} \times \ddot{\mathbf{q}} = \mathbf{0}. \tag{44}$$

The second equation yields two integrals $\dfrac{2}{5}r\boldsymbol{\omega} + \mathbf{e}_3^{(0)} \times \dot{\mathbf{q}} = \mathbf{a} = \boldsymbol{const}$ and $\boldsymbol{\omega}.\mathbf{e}_3^{(0)} = \boldsymbol{\omega}_0 \cdot \mathbf{e}_3^{(0)} = const$, where $\mathbf{a} = \dfrac{2}{5}r\boldsymbol{\omega}_0 + \mathbf{e}_3^{(0)} \times \dot{\mathbf{q}}_0 = \dfrac{7}{5}r\boldsymbol{\omega}_0 - r(\boldsymbol{\omega}_0 \cdot \mathbf{e}_3^{(0)})\mathbf{e}_3^{(0)} - -\Omega(t_0)\mathbf{q}_0$. Here $\boldsymbol{\omega}_0$ is the initial value of the absolute angular velocity $\boldsymbol{\omega}$ and $\mathbf{q}_0$ is the initial position vector of the contact point Q. Substituting $\boldsymbol{\omega} = \dfrac{5}{2r}\mathbf{a} - \dfrac{5}{2r}\mathbf{e}_3^{(0)} \times \dot{\mathbf{q}}$ in the first equation we obtain the linear differential equation

$$\dot{\mathbf{q}} = \frac{2\Omega}{7}\mathbf{e}_3^{(0)} \times \mathbf{q} + \mathbf{b}, \quad \mathbf{b} = -\frac{5}{7}\mathbf{e}_3^{(0)} \times \mathbf{a}. \tag{45}$$

If $\Omega(t)$=Ω=*const.*, then the derivative of $\mathbf{q}$ can be represented in the form

$$\dot{\mathbf{q}} = \frac{2\Omega}{7}\mathbf{e}_3^{(0)} \times \left(\mathbf{q} - \frac{5}{2\Omega}\mathbf{a}\right) = \frac{2\Omega}{7}\mathbf{e}_3^{(0)} \times (\mathbf{q} - \mathbf{s})$$

where $\mathbf{s} = \frac{5}{2\Omega}\left[\frac{7}{5} r\boldsymbol{\omega}_0 \cdot (\mathbf{e}_1^{(0)}\mathbf{e}_1^{(0)} + \mathbf{e}_2^{(0)}\mathbf{e}_2^{(0)}) - \Omega \mathbf{q}_0\right]$ is the projection of the vector $\frac{5}{2\Omega}\mathbf{a}$ on the horizontal plane. This means that the derivative of the vector $\mathbf{x}=\mathbf{q}-\mathbf{s}$ is equal to $\dot{\mathbf{x}} = \frac{2\Omega}{7}\mathbf{e}_3^{(0)} \times \mathbf{x}$, i.e., the vector $\mathbf{x}$ is spinning with a constant angular velocity $\frac{2\Omega}{7}$ around the vertical axis and its length is

$$|\mathbf{x}| = |\mathbf{x}_0| = |\mathbf{q}_0 - \mathbf{s}| = \left|\frac{7}{2\Omega} r\boldsymbol{\omega}_0 \cdot (\mathbf{e}_1^{(0)}\mathbf{e}_1^{(0)} + \mathbf{e}_2^{(0)}\mathbf{e}_2^{(0)}) - \frac{5}{2}\mathbf{q}_0\right|.$$

Thus, the trajectory of the contact point Q is a circle with a centre S, $\overline{OS}=\mathbf{s}$ and a radius $|\mathbf{x}_0|$ (Fig. 1).

In the case $\Omega=\Omega(t)$ the solution $\mathbf{q}$ can be represented as $\mathbf{q} = \mathbf{x} + \mathbf{y}$, where $\mathbf{x}$ is the solution of the homogenous equation $\dot{\mathbf{x}} = \frac{2\Omega}{7}\mathbf{e}_3^{(0)} \times \mathbf{x}$ satisfying the initial condition $\mathbf{x}(0) = \mathbf{q}_0$ and $\mathbf{y}$ is a particular solution of (45).

It always can be written $\mathbf{x} = \mathbf{A} \cdot \mathbf{q}$, where $\mathbf{A}=\mathbf{A}(t)$, $\mathbf{A}(0)=\mathbf{E}_2$, is a tensor of second order which transforms the vector $\mathbf{q}_0$ into the vector $\mathbf{x}$. From $\mathbf{x} \cdot \dot{\mathbf{X}} = 0$ it follows that the tip of the vector $\mathbf{x}$ moves in a circle while the vector rotates with the variable angular velocity $\frac{2\Omega}{7}$. This means, that $\mathbf{A}$ is a tensor describing the rotation around the axis $\mathbf{e}_3^{(0)}$. This tensor can be represented in the form (Lilov 1993) $\mathbf{A} = -\cos\varphi \tilde{\mathbf{e}}_3^{(0)} \cdot \tilde{\mathbf{e}}_3^{(0)} + \sin\varphi \tilde{\mathbf{e}}_3^{(0)} + \mathbf{e}_3^{(0)}\mathbf{e}_3^{(0)}$, where $\varphi = \frac{2}{7}\int_0^t \Omega(\tau)d\tau$ and its derivative is $\dot{\mathbf{A}} = \frac{2\Omega}{7}\mathbf{e}_3^{(0)} \times \mathbf{A}$.

Let us substitute in Eq.(45) $\mathbf{q} = \mathbf{y} = \mathbf{A} \cdot \mathbf{u}$, $\mathbf{u} = \mathbf{u}(t)$, $\dot{\mathbf{y}} = \dot{\mathbf{A}} \cdot \mathbf{u} + \mathbf{A} \cdot \dot{\mathbf{u}}$. The derivative $\dot{\mathbf{u}}$ satisfies the equation $\mathbf{A} \cdot \dot{\mathbf{u}} = \mathbf{b}$ and, thus, $\mathbf{u} = \int_0^t \mathbf{A}^{-1}(\tau)d\tau \cdot \mathbf{b}$. Taking into account that $\mathbf{A}^{\mathrm{T}} \cdot \tilde{\mathbf{e}}_3^{(0)} = -\frac{7}{2\Omega}\dot{\mathbf{A}}^{\mathrm{T}}$, we can write down $\mathbf{u} = \int_0^t \mathbf{A}^{\mathrm{T}}(\tau)d\tau \cdot \mathbf{b} =$

$$= \frac{5}{2}\int_0^t \frac{\dot{\mathbf{A}}^{\mathrm{T}}(\tau)}{\Omega(\tau)}d\tau \cdot \mathbf{a}.$$

In this way, if $\mathbf{q}_0$ and $\boldsymbol{\omega}_0$ are the initial values of $\mathbf{q}$ and $\boldsymbol{\omega}$ the solution of the Eqs. (44) is

$$\mathbf{q} = \mathbf{A} \cdot \mathbf{q}_0 + \frac{5}{2}\mathbf{A} \cdot \int_0^t \frac{\dot{\mathbf{A}}^{\mathrm{T}}(\tau)}{\Omega(\tau)}d\tau \cdot \mathbf{a}, \quad \boldsymbol{\omega} = \boldsymbol{\omega}_0 + \frac{5}{2r}(\Omega\mathbf{q} - \Omega_0\mathbf{q}_0)$$

5.3. System described by nonrotational second-ordered tensor. A system of 12 thin homogenous heavy rods which are the edges of a parallelepiped, is moving without friction on a horizontal plane (Fig. 2). The rods are coupled by ball-and-socket

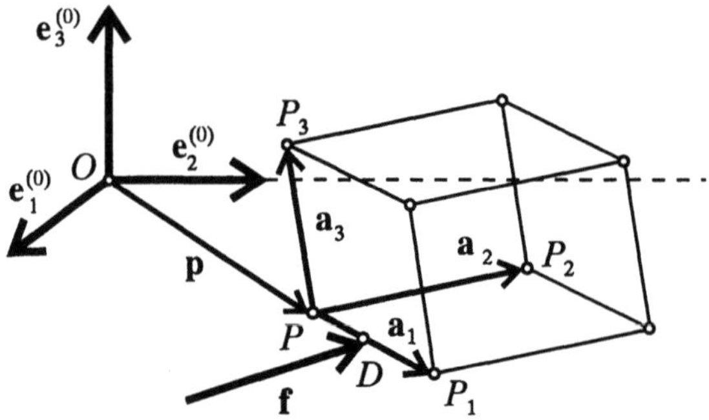

Fig. 2

joints. The parallel rods are supposed to be identical. The mass of the rod PP_i is m_i (i=1,2,3) and the vectors $\overline{PP_i}$ are denoted with $\mathbf{a}_i$. In the point D, $\overline{PD} = l\mathbf{a}_1$, an external horizontal force is applied. Three couples of internal forces are acting between the rods PP_i and PP_j ($i \neq j$, i, j = 1,2,3). These forces restore the initial configuration and are supposed to depend on the disturbance of the initial distance between the end points of the vectors $\mathbf{a}_i$ and $\mathbf{a}_j$.

As the representation of the position-vector of a typical system point M, in the basis $\mathbf{a}_1$, $\mathbf{a}_2$, $\mathbf{a}_3$, does not depend on the time, the vectors $\mathbf{X}_k$=$\mathbf{a}_k$ (k=1,2,3) and the vector $\mathbf{X}_4$=$\mathbf{p}$=$\overline{OP}$ may be chosen as tensor generalized parameters. The tensor generalized parameters and their variations satisfy the relationships

$$\mathbf{a}_1 \cdot \mathbf{a}_1 = const, \;\; \mathbf{a}_1 \cdot \mathbf{e}_3^{(0)} = 0, \quad \Rightarrow \quad \delta\mathbf{a}_1 = \delta x_1 \mathbf{e}_3^{(0)} \times \mathbf{a}_1,$$

$$\mathbf{a}_2 \cdot \mathbf{a}_2 = const, \;\; \mathbf{a}_2 \cdot \mathbf{e}_3^{(0)} = 0, \quad \Rightarrow \quad \delta\mathbf{a}_2 = \delta x_2 \mathbf{e}_3^{(0)} \times \mathbf{a}_2,$$

$$\mathbf{a}_3 \cdot \mathbf{a}_3 = const \quad \Rightarrow \quad \delta\mathbf{a}_3 = \delta\boldsymbol{\xi} \cdot \widetilde{\mathbf{a}}_3,$$

$$\mathbf{p} \cdot \mathbf{e}_3^{(0)} = 0, \quad \Rightarrow \quad \delta\mathbf{p} = \delta\boldsymbol{\eta} \cdot \widetilde{\mathbf{e}}_3^{(0)}.$$

Here the scalars δx_1, δx_2 and the vectors $\delta\boldsymbol{\xi}$, $\delta\boldsymbol{\eta}$ are arbitrary. In this way, we ought to substitute $\mathbf{A}_{11} = \mathbf{e}_3^{(0)} \times \mathbf{a}_1$, $\mathbf{A}_{22} = \mathbf{e}_3^{(0)} \times \mathbf{a}_2$, $\mathbf{A}_{33} = \widetilde{\mathbf{a}}_3$, $\mathbf{A}_{44} = \widetilde{\mathbf{e}}_3^{(0)}$ in the equations (34). Having in mind, that (see Eqs. (19) and (16))

$$\frac{\partial \mathbf{W}}{\partial \dot{\mathbf{X}}_k}=\frac{\partial \mathbf{W}}{\partial \dot{\mathbf{a}}_k}=\left(\hat{\mathbf{a}}_k{}^{[1]}\mathbf{E}_2\right)^{[1]}=\left(\hat{\mathbf{a}}_k\mathbf{E}_2\right)^{[1]},\qquad {}^{[2]}\frac{\partial \mathbf{W}}{\partial \dot{\mathbf{X}}_k}=\mathbf{E}_2\hat{\mathbf{a}}_k,\quad (k{=}1,2,3)$$

$$\frac{\partial \mathbf{W}}{\partial \dot{\mathbf{X}}_4}=\frac{\partial \mathbf{W}}{\partial \dot{\mathbf{p}}}=\mathbf{0}_3,\quad \frac{\partial \mathbf{r}_P}{\partial \mathbf{X}_k}=\mathbf{0}_2,\quad (\mathrm{k}=1,2,3),\quad \frac{\partial \mathbf{r}_P}{\partial \mathbf{X}_4}=\frac{\partial \mathbf{p}}{\partial \mathbf{p}}=\mathbf{E}_2$$

the equations of motion take the form

$$\hat{\mathbf{a}}_1\cdot\left(\int_S \boldsymbol{\rho}\ddot{\boldsymbol{\rho}}dm+4m\boldsymbol{\rho}_C\ddot{\mathbf{p}}-\mathbf{M}_P\right)\cdot\mathbf{e}_3^{(0)}\times\mathbf{a}_1=0,$$

$$\hat{\mathbf{a}}_2\cdot\left(\int_S \boldsymbol{\rho}\ddot{\boldsymbol{\rho}}dm+4m\boldsymbol{\rho}_C\ddot{\mathbf{p}}-\mathbf{M}_P\right)\cdot\mathbf{e}_3^{(0)}\times\mathbf{a}_2=0,$$

$$\hat{\mathbf{a}}_3\cdot\left(\int_S \boldsymbol{\rho}\ddot{\boldsymbol{\rho}}dm+4m\boldsymbol{\rho}_C\ddot{\mathbf{p}}-\mathbf{M}_P\right)\times\mathbf{a}_3=\mathbf{0},$$

$$\left[4m(\ddot{\mathbf{p}}+\ddot{\boldsymbol{\rho}}_C)-\mathbf{f}\right]\times\mathbf{e}_3^{(0)}=\mathbf{0},$$

where $4m=4(m_1+m_2+m_3)$ is the mass of the system and

$$\boldsymbol{\rho}_C=\frac{1}{2}(\mathbf{a}_1+\mathbf{a}_2+\mathbf{a}_3).$$

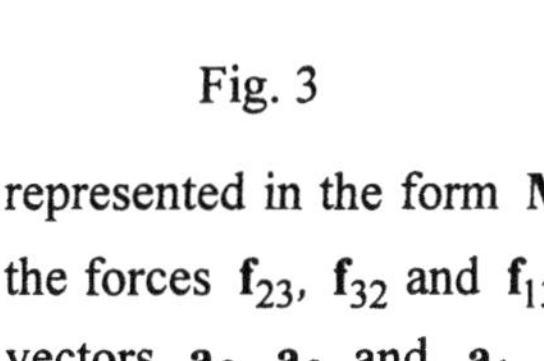

Fig. 3

Let us suppose that the forces preventing the changes of the angle between the vectors $\mathbf{a}_1$ and $\mathbf{a}_2$ are represented in the form $\mathbf{f}_{12}=\tau_{12}(\mathbf{a}_2-\mathbf{a}_1)$, $\mathbf{f}_{21}=\tau_{12}(\mathbf{a}_1-\mathbf{a}_2)$ and are applied at the points P_1 and P_2, respectively where $\tau_{12}=\tau_{12}(\mathbf{a}_1,\mathbf{a}_2)$ is a known function (Fig.3). Thus, the tensor $\mathbf{M}_{12}=\overline{PP_1}\,\mathbf{f}_{12}+\overline{PP_2}\,\mathbf{f}_{21}=\mathbf{a}_1\mathbf{f}_{12}+\mathbf{a}_2\mathbf{f}_{21}$ can be represented in the form $\mathbf{M}_{12}=\tau_{12}\left(\mathbf{a}_1\mathbf{a}_2+\mathbf{a}_2\mathbf{a}_1-\mathbf{a}_1\mathbf{a}_1-\mathbf{a}_2\mathbf{a}_2\right)$. Analogously, for the forces $\mathbf{f}_{23}$, $\mathbf{f}_{32}$ and $\mathbf{f}_{13}$, $\mathbf{f}_{31}$, preventing the changes of the angles between the vectors $\mathbf{a}_2$, $\mathbf{a}_3$ and $\mathbf{a}_1$, $\mathbf{a}_3$ and the corresponding tensors $\mathbf{M}_{23}$ and $\mathbf{M}_{31}$ we have

$$\mathbf{f}_{23}=\tau_{23}(\mathbf{a}_3-\mathbf{a}_2),\ \mathbf{f}_{32}=\tau_{23}(\mathbf{a}_2-\mathbf{a}_3),$$

$$\mathbf{f}_{31}=\tau_{31}(\mathbf{a}_1-\mathbf{a}_3),\quad \mathbf{f}_{13}=\tau_{31}(\mathbf{a}_3-\mathbf{a}_1)$$

$$\mathbf{M}_{23}=\tau_{23}\left(\mathbf{a}_2\mathbf{a}_3+\mathbf{a}_3\mathbf{a}_2-\mathbf{a}_2\mathbf{a}_2-\mathbf{a}_3\mathbf{a}_3\right),$$

$$\mathbf{M}_{31}=\tau_{31}\left(\mathbf{a}_3\mathbf{a}_1+\mathbf{a}_1\mathbf{a}_3-\mathbf{a}_3\mathbf{a}_3-\mathbf{a}_1\mathbf{a}_1\right),$$

where τ_{12}, τ_{23}, τ_{31} are known functions of the vectors $\mathbf{a}_1$, $\mathbf{a}_2$, $\mathbf{a}_3$. In this way the tensor $\mathbf{M}_P$ in Eq. (31) is

$$\mathbf{M}_P = \int_S \boldsymbol{\rho} d\mathbf{f} = -4mg\rho_C \mathbf{e}_3^{(0)} + l\mathbf{a}_1\mathbf{f} + \mathbf{M}_{12} + \mathbf{M}_{23} + \mathbf{M}_{31}.$$

Applying Eq. (17) and the relations $\mathbf{e}_3^{(0)} \cdot \mathbf{a}_1 = 0$, $\mathbf{e}_3^{(0)} \cdot \mathbf{a}_2 = 0$, $\mathbf{e}_3^{(0)} \cdot \mathbf{f} = 0$, we obtain

$$\hat{\mathbf{a}}_1 \cdot \mathbf{M}_P \cdot (\mathbf{e}_3^{(0)} \times \mathbf{a}_1) = (l\mathbf{f} + \tau_{12}\mathbf{a}_2 + \tau_{31}\mathbf{a}_3) \cdot (\mathbf{e}_3^{(0)} \times \mathbf{a}_1)$$

$$\hat{\mathbf{a}}_2 \cdot \mathbf{M}_P \cdot (\mathbf{e}_3^{(0)} \times \mathbf{a}_2) = (\tau_{12}\mathbf{a}_1 + \tau_{23}\mathbf{a}_3) \cdot (\mathbf{e}_3^{(0)} \times \mathbf{a}_2) \tag{46}$$

$$\hat{\mathbf{a}}_3 \cdot \mathbf{M}_P \times \mathbf{a}_3 = (-2mg\mathbf{e}_3^{(0)} + \tau_{23}\mathbf{a}_2 + \tau_{31}\mathbf{a}_1) \times \mathbf{a}_3.$$

The representation of the tensor $\int_S \boldsymbol{\rho}\ddot{\boldsymbol{\rho}} dm$ is given by Eq. (30). The coordinates of the tensor $\mathbf{I} = \int_S \boldsymbol{\rho}\boldsymbol{\rho} dm$ in the basis $\underline{\mathbf{a}}$ can be obtained using the relations

$$\mathbf{I} = \sum_{k=1}^{12} \mathbf{I}_k, \quad \mathbf{I}_k = \int_{S_k} \boldsymbol{\rho}\boldsymbol{\rho} dm, \quad \boldsymbol{\rho} = \overline{PM}, \quad M \in S_k,$$

$$\mathbf{I}_k = m_k \mathbf{r}_{C_k}\mathbf{r}_{C_k} + \mathbf{I}_{C_k}, \quad \mathbf{r}_{C_k} = \overline{PC_k}$$

where m_k and C_k are the mass and the mass-centre of the rod number k (k=1,...,12) and $\mathbf{I}_k$ and $\mathbf{I}_{C_k}$ are the tensors of the rods with respect to the point P and to the point C_k, respectively. For the tensor $\mathbf{I}_{C_k}$ (k=1,2,3) with respect to the mass-centre C_k of the rod we have the expression $\mathbf{I}_{C_k} = \frac{m_k}{12}\mathbf{a}_k\mathbf{a}_k$ (k=1,2,3). Applying the last formulae several times, we obtain

$$\mathbf{I} = (\mathbf{a}_1, \mathbf{a}_2, \mathbf{a}_3) \begin{pmatrix} 2m - \frac{2m_1}{3} & m & m \\ m & 2m - \frac{2m_2}{3} & m \\ m & m & 2m - \frac{2m_3}{3} \end{pmatrix} \begin{pmatrix} \mathbf{a}_1 \\ \mathbf{a}_2 \\ \mathbf{a}_3 \end{pmatrix}.$$

Hence,

$$\int_s \rho\ddot{\rho}dm = (\mathbf{a}_1, \mathbf{a}_2, \mathbf{a}_3)\begin{pmatrix} 2m - \frac{2m_1}{3} & m & m \\ m & 2m - \frac{2m_2}{3} & m \\ m & m & 2m - \frac{2m_3}{3} \end{pmatrix}\begin{pmatrix} \ddot{\mathbf{a}}_1 \\ \ddot{\mathbf{a}}_2 \\ \ddot{\mathbf{a}}_3 \end{pmatrix}. \tag{47}$$

Taking into account the Eqs. (46) and (47) we reduce the motion equations to the vector form

$$\left[\left(2m - \frac{2m_1}{3}\right)\ddot{\mathbf{a}}_1 + m\ddot{\mathbf{a}}_2 + m\ddot{\mathbf{a}}_3 + 2m\ddot{\mathbf{p}} - l\mathbf{f} - \tau_{12}\mathbf{a}_1 - \tau_{31}\mathbf{a}_3\right]\cdot \mathbf{c}_3^{(0)} \times \mathbf{a}_1 = 0$$

$$\left[m\ddot{\mathbf{a}}_1 + \left(2m - \frac{2m_2}{3}\right)\ddot{\mathbf{a}}_2 + m\ddot{\mathbf{a}}_3 + 2m\ddot{\mathbf{p}} - \tau_{12}\mathbf{a}_2 - \tau_{23}\mathbf{a}_3\right]\cdot \mathbf{e}_3^{(0)} \times \mathbf{a}_2 = 0$$

$$\mathbf{a}_3 \times \left[m\ddot{\mathbf{a}}_1 + m\ddot{\mathbf{a}}_2 + \left(2m - \frac{2m_3}{3}\right)\ddot{\mathbf{a}}_3 + 2m\ddot{\mathbf{p}} + 2mg\mathbf{e}_3^{(0)} - \tau_{23}\mathbf{a}_2 - \tau_{31}\mathbf{a}_1\right] = \mathbf{0}$$

$$\mathbf{e}_3^{(0)} \times \left[4m\ddot{\mathbf{p}} + 2m(\ddot{\mathbf{a}}_1 + \ddot{\mathbf{a}}_2 + \ddot{\mathbf{a}}_3) - \mathbf{f}\right] = \mathbf{0}\,.$$

The scalar equations can be obtained by introducing suitable scalar variables determining the vectors $\mathbf{a}_1$, $\mathbf{a}_2$, $\mathbf{a}_3$ and $\mathbf{p}$.

6. Conclusions

The main purpose of formulating dynamic equations of motion in terms of tensor variables is to retain the natural mechanical description of the system and by this to present the equations in a more compact and physically clear form. This makes the tensor equations more effective than the traditional scalar form of motion equations especially in cases when the mechanical system can be separated into parts with identical types of behaviour, such as in multibody systems. The equations can find application also in analytical investigations for determining, for instance, the steady-state motions and their stability, i.e., for investigations of the qualitative properties of the mechanical system itself. It is hoped that the new form of motion equations will be recognized as a natural method for obtaining the motion equations.

References

Barhorst AA, Everett LJ (1995) "Modeling Hybrid Parameter Multiple Body Systems: a Different Approach". *International J. of Non-Linear Mech.* **30** (1), p. 1-21

Boltzmann L (1902) "Über die Form der Lagrangeschen Gleichungen für nichtholonome generalizierte Koordinaten", Sitzungberichte der Mathematisch-Naturwissenschaftlichen Akademie der Wissenschaften zu Wien, **111**, Abteilung IIa, Heft 1-2, 1603-1614

Essén H (1994) "On the Geometry of Nonholonomic Dynamics". *ASME, Journal of Applied Mechanics* **61**, p. 689-694

Everett LJ, McDermott MJ (1986) "The Use of Vector Techniques in Variational Problems", *ASME, J. of Dynamic Systems, Measurement, and Control* **108**, p. 141-145

García de Jalón J, Unda J, Avello A (1986) "Natural Coordinates for Computer Analysis of Multibody Systems". *Comput. Meth. Appl. Mech. and Eng.* **56**(3), p. 309-327

Gersten J, Soodak H, Tiersten MS (1991) "Ball Moving on Stationary or Rotating Horizontal Plane". *American Journal of Physics* **60**, p. 43-47

Gorr GV, Kudryaschova F, Stepanova LA (1978) "*Classical Problems of Dynamics of Rigid Body*". Naukova dumka, Kiev (in Russian)

Hamel G (1904) "Die Lagrange-Eulersche Gleichungen der Mechanik", *Zeitschrift für Mathematik und Physik* **50**, p.1-57

Kharlamov PV (1972) "On the motion equations of rigid bodies", *Mech. of rigid bodies*, Kiev, **4**, p. 52-73 (in Russian)

Lagrange JL (1788) "*Mechanique analytique*", Paris

Lebedev PA (1986) "Vector Algorithm's of Explicit Solution of Problems for Positions and Displacements of Rigid-Body Systems". *Conference of theor. and appl. mech.*, Taschkent , 24-30 Sept., 1986, (in Russian)

Lévy-Leblond JM (1986) "Le billiard d'ANAIS", *European Journal of Physics* **7**, p. 252-258

Lewis AD, Murray RM (1994) "Variational Principles for Constrained Systems: Theory and Experiment". *International Journal of Non-Linear Mechanics* **30** (6), p. 793-815

Lilov L (1993) "*Modeling of Multibody Systems*", Nauka, Moscow (in Russian)

Lilov L, Vasileva N (1984) "Steady-state Motions of a Tree-like System of Lagrange's Gyroscopes", *Theor. and Appl. mech.* Sofia, **15**(3), p. 24-34 (in Russian)

Maggi GA (1901) "Di alcune nuove forma della equazioni della dinamica applicabile ai sistemi anolonomi", *Atti della Reale Accademia Nazionale dei Linei, Rendiconti Classe fisiche e mathematiche*, Ser.5, v. 10, fasc.2, p. 287-291

Milne-Thomson LM (1938) "*Theoretical Hydrodynamics*". Macmillan, London

Neimark JI, Fufaev NA (1967) "*Dynamics of Nonholonomic Systems*" . Nauka, Moscow (in Russian) (Translated version in English, American Mathematical Society, Providence, RI, 1972)

Papastavridis JG (1991) "On Energy Rate Theorems for Linear First-Order Nonholonomic Systems". *ASME, Journal of Applied Mechanics* **58**, p. 536-544

Papastavridis JG (1994) On the Boltzmann-Hamel Equations of Motion: a Vectorial Treatment". *ASME, Journal of Applied Mechanics* **61**, p. 453-459

Quinn RD (1990) "Equations of Motion for Structures in Terms of Quasi-Coordinates", *ASME, Journal of Applied Mechanics* **57**, p. 745-749

Vasileva N, Lilov L (1988) "Tensor Form of Lagrange's Equations". *Annual of Sofia University* , book 2 - Mechanique **82**, p. 249-286 (in Russian)

Wittenburg J (1977) "*Dynamics of Systems of Rigid Bodies*". B.G. Teubner, Stuttgart

Wittenburg J, Lilov L (1975) "Relative Equilibrium Positions and Their Stability for a Multi-Body Satellite in a Circular Orbit", *Ingenieur Archiv* **44**(4), p. 269-279

Efficient Description and Geometrical Interpretation of the Dynamics of Constrained Systems

Dedicated to my Friend John Tinsley Oden on the Occasion of his Sixtieth Birthday

Hans Brauchli

Institute of Mechanics, ETH-Zentrum, 8092 Zürich, Switzerland
e-mail: brauchli@ifm.mavt.ethz.ch

Summary. The methodology of multibody dynamics is briefly discussed. As an example of a descriptor formulation, the mass-orthogonal projection method is presented and it's numerical efficiency illustrated by an example of a dynamic optimization problem. Typically, in a descriptor formulation, the numerical efficiency depends on the particular choice of the equations in the redundant variables. Also, the use of sparse-matrix techniques may be advantageous. In a second part, the geometric interpretation of dynamics is reviewed. As an example, the topology of the configuration space of a four-bar linkage is discussed.

Keywords. Mass-orthogonalty, efficiency, optimization, curvature, topology, bifurcation, singularity.

1 Introduction

The last twenty years saw a big surge in computer-aided analysis techniques of multibody systems. Applications range from the classical field of machinery to vehicle dynamics, space research, robotics and others. A variety of methods for the description of multibody systems is now available; for example, Schiehlen (1990) lists twenty different computer codes for the simulation of such systems. In today's applications, the simulation of a mechanical system will often be performed in a broader environment, providing just one step in a more complex task. One example for this is the optimization of transient dynamical processes, where typically a simulation has to be repeated for a few thousand times. Even with the best and fastest computers available today, the question of numerical efficiency is then of consequence. On the other hand, applications of multibody dynamics are now often performed on a workstation or a personal computer, so the question of efficiency comes up again. For the user, however, the numerical efficiency may be less important than the overall efficiency: how long does it take to model a new mechanical system on the computer and to have results available?

It is not a simple task to put the developments in the field of multibody dynamics over the last twenty years in perspective. A concise review of this period would certainly be very welcome. The first Symposium on *Multibody Dynamics* was chaired and organized in 1977 by Kurt Magnus (1978), who may also be considered to be the mentor of most researchers in this field in Germany. Contributions were made, among others, by J.V. Breakwell, G.E. Fleischer, C.S. Hsu, J. Inoue, A. Ishlinskii, T.R. Kane, A.J. Kerr, D. Klimov, W. Kortüm, J. Lückel, J.W. Lund, P. Meinke, L. Meirovitch, V.J. Modi, R.E. Roberson, W.O. Schiehlen and J. Wittenburg. Many of the now well known researchers were in the audience. Since 1977 many important meetings have been held. Among the most memorable ones are the Symposium on *Dynamics of Multibody Systems* in 1985 chaired by Bianchi and Schiehlen (1986), the International Symposium 1992 concluding a five-year research project of the DFG (Schiehlen 1993), and the NATO-ASI Symposium 1993 chaired by Pereira and Ambrósio (1994, 1995). A remarkable fact is the sucess in multibody dynamics of researchers coming from the field of Finite Elements, such as García de Jalón (1981) and Géradin (1994).

For somebody educated in the spirit of classical dynamics it is surprising that the community of researchers in computer-aided analysis of multibody systems shows some reluctance to use the apparatus of analytical dynamics. In the first book in this newly emerging field, Wittenburg (1997) states Lagrange's equations only to remark that *a substantial amount of labor is required to formulate the Lagrangian L and its derivatives.* One of the early codes in this field, NEWEUL by Kreuzer and Schiehlen (Schiehlen 1990, pp.181-202), is simply based on the laws of linear and angular momentum. Another example is the book of Kane and Levinson (1985). As a rule, very few authors take advantage of the basic concepts of analytical dynamics such as canonical momenta, or the geometric structure of the configuration space or phase space. Noteworthy exceptions are Meirovitch (1970), Renaud (1975), Hiller (1983) and Nikravesh (1988).

The computational methods used for mechanisms are not basically different from those for multibody systems. However, for mechanisms, some problems are relevant which may be of minor interest in general multibody systems. One of these is the problem of inverse dynamics, i.e. the determination of the driving forces or driving torques for manipulators necessary to produce a prescribed motion. Also, some questions of kinematics may play a dominant role: the determination of optimal trajectories, or of trajectories avoiding obstacles. One topic intimately related to kinematics will be addressed below: the geometry and the connectivity of the configuration space.

In a first section of this paper, we will briefly review the methodology of multibody dynamics. Also, we point out that *there is* a form of Lagrange's equations useful for systems with tree-structure, i.e. most manipulators. We then introduce the concept of mass-metric and mass-orthogonality and describe the mass-orthogonal projection method. We claim that this method is a good candidate to deal with the problem of overall-efficiency. It also yields a rather simple approach to the problem of inverse dynamics.

In a second section the problem of numerical efficiency, in the setting of the projection method, is addressed. There are three main issues: the choice of the formulation of the equations; the use of sparse matrix techniques; and finally, the choice of integration schemes. The optimization of a low-voltage circuit breaker is also included as an application.

In a third section we discuss the relation of geometry to dynamics. Conceptually, this presentation is based on a beautiful paper *On the Geometry of Dynamics* by Synge (1927), which appeared seventy years ago. The basis is the geometry of the configuration space given by the mass-metric: according to Jacobi's principle, the geodesics in this metric correspond to inertial orbits of the mechanical system. Synge relates the stability of motion to the curvature in this geometry. In the setting of the projection method, the configuration space of the constrained system is embedded in the space of the free system. There are now two curvatures: the *intrinsic curvature* of the configuration space and the *extrinsic curvature* of the embedding. The intrinsic curvature is related to the stability of orbits of the constrained system, while the extrinsic curvature is related to the constraint reactions. Both can be determined with the equations of the projection method. They are also related to each other by a generalization of the *teorema egregium* of Gauss. We also consider a question of global geometry, the topology of the configuration space. In two dimensions the genus of the configuration space can be easily determined using a theorem of Poincaré. As an example we include a four-bar linkage, a mechanism with two degrees of freedom. The genus of the configuration space varies from 0 to 4, depending on geometric parameters. An important issue, of course, is to understand the dynamical behavior of the system near a bifurcation, where a singularity occurs. This singularity can be described by a localized curvature. It's effect on the orbits is to produce a discontinuity in the velocity, which may be understood intuitively as an impact caused by localized curvature. Although as a rule singularities should be avoided, their understanding may be helpful for the designer of mechanisms.

2 Descriptions of Multibody Systems

2.1 Methodology of Multibody Dynamics

We briefly discuss some criteria for classifying the methods of multibody dynamics, see e.g. (Bremer and Pfeiffer 1992). The most important distinction is between *minimal formulations* and *descriptor formulations.* In minimal formulations a minimal set of coordinates is used for describing the system. The governing equations appear as ordinary differential equations (ODE). Their application is restricted to systems with some kind of simple structure, say tree-structure. Minimal formulations may be very useful for real-time simulation or simulations on small computers. In descriptor formulations the system is described by a set of redundant coordinates and constraints are imposed. The system is then governed by differential algebraic equations (DAE). Descriptor formulations have the advantage that constraint reactions can also be computed. Furthermore, the modeling of the mechanical system is greatly simplified, since it may be assembled based on simpler constituents, e.g. rigid bodies. Therefore, descriptor formulations are now used very widely.

We have already mentioned that the governing equations may appear as ODEs or DAEs. About ten years ago, obtaining ODEs was important for numerical efficiency. But the development of good integration schemes for DAEs - see e.g. (Petzold 1994) - has made the DAE-formulations competitive.

In many codes for multibody systems, e.g. NEWEUL, the governing equations are produced by computer algebra. This has the disadvantage that the governing equations may be rather involved and difficult to handle. The derivation of the equations of the first variation, e.g., is not an easy task. As a rule, computer algebra should be used only as a last resort.

2.2 The Equations of Lagrange

A holonomic and scleronomic system with tree-structure, i.e, without kinematical loops, can be described by a set of n generalized coordinates q^k, $k = 1, \ldots, n$. Generalized forces Q_k are defined through the virtual work $\delta W = Q_k \, \delta q^k$. Summation over repeated indices is understood. The equations of Lagrange of the second kind read

$$\left(\frac{\partial T}{\partial \dot{q}_k}\right)^{\cdot} - \frac{\partial T}{\partial q_k} = Q_k \,. \tag{1}$$

Here,

$$T(q^k, \dot{q}^k) = \frac{1}{2} \, m_{ij}(q^k) \, \dot{q}^i \, \dot{q}^j \tag{2}$$

denotes the kinetic energy as a function of the generalized coordinates q^k and generalized velocities $\dot{q}^k$, and $m_{ij}(q^k)$ stands for the mass matrix, which will in general be a function of the generalized coordinates q^k. The *substantial amount of labor* required to make these equations explicit obviously stems from the first term on the left hand side of eq. (1). Canonical momenta, also called generalized momenta, are defined as

$$p_k = \frac{\partial T}{\partial \dot{q}_k} \,. \tag{3}$$

Since the mass matrix has to be positive definite, eq. (3) defines a regular transformation between generalized momenta and velocities. Therefore, we may use the p_k and q^k as independent variables and, introducing the short-hand notation

$$\frac{\partial m_{ij}}{\partial q_k} = m_{ij,k} \,, \tag{4}$$

write eqs. (1) and (3) in the form

$$\dot{p}_k = Q_k + \frac{\partial T}{\partial q_k} = Q_k + \frac{1}{2}\, m_{ij,k}\, \dot{q}^i\, \dot{q}^j \,, \tag{5}$$

$$m_{kj}\, \dot{q}^j = p_k \,. \tag{6}$$

Eqs. (5) and (6) represent a system of first-order differential equations upon solving eq. (6) for the velocities $\dot{q}^k$, which then have to be inserted in eq. (5).

The equations of the first variation, useful in some applications, also take a convenient form:

$$\delta\dot{p}_k = Q_{k,\ell}\, \delta q^\ell + m_{ij,k}\, \dot{q}^i\, \delta\dot{q}^j + \frac{1}{2}\, m_{ij,k\ell}\, \dot{q}^i\, \dot{q}^j\, \delta q^\ell \,, \tag{7}$$

$$m_{k\ell}\, \delta\dot{q}^\ell = \delta p_k - m_{k\ell,j}\, \dot{q}^\ell\, \delta q^j \,. \tag{8}$$

Eqs. (7) and (8) have to be solved in conjunction with eqs. (5) and (6). Again, the two linear systems (6) and (8) have to be solved first for $\dot{q}^k$ and then for $\delta\dot{q}^k$.

For modeling nonholonomic systems, or for using nonholonomic velocity parameters, a similar procedure can be applied to the equations of Euler-Lagrange (Brauchli 1991) to obtain the analogues of eqs. (5) and (6). The derivation of the corresponding equations of the first variation, however, may be more involved.

2.3 Mass-Metric and Mass-Orthogonality

Following Synge (1927), we may endow the configuration space of a constrained mechanical system with a metric

$$ds^2 = m_{k\ell}\, dq^k\, dq^\ell \ , \tag{9}$$

which Synge calles the *kinematical line element.* This metric is given by the kinetic energy of the system. Jacobi's principle then states that the inertial orbits of the mechanical system are in a one-to-one correspondence with the geodesics of this metric defined by the variational principle

$$\delta \int ds \ = 0 \ . \tag{10}$$

There is also another form of Jacobi's principle, taking into account conservative forces, but since the irksome term in Lagrange's equations stems from the kinetic energy, understanding inertial motions in terms of geometry is sufficient for our purpose. We may then introduce a scalar product

$$\langle u, v \rangle = m_{k\ell}\, u^k\, v^\ell \tag{11}$$

of two generalized velocities u and v of the system in a given configuration, and define two such velocities to be mass-orthogonal, if they satisfy

$$\langle u, v \rangle = 0 \ , \tag{12}$$

that is, if their scalar product vanishes.

2.4 Projections

We now view the mechanical system as a free system without constraints and impose kinematic constraints

$$E^T u = 0 \tag{13}$$

with an $n \times n_c$ matrix E of rank n_c to describe the constrained system. Here, the degree of freedom of the free system is denoted by n, the degree of freedom of the constrained system by n_a and the number of constraints by n_c. Naturally,

$$n_a + n_c = n \ . \tag{14}$$

The solution space of eq. (13) is given by the *admissible velocities* u_α of the constrained system. As shown in (Brauchli 1991), the admissible velocities can be obtained from the velocities of the free system by a projection α,

$$u_\alpha = \alpha\, u \ . \tag{15}$$

A complementary projection β will bring forth the *inadmissible velocities*

$$u_\beta = \beta\, u \ . \tag{16}$$

The two projections are uniquely defined by a condition of mass-orthogonality, namely,

$$\langle u_\alpha, u_\beta \rangle = 0 \, . \tag{17}$$

These projections are given by the formulas

$$\alpha = \mathbf{1} - AEG^{-1}E^T \, , \; \beta = AEG^{-1}E^T \, , \tag{18}$$

where

$$G = E^T AE \tag{19}$$

is the *fundamental matrix* and $A = M^{-1}$ the inverse mass matrix, while $\mathbf{1}$ denotes the identity matrix.

The projections α and β act on all physical entities of the system. For forces and momenta, however, the transposed projections have to be used, since they belong to spaces dual to the velocities. We have, therefore,

$$Q_\alpha = \alpha^T Q \, , \; Q_\beta = \beta^T Q \tag{20}$$

and

$$p_\alpha = \alpha^T p \, , \; p_\beta = \beta^T p \, . \tag{21}$$

As a consequence of mass-orthogonality, $p_\beta = 0$ in the constrained system, which can be used as the starting point to derive the equations of motion (Brauchli 1991). The mass matrix of the constrained system,

$$M_{\alpha\alpha} = \alpha^T M \, \alpha \, , \tag{22}$$

is a singular $n \times n$ matrix of rank n_a. Its generalized inverse (Penrose 1955) is given by

$$A_{\alpha\alpha} = \alpha \, A \, \alpha^T \, . \tag{23}$$

In fact

$$A_{\alpha\alpha} \, M_{\alpha\alpha} = \alpha \, , \tag{24}$$

and α can be viewed as the identity matrix in the space of admissible velocities.

2.5 The Mass-Orthogonal Projection Method

There are many ways to derive the governing equations of a constrained system. In (Brauchli 1991) a geometric approach was used. An alternative possibility is to use Lagrangian multipliers (Bach 1993). The resulting equations can be cast in many different forms, since they are stated in the state space of the free system, but their right-hand sides are uniquely defined for the constrained system only. One way to express the governing equations is

$$\dot{p} = Q_\alpha + \frac{\partial T}{\partial q^k} - \omega\, A p + \dot{M}_{\beta\alpha}\, A p\,, \tag{25}$$

$$\dot{q} = u = A_{\alpha\alpha}\, p\,. \tag{26}$$

These equations make sense only for initial data satisfying the kinematic *and* geometric constraints of the system. In order to state eqs. (25) and (26) explicitly, one needs the matrices E, $\dot{E}$, M, and $\dot{M}$, and the matrices M and G have to be inverted. The new terms appearing in eq. (25) are

$$\omega = \Omega\, \alpha + \alpha^T \Omega^T\,, \tag{27}$$

$$\Omega = E\, G^{-1} \dot{E}^T \tag{28}$$

and

$$\dot{M}_{\beta\alpha} = \beta^T \dot{M}\, \alpha\,. \tag{29}$$

The constraint reactions can also be determined as functions of position and velocities, i.e.,

$$Q_\beta = -\omega\, A p + \dot{M}_{\beta\alpha}\, A p + \frac{\partial T}{\partial q^k}\,. \tag{30}$$

2.6 Inverse Dynamics

There is a simple approach to the problem of inverse dynamics, if the equations of the projection method are used with a constant mass matrix. They then take the form (Brauchli and Weber 1991)

$$M\, \ddot{q} = Q_\alpha - \Omega(\dot{q})\, \dot{q}\,. \tag{31}$$

For determining the driving forces necessary for a given motion of the system, one simply has to apply the principle of virtual work to eq. (31) with an admissible displacement δq_α. One can show that the contribution of the second term on the right hand side of eq. (31) vanishes. Therefore,

$$\delta W = Q_\alpha^T\, \delta q_\alpha = \ddot{q}^T M\, \delta q_\alpha \tag{32}$$

yields the desired driving forces. This method has been applied to the Delta-robot of R. Clavel (1988), a manipulator with many closed loops (Devaquet and Brauchli 1992).

3 Numerical Efficiency

3.1 Choice of Formulation and Linear Algebra

As mentioned previously, the governing equations of the projection method are not uniquely determined. Two families were proposed by (Sofer, Melliger, and Brauchli 1992, Sofer, Brauchli, and Melliger 1993, Sofer, Bach, and Brauchli 1995) and tested for efficiency, namely,

$$F_1: \quad \dot{q} = A\,p_\alpha + X\,p_\beta, \quad \dot{p} = Q_\alpha - \Omega\,A\,p_\alpha + Y\,p_\beta\,, \tag{33}$$

$$F_2: \quad \dot{q} = A\,p_\alpha, \quad \dot{p} = Q_\alpha + \alpha^T \Omega^T A\,p + \beta^T x\,. \tag{34}$$

X and Y can be chosen arbitrarily in the first family, x in the second. The best performance was found for

$$\dot{q} = A\,p_\alpha, \quad \dot{p} = Q_\alpha + \alpha^T \Omega^T A\,p - \Omega\,A\,p_\alpha\,, \tag{35}$$

which belongs to both families. This form was then chosen to be implemented in DYNAMITE (Bach, Brauchli, Melliger, and Hillmer 1993), a code used at the ETH, the Federal Institute of Technology of Switzerland.

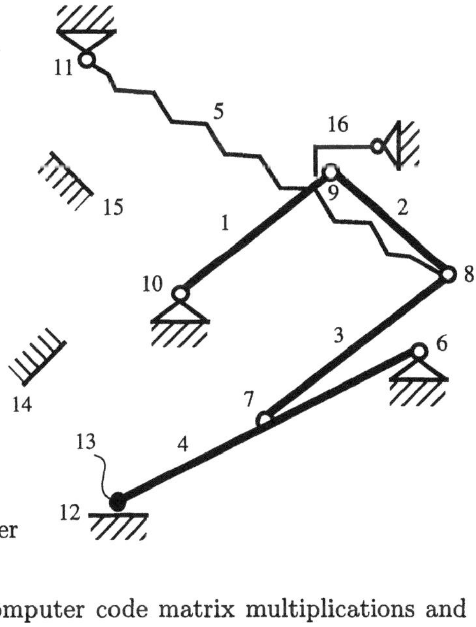

Fig. 1 Circuit breaker

Naturally, in a computer code matrix multiplications and matrix inversions have to be avoided and linear system solvers are to be used instead. In typical examples, the M and E matrices have a sparse structure. Even for an occupancy of about 80% the application of sparse-matrix techniques may reduce the cpu-time used by a factor of about 4. This is so because the sparsity structure is constant and has to be analyzed only once. An alternative would be to use parallel processing, but the two cannot be combined. We found the use of sparse-matrix techniques more advantageous. Many of my Ph.D.-students (Weber 1981, Weber 1986, Brauchli and Weber 1986, Devaquet 1993, Melliger 1994, Sofer, Bach, and Brauchli 1995, Schulz 1997, Chen and Brauchli 1997) have made significant contributions to the projection method and its implementation in DYNAMITE.

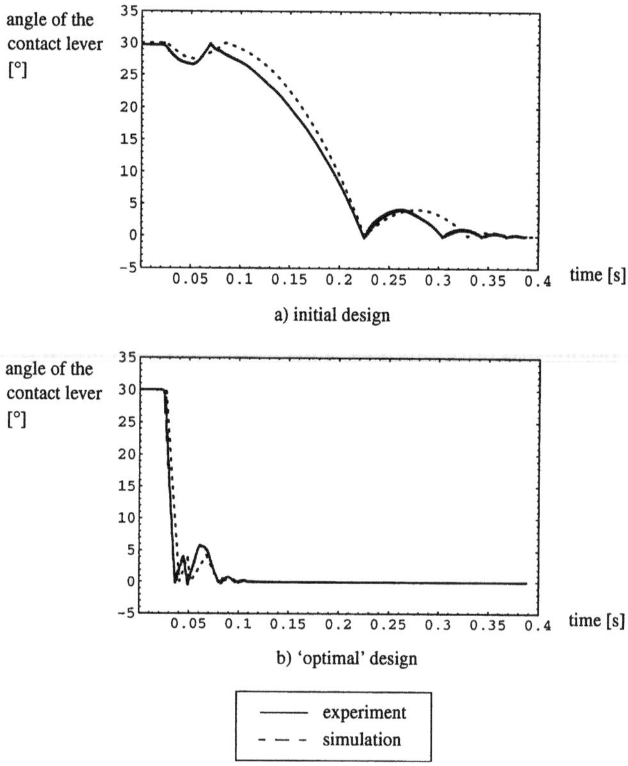

Fig. 2 Comparison between measurement and computation

3.2 Optimization

As an industrial example (Schulz and Brauchli 1996, Schulz 1997, Schulz and Brauchli 1997, Schulz, Mücke, and Walser 1997) we present the optimization of a low voltage curcuit breaker for ABB. In Fig. 1 a simplified model of the circuit breaker is shown. The contact at point 12 has to open as fast as possible and without recoil, when the lever 16 is moved. Several optimization strategies were compared (Obrist 1996). Typically, in problems of optimization of transient dynamical phenomena, the cost-function is very rough and the classical search methods are inefficient. They have to be supplemented by random methods. We have used multilevel single linkage and simulated annealing. The results of the computer-simulation were verified by an experiment (Schulz, Mücke, and Walser 1997). These results are shown in Fig. 2. With the optimal design, the cost-function could be reduced by a factor of 16.

4 Geometry of Dynamics

4.1 Curvature

We introduced the mass metric

$$ds^2 = m_{k\ell}\, dq^k\, dq^\ell \,, \tag{36}$$

in Section 2.3. and mentioned that the geodesics in this metric correspond to the inertial orbits in the mechanical system. If we use a constant mass matrix, the equations for an inertial motion take the form

$$M\,\ddot{q} + \Omega(\dot{q})\,\dot{q} = 0 \,. \tag{37}$$

On the other hand, the differential equation of a geodesic is given by

$$\ddot{q}^k + \Gamma^k_{ij}\,\dot{q}^i\,\dot{q}^j = 0 \,, \tag{38}$$

where Γ^k_{ij} denote the Christoffel symbols. However, in spite of the similarity of eqs. (37) and (38), the situation is completely different: eq. (37) describes the embedded system, and, as can be seen from eq. (30), Ω is related to the constraint reactions. Therefore, Ω is a measure of the curvature of the embedding and corresponds to the *second fundamental form* of elementary surface geometry. The Γ^k_{ij} coefficients, on the other hand, belong to the inner geometry of the configuration space. This relation can be made clear by stating the equations of the first variation. From eq. (37) we obtain (Brauchli and Weber 1991) for an inertial motion

$$\delta\ddot{q} + A\,\alpha^T\,\Omega^T(\delta q)\,A\,\Omega(\dot{q})\,\dot{q} - A\,\alpha^T\,\Omega^T(\dot{q})A\,\Omega(\delta q)\,\dot{q} = 0 \,, \tag{39}$$

while the corresponding equations in classical differential geometry read

$$\delta\ddot{q}^k + R^k_{ij\ell}\,\delta\dot{q}^i\,\delta q^j\,\dot{q}^\ell = 0. \tag{40}$$

So the Ω matrices are related to the intrinsic curvature through (39) as well as to the extrinsic curvature through eq. (37).

4.2 Topology of the Configuration Space

Typically, in mechanisms, the configuration space is a closed surface or manifold, which may or may not have boundaries. It is interesting to investigate its connectivity and related effects on the dynamics of the system. As an example we consider a planar four-bar linkage as shown in Fig. 3. The inclination of the four moving rods can be described by four angles, but these are not independent, the degree of freedom of this system being 2. Therefore, we have a two-dimensional surface as a configuration space. We use a geometrical parameter Λ for the respective lenghts of the rigid links,

$$\ell_1 = \ell \, , \, \ell_2 = \Lambda \ell \, , \, \ell_3 = \Lambda \ell \, , \, \ell_4 = \ell \, , \tag{41}$$

and a parameter a for the distance of the two fixed joints. Depending on these two parameters, this surface exhibits a different topology. We use an admissible motion as a tangential vector field of the configuration space with isolated singularities (Brauchli 1997). By a theorem of Poincaré the Euler-characteristic χ of the surface may then be computed as the index-sum over these singularities. The *genus* g of the surface indicates the number of holes. It is related with the Euler characteristic by

$$\chi = 2 - 2g \, . \tag{42}$$

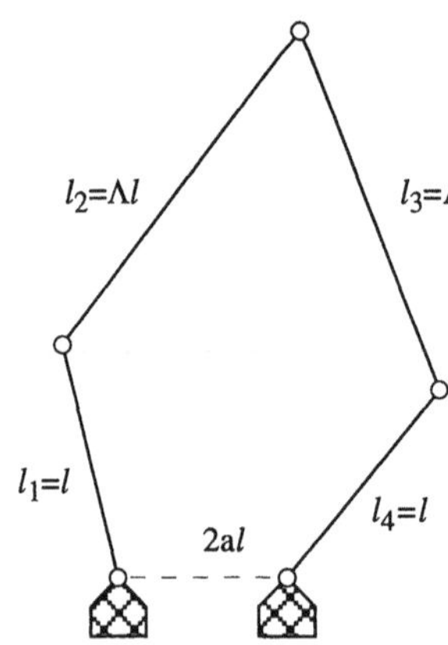

Fig. 3 Four-bar linkage

The corresponding bifurcation diagram is shown in Fig. 4. There are two regions, III and VII, where the surface consists of two components and both are tori ($\chi = 0$, $g = 1$). If the centers of all joints are aligned, the surface is a topological sphere ($\chi = 2$, $g = 0$), region I. The maximum number of wholes is taken in region V ($\chi = -6$, $g = 4$). In the other regions all intermediate values occur, as shown in Table 1.

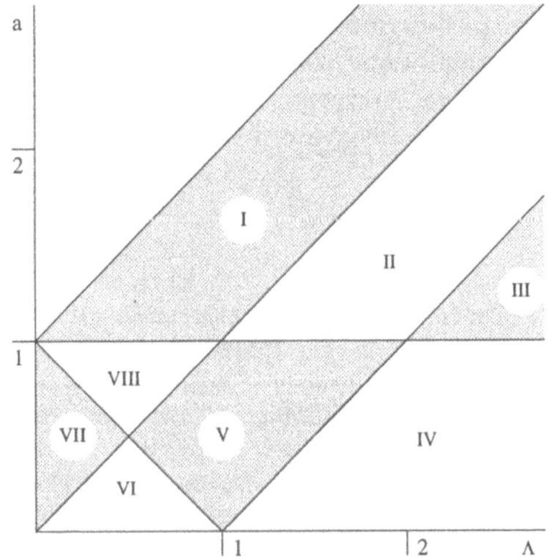

Fig. 4 Bifurcation diagram of four-bar linkage

For certain values of the geometric parameters a bifurcation occurs. From the theorem on the curvatura integra it can be understood that this is coupled with a jump in the curvature of 4π at a singular point. In the vicinity of this singularity the configuration space can be approximated by a cone, and the dynamics of the inertial motion locally admits an integral. Intuitively, this can be understood as a localized curvature producing an impact on the mechanical system.

domains	g	χ	K
I	0	+ 2	1
II, VIII	2	− 2	1
III, VII	1	0	2
IV, VI	3	− 4	1
V	4	− 6	1

Table 1: genus g, Euler-characteristic χ and number of components K

5 Conclusions

The question of numerical efficiency of the different methods for describing multibody systems and mechanisms is still important. For overall efficiency preference is given to descriptor formulations, while e.g. for real-time simulations minimal formulations will be chosen. One of the most intriguing questions is the geometric interpretation of dynamics. It may be particularly helpful in the study of the behaviour of mechanisms near a singularity.

References

Bach, D.N. (1993), *Ein Beitrag zur Starrkörpermechanik*, Diss. ETH Nr. 10367.

Bach, D., Brauchli, H., Melliger, O., Hillmer P. (1993), *DYNAMITE: Multibody Dynamics via Projection Methods*, in (Schiehlen 1993), pp. 355-360.

Bestle, D., Schiehlen W. (eds.), (1996), *IUTAM Symposium on Optimization of Mechanical Systems*, Kluwer Academic Publishers, Dordrecht.

Bianchi, G., Schiehlen, W. (eds.), (1986), *Dynamics of Multibody Systems*, IUTAM/ IFToMM Symposium Udine 1985, Springer-Verlag, Berlin.

Brauchli, H. (1991), *Mass-Orthogonal Formulation of Equations of Motion for Multibody Systems*, J. Appl. Math. Phys. (ZAMP) **42**, pp. 169-182.

Brauchli, H. (1997), *Konfigurationsräume einfacher Mechanismen*, to appear in Elemente der Mathematik.

Brauchli H., Weber, R. (1991), *Dynamical Equations in Natural Coordinates*, Comp. Meth. Appl. Eng. **91**, pp. 1403-1414.

Bremer, H., Pfeiffer, F. (1992), *Elastische Mehrkörpersysteme*, Teubner Studienbücher.

Chen, T., Brauchli, H. (1997), *On the Geometry, Dynamics and Stability of Non-Holonomic Systems*, presented at the NATO advanced Study Institute on Computational Methods in Mechanisms, Varna, Bulgaria.

Clavel, R. (1988), *Delta, a Fast Robot with Parallel Geometry*, Proc. Internat. Symposium on Industrial Robots, pp. 91-100.

Devaquet, G. (1993), *Die Gleichungen der ersten Variation redundant modellierter Starrkörpersysteme*, Diss. ETH Nr. 10395.

Devaquet, G., Brauchli, H. (1992), *A Simple Model for the DELTA-Robot*, Robotersysteme **8**, pp. 193-199.

Hiller, M. (1983), *Mechanische Systeme - eine Einführung in die analytische Mechanik und Systemdynamik*, Springer-Verlag, Berlin.

García de Jalón, J., Cuadrado, J., Avello, A., Jiménez, J.M. (1994), *Kinematic and Dynamic Simulation of Flexible Systems with Fully Cartesian Coordinates*, in (Pereira and Ambrósio 1994), pp. 285-323.

García de Jalón, J., Serna M.A., Avilés R (1981), *A Computer Method for Kinematic Analysis of Lower Pair Mechanisms. Part I: Velocities and Accelerations, Part II: Position Problems*, Mechanism and Machine Theory **16**, pp. 543-566.

Géradin, M., Cardona, A., Doan D.B., Duysens J. (1994), *Finite Element Modeling Concepts in Multibody Dynamics*, in (Pereira and Ambrósio 1994), pp. 233-284.

Kane, T.R., Levinson, D.A. (1985), *Dynamics: Theory and Applications*, McGraw-Hill, New York.

Magnus K. (ed.), (1978), *Dynamics of Multibody Systems*, IUTAM-Symposium Munich/Germany, August 29 - September 3, 1977, Springer-Verlag, Berlin.

Meirovitch, L. (1970), *Methods of Analytical Dynamics*, McGraw-Hill, New York.

Melliger, O.F. (1994), *Numerisch effiziente Methoden der Mehrkörperdynamik*, Diss. ETH Nr. 10755.

Nikravesh, P.E. (1988), *Computer-Aided Analysis of Mechanical Systems*, Prentice-Hall, Englewood-Cliffs.

Obrist, D. (1996), *Anwendung globaler Optimierungsverfahren auf die Optimierung der Dynamik eines Mechanismus mit einseitigen Bindungen und Stössen*, Semesterarbeit, Institut für Mechanik, ETH-Zürich.

Penrose, R. (1955), *A Generalized Inverse for Matrices*, Proc. Cambr. Phil. Soc. **51**, pp. 406-413.

Pereira, M.S., Ambrósio J.A.C. (eds.), (1994), *Computer-Aided Analysis of Rigid and Flexible Mechanical Systems*, NATO ASI Series E: Applied Sciences - Vol. 268, Kluwer Academic Publishers, Dordrecht.

Pereira, M.S., Ambrósio J.A.C. (eds.), (1995), *Computational Dynamics in Multibody Systems*, Kluwer Academic Publishers, Dordrecht.

Petzold, L.R. (1994), *Computational Challenges in Mechanical Systems Simulation*, in (Pereira and Ambrósio 1994), pp. 483-499.

Renaud, M. (1975), *Contribution à l'Etude de la Modélisation des Systèmes Mécaniques Articulés*, Diss. Univ. Toulouse.

Schiehlen, W.O. (1990), *Multibody Systems Handbook*, Springer-Verlag, Berlin.

Schiehlen, W. (ed.), (1993) *Advanced Multibody System Dynamics, Simulation and Software Tools*, Kluwer Academic Publishers, Dordrecht.

Schulz, M. (1997), *Optimierung von Mechanismen mit einseitigen Bindungen und Kollisionen*, Diss. ETH Nr. 12041.

Schulz, M., Brauchli, H. (1996), *Simulation, Sensitivity Analysis and Optimization of Constrained Multibody Systems with Impacts Based on Mass-Orthogonal Projections*, in (Bestle and Schiehlen 1996), pp. 261-268.

Schulz, M., Mucke, R., Walser, H.-P. (1997), *Optimisation of Mechanisms with Collisions and Unilateral Constraints*, Multibody System Dynamics **1**, pp. 223-240.

Schulz, M., Brauchli, H. (1997), *Optimisation of Mechanisms and the Effect of Collisions*, presented at the NATO advanced Study Institute on Computational Methods in Mechanisms, Varna, Bulgaria; submitted for publication.

Sofer M.M. (1991), *On Equilibrium, Stability and Nonlocality in Elasticity Theory*, Diss. ETH Nr. 9420.

Sofer, M., Melliger, O., Brauchli, H. (1992), *Numerical behaviour of different formulations for multibody dynamics*, Numerical Methods in Engineering '92, Ch. Hirsch et al (eds.), Elsevier Science Publishers, Amsterdam.

Sofer, M., Brauchli, H., Melliger O. (1993), *ODE Formulations for Multibody Dynamics: Numerical Aspects*, in (Schiehlen 1993), pp. 397-402.

Sofer, M., Bach, D., Brauchli, H. (1995), *Dynamics of Constrained Systems Based on Mass-Orthogonal Projections*, in (Pereira and Ambrósio 1995), pp. 1-13.

Synge, J.L. (1927), *On the Geometry of Dynamics*, Phil. Trans. A **226**, pp. 31-106.

Weber, R.W. (1981), *Kanonische Theorie nichtholonomer Systeme*, Diss. ETH Nr. 6876.

Weber, R. (1986), *Hamiltonian Systems with Constraints and their Meaning in Mechanics*, Arch. Rational Mech. Anal. **91**, pp. 309-335.

Wittenburg, J. (1977) *Dynamics of Systems of Rigid Bodies*, Teubner, Stuttgart.

A Generic Numerical Method for Mechanical System Kinematics and Dynamics Modeling

Evtim V. Zakhariev

Institute of Mechanics, Bulgarian Academy of Sciences, Acad. G. Bonchev Street, bl. 4, BG-1113 Sofia, Bulgaria. evtimvz@bgcict.acad.bg

Summary. The theory and basic numerical procedures of unified approach for kinematic analysis and optimization, as well as for force analysis and modeling of the nonlinear dynamics of spatial poly-contour mechanical systems are presented. Matrix transformation methods and Newton-Euler equations are used for deriving the kinematic and dynamic equations of motion. The resulting nonlinear kinematic model is common for the analysis and synthesis problems. Nonlinear programming techniques are used for solution of the kinematic equations. A new method of treating the nonlinear explicit form dynamic equations is presented. Generalized forces, reactions, internal forces, etc., obeying principles of proportionality and superposition, are derived as a function of generalized velocities and accelerations. The nonlinear model is focused on solution of special problems, including, for example, friction forces in the pairs. On the basis of the discretized nonlinear dynamic equations, a new numerical method for mechanical system motion control and optimization is suggested.

Keywords. Multibody systems, kinematics, nonlinear dynamics, friction, motion control

1. Introduction

In the recent past a large number of papers and books have developed methods for computer simulation of complex mechanical systems. The books of Wittenburg (1977), Robertson and Schwertassek (1988), Haug (1989), Shabana (1989), Garcia de Jalon and Bayo (1993), Schiehlen (1993), Lilov (1993), and Angeles and Kecskemethy (1995) have summarized the results of many research activities that have built an extensive basis for future studies.

The choice of kinematic parameters and characteristics of motion has significant influence on the effectiveness of analysis and synthesis. Several methods of kinematic analysis are available, most of which are based on screw theory and coordinate transformation matrices (Denavit and Hartenberg 1950; Freudenstain 1962; Bottema and Roth 1970; Angeles 1982). A detailed study would show their different descriptions but very similar algorithms and computer programs. The aim is computational efficiency and universality of the algorithms. As a whole, mathematical models for analysis and optimal synthesis are very similar, since the same equations of kinematic restrictions are being solved. The Newton method for numerical solution of systems of nonlinear equations has been successfully applied for the purposes of analysis, while methods of mechanism optimization are effective for the purposes of complex spatial mechanism synthesis. This is true since analytical and geometrical approaches are impractical and the experience of researchers is required to find feasible solutions. Recently developed and contemporary methods that deal with a large variety of optimal kinematic synthesis problems are based on consecutive analyses. The first treatments concerning optimal synthesis of mechanisms using optimization techniques are form the 1960-s; e.g., by Roth and Freudenstain (1963) and Han (1966). Since Fox and Willmert (1967) introduced nonlinear programming to optimization of a planar four bar mechanism, optimization methods for synthesis of both planar and spatial mechanisms have been developed in recent years. Detailed review on publications in this field is presented by Dimentberg et al. (1983). Many researches (Gupta 1973; Soylemez and Freudenstain 1982; etc.) have presented methods for optimization of spatial mechanisms. Angeles and Liu (1990) applied the method of least-squares with constrains for optimal synthesis of spherical path generators. Recent developments (for review see de Jalon et al. 1993; Hansen and Tortorelli 1996) have presented general methods for optimal synthesis of spatial mechanisms using Cartesian (natural) coordinates and nonlinear programming techniques.

The kinematic characteristics of motion are included in every kind of dynamics models, no matter whether they are presented explicitly or numerically. A fundamental difference between existing methods is the coordinates used in kinematic and dynamic models. Most often, Cartesian and generalized coordinates are applied. When generalized coordinates are used most often with Lagrange and Newton-Euler equations, kinematic characteristics cannot be explicitly defined and their numerical realization is required. Methods of dynamics modeling that are based on differential - algebraic equations (DAE) demand that kinematic characteristics to be presented in the constraints, which are kinematic equations of the system. When utilizing Cartesian coordinates, the kinematic model is characterized by nonlinear equations with a large number of parameters, while the dynamic model is, in general, a second order DAE that is much more complex than an ordinary differential equations (ODE). The stability of numerical integration methods for solving DAE is difficult to analyze since the kinematic restrictions are algebraic components of the second order differential dynamics equations that should be solved simultaneously. When using generalized coordinates, the initial calculations and computer algorithms are cumbersome. However, the final result for the dynamics model is an ODE, which leads to stable numerical solution for even nonlinear dynamic equations.

The inclusion of a general rigid-body dynamic model in the process of design and control improves mechanical system behavior and performance fidelity. However, friction imposes additional non-linearity in the dynamic equations of connected moving bodies. Force analysis for spatial mechanisms with open or one closed contour including friction in the joints was successfully achieved by Keler (1973). An iterative procedure for friction force presentation as external forces is applied in an algorithm for closed loop mechanism analysis. For the same problem, Zakhariev (1995) derived quasi-static nonlinear equations for the reactions (including friction forces) and generalized forces and numerically solved them, obtaining better convergence. Haug et al. (1986) presented unified formulation of the equations of motion for constrained multibody systems with Coulomb friction, employing Cartesian generalized coordinates and Lagrange multipliers. The dynamics model so obtained is in the form of a DAE.

In this chapter generalized coordinates are used for kinematic model description. The method of Denavit and Hartenberg for presentation of kinematic parameters is modified and applied for poly-contour spatial mechanisms and adjacent bodies with many joints. A procedure for orientation of joint and body coordinate systems and estimation of independent design parameters is suggested. The kinematic model is generalized for the purposes of analysis and synthesis. Relative and absolute characteristics of motion are numerically obtained and involved in the next stages of the simulation process. Newton-Euler equations are applied for dynamics problem solution. A recurrent algorithm for force analysis is suggested, for use in deriving of the nonlinear dynamics model. The explicit form dynamic equations of a multibody system are obtained using a minimal set of generalized coordinates, with no kinematic constraints. In the method presented, reactions in the joints are obtained as intermediate values and their functional dependence with respect to the generalized velocities and accelerations is derived numerically. The method is based on principles of proportionality and superposition. Some special dynamic effects such as friction in the pairs are effectively modeled.

The nonlinear dynamic model presented here is involved in a method for motion control and optimization of poly-contour spatial mechanical systems. The control algorithm could be described with continuous targeting of the effector to its final positions. The trajectory is discretized in small intervals (nodes) where the robot motion is assumed to be with constant general coordinate accelerations. The mathematical description of the problem for optimal motion planning is presented as nonlinear programming problem. The characteristics of motion and discretized generalized forces in every node are parameters of the optimization problem. The discretized equations of motions and dynamic equations are equality constraints for the programming problem.

An example of motion of a double pendulum mechanism with one open branch and a closed contour taking into account friction in the pairs is presented.

2. Representation of Mechanism Parameters

2.1. Topological Description

Representation of a mechanism structure in a comprehensive manner using the computer is the initial step for its analysis and synthesis. The most developed and contemporary approach of description of mechanism topology is based on graph theory (Wittenburg 1993). In Fig. 2.1a, a chain with closed loops is shown. The

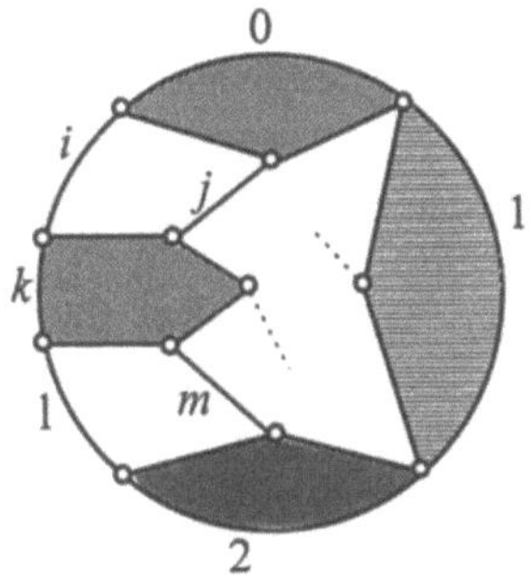

a

$$\text{Joint No.}\quad 1\ 2\ 3\ 4\ 5\ 6\ 7\ 8\ 9\ 10\ \ldots\ldots$$

$$IS=\begin{bmatrix} 0 & 0 & 0 & 1 & 2 & 2 & k & k & k & k & \ldots\ldots \\ 1 & i & j & 2 & m & l & i & j & l & m & \ldots\ldots \end{bmatrix}$$

b

$$IR=\begin{bmatrix} 2 & 9 & 2 & \ldots\ldots \\ -7 & -6 & -7 & \ldots\ldots \\ 8 & 5 & 9 & \ldots\ldots \\ -3 & -10 & -6 & \ldots\ldots \\ * & * & -4 & \ldots\ldots \\ * & * & -1 & \ldots\ldots \end{bmatrix}$$

c

Fig. 2.1 Structure representation of a poly-contour chain

links are numbered consequently from 0 to $\underline{n}$ (chain with $\underline{n}+1$ links). For a pair with index *ij*, the links numbered *i* and *j* are adjacent. For a mechanism with $\underline{p}$ pairs, a $2\times\underline{p}$ initial structure matrix ***IS*** (Fig. 2.1b) describes the chain structure. For this matrix (Wittenburg 1993), the *k*th column describes the pair *ij* (numbered as *k*th) in which column the first element ***IS***$(1,k)$ is the number of *i*th link and the second element ***IS***$(2,k)$ is the number of *j*th link indicating that they are adjacent. This matrix supplies the computer program with information on the number and kind of links and the number of pairs. It is enough that the graph and its incidence matrix are defined (Wittenburg 1993; McPhee 1996). If Cartesian (natural) coordinates are used for mechanical system modeling, the numerical structure data for incidence matrix elements are the numerical basis for defining kinematic restrictions. If generalized coordinates are applied, the restrictions present the connectivity of links in chains (closed contours and branches), so additional structure information must be given. In Fig. 2.1c, the matrix ***IR*** of contours and branches is shown, obtained with input values of elements of matrix ***IS***. The *i*th column (contour) contains $\underline{p}_i$ elements, where $\underline{p}_i$ is the number of joints in the *i*th contour (i = 1, 2,..., $\underline{r}$). The values of elements of matrix ***IR*** define the joint number in the initial structure matrix. The consequence of the elements of the *k*th column indicates the place of the corresponding joint in the matrix equations for the coordinate system transformation of links pertaining to the *k*th chain. For example, if ***IR***$(i,k) = m$ then the *m*th joint participate in the corresponding *k*th contour and the coordinate transformation is from link ***IS***$(2,m)$ to link ***IS***$(1,m)$. If ***IS***$(i,k) = -m$, then the coordinate transformation is from link ***IS***$(1,m)$ to link ***IS***$(2,m)$.

2.2. Kinematic Parameters and Characteristics of Motion

A main problem of kinematic analysis and synthesis is identification of the independent characteristics of motion and design parameters. Angeles (1993) presented a detailed study on joint parameters and motions and their description through screw theory. It is known that the parameters of Denavit and Hartenberg (1950) are the minimal number of parameters that give the relative position of two axes connecting adjacent links in space. The application of these parameters has been explored for open and one contour spatial chains. Such a procedure for poly-contour spatial mechanisms can create problems, due to inclusion of dependent parameters in the analysis and design processes.

Every link has unique body fixed coordinate system defined. Two adjacent links undergo relative motion restricted by the geometry of the contacting surfaces of the corresponding pair. Relative motions in occurring conventional joints, single degree of freedom (sdf) or many degree of freedom (mdf), are mostly along the axes of a spatial Cartesian coordinate system that defines the position of the contact surfaces with respect to the body reference frames. These are the parameters of the relative position of the joint coordinate systems that present kinematic parameters of the links. The aim of this section is to describe simple rules for joint coordinate system orientation, in such a manner that these parameters are independent motion and design variables.

In Fig. 2.2a, the *k*th link with n_k joints (multiple link with more than three adjacent links) of a spatial mechanism is shown. In every pair, a coordinate system is located with an index of the corresponding pair ($O_{kl}X_{kl}Y_{kl}Z_{kl}$ in the *kl*th joint) and with axes set up along geometrical axes of the joint. The motions in each joint are defined by a 6 × 1 motion matrix Δ_{kl} (for the *kl*th joint) in which the first three elements define translations along the joint coordinate system axes. The remaining three elements correspond to possible small rotations in the joint. These values for all joints of the system, are ordered in a 6 × $\underline{p}$ matrix Δ (Fig. 2.2c, d) that is of similar order to the initial structure matrix ***IS*** (Fig. 2.2b). A matrix ***IΔ*** (Fig. 2.2e) with elements 1 or 0 determines whether the corresponding elements are active or not; i.e., if free motion along the corresponding axis occurs or not.

Only one coordinate system is firmly fixed to the kth link and arbitrarily located in it. In order to eliminate dependent design parameters in Fig. 2.2a, this coordinate system is fixed in the *kl*th joint (the first of the *k*th link). Here sdf joints are discussed, since each mdf joint could be presented as a consequence of binary links and sdf joints (Woo and Freudenstain 1970). The *Z* axes representing degenerate joint contact surfaces are set up along the joint axes. Obviously, the parameters shown as a_{ij}, s_{ij}, φ_{ij}, θ_{ij} for the *ij*th pair are similar to these of Denavit and Hartenberg. All of them are design variables, not characteristics of motion. In Fig. 2.2c, the dependent parameters are zero, because the first joint

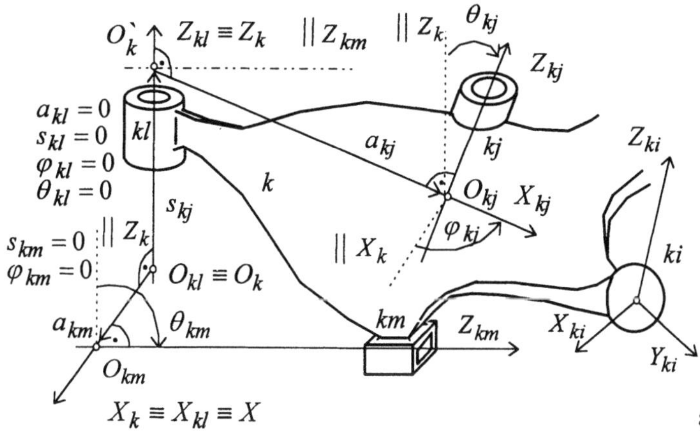

a

$$IS = \begin{bmatrix} .. & .. & k & k & k & k & .. & .. \\ .. & .. & l & m & i & j & .. & .. \end{bmatrix}$$

b

$$\Delta = \begin{bmatrix} .. & .. & \Delta_{kl} & \Delta_{km} & \Delta_{ki} & \Delta_{kj} & .. & .. \end{bmatrix}$$

c

$$\Delta = \begin{bmatrix} .. & .. & \Delta x_{kl} & \Delta x_{km} & \Delta x_{ki} & \Delta x_{kj} & .. & .. \\ .. & .. & \Delta y_{kl} & \Delta y_{km} & \Delta y_{ki} & \Delta y_{kj} & .. & .. \\ .. & .. & \Delta z_{kl} & \Delta z_{km} & \Delta z_{ki} & \Delta z_{kj} & .. & .. \\ .. & .. & \theta x_{kl} & \theta x_{km} & \theta x_{ki} & \theta xj & .. & .. \\ .. & .. & \theta y_{kl} & \theta y_{km} & \theta y_{ki} & \theta y_{kj} & .. & .. \\ .. & .. & \theta z_{kl} & \theta z_{km} & \theta z_{ki} & \theta z_{kj} & .. & .. \end{bmatrix}$$

d

$$I\Delta = \begin{bmatrix} .. & .. & 0 & 0 & 0 & 0 & .. & .. \\ .. & .. & 0 & 0 & 0 & 0 & .. & .. \\ .. & .. & 0 & 1 & 0 & 0 & .. & .. \\ .. & .. & 0 & 0 & 1 & 0 & .. & .. \\ .. & .. & 0 & 0 & 1 & 0 & .. & .. \\ .. & .. & 1 & 0 & 1 & 1 & .. & .. \end{bmatrix}$$

e

Fig. 2.2 The *k*th link and its numerical representation

coordinate system (index *kl*) coincides with the link coordinate system (index *k*), the axis X_{kl} coincides with X_{km} (*km*th is the second joint of the *k*th link), and they are along the shortest line between the axes Z_{kl} and Z_{km}. The positions of the other joint coordinate systems are shown with the full number of Denavit and Hartenberg parameters. It is known (Angeles 1993) that the axis of a translation is not fixed (free vector) in the relative space and only its direction is significant regarding the relative motion of connected bodies. This means that the parameters of a translational pair, which define its position with respect to other joints of the link, are dependent. For example, the translational axis *km* in Fig. 2.2 could be located at point $\grave{O}_k$, with axis $\grave{Z}_{km}$ parallel to Z_{km}.

For a sdf pair connecting two links, eight design parameters exist; four for each of adjacent link. In Fig. 2.3, two links (*k*th and *l*th) are displayed. The *kl* and *lk* coordinate systems in the joint *kl* are shown, and their relative position is pointed

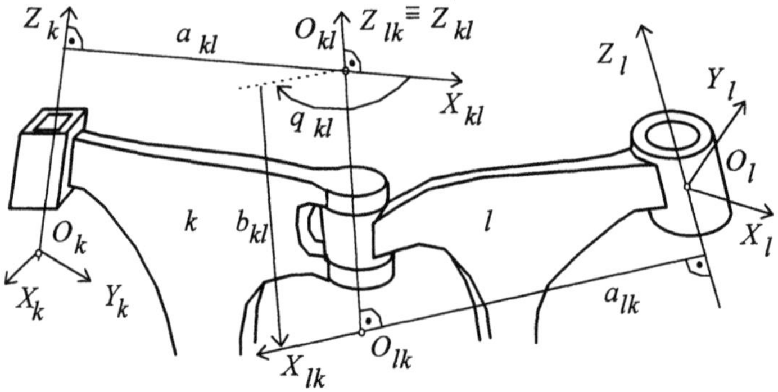

Fig. 2.3 Two adjacent links

out. If joint *kl* is rotational, then the angle between the X_{kl} and X_{lk} axes is a characteristic of motion (joint coordinate q_{kl}), and the distance b_{kl} between the coordinate system origins O_{kl} and O_{lk} is a kinematic (design) parameter of the chain, and vice versa for a translational pair. The characteristics of motion in all joints of the chain are arranged in a $\underline{p} \times 1$ matrix $\boldsymbol{Q}_{\underline{p}} = \begin{bmatrix} q_1 & q_2 & \cdots & q_{\underline{p}} \end{bmatrix}^{\mathrm{T}}$. All parameters of a mechanism are arranged in a $9 \times \underline{p}$ matrix $\boldsymbol{A}$ of the parameters located by the $9 \times \underline{p}$ matrix $\boldsymbol{IA}$. The first four elements of the *i*th column (joint ordered as *i* in matrix $\boldsymbol{IS}$) of matrix $\boldsymbol{A}$ correspond to the design parameters of the *i*th joint with respect to the $\boldsymbol{IS}(1,i)$th link. The next element of the column corresponds to the parameter b_{kl}. The last four elements of the *i*th column correspond to the design parameters of the *i*th joint with respect the $\boldsymbol{IS}(2,i)$th link. The elements of $\boldsymbol{IA}$ are 1 or 0 if the corresponding elements (design parameters) of matrix $\boldsymbol{A}$ are respectively variable or constant. If there are not special considerations, the elements corresponding to the parameters of the joint, in which the link coordinate system is set up, are zero, as are parameters s and φ for the joints arranged as number two in the link. The elements of matrices $\boldsymbol{A}$ and $\boldsymbol{IA}$ for the link and joints in Fig. 2.2 are as follows:

$$\boldsymbol{A} = \left[\begin{array}{cccccccc} \cdot\cdot & \cdot\cdot & a_{kl}=0 & a_{km} & a_{ki} & a_{kj} & \cdot\cdot & \cdot\cdot \\ \cdot\cdot & \cdot\cdot & s_{kl}=0 & s_{km}=0 & s_{ki} & s_{kj} & \cdot\cdot & \cdot\cdot \\ \cdot\cdot & \cdot\cdot & \varphi_{kl}-0 & \varphi_{km}=0 & \varphi_{ki} & \varphi_{kj} & \cdot\cdot & \cdot\cdot \\ \cdot\cdot & \cdot\cdot & \theta_{kl}=0 & \theta_{km} & \theta_{ki} & \theta_{kj} & \cdot\cdot & \cdot\cdot \\ \hline \cdot\cdot & \cdot\cdot & b_{kl} & b_{mk} & b_{ki} & b_{kj} & \cdot\cdot & \cdot\cdot \\ \hline \cdot\cdot & \cdot\cdot & a_{lk} & a_{mk} & a_{ik} & a_{jk} & \cdot\cdot & \cdot\cdot \\ \cdot\cdot & \cdot\cdot & s_{lk} & s_{mk} & s_{ik} & s_{jk} & \cdot\cdot & \cdot\cdot \\ \cdot\cdot & \cdot\cdot & \varphi_{lk} & \varphi_{mk} & \varphi_{ik} & \varphi_{jk} & \cdot\cdot & \cdot\cdot \\ \cdot\cdot & \cdot\cdot & \theta_{lk} & \theta_{mk} & \theta_{ik} & \theta_{jk} & \cdot\cdot & \cdot\cdot \end{array}\right]; \quad \boldsymbol{IA} = \left[\begin{array}{cccccccc} \cdot\cdot & \cdot\cdot & 0 & 0 & 1 & 1 & \cdot\cdot & \cdot\cdot \\ \cdot\cdot & \cdot\cdot & 0 & 0 & 1 & 0 & \cdot\cdot & \cdot\cdot \\ \cdot\cdot & \cdot\cdot & 0 & 0 & 1 & 1 & \cdot\cdot & \cdot\cdot \\ \cdot\cdot & \cdot\cdot & 0 & 1 & 1 & 1 & \cdot\cdot & \cdot\cdot \\ \hline \cdot\cdot & \cdot\cdot & 1 & 1 & 1 & 1 & \cdot\cdot & \cdot\cdot \\ \hline \cdot\cdot & \cdot\cdot & 1 & 1 & 1 & 1 & \cdot\cdot & \cdot\cdot \\ \cdot\cdot & \cdot\cdot & 1 & 1 & 1 & 1 & \cdot\cdot & \cdot\cdot \\ \cdot\cdot & \cdot\cdot & 1 & 1 & 1 & 1 & \cdot\cdot & \cdot\cdot \\ \cdot\cdot & \cdot\cdot & 1 & 1 & 1 & 1 & \cdot\cdot & \cdot\cdot \end{array}\right].$$

In matrix $\boldsymbol{A}$ it can be seen that joint kl is first and joint km is second for the kth link, because the parameters $a_{kl}, s_{kl}, \varphi_{kl}, \theta_{kl}, s_{km}, \varphi_{km}$ are zero and their corresponding elements in matrix $\boldsymbol{IA}$ are also zero.

The parameters of two adjacent links, which are connected by a translational pair, defining their relative position (in this case kth and mth links) are dependent. Some of them are constant and so evaluated as to maintain only the relative position of other joint axes. For example, the parameters a_{km} and s_{kj} in Fig. 2.2 are not variable, which is expressed as zero values of the corresponding elements in matrix $\boldsymbol{IA}$. Of course a fictitious location of a translational pair is most often a bad solution, as far as the design method is concerned. It is preferably the corresponding parameters to be restricted in certain boundaries. Considering the links of a mechanism and taking into account the connectivity; i.e., the parameters b_{kl} of the relative movement, the number of design parameters $\underline{N}$ is

$$\underline{N} = \sum_{j} n_j (4.j - 6) + p_5 \tag{2.1}$$

where p_5 is the number of joints of class 5 (all joints for a chain with sdf pairs). This number could be decreased with the number of dependent parameters of translational pairs. When function generating mechanisms are considered, the number of parameters should be reduced - normalized lengths of the links are applied. If path generating mechanism is examined the parameters must be increased with the path generating point parameters and the coordinates of base joints.

3. Kinematic Constraints of Spatial Mechanisms

The relative positions of the coordinate systems of links and joints, defined according to the rules suggested above, are presented by 4 $\times$ 4 transformation matrices (Denavit and Hartenberg 1950). The matrix of coordinate system transformation of body j with respect to body i is $\boldsymbol{T}_{i,j} = \begin{bmatrix} \tau_{i,j} & \rho_{i,O_j} \\ \boldsymbol{0}^3 & 1 \end{bmatrix}$ where

$$\boldsymbol{T}_{i,j}^{-1} = \begin{bmatrix} \tau_{i,j}^{\mathrm{T}} & -\tau_{i,j}^{\mathrm{T}} \cdot \rho_{i,O_j} \\ \boldsymbol{0}^3 & 1 \end{bmatrix} \tag{3.1}$$

$\tau_{i,j}$ is matrix of rotation and $\tau_{i,j}^{\mathrm{T}}$ is its matrix transpose; the superscript (-1) denotes matrix inverse; ρ_{i,O_j} is a 3 $\times$ 1 matrix, the vector of coordinates of jth coordinate system origin with respect to *i*th coordinate system; and 0^3 is 1×3 zero matrix.

For every pair, three matrices are defined. For example, the *ij*th pair is defined by the matrices $T_{i,ij}$, $T_{ij,ji}$, and $T_{j,ji}$. The matrix $T_{i,ij}$ presents the position and orientation of the *ij*th coordinate system (in *ij*th pair) with respect to the coordinate system of the *i*th link. The matrix $T_{ij,ji}$ presents the position and orientation of the *ji*th coordinate system with respect to the *ij*th one. The values of elements of these matrices do not depend on the kind of the relative parameters; Euler, Denavit and Hartenberg, etc. Once these parameters are evaluated and matrix elements calculated, only the values of the elements of matrix $T_{ij,ji}$ change during system motion. It is preferable that the parameters of Denavit and Hartenberg be used (see the aforementioned approach) in kinematic synthesis, while this is not mandatory in kinematic analysis. The matrix $T_{i,j}$ of relative position of the link *j* with respect to link *i* is

$$T_{i,j} = T_{i,ij} \cdot T_{ij,ji} \cdot T_{j,ji}^{-1} \tag{3.2}$$

The matrices $T_{i,ij}$ and $T_{j,ji}$ depend on the design parameters of their own links. The elements of matrix $T_{ij,ji}$ depend on a design parameter and a characteristic of motion that defines the relative displacement of the adjacent links in joint *ij* (see sect. 2.2). For this matrix, the following equation holds:

$$T_{ij,ji} = T_{ij,ji}^{\varphi} \cdot T_{ij,ji}^{s} = \begin{bmatrix} cos(q_{ij}^{\varphi}) & sin(q_{ij}^{\varphi}) & 0 & 0 \\ -sin(q_{ij}^{\varphi}) & cos(q_{ij}^{\varphi}) & 0 & 0 \\ 0 & 0 & 1 & 0 \\ 0 & 0 & 0 & 1 \end{bmatrix} \cdot \begin{bmatrix} 1 & 0 & 0 & 0 \\ 0 & 1 & 0 & 0 \\ 0 & 0 & 1 & q_{ij}^{s} \\ 0 & 0 & 0 & 1 \end{bmatrix}$$

$$= T_{ij,ji}^{s} \cdot T_{ij,ji}^{\varphi} \tag{3.3}$$

If the joint is rotational, so matrix $T_{ij,ji}^{s}$ is first in the matrix equation of Eq. 3.3 and vice versa if the joint is translational; i.e., the matrix containing the characteristic of motion is the second in Eq. 3.3. The matrix $T_{i,ij}$ could be presented as its multiplication by the first matrix of $T_{ij,ji}$, which in turn is presented by its second matrix. Thus, the matrix $T_{ij,ji}$ contains only the characteristic of motion in the corresponding joint.

When the elements of ***IS***, corresponding to joint k (ordered in the kth column of ***IS***), are ***IS***$(1,k) = i$ and ***IS***$(2,k) = j$, then $\boldsymbol{T}_{i,j} = \boldsymbol{T}_k$. The partial derivatives of $\boldsymbol{T}_k$ for the design parameters are

$$\frac{\partial \boldsymbol{T}_k}{\partial a_{m\backslash k}} = \frac{\partial \boldsymbol{T}_{i,ij}}{\partial a_{m\backslash k}} . \boldsymbol{T}_{ij,ji} . \boldsymbol{T}_{j,ji}^{-1} \quad , m = 1, 2, 3, 4, 5 \tag{3.4}$$

$$\frac{\partial \boldsymbol{T}_k}{\partial a_{m\backslash k}} = \boldsymbol{T}_{i,ij} . \boldsymbol{T}_{ij,ji} . \frac{\partial \boldsymbol{T}_{j,ji}^{-1}}{\partial a_{m\backslash k}} \quad ,m=6,7,8,9 \tag{3.5}$$

The partial derivatives of $\boldsymbol{T}_k$ for the characteristics of motions are

$$\frac{\partial \boldsymbol{T}_k}{\partial q_k} = \boldsymbol{T}_{i,ij} . \frac{\partial \boldsymbol{T}_{ij,ji}}{\partial q_k} . \boldsymbol{T}_{j,ji}^{-1} \tag{3.6}$$

$$\frac{\partial^2 \boldsymbol{T}_k}{\partial q_k^2} = \boldsymbol{T}_{i,ij} . \frac{\partial \boldsymbol{T}_{ij,ji}^2}{\partial q_k^2} . \boldsymbol{T}_{j,ji}^{-1} \tag{3.7}$$

$$\frac{\partial \boldsymbol{T}_k}{\partial q_i} = \boldsymbol{0}_4^4 , \ i = 1, 2, \dots , k\text{-}1, k\text{+}1, \dots, \underline{p} ; \frac{\partial^2 \boldsymbol{T}_k}{\partial q_i \partial q_j} = \boldsymbol{0}_4^4 \tag{3.8}$$

The kinematic constraints for a mechanism, as well as the nonlinear function for the parameters and characteristics of motion, are as follows

$$\boldsymbol{H}^{(l)} = \boldsymbol{T}_{(1)}^{(l)} . \boldsymbol{T}_{(2)}^{(l)} ..... \boldsymbol{T}_{(\underline{p}_l)}^{(l)} - \boldsymbol{E}_4 = \boldsymbol{0}_4^4 , \ l = 1, 2, \dots, \underline{r} \tag{3.9}$$

where $\boldsymbol{0}_4^4$ is the 4×4 zero matrix; $\boldsymbol{E}_4$ is the 4×4 identity matrix; $\underline{p}_l$ is the number of elements in the lth column of ***IR*** ; i.e. the number of joints in chain l. The matrices $\boldsymbol{T}_{(n)}^{(l)}$, $n = 1, 2, \dots, \underline{p}_l$ are identified as follows:

$k = abs(\boldsymbol{IR}(n,l))$;

$\boldsymbol{T}_{(n)}^{(l)} = \boldsymbol{T}_k$, if $\boldsymbol{IR}(n,l) > 0$;

$\boldsymbol{T}_{(n)}^{(l)} = \boldsymbol{T}_k^{-1}$, if $\boldsymbol{IR}(n,l) < 0$.

If chain l is an open branch, the last matrix represents the relative position of the end link of the corresponding branch l with respect to the initial one, both connected by an mdf (in this case 6) joint. The elements of such a matrix are characteristics of motion and are treated as joint coordinates.

The matrix-vector of joint coordinates

$$\boldsymbol{Q}_{\underline{p}}(t)=\begin{bmatrix}\boldsymbol{Q}\\ \boldsymbol{Q}_{\underline{d}}\end{bmatrix}=\begin{bmatrix}\begin{bmatrix}q_1(t) & q_2(t) & \dots & q_{\underline{g}}(t)\end{bmatrix}^{\mathrm{T}}\\ \begin{bmatrix}q_{\underline{g}+1}(q_1,q_2,\dots,q_{\underline{g}}) & \dots & q_{\underline{p}}(q_1,q_2,\dots,q_{\underline{g}})\end{bmatrix}^{\mathrm{T}}\end{bmatrix} \tag{3.10}$$

is composed of the $\underline{g}\times 1$ column $\boldsymbol{Q}$ of the generalized (controlled or driven) joint movements (coordinates) and the $\underline{d}\times 1$ sized ($\underline{d}=\underline{p}-\underline{g}$) column $\boldsymbol{Q}_{\underline{d}}$ of the dependent joint movements (coordinates). For a common (not overconstrained) mechanism, six equations in the lth matrix equation of Eq. 3.9 are independent; e.g., three linear independent elements of the matrix of rotation (for example the diagonal elements) and the first three elements of the last column of $\boldsymbol{H}^{(l)}$. For $l=1, 2, \dots, \underline{r}$, the nonlinear system

$$\boldsymbol{H}(\boldsymbol{A},\boldsymbol{Q}_{\underline{p}})=\begin{bmatrix}h_1 & h_2 & \dots & h_{6.\underline{r}}\end{bmatrix}^{\mathrm{T}}=\begin{bmatrix}h^{(1)}_{1\backslash 1} & h^{(1)}_{2\backslash 2} & h^{(1)}_{3\backslash 3} & h^{(1)}_{1\backslash 4} & h^{(1)}_{2\backslash 4}\\ h^{(1)}_{3\backslash 4} & h^{(2)}_{1\backslash 1} & h^{(2)}_{2\backslash 2} & \dots & h^{(\underline{r})}_{3\backslash 3} & h^{(\underline{r})}_{1\backslash 4} & h^{(\underline{r})}_{2\backslash 4} & h^{(\underline{r})}_{3\backslash 4}\end{bmatrix}^{\mathrm{T}}=\boldsymbol{0}_{6.\underline{r}} \tag{3.11}$$

is composed of $6.\underline{r}=\underline{d}$ independent equations of Eq. 3.9. In Eq. 3.11, $\boldsymbol{0}_{6.\underline{r}}$ is the $6.\underline{r}\times 1$ zero matrix. Assuming Eqs. 3.4, 3.5, 3.6, 3.9, and 3.11 the $9\times\underline{p}$ block-matrix $\dfrac{\partial\boldsymbol{H}}{\partial\boldsymbol{A}}$ of partial derivatives of $\boldsymbol{H}$ with respect to the design parameters is calculated, where its ijth block-element is $\left\langle\dfrac{\partial\boldsymbol{H}}{\partial\boldsymbol{A}}\right\rangle_{i\backslash j}=\dfrac{\partial\boldsymbol{H}}{\partial a_{i\backslash j}}$; $i=1, 2, \dots, 9$; $j=1, 2, \dots, \underline{p}$. The same matrix is presented as a cubic-matrix with element ij as follows: $\left\langle\dfrac{\partial\boldsymbol{H}}{\partial\boldsymbol{A}}\right\rangle_{i\backslash j\backslash k}=\dfrac{\partial h_i}{\partial a_{j\backslash k}}$; $i=1, 2, \dots, \underline{d}$; $j=1, 2, \dots, 9$; $k=1, 2, \dots, \underline{p}$. In the same way, the $6.\underline{r}\times\underline{p}$ matrix $\dfrac{\partial\boldsymbol{H}}{\partial\boldsymbol{Q}_{\underline{p}}}$, from which the matrices $\dfrac{\partial\boldsymbol{H}}{\partial\boldsymbol{Q}}$ and $\dfrac{\partial\boldsymbol{H}}{\partial\boldsymbol{Q}_{\underline{d}}}$ are composed, could be calculated. The matrix equations of Eq. 3.11 composed from Eq. 3.9, present an inexplicit function of the motion characteristics depending on the design parameters. With prescribed values of the design parameters and generalized (controlled) coordinates, solving Eq. 3.11 for the dependent joint motions yields the desired results.

3.1. Kinematic Analysis

Since the Eqs. 3.9 and 3.11 are highly nonlinear, the iterative Newton-Raphson method is applied for their solution; i.e., with initial values of the dependent joint coordinates $\tilde{\underline{Q}}_d$ for which Eqs. 3.11 are not satisfied, a new solution $\underline{Q}_d$ is obtained from the linear system

$$\underline{H}(\tilde{\underline{Q}}_d) + \frac{\underline{H}(\tilde{\underline{Q}}_d)}{\partial \underline{Q}_d}(\tilde{\underline{Q}}_d - \underline{Q}_d) = \underline{0}_{6.r}\,. \tag{3.12}$$

If $\underline{Q}_d$ does not satisfy Eq. 3.11 within prescribed limits, Eqs. 3.12 are again solved with the new values of $\tilde{\underline{Q}}_d = \underline{Q}_d$ obtained in the foregoing iteration. Once $\underline{Q}_d$ is calculated and, with given generalized velocities $\dot{\underline{Q}}$, the matrix of the dependent joint velocities is obtained from the linear system

$$\frac{\partial \underline{H}}{\partial \underline{Q}_p} \cdot \begin{bmatrix} \dot{\underline{Q}} \\ \dot{\underline{Q}}_d \end{bmatrix} = \underline{0}_{6.r} \tag{3.13}$$

Computing $\underline{Q}_d$ and $\dot{\underline{Q}}_d$, the generalized acceleration $\ddot{\underline{Q}}$ is obtained from

$$\frac{\partial \underline{H}}{\partial \underline{Q}_p} \cdot \begin{bmatrix} \ddot{\underline{Q}} \\ \ddot{\underline{Q}}_d \end{bmatrix} + \dot{\underline{Q}}_p^{\mathrm{T}} \cdot \frac{\partial^2 \underline{H}}{\partial \underline{Q}_p \partial \underline{Q}_p} \cdot \dot{\underline{Q}}_p = \underline{0}_{6.r} \tag{3.14}$$

The matrix $\ddot{\underline{Q}}_d$ of the dependent joint accelerations can thus be calculated. With

$$\left\langle \frac{\partial^2 \underline{H}}{\partial \underline{Q}_p \partial \underline{Q}_p} \right\rangle_{i\backslash j\backslash k} = \frac{\partial h_k}{\partial q_i \partial q_j}\,;\ i, j = 1, 2, \ldots, \underline{p};\ k = 1, 2, \ldots, \underline{d} \tag{3.15}$$

element *ijk* of the corresponding cubic-matrix in Eq. 3.14 is obtained.

If the relative characteristics of motion are known, the characteristics of motion of the links with respect to the inertia reference frame (index "0") could be calculated, knowing the absolute position of one link, most often the base link. Starting with link i for which the matrices $\underline{T}_{0,i}$, $\dot{\underline{T}}_{0,i}$ and $\ddot{\underline{T}}_{0,i}$ are known, motion characteristics of link j adjacent to link i with respect to the inertia reference frame, are calculated as follows:

$$\boldsymbol{T}_{0,j} = \boldsymbol{T}_{0,i}.\boldsymbol{T}_{i,j} \tag{3.16}$$

$$\dot{\boldsymbol{T}}_{0,j} = \dot{\boldsymbol{T}}_{0,i}.\boldsymbol{T}_{i,j} + \boldsymbol{T}_{0,i}.\dot{\boldsymbol{T}}_{i,j} \tag{3.17}$$

$$\ddot{\boldsymbol{T}}_{0,j} = \ddot{\boldsymbol{T}}_{0,i}.\boldsymbol{T}_{i,j} + 2.\dot{\boldsymbol{T}}_{0,i}.\dot{\boldsymbol{T}}_{i,j} + \boldsymbol{T}_{0,i}.\ddot{\boldsymbol{T}}_{i,j} \qquad 3.18)$$

where 3×1 vectors for position ρ_{0,M_i} , velocity $\boldsymbol{v}_{0,M_i}$, and acceleration $\boldsymbol{a}_{0,M_i}$ of point M_i; as well as angular velocity $\omega_{0,i}$ and angular acceleration $\varepsilon_{0,i}$ of ith coordinate system with respect to the base (0th link) are as follows:

$$\begin{bmatrix} \rho_{0,M_i} \\ 1 \end{bmatrix} = \boldsymbol{T}_{0,i} \cdot \begin{bmatrix} \rho_{i,M_i} \\ 1 \end{bmatrix} \tag{3.19}$$

$$\begin{bmatrix} \boldsymbol{v}_{0,M_i} \\ 0 \end{bmatrix} = \dot{\boldsymbol{T}}_{0,i} \cdot \begin{bmatrix} \rho_{i,M_i} \\ 1 \end{bmatrix} \tag{3.20}$$

$$\begin{bmatrix} \boldsymbol{a}_{0,M_i} \\ 0 \end{bmatrix} = \ddot{\boldsymbol{T}}_{0,i} \cdot \begin{bmatrix} \rho_{i,M_i} \\ 1 \end{bmatrix} \tag{3.21}$$

$$\hat{\omega}_{0,i} = \dot{\tau}_{0,i} \cdot \tau_{0,i}^{\mathrm{T}} \tag{3.22}$$

$$\hat{\varepsilon}_{0,i} = \ddot{\tau}_{0,i} \cdot \tau_{0,i}^{\mathrm{T}} + \dot{\tau}_{0,i} \cdot \dot{\tau}_{0,i}^{\mathrm{T}} \tag{3.23}$$

where $\hat{\omega}_{0,i}$ and $\hat{\varepsilon}_{0,i}$ are skew symmetric matrices associated with $\omega_{0,i}$ and $\varepsilon_{0,i}$, respectively. In the same way the relative positions, velocities, and accelerations of two moving links could be calculated. This problem is of importance for adjacent links connected by a higher pair.

3.2. Optimization Techniques in Kinematic Synthesis

The problem of kinematic synthesis, can be defined as follows: For a given function of the generalized coordinates $\boldsymbol{Q}(t)$, the design parameters $\boldsymbol{A}$ are to be found, to satisfy the desired motion $\underline{\tilde{\boldsymbol{Q}}_d}(\boldsymbol{Q}(t))$, i.e.,

$$H\left[\tilde{Q}_{\underline{d}}(Q), A\right] = 0_{\underline{d}} \tag{3.24}$$

or, in discretized form defined by n positions of the system,

$$H^i(\tilde{Q}^i_{\underline{d}}, Q^i, A) = 0_{\underline{d}}, i=1,2,...,n \tag{3.25}$$

In the practice of spatial mechanism synthesis only the discretized form has been applied, since the function of Eq. 3.24 is highly nonlinear. If the number of equations is equal to or less than the number of design parameters, Eq. 3.25 could be solved by means of numerical procedures for solution of nonlinear equations. Direct search synthesis methods are available mostly for simple one contour plane and some spatial mechanisms. For polycontour mechanisms, only optimization techniques are applicable. In general, the process of kinematic synthesis is presented as a nonlinear programming problem; i.e., to minimize the objective function

$$Z = \frac{1}{2}\sum_{i=1}^{n}(\tilde{Q}^i_{\underline{d}} - Q^i_{\underline{d}})^{\mathrm{T}}.M^i.(\tilde{Q}^i_{\underline{d}} - Q^i_{\underline{d}}), \tag{3.26}$$

subject to the equality constrains of Eq. 3.25 and discretized and/or parametric inequality constrains. Here M^i is a $\underline{d} \times \underline{d}$ diagonal weight coefficient matrix.

The inequality constraints are most often prescribed limits of link lengths, joint positions, force transmission properties, etc. An important condition for mechanism synthesis is the case of a mechanism with a closed chain. This means that the matrix Eq. 3.25 are to be satisfied for every value of $Q^i(t_i)$, within the prescribed range $t_1 \leq t_i \leq t_2$, without wedging. This condition is mathematically expressed with equality constraints, discretized Eqs. 3.25 for a finite number of adjacent mechanism positions. Conditions for the existence of closed chains are numerically expressed as parametric inequality constraints; i.e.,

$$det\left(\frac{\partial H(t)}{\partial Q_{\underline{d}}}\right) > 0 \tag{3.27}$$

As could be seen above, the objective function is composed of the characteristics of motion, it contains no constructive parameters. In many research activities, the equality constraints are included in the objective function (Himmelblau 1972), and its minimization is carried out simultaneously with the solution of the equations. In mechanism synthesis, especially when spatial mechanisms must be optimized, the varying parameters are large in number and, in this case, the optimization process could be ineffective.

For the mathematical model presented, the method of conjugate gradients is applied. Since, for the discrete mechanism positions, the characteristics of motion are not explicit functions of the design parameters, the gradient of the objective function is as follows:

$$\frac{\partial Z}{\partial a_{j\backslash k}} = -\sum_{i=1}^{n}\left(\underline{\tilde{Q}}_d^i - \underline{Q}_d^i\right)^{\mathrm{T}} . M^i . \frac{\partial \underline{Q}_d^i}{\partial a_{j\backslash k}} \tag{3.28}$$

where

$$\frac{\partial \underline{Q}_d^i}{\partial a_{j\backslash k}} = -\left(\frac{\partial H^i}{\partial \underline{Q}_d^i}\right)^{-1} . \frac{\partial H^i}{\partial a_{j\backslash k}} \tag{3.29}$$

In this case, for n mechanism positions, n sets of the matrix equations of Eq. 3.29 must be solved. In implementing the optimization procedure based on this algorithm proposed, the main computational burden is the general purpose program for kinematic analysis. Since the gradient of the objective function is calculated, one step of its minimization could be implemented by changing the design parameters along the vector-gradient direction. In this case, the equality constraints are not violated during the iterative optimization procedure. Gradient projection techniques are appropriate for satisfying the inequality constraints.

4. Dynamics of Spatial Mechanisms

4.1. Force Analysis

The forces acting on the links of a mechanical system include gravitational forces, spring forces, and resistance forces of the environment. When kinematic characteristics are known, the appropriate approach for direct dynamics solution is to use Newton-Euler dynamic equations. These are interpreted as quasi-static conditions for equilibrium of the forces loading the bodies, where the inertia forces that depend on the mass and motion characteristics of the links are treated as external forces. For link i, inertia forces (3×1 matrix-vector $\boldsymbol{\Phi f}_{i,C_i}$) and moments ($3 \times 1$ matrix-vector $\boldsymbol{\Phi m}_{i,C_i}$) reduced to the center of gravity C_i are determined by the matrix equations

$$\Phi_{i,C_i} = \begin{bmatrix} \Phi \boldsymbol{f}_{i,C_i} \\ \Phi \boldsymbol{m}_{i,C_i} \end{bmatrix} = \begin{bmatrix} -\boldsymbol{M}_i . \boldsymbol{a}_{i,C_i} \\ -\boldsymbol{J}_{i,C_i} . \varepsilon_{i,i} - \hat{\omega}_{i,i} . \boldsymbol{J}_{i,C_i} . \omega_{i,i} \end{bmatrix}$$

$$= \begin{bmatrix} \Phi fx_{i,C_i} & \Phi fy_{i,C_i} & \Phi fz_{i,C_i} & \Phi mx_{i,C_i} & \Phi my_{i,C_i} & \Phi mz_{i,C_i} \end{bmatrix}^{\mathrm{T}} \tag{4.1}$$

where $\boldsymbol{M}_i$ is the mass of link i and $\boldsymbol{J}_{i,C_i}$ is the 3×3 inertia tensor of link i, relative to its center of gravity. When all external forces acting on a body are known, including the inertia forces and the reactions in the connections, then this body is considered to be in dynamic equilibrium. Figure 4.1 shows link i in quasi-static equilibrium with the contacting links j and k and performing complex spatial motion; inertia forces and moments in the center of gravity C_i; external forces $\boldsymbol{Pf}_{i,M_i}$ and moments $\boldsymbol{Pm}_{i,M_i}$ reduced to point M_i, where

$$\boldsymbol{P}_{i,M_i} = \begin{bmatrix} \boldsymbol{Pf}_{i,M_i} \\ \boldsymbol{Pm}_{i,M_i} \end{bmatrix} = \begin{bmatrix} Pfx_{i,M_i} & Pfy_{i,M_i} & Pfz_{i,M_i} & Pmx_{i,M_i} & Pmy_{i,M_i} & Pmz_{i,M_i} \end{bmatrix}^{\mathrm{T}};$$

and reaction forces $\boldsymbol{Rf}_{im_k}$ and moments $\boldsymbol{Rm}_{im_k}$, $k = 1, 2, \ldots, n_i$, in the im_k th joint, where

$$\boldsymbol{R}_{im_k} = \begin{bmatrix} \boldsymbol{Rf}_{im_k} \\ \boldsymbol{Rm}_{im_k} \end{bmatrix} = \begin{bmatrix} Rfx_{im_k} & Rfy_{im_k} & Rfz_{im_k} & Rmx_{im_k} & Rmy_{im_k} & Rmz_{im_k} \end{bmatrix}^{\mathrm{T}}.$$

In this notation, n_i is the number of the links adjacent to link i. In Fig. 4.1, $n_i = 2$, $m_1 = j$, and $m_2 = k$. For a mechanical system with $\underline{n}$ links, the following linear system of matrix equations is valid:

$$\sum_{k=1}^{n_i} \overset{*}{\rho}_{i,O_{im_k}} \cdot \overset{*}{\tau}_{i,im_k} \cdot \boldsymbol{R}_{im_k} + \overset{*}{\rho}_{i,C_i} \cdot \boldsymbol{\Phi}_{i,C_i} + \overset{*}{\rho}_{i,M_i} \cdot \boldsymbol{P}_{i,M_i} = \boldsymbol{0}_6, \; i=1,2,\ldots,\underline{n} \quad (4.2)$$

$$\boldsymbol{R}_{ij} = -\overset{*}{\rho}_{ij,O_{ji}} \cdot \overset{*}{\tau}_{ij,ji} \cdot \boldsymbol{R}_{ji} \quad (4.3)$$

Here $\boldsymbol{R}_{ij}$ denotes the matrix of reactions in joint ij, with components in the joint ij coordinate system. In Eqs. 4.2 and 4.3 $\rho^* = \begin{bmatrix} \boldsymbol{E}_3^3 & \boldsymbol{0}_3^3 \\ \hat{\rho} & \boldsymbol{E}_3^3 \end{bmatrix}$ and $\tau^* = \begin{bmatrix} \tau & \boldsymbol{0}_3^3 \\ \boldsymbol{0}_3^3 & \tau \end{bmatrix}$.

If an open chain is to be analyzed the Eqs. (4.2 and 4.3 could be solved separately for each link, starting with the end-effector; i.e., for $i=\underline{n}$, $\underline{n}$-1, ... , 1

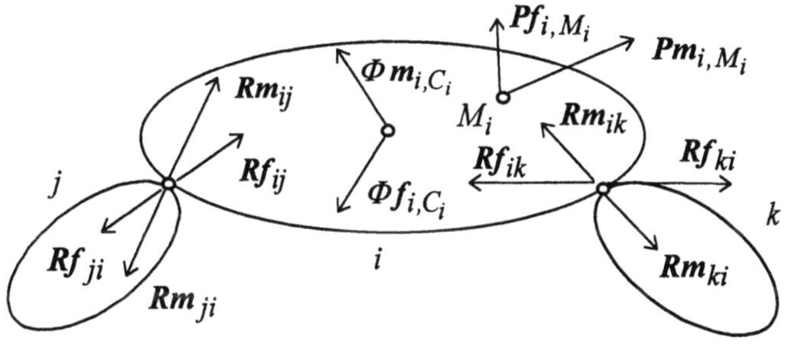

Fig. 4.1 Link *i* in quasi-static equilibrium

which is to the most often used numbering of open chains. Then, for link i in notation R_{im_k} k=1, 2, and the indices are $m_1 = i-1$, $m_2 = i+1$. When $i=\underline{n}$ $R_{im_2} = R_{\underline{n}(\underline{n}+1)} = 0_6$, and in every step for $i<\underline{n}$ $R_{i(i+1)}$ is known from the foregoing iteration. This means that $\underline{n}$ times a system of six equations must be solved. Controlled open chains have sdf driven joints, and Rmz_{ij} and Rfz_{ij} correspond to the generalized forces of the rotational or translational pairs, respectively. These are forces for driven pairs in closed contours. If the mechanism to be analyzed is built up as a multiloop chain, Eqs. 4.2 and 4.3 must be solved simultaneously for all links. When a mechanical system is not overconstrained, and assuming the reactions along the joint axes where motion occurs are equal zero, the system of Eqs. 4.2 and 4.3 has a unique solution.

4.2. Closed Form Dynamic Equations

One is often interested in obtaining better insight into the structure of the dynamic equations. For example, what is the form of the gravity terms? How does the magnitude of gravity effect compare with the magnitude of the internal effects? To investigate these questions, to model the motion of the system loaded with forces, and to find the answers to other questions, the dynamic equations must be written in closed form; i.e.,

$$S = A(Q).\ddot{Q} + B(Q,\dot{Q}) \tag{4.4}$$

or as configuration space dynamic equations

$$S = A(Q).\ddot{Q} + C(Q).\langle\dot{Q}\dot{Q}\rangle + D(Q).\langle\dot{Q}^2\rangle + G(Q) \tag{4.5}$$

where $S = \begin{bmatrix} s_1 & s_2 & \dots & s_{\underline{g}} \end{bmatrix}^T$ is the $\underline{g} \times 1$ matrix-column of the generalized forces, $A(Q)$ and $D(Q)$ are $\underline{g} \times \underline{g}$ matrices; $B(Q,\dot{Q})$ and $G(Q)$ are $\underline{g} \times 1$ columns; and $C(Q)$ is a $\underline{g}.(\underline{g}-1)/2 \times \underline{g}$ matrix. Here $\langle\dot{Q}^2\rangle = \begin{bmatrix} \dot{q}_1^2 & \dot{q}_2^2 & \dots & \dot{q}_{\underline{g}}^2 \end{bmatrix}^T$;

$\langle\dot{Q}\dot{Q}\rangle = \begin{bmatrix} \dot{q}_1 \cdot \dot{q}_2 & \dot{q}_1 \cdot \dot{q}_3 & \dots & \dot{q}_1 \cdot \dot{q}_{\underline{g}} & \dot{q}_2 \cdot \dot{q}_3 & \dots & \dot{q}_{\underline{g}-1} \cdot \dot{q}_{\underline{g}} \end{bmatrix}^T$.

There are many methods of numerical integration. The simplest numerical integration scheme is that of Euler; i.e., from an initial state Q, $\dot{Q}$ of the system at time $t = 0$ and with given driving forces, the next position at time $t + \Delta t$ is determined from:

$$Q(t+\Delta t)=Q(t)+\dot{Q}(t).\Delta t+\frac{1}{2}.\ddot{Q}(t).\Delta t^2 \tag{4.6}$$

$$\dot{Q}(t+\Delta t)=\dot{Q}(t)+\ddot{Q}(t).\Delta t \tag{4.7}$$

where Eqs. 4.4 or 4.5 are to be solved in order that $\ddot{Q}$ can be calculated. In this numerical procedure, the most time consuming problem is calculations of the coefficients of matrices $\boldsymbol{A}$ and $\boldsymbol{B}$, since they are highly nonlinear in $\boldsymbol{Q}$ and $\dot{\boldsymbol{Q}}$. In Sect. 4.1 the numerical procedure for solution of the first dynamics problem on the basis of Newton-Euler equations is presented. Here, the quasi-static approach (Walker and Orin 1982; Zakhariev 1988) is suggested, to be applied iteratively for solution of the second dynamics problem. This procedure is comprised of the following steps:

Step A: The reactions in the joints are calculated for the actual generalized coordinates $\boldsymbol{Q}$ and velocities $\dot{\boldsymbol{Q}}$ and for $\ddot{\boldsymbol{Q}}=0$. With these conditions

$$\boldsymbol{B}(\boldsymbol{Q},\dot{\boldsymbol{Q}})=\boldsymbol{S} \tag{4.8}$$

where the vector of generalized forces $\boldsymbol{S}$ is calculated by solving the direct dynamics problem and Newton-Euler equations; i.e., the matrix $\boldsymbol{B}$ is calculated numerically.

Step B: For the same mechanical system configuration and ignoring all external forces and gravity of the links, the reactions in the joints are calculated with the conditions $\dot{\boldsymbol{Q}}=0$ and $\ddot{q}_i=1$, $\ddot{q}_j=0$, $j=1, 2, \ldots, i\text{-}1, i\text{+}1, \ldots, \underline{g}$. For the generalized forces $\boldsymbol{S}^{(i)}$ obtained and solving the first problem of dynamics, the following matrix equation is valid:

$$\boldsymbol{A}.\ddot{\boldsymbol{Q}}=\begin{bmatrix} * & * & \ldots & a(1,i) & \ldots & * \\ * & * & \ldots & a(2,i) & \ldots & * \\ \ldots & \ldots & \ldots & \ldots & \ldots & \ldots \\ * & * & \ldots & a(i,i) & \ldots & * \\ \ldots & \ldots & \ldots & \ldots & \ldots & \ldots \\ * & * & \ldots & a(\underline{g},i) & \ldots & * \end{bmatrix}\cdot\begin{bmatrix} 0 \\ 0 \\ \ldots \\ 1 \\ \ldots \\ 0 \end{bmatrix}=\boldsymbol{S}^{(i)} \tag{4.9}$$

Hence, the vector $\boldsymbol{S}^{(i)}$ is equal to the ith column of $\boldsymbol{A}(\boldsymbol{Q})$. In this way, consequently, for $i=1, 2, \ldots, \underline{g}$ matrix $\boldsymbol{A}(\boldsymbol{Q})$ can be computed.

If configuration space dynamic equations are to be derived, the procedure proceeds with the following steps.

Step C: The reactions in the joints are calculated for the actual values of the generalized coordinates $\boldsymbol{Q}$ and the restrictions $\dot{\boldsymbol{Q}}=0$ and $\ddot{\boldsymbol{Q}}=0$. With these conditions,

$$G(Q) = S \tag{4.10}$$

where the generalized forces S are calculated by solving the direct dynamics problem..

Step D: The coefficients of matrix $D(Q)$ are computed by solving the first problem of dynamics for the generalized coordinates Q and ignoring all external forces and gravity terms, with the restrictions $\ddot{Q}$ = 0, $\dot{q}_i$ = 1, and $\dot{q}_j$= 0, j = $1, 2, ..., j\text{-}1, j+1, ... , \underline{g}$. For the generalized forces so derived, the following equation holds:

$$D.\left\langle \dot{Q}^2 \right\rangle = \begin{bmatrix} * & * & \dots & d(1,i) & \dots & * \\ * & * & \dots & d(2,i) & \dots & * \\ \dots & \dots & \dots & \dots & \dots & \dots \\ * & * & \dots & d(i,i) & \dots & * \\ \dots & \dots & \dots & \dots & \dots & \dots \\ * & * & \dots & d(\mathbf{g},i) & \dots & * \end{bmatrix} \cdot \begin{bmatrix} 0 \\ 0 \\ \dots \\ 1 \\ \dots \\ 0 \end{bmatrix} = S^{(i)} \tag{4.11}$$

Thus, the ith column of matrix D is computed.

Step D: For the configuration Q, ignoring all gravity terms and with the conditions $\ddot{Q}$ = 0, $\dot{q}_i$ = 1, $\dot{q}_j$ =1, and $\dot{q}_k$= 0, k =1,2,...,i-1,i+1, ... ,j-1,j+1,..., $\underline{g}$, the coefficients of the matrix $C(Q)$ are computed solving the direct dynamics problem. For the generalized forces $S^{(m)}$ (m is the number of the element $\dot{q}_i \cdot \dot{q}_j$ in the matrix $\left\langle \dot{Q}\dot{Q} \right\rangle$), the following matrix equation is valid:

$$C.\left\langle \dot{Q}\dot{Q} \right\rangle + D.\left\langle \dot{Q}^2 \right\rangle = \begin{bmatrix} * & * & \dots & c(1,m) & \dots & * \\ * & * & \dots & c(2,m) & \dots & * \\ \dots & \dots & \dots & \dots & \dots & \dots \\ * & * & \dots & c(i,m) & \dots & * \\ \dots & \dots & \dots & \dots & \dots & \dots \\ * & * & \dots & c(\underline{g},m) & \dots & * \end{bmatrix} \cdot \begin{bmatrix} 0 \\ 0 \\ \dots \\ 1 \\ \dots \\ 0 \end{bmatrix} + D.\left\langle \dot{Q}^2 \right\rangle = S^{(m)} \tag{4.12}$$

where the ith and jth elements of $\left\langle \dot{Q}^2 \right\rangle$ are 1 and all others are equal zero. As it is seen from Eq. 4.12 the mth column of C is evaluated, since matrix D is defined during the previous step C.

4.3. Dynamics of Mechanisms Including Friction in the Pairs

It is assumed that the reaction $\boldsymbol{R}_{ij}$ and $\boldsymbol{R}_{ji}$ in the *ij*th pair are applied at the joint coordinate system origins O_{ij} and O_{ji}, respectively. The real reactions in the pairs are always applied at some points of the joint contact surfaces, and they are composed of normal forces and friction forces. Coulomb friction is taken into consideration, since its discontinuous nature and nonlinear velocity dependent nature is appropriate model for the reality. The numerical algorithm of force analysis imposes the necessity of discretized representation of the friction coefficient function.

If friction is not taken into consideration, the components of the normal reactions ($k_{ij} \times 1$ matrix-column N_{ij}) in a pair could be reduced to the joint coordinate system origin; i.e.,

$$\boldsymbol{R}_{ij} = \boldsymbol{TN}_{ij} . \boldsymbol{N}_{ij} \tag{4.13}$$

where $\boldsymbol{TN}_{ij}$ is a $6 \times k_{ij}$ transformation matrix and k_{ij} is the class of the *ij*th pair. For a sdf joint k_{ij}=5. For cylindrical and slotted sphere pairs, k_{ij}=4, and for a sphere pair, k_{ij}=3. If friction is assumed,

$$\boldsymbol{R}_{ij} = \boldsymbol{TN}_{ij} . \boldsymbol{N}_{ij} + \boldsymbol{t}_{ij}\left(\boldsymbol{N}_{ij}\right) \tag{4.14}$$

where $\boldsymbol{t}_{ij}$ is a 6×1 matrix-column, whose elements are nonlinear functions of the components of the normal reactions in the joint contact points.

4.3.1. Dynamics of Open Chains

If the kinematic characteristics of motion are known, the solution of the direct problem of dynamics consists of solving Eqs. 4.2 and 4.3 for every link and joint, starting from the end effector (numbered with *n*) and finishing at the base (numbered with 0). For a pair (with index *ij*) the six equations of Eq. 5.11 could be decomposed to k_{ij} linear equations for k_{ij} unknown components of the normal reactions. The column of friction forces $\boldsymbol{t}_{ij}$ is composed of 6- k_{ij} elements, which are not equal zero, and could be calculated explicitly if the normal reactions are calculated.

From the structure of the matrices $\boldsymbol{TN}_{ij}$ and $\boldsymbol{t}_{ij}$, it can be seen that the principles of superposition and proportionality are valid for the normal reactions when friction occurs. These principles are basic for the numerical algorithm of deriving the explicit form dynamic equations including friction in the pairs. The equations below are derived as closed form dynamic equations, since the procedure for the configuration space dynamic equations is the same. According to this procedure, for every pair ij two matrices are calculated; $k_{ij} \times 1$ matrix $N_{ij}^{(B)}$ and $k_{ij} \times \underline{g}$ matrix $N_{ij}^{(A)}$. The elements of $N_{ij}^{(B)}$ are the normal reactions of the ijth pair calculated with the restrictions in Step A, Eq. 4.8. The elements of the kth ($k = 1, 2, \ldots, \underline{g}$) column of $N_{ij}^{(A)}$ are calculated under the restrictions imposed in Step B, Eq. 4.9, on the kth acceleration. For the normal reactions so derived, the following equation holds:

$$N_{ij} = N_{ij}^{(A)}.\ddot{Q} + N_{ij}^{(B)} \tag{4.15}$$

Since the nonlinear function $t_{ij} = t_{ij}\left(N_{ij}\right)$ is known (for review see Zakhariev 1995) and from Eq. 4.15, the nonlinear function

$$t_{ij} = t_{ij}\left(\dot{Q},\ddot{Q}\right) \tag{4.16}$$

can be numerically derived.

Open chains are build up with sdf joints, so only one element of every joint matrix t_{ij} does not equal zero; i.e., this is the friction force or moment in the joint. The explicit form dynamic equations of Eq. 4.4 assuming friction in the pairs, are as follows:

$$H(Q,\dot{Q},\ddot{Q}) = A(Q).\ddot{Q} + B(Q,\dot{Q}) + t(Q,\dot{Q},\ddot{Q}) + U = 0_{\underline{g}} \tag{4.17}$$

where $t = \begin{bmatrix} t_1 & t_2 & \ldots & t_{\underline{g}} \end{bmatrix}^{\mathrm{T}}$ is a $\underline{g} \times 1$ matrix-column of friction forces composed from the non zero element of the matrix t_{ij} for all joints, numbered as in the initial structure matrix $\boldsymbol{IS}$ (see Sect. 2.1); i.e.,

$$t_k = \left(\frac{\partial \Delta_{ij}}{q_k}\right).t_{ij} \tag{4.18}$$

Finally $U = \begin{bmatrix} u_1 & u_2 & \ldots & u_{\underline{g}} \end{bmatrix}^{\mathrm{T}}$ is the $\underline{g} \times 1$ column of controlled forces. In Eq. 4.18, the fact that $\Delta_{ij} = -\Delta_{ji}$ must be taken into consideration when computations are implemented. For example, for a rotational pair of an open chain $\dfrac{\partial \Delta_{ij}}{\partial q_k} = \dfrac{\partial \Delta_k}{\partial q_k} = \begin{bmatrix} 0 & 0 & 0 & 0 & 0 & 1 \end{bmatrix}^{\mathrm{T}}$, and if U is given, the solution of Eq. 4.17

for $\ddot{Q}$ could be achieved numerically by the iterative procedure of Newton-Raphson, where

$$\frac{\partial H}{\partial \ddot{Q}} = A(Q) + \frac{\partial t}{\partial \ddot{Q}} \tag{4.19}$$

$$\frac{\partial t_{ij}}{\partial \ddot{Q}} = \frac{\partial t_{ij}}{\partial N_{ij}} \cdot \frac{\partial N_{ij}}{\partial \ddot{Q}} = \frac{\partial t_{ij}}{\partial N_{ij}} . N_{ij}^{(A)} \tag{4.20}$$

4.3.2. Dynamics of Closed Chains

The solution of the inverse problem of dynamics consists of applying the procedure for open chains suggested above, where for every real joint ij two matrices are calculated, the $k_{ij} \times 1$ matrix $N_{ij}^{(B)}$ and the $k_{ij} \times \underline{g}$ matrix $N_{ij}^{(A)}$.

For normal reactions so defined, Eq. 4.15 holds and friction forces are calculated from Eq. 4.16. Then, the elements of matrix $t(\dot{Q},\ddot{Q})$ in Eq. 4.17 are calculated as

$$t_k = \sum_{(ij)} \left(\frac{\partial \Delta_{ij}}{\partial q_k} \right)^{\mathrm{T}} . t_{ij} \text{ , } k=1, 2, \ldots, \underline{g} \tag{4.21}$$

where the sum is over all real joints. Knowing the elements of matrix D, the nonlinear dynamic equations of Eq. 4.17 are solved by the Newton-Raphson procedure, where Eqs. 4.20, 4.21 are taken into account.

4.3.3. Stick-Slip Effect in Joints

The stick-slip process is a result of the discontinuous nature of friction, as well as of the Coulomb friction model discussed here. The numerical procedure suggested here could be implemented for solution of the first dynamics problem, including the stick-slip process. However, the second problem of dynamics, if zero velocity in a pair occurs, cannot be solved uniquely. A numerical procedure for solution of the following problem is thus presented: Integrate the explicit form dynamic equations at the occurrence of zero velocity in a pair of a moving chain, with a given vector U of the driving forces.

Movement in this pair is dependent on the sign of the friction force and is determined by the sign of acceleration in the pair. If this pair is controlled (its movement is a generalized coordinate), then the following procedure is applied:

(a) The kinematic chain is assumed to have $\underline{g}$-1 dof , and the pair in which zero velocity occurs (let it be the *i*th) is assumed to be fixed. For this configuration of the chain, the explicit form dynamic equations; i.e.,

$$\underline{\boldsymbol{A}}.\underline{\ddot{\boldsymbol{Q}}}+\underline{\boldsymbol{B}}+\underline{\boldsymbol{t}}+\underline{\boldsymbol{U}}=\boldsymbol{0}_{\underline{g}-1} \tag{4.23}$$

are solved for $\underline{\ddot{\boldsymbol{Q}}}$, where the underlined notation implies that the *i*th row and column are deleted in the corresponding matrices in Eq. 4.17.

(b) From $\underline{\ddot{\boldsymbol{Q}}}$ so obtained, the matrix of accelerations $\ddot{\boldsymbol{Q}}$ is assumed to have its *i*th element equal to zero. With this $\ddot{\boldsymbol{Q}}$, the matrix of the generalized forces $\boldsymbol{S}$ and $\boldsymbol{t}$ are calculated with coefficient of friction for zero velocity of the *i*th generalized movement.

(c) For the *i*th generalized force s_i and absolute value of the friction force $|t_i|$ so calculated, the following conditions hold: if

$$-|t_i|\leq u_i+s_i\leq|t_i| \tag{4.24}$$

there will be no motion in the joint and $\underline{\ddot{\boldsymbol{Q}}}$ is the real solution; if

$$u_i+s_i<-|t_i| \tag{4.25}$$

the sign of friction force t_i is (+) and if

$$u_i+s_i>|t_i| \tag{4.26}$$

the sign of friction force t_i is (-).

The final step is to derive the real value of $\ddot{\boldsymbol{Q}}$ by solving Eq. 4.17 with full size of matrices and with the sign of friction forces defined above.

4.4. Control and Optimization of Motion

The explicit form dynamic equations of Eqs. 4.4 and 4.5 could be discretized for n positions at time $t_0\leq t_i\leq t_n$,

$$\boldsymbol{S}_i=\boldsymbol{A}_i.\ddot{\boldsymbol{Q}}_i+\dot{\boldsymbol{Q}}_i^{\mathrm{T}}.\boldsymbol{B}_i.\dot{\boldsymbol{Q}}_i+\boldsymbol{G}_i\ ,\ i=0,1,\ldots,n\text{-}1 \tag{4.27}$$

where $\boldsymbol{B}$ is a $\underline{g}\times\underline{g}\times\underline{g}$ cubic matrix in which the *i*th column of $\boldsymbol{C}$ is the diagonal of the *i*th matrix (*j,j,i*th elements, j=1,2, ..., $\underline{g}$) of $\boldsymbol{B}$ and the elements of the *i*th column of $\boldsymbol{D}$ is the upper triangular part of the *i*th matrix of $\boldsymbol{B}$ ordered in rows. Together with the interpolation formulas of the numerical integration algorithm, for example the simple method of Euler,

$$\boldsymbol{Q}_{i+1} = \boldsymbol{Q}_i + \dot{\boldsymbol{Q}}_i . \Delta t_i + \ddot{\boldsymbol{Q}}_i . \frac{\Delta t_i^2}{2} \tag{4.28}$$

$$\dot{\boldsymbol{Q}}_{i+1} = \dot{\boldsymbol{Q}}_i + \ddot{\boldsymbol{Q}}_i . \Delta t_i \, , \; i = 0,\, 1,\, ...,\, n\text{-}1 \tag{4.29}$$

Eqs. 4.27, 4.28, and 4.29 fully describe numerical dependencies between the mass parameters, characteristics of motion, and driving forces throughout the trajectory of movement. The accuracy is better as the increment of time $\Delta t_i = t_{i+1} - t_i$, $i = 0, 1, ... n\text{-}1$, is reduced.

The controlled system path planning algorithm (Zakhariev at al. 1996) is as follows: Starting from position $\boldsymbol{Q}_0$, with velocity $\dot{\boldsymbol{Q}}_0$, the next state of the system $\boldsymbol{Q}_n$, $\dot{\boldsymbol{Q}}_n$ is to be achieved at time Δt. If no strategy of movement between the positions 0 and n, or outlying points is known, Δt is divided into n small increments of time Δt_i, $i=1,2,...,n$, where the new values $\ddot{\boldsymbol{Q}}_i$ of the accelerations are computed according to the values $\boldsymbol{Q}$, $\dot{\boldsymbol{Q}}$ of the feedback control system sensors. If Δt_i is fixed, the strategy of movement is assumed to be carried out with at least two values of the accelerations $\ddot{\boldsymbol{Q}}_i$, $\ddot{\boldsymbol{Q}}_i^-$, and $\ddot{\boldsymbol{Q}}_i^+$ at time $\Delta t_i/2$, where the movement is assumed to occur with constant acceleration. From Eqs. 4.28 and 4.29 the accelerations $\ddot{\boldsymbol{Q}}_i$ are calculated and from Eq. 4.27 the generalized forces $\boldsymbol{S}_i$ are computed. For the problem so defined, the main aim is to maintain constant accelerations $\ddot{\boldsymbol{Q}}_i$. Then, in m_i positions, proportional of the frequency of the controlling system f ($m_i = \Delta t_i . f$), the characteristics of motion of the generalized coordinates at time t, $t_i \le t \le t_{i+1}$ must be

$$\ddot{\boldsymbol{Q}}_i^j = \ddot{\boldsymbol{Q}}_i = const \tag{4.30}$$

$$\dot{\boldsymbol{Q}}_i^j = \dot{\boldsymbol{Q}}_i + \frac{j}{f} \ddot{\boldsymbol{Q}}_i \tag{4.31}$$

$$\boldsymbol{Q}_i^j = \boldsymbol{Q}_i + \frac{j}{f} . \dot{\boldsymbol{Q}}_i + \frac{1}{2}\left(\frac{j}{f}\right)^2 . \ddot{\boldsymbol{Q}}_i \, , \; j = 1,2,...,m_i \tag{4.32}$$

When the control system creates these motion characteristics at time $t_i \le t \le t_{i+1}$, the integration scheme of Eqs. 4.28 and 4.29 is satisfied and convergence is achieved. For this purpose, a feedback control scheme depending on acceleration errors and modification of the generalized forces $\Delta \boldsymbol{S}_i$ is suggested. The method is realized assuming constant matrices $\boldsymbol{A}_i$, $\boldsymbol{B}_i$, and $\boldsymbol{G}_i$ at time $t_i \leq t \leq t_{i+1}$. It must be pointed out that no convergence to any position is needed, but only keeping the generalized accelerations at the prescribed values.

When prescribed system positions are relatively outlying, the strategy of the control algorithm presented provides a movement with almost constant generalized accelerations that are limited by the prescribed velocities of the state examined. This path could not be optimal for popular criteria such as minimal time of path motion or energy consumption. In Eqs. 4.27, 4.28, and 4.29 the trajectory between end positions $\boldsymbol{Q}_0$ and $\boldsymbol{Q}_n$ is discretized at n-1 intermediate positions $\boldsymbol{Q}_i$, i=1, 2,..., n-1, of the system. For these positions, 2.(n-1) motion characteristics $\boldsymbol{Q}_i$, $\dot{\boldsymbol{Q}}_i$, i=1, 2,...,n-1, are defined as parameters. For the same state, parameters are also 3.n values of $\ddot{\boldsymbol{Q}}_i, \boldsymbol{S}_i, \Delta t_i$, i=0, 1,...,n-1. If we assume equal increments of time $\Delta t_i = \Delta t = const$ and taking into account 3.n sized nonlinear equation system of Eqs. 4.27, 4.28, and 4.29, the number of independent parameters is n; e.g., n-1 values of $\boldsymbol{Q}_i$, i=1, 2,..., n-1, and Δt. When a general scalar function is defined as

$$Z = \sum_{i=0}^{n-1} F\left(\boldsymbol{Q}_i, \dot{\boldsymbol{Q}}_i, \boldsymbol{S}_i\right) . \Delta t \tag{4.33}$$

it could be optimized for the parameters of the synthesis problem $\boldsymbol{Q}_i$, i=1, 2,..., n-1, and Δt. Equations 4.27, 4.28, and 4.29 are equality constraints. Restrictions on generalized velocities and forces are inequality constraints in the nonlinear programming problem so defined.

5. Example

In Fig. 5.1, an example of a double pendulum with one dof spatial closed chain and one dof open branch is shown. The basic kinematic measures and inertia parameters are identified in the figure - all in SI UNITS. The chain is moving initially with the position q_1 = 1.5708, $\dot{q}_1$ = 0, q_8 = 1.5708, $\dot{q}_8$ = 0. Free motion occurs, starting from a horizontal position of the links in joints corresponding to the generalized coordinates. The coefficient of friction in the pairs is calculated numerically. The mechanism is very sensitive for self-locking. In Figs. 5.2 and 5.3 the characteristics of motion are shown.

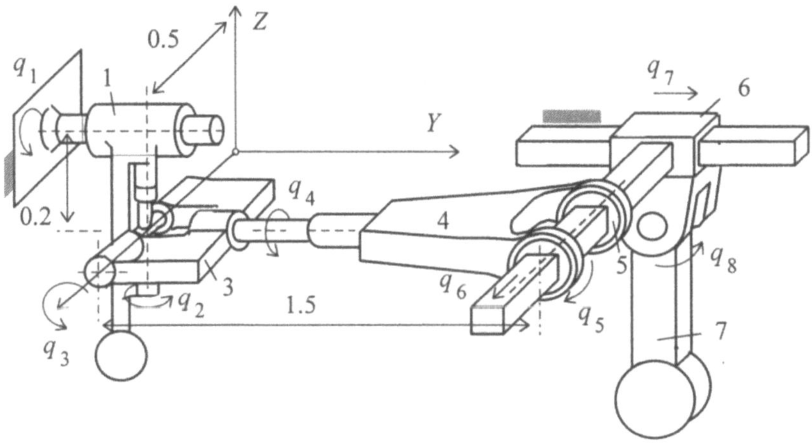

$$M_1 = 5.\ ;\ \rho_{1,C_1} = \begin{bmatrix} 0. \\ 0.4 \\ 0. \end{bmatrix};\ \boldsymbol{J}_{1,C_1} = \begin{bmatrix} 1.2 & 0. & 0. \\ 0. & 0.3 & 0. \\ 0. & 0. & 1.2 \end{bmatrix};$$

$$M_2 = 2.\ ;\ \rho_{2,C_2} = \begin{bmatrix} 0. \\ 0. \\ 0. \end{bmatrix};\ \boldsymbol{J}_{2,C_2} = \boldsymbol{0}_3^3;$$

$$M_3 = 2.\ ;\ \rho_{3,C_3} = \begin{bmatrix} 0. \\ 0. \\ 0. \end{bmatrix};\ \boldsymbol{J}_{3,C_3} = \boldsymbol{0}_3^3;$$

$$M_4 = 20.\ ;\ \rho_{4,C_4} = \begin{bmatrix} 0. \\ 0. \\ 0.8 \end{bmatrix};\ \boldsymbol{J}_{4,C_4} = \begin{bmatrix} 3. & 0. & 0. \\ 0. & 3. & 0. \\ 0. & 0. & 0.3 \end{bmatrix};$$

$$M_5 = 5.\ ;\ \rho_{5,C_5} = \begin{bmatrix} 0. \\ 0. \\ 0. \end{bmatrix};\ \boldsymbol{J}_{5,C_5} = \boldsymbol{0}_3^3;$$

$$M_6 = 30.\ ;\ \rho_{6,C_6} = \begin{bmatrix} 0. \\ 0. \\ 0. \end{bmatrix};\ \boldsymbol{J}_{6,C_6} = \boldsymbol{0}_3^3;$$

$$M_7 = 10.\ ;\ \rho_{7,C_7} = \begin{bmatrix} -0.2 \\ 0. \\ 0. \end{bmatrix};\ \boldsymbol{J}_{7,C_7} = \begin{bmatrix} 0.2 & 0. & 0. \\ 0. & 1. & 0. \\ 0. & 0. & 1. \end{bmatrix}.$$

Fig. 5.1 Double pendulum mechanism with one closed contour and one open branch

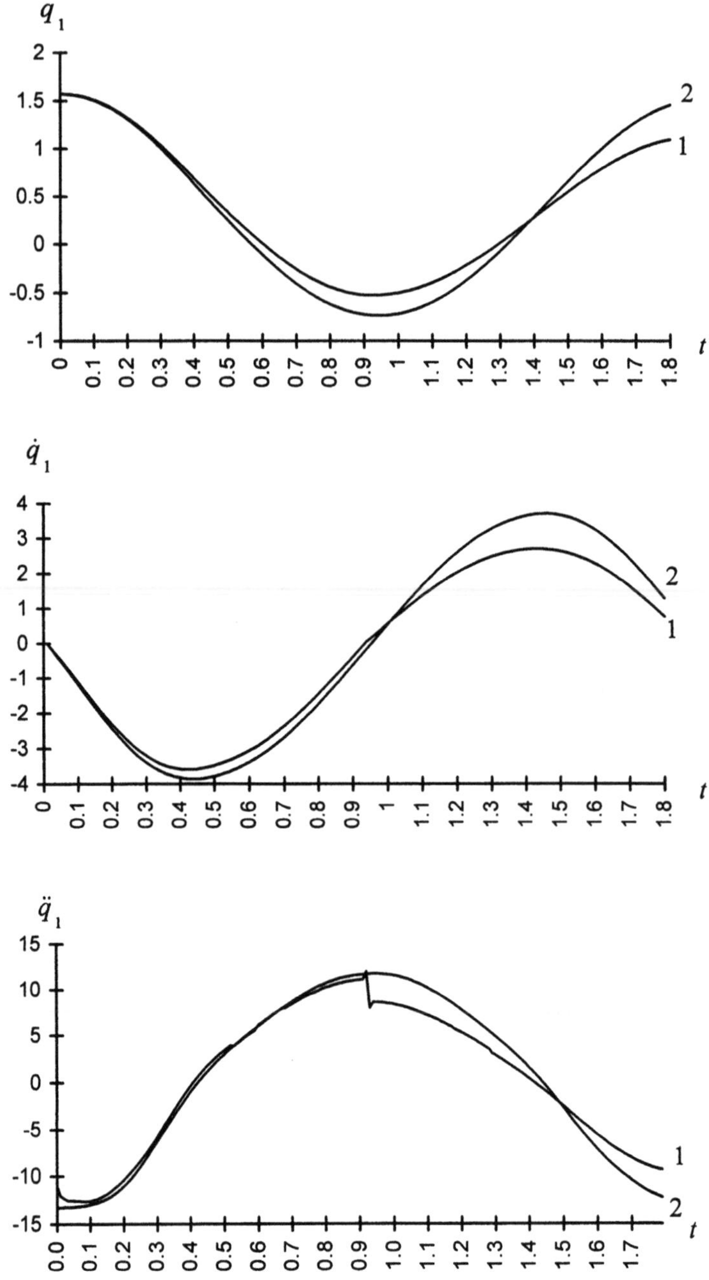

Fig. 5.2 Numerical results for q_1 with (1) and without (2) friction in the joints

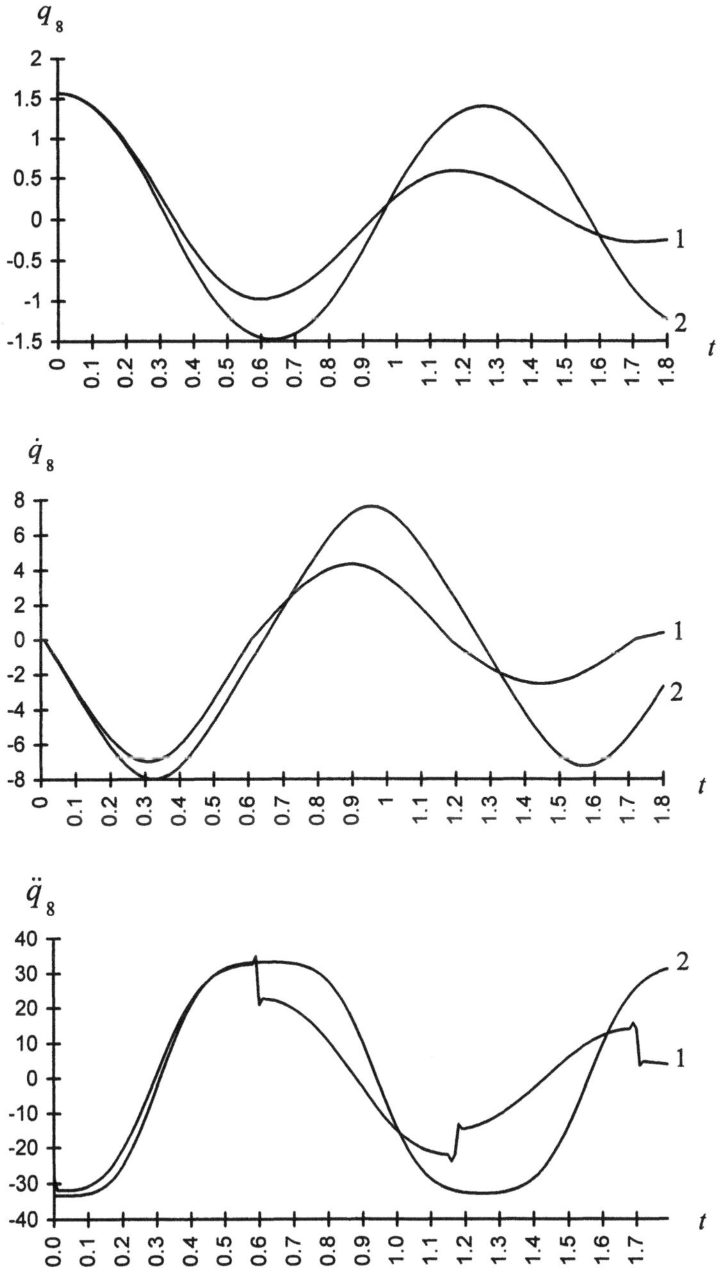

Fig. 5.3 Numerical results for q_8 with (1) and without (2) friction in the joints

Acknowledgments

The work reported here was possible under the support of National Foundation "Scientific Investigations" of Bulgaria, contract MM 526/95. NATO SEAD's financial support is gratefully acknowledged. The author would like to thank Prof. Edward Haug for his help.

References

Angeles J (1982) Spatial Kinematic Chains, Springer-Verlag, Berlin

Angeles J (1994) On Twist and Wrench Generators and Annihilators, in Computer-Aided Analysis of Rigid and Flexible Mechanical Systems. M.S. Pereira and J.A.C. Ambrosio (ed.), NATO ASI Series, E, Vol. 268, Kluwer Academic Publishers, Dordrecht: 379-411

Angeles J, Kecskemethy A (1995) Kinematics and Dynamics of Multibody Systems. Springer-Verlag, Berlin

Angeles J, Liu Z (1990) A constrained Least square method for the optimization of spherical four-bar path generators, Proc. ASME 21st Biennial Mechanisms Conference. DE -Vol. 25, September 17-19, Chicago: 289-294

Bottema O, Roth B (1970) Theoretical Kinematics, Applied Mathematics and Mechanics. North-Holland, Amsterdam

Denavit JJ, Hartenberg RSA (1950) Kinematic notation for lower pair mechanisms based on matrices. Journal of Engineering for Industry, 22(2): 215-221

Dimentberg FM, Sarkisian JL, Uskov MK (1983) Spatial Mechanisms. A Survey of the contemporary investigations. Moscow, Nauka, in Russian

Fox FL, Willmert KD (1967) Optimum Design of Curve Generating Linkages with Inequality Constraints. Journal of Engineering for Industry, 89: 144-152

Freundenstein F (1962) On the variety of motions generated by mechanisms. Journal of Engineering for Industry, 84(1): 156-160

García de Jalón J, Bayo E (1993) Kinematic and Dynamic Simulation of Multibody Systems – The Real-Time Challenge. Springer-Verlag, New York

García de Jalón J, Cuadrado J, Avello A, Jiménez JM (1993) Kinematic and Dynamic Simulation of Rigid and Flexible Systems With Fully Cartesian Coordinates. In: Computer-Aided Analysis of Rigid and Flexible Mechanical Systems, MS Pereira and JAC Ambrosio (eds), NATO ASI Series E, Vol. 268, Kluwer Academic Publishers, Dordrecht: 285-324

Gupta VK (1973) Computer Aided Synthesis of Mechanisms Using Nonlinear Programming. Journal of Engineering for Industry, 95: 339-344

Han C-Y (1966) A General Method for the Optimal Design of Mechanisms. Journal of Mechanisms, Vol. 1: 301-313

Hansen JM, Tortorelli DA (1996) An Efficient Method for Synthesis of Mechanisms Using an Optimization Method. IUTAM Symposium on Optimization of Mechanical Systems, D. Bestle and W. Schiehlen (ed), Kluwer Academic Publishers, Dordrecht: 129-138

Haug EJ (1989) Computer-aided Kinematics and Dynamics of Mechanical Systems. Allyn and Bacon, Boston

Haug EJ, Wu SC, Yang SM (1986) Dynamics of Mechanical Systems with Coulomb Friction, Stiction, Impact and Constraint Addition-Deletion – Part I: Theory. Mechanism and Machine Theory, 21(5): 401-406.

Himmelblau DM (1972) Applied Nonlinear Programming. McGraw-Hill

Keler ML (1973) Kinematics and statics including friction in single-loop mechanisms by screw calculus and dual vectors. Journal of Engineering for Industry, 95: 471 - 479

Lilov L (1993) Modelling of Multibody Systems. Nauka, Moscow (in Russian)

McPhee JJ (1996) On the Use of Linear Graph Theory in Multibody System Dynamics, Nonlinear Dynamics, 9: 73-90

Roberson RE, Schwertassek R (1988) Dynamics of Multibody Systems. Springer-Verlag, Berlin

Roth B, Freudenstein F (1963) Synthesis of Path Generating Mechanisms by Numerical Methods. Journal of Engineering for Industry, 85(3): 298-306

Schiehlen W (ed) (1993) Advanced Multibody System Dynamics, Solid Mechanics and its Applications. Kluwer Academic Publishers, Dordrecht

Shabana AA (1989) Dynamics of Multibody Systems. John Wiley & Sons, New York

Soylemez E, Freundenstein F (1982) Transmission Optimisation of spatial four-link mechanisms. Mechanism and Machine Theory, 17: 263-283

Walker MW, Orin DE (1982) Efficient Dynamic Computer Simulation of Robotic Mechanisms, Journal of Dynamic Systems, Measurement and Control, 104: 205-211

Wittenburg J (1977) Dynamics of Systems of Rigid Bodies. B.G. Teubner, Stuttgart

Wittenburg J (1993) Topological description of articulated systems. In: Computer-Aided Analysis of Rigid and Flexible Mechanical Systems, MS Pereira and JAC Ambrosio (eds), NATO ASI Series E, Vol. 268, Kluwer Academic Publishers, Dordrecht: 159-196

Woo L, Freundenstein F (1970) Application of line geometry to theoretical kinematics and the kinematic analysis of mechanical systems, Journal of Mechanisms. 5: 417-460

Zakhariev EV (1988) Design and Optimization of the Parameters of the Industrial Manipulators and Robots. Ph.D. thesis. Technical University Sofia

Zakhariev EV (1995) Solution of a nonlinear direct problem of multibody system dynamics. Mechanics of Machines, Technical University Varna, 11(3): 68-72

Zakhariev EV, Delchev KK, Karastanev SB (1996) Nonlinear Dynamics Modeling and Control of Multibody Systems. Proceedings of the 10th European Simulation Multiconference, Budapest, Hungary, June 2-6: 913-917

Implicit Integration of the Equations of Multibody Dynamics

E.J. Haug*, D. Negrut*, and M. Iancu**

Department of Mechanical Engineering* and Department of Mathematics**, The University of Iowa, Iowa City, Iowa 52242, USA, email: haug@ccad.uiowa.edu

Summary. Implicit integration approaches based on generalized coordinate-partitioning of the differential-algebraic equations of motion of multibody dynamics are presented. The methods developed are intended for simulation of stiff mechanical systems using the well-known Newmark integration method from structural dynamics and more recent implicit Runge-Kutta methods. Newmark integration formulas for second-order differential equations are first used to define independent generalized coordinates and their first time-derivative as functions of independent accelerations. The latter are determined as the solution of discretized equations obtained using the descriptor form of the equations of motion. Dependent variables in the formulation, including Lagrange multipliers, are determined to satisfy all the kinematic and kinetic equations of multibody dynamics. Next, the process is extended to implicit Runge-Kutta methods with variable step size and error control. The approach is illustrated by solving the constrained equations of motion for mechanical systems that exhibit stiff behavior. Results show that the approach is robust and has the capability to integrate differential-algebraic equations of motion for stiff multibody dynamic systems

Keywords. Implicit numerical integration, differential-algebraic equation integrators, multibody dynamics integration.

1. Introduction

In this chapter, $\mathbf{q} = [q_1, q_2, \ldots, q_n]^T$ denotes the vector of generalized coordinates that define the state of a multibody system (Haug 1989). Generalized coordinates for rigid bodies are Cartesian coordinates the origins of body reference frames and Euler parameters (Haug, 1989) that define their orientations. Joints connecting bodies in a mechanical system restrict their relative motion and impose constraints on their generalized coordinates. In most cases, kinematic constraints are algebraic equations involving generalized coordinates, of the form

$$\mathbf{\Phi}(\mathbf{q}) \equiv \left[\Phi_1(\mathbf{q}), \Phi_2(\mathbf{q}), \ldots, \Phi_m(\mathbf{q})\right]^T = \mathbf{0} \tag{1}$$

Differentiating eq.(1) with respect to time yields the kinematic velocity equation,

$$\mathbf{\Phi_q}(\mathbf{q})\dot{\mathbf{q}} = \mathbf{0} \tag{2}$$

where subscript denotes partial differentiation; i.e., $\mathbf{\Phi_q} = \left[\partial\Phi_i/\partial q_j\right]$, and an over dot denotes differentiation with respect to time. Finally, differentiating eq.(2) with respect to time yields the kinematic acceleration equation,

$$\mathbf{\Phi_q}(\mathbf{q})\ddot{\mathbf{q}} = -\left(\mathbf{\Phi_q}\dot{\mathbf{q}}\right)_\mathbf{q}\dot{\mathbf{q}} \equiv \boldsymbol{\tau}(\mathbf{q},\dot{\mathbf{q}}) \tag{3}$$

Equations (1) through (3) characterize the admissible motion of the mechanical system.

The state of a mechanical system changes in time under the influence of both applied and constraint forces. The Lagrange multiplier form of the constrained equations of motion (Haug 1989) for mechanical systems is

$$\mathbf{M}(\mathbf{q})\ddot{\mathbf{q}} + \mathbf{\Phi_q^T}(\mathbf{q})\boldsymbol{\lambda} = \mathbf{Q}^A(\mathbf{q},\dot{\mathbf{q}},t) \tag{4}$$

where $\mathbf{M}(\mathbf{q})$ is the system mass matrix that may depend on generalized coordinates, $\boldsymbol{\lambda}$ is the vector of Lagrange multipliers that account for workless constraint forces, and $\mathbf{Q}^A(\mathbf{q},\dot{\mathbf{q}},t)$ is the vector of generalized applied forces that may depend on generalized coordinates, their first time derivatives, and time.

Equations (1- 4) comprise a system of differential-algebraic equations (DAE). It is known (Petzold 1982) that DAE are not ordinary differential equations (ODE). While analytically satisfying eqs.(1) and (4) assures that eqs.(2) and (3) are also satisfied, when the problem is solved numerically this ceases to be the case. In general, the task of obtaining a numerical solution of the DAE of eqs.(1) and (4), especially for stiff systems that require implicit methods, is substantially more difficult and prone to intense numerical computation than one of solving an ODE (Hairer and Wanner 1996). Implicit numerical integration methods based on Newmark integration formulas that are used extensively in solving large-scale ordinary differential equations arising in structural dynamics (Hughes 1987) are employed here. More recent implicit Runge-Kutta methods with attractive variable-step-size error control (Hairer and Wanner 1996) are also employed. For a review of the literature on numerical integration methods for the solution of the DAE of multibody dynamics, the reader is referred to a recent paper by Haug, Negrut, and Iancu (1997).

2. Differential-Algebraic Equations Of Multibody Dynamics

Equations (1) and (4) may be written in matrix form as

$$\begin{bmatrix} \mathbf{M}(\mathbf{q}) & \mathbf{\Phi_q^T} \\ \mathbf{\Phi_q} & \mathbf{0} \end{bmatrix} \begin{bmatrix} \ddot{\mathbf{q}} \\ \boldsymbol{\lambda} \end{bmatrix} = \begin{bmatrix} \mathbf{Q}^A(\mathbf{q},\dot{\mathbf{q}},t) \\ \boldsymbol{\tau}(\mathbf{q},\dot{\mathbf{q}}) \end{bmatrix} \tag{5}$$

which is called the "descriptor form" of the equations of motion. One way to solve the DAE of eqs.(1), (2), and (5) is to reduce it to an equivalent ODE. This can be done by selecting a subset of independent generalized coordinates from the generalized coordinate vector $\mathbf{q}$. All dependent variables in the equations of motion are expressed in terms of independent coordinates, to obtain a state space ODE (SSODE).

In order to determine a partitioning of the generalized-coordinate vector $\mathbf{q}$ into dependent and independent coordinate vectors $\mathbf{u}$ and $\mathbf{v}$, respectively, a set of consistent generalized coordinates $\mathbf{q}^0$, i.e., satisfying eq.(1), is first determined. In this configuration, the constraint Jacobian matrix is evaluated and numerically factored, using the Gauss-Jordan algorithm (Atkinson 1989),

$$\mathbf{\Phi_q}(\mathbf{q}_0) \rightarrow \left[\mathbf{\Phi_u}(\mathbf{q}_0) \,\middle|\, \mathbf{\Phi_v}(\mathbf{q}_0)\right] \tag{6}$$

The order of appearance of generalized coordinates associated with columns of the resulting matrix yields a nonsingular subjacobian with respect to $\mathbf{u}$; i.e.,

$$\det\left(\mathbf{\Phi_u}(\mathbf{q}_0)\right) \neq 0 \tag{7}$$

This can always be done if the constraint equations are independent.

Having partitioned the generalized coordinates, eqs.(1) and (2) can be written in the partitioned form

$$\mathbf{\Phi}(\mathbf{u}, \mathbf{v}) = \mathbf{0} \tag{8}$$

$$\mathbf{\Phi_u}(\mathbf{u}, \mathbf{v})\dot{\mathbf{u}} + \mathbf{\Phi_v}(\mathbf{u}, \mathbf{v})\dot{\mathbf{v}} = \mathbf{0} \tag{9}$$

The condition of eq.(7) and the implicit-function theorem (Corwin and Szczarba 1982) guarantee that eq.(8) can be solved for $\mathbf{u}$ as a function of $\mathbf{v}$, in some neighborhood of $\mathbf{v}_0$, yielding

$$\mathbf{u} = \mathbf{g}(\mathbf{v}) \tag{10}$$

where the function $\mathbf{g}(\mathbf{v})$ has as many continuous derivatives as does the constraint function $\mathbf{\Phi}(\mathbf{q})$.

3. Newmark Integration: Descriptor Form

3.1 Formulation

The numerical integration of the DAE of eqs.(1-4) requires discretization of the equations using a numerical integration formula. Since only the independent coordinates $\mathbf{v}$ may be integrated, based on state-space reduction considerations (Haug 1989), discretization via integration formulas is applied only to these coordinates. All other coordinates and their time-derivatives, as well as the

Lagrange multipliers, become dependent variables in the discretized equations. While numerous ODE integrators, both explicit and implicit, can be applied to integrate independent coordinates, attention in this section is limited to the implicit Newmark family of integrators (Hughes 1987).

Applied to the independent coordinates, the acceleration form of the Newmark equations is

$$\dot{\mathbf{v}}_{n+1} = \tilde{\dot{\mathbf{v}}}_{n+1} + \gamma h \ddot{\mathbf{v}}_{n+1} \tag{11}$$

$$\mathbf{v}_{n+1} = \tilde{\mathbf{v}}_{n+1} + \beta h^2 \ddot{\mathbf{v}}_{n+1} \tag{12}$$

where

$$\tilde{\dot{\mathbf{v}}}_{n+1} = \dot{\mathbf{v}}_n + (1-\gamma) h \ddot{\mathbf{v}}_n \tag{13}$$

$$\tilde{\mathbf{v}}_{n+1} = \mathbf{v}_n + h\dot{\mathbf{v}}_n + \left(\frac{h^2}{2}\right)(1-\gamma)\ddot{\mathbf{v}}_n \tag{14}$$

in which subscripts denote the time step, β and γ are integration constants, and h is the integration step size.

As used in structural dynamics, eqs.(11) and (12) determine independent velocity and position as functions of independent acceleration at time step n + 1, given a knowledge of independent position, velocity, and acceleration at time step n. This class of implicit algorithms has good stability properties (Hughes 1987), so it is used here to treat multibody dynamics applications in which stiffness occurs.

For the remainder of this section, subscript n + 1 that defines the unknowns evaluated at time step n + 1 is suppressed, for notational convenience. In order to discretize the DAE to be integrated, eqs.(11) and (12) are substituted into the kinematic constraint equations and the equations of motion. Recalling that eqs.(8) and (9) determine $\mathbf{u}$ and $\dot{\mathbf{u}}$ as functions of $\mathbf{v}$ and $\dot{\mathbf{v}}$, $\mathbf{u}$ and $\dot{\mathbf{u}}$ are in turn functions of the independent accelerations. The descriptor form of the DAE of eq.(5) at time step n + 1 may thus be regarded as nonlinear equations involving all accelerations and Lagrange multipliers. In other words, the solution of Eq. (5) is sought, with $\mathbf{u}$ and $\dot{\mathbf{u}}$ obtained from eqs.(8) and (9), and $\mathbf{v}$ and $\dot{\mathbf{v}}$ obtained using eqs.(11) and (12). Therefore, in terms of the number of variables to be determined via the iterative solution of the discretized equation of motion, this approach is situated between the two extreme approaches proposed by Haug, Negrut, and Iancu (1997), on one hand, and Haug and Yen (1992) and Potra (1993) on the other.

To solve the discretized form of eq.(5), a Jacobian with respect to acceleration and Lagrange-multiplier variables must thus be evaluated. In preparation for this calculation, the chain rule of differentiation is applied to obtain derivatives of generalized coordinates and their first time derivative with respect to unknown accelerations.

Differentiating eqs.(11) and (12) with respect to independent accelerations yields

$$\dot{\mathbf{v}}_{\ddot{\mathbf{v}}} = \gamma h \mathbf{I} \tag{15}$$

$$\mathbf{v}_{\ddot{\mathbf{v}}} = \beta h^2 \mathbf{I} \tag{16}$$

Next, differentiating eq.(8) with respect to independent accelerations and using eq.(16) yields

$$\mathbf{\Phi}_{\mathbf{u}} \mathbf{u}_{\ddot{\mathbf{v}}} = -\mathbf{\Phi}_{\mathbf{v}} \mathbf{v}_{\ddot{\mathbf{v}}} = -\beta h^2 \mathbf{\Phi}_{\mathbf{v}} \tag{17}$$

Since eq.(7) implies that the coefficient matrix on the left of eq.(17) is nonsingular,

$$\mathbf{u}_{\ddot{\mathbf{v}}} = -\beta h^2 \mathbf{\Phi}_{\mathbf{u}}^{-1} \mathbf{\Phi}_{\mathbf{v}} \equiv \beta h^2 \mathbf{H} \tag{18}$$

where the matrix $\mathbf{H}$ is obtained by solving the system of equations

$$\mathbf{\Phi}_{\mathbf{u}} \mathbf{H} = -\mathbf{\Phi}_{\mathbf{v}} \tag{19}$$

A Boolean matrix $\mathbf{P} \in \mathbf{R}^{\mathbf{ndof} \times \mathbf{n}}$, where $\mathrm{ndof} = \mathrm{n} - \mathrm{m}$, is defined such that row i of this matrix has only zeros, with the exception of the kth column, so q_k occupies the ith entry of the independent vector $\mathbf{v}$. Then,

$$\mathbf{v} = \mathbf{P}\mathbf{q} \tag{20}$$

In order to simplify the presentation, it is assumed that the generalized coordinates have been reordered so that $\mathbf{q} = \left[\mathbf{u}^{\mathrm{T}}, \mathbf{v}^{\mathrm{T}}\right]^{\mathrm{T}}$. Then, matrix $\mathbf{P}$ assumes the form $\mathbf{P} = [\mathbf{0}, \mathbf{I}]$, where $\mathbf{I}$ is the $\mathrm{ndof} \times \mathrm{ndof}$ identity matrix. Recalling that $\mathbf{P}$ is constant, $\ddot{\mathbf{v}} = \mathbf{P}\ddot{\mathbf{q}}$ and

$$\ddot{\mathbf{v}}_{\ddot{\mathbf{q}}} = \mathbf{P} \tag{21}$$

Combining the results of eqs.(16) and (18),

$$\mathbf{q}_{\ddot{\mathbf{v}}} = \beta h^2 \begin{bmatrix} \mathbf{H} \\ \mathbf{I} \end{bmatrix} \tag{22}$$

Using the chain rule of differentiation with eq.(21),

$$\mathbf{q}_{\ddot{\mathbf{q}}} = \mathbf{q}_{\ddot{\mathbf{v}}} \ddot{\mathbf{v}}_{\ddot{\mathbf{q}}} = \beta h^2 \begin{bmatrix} \mathbf{HP} \\ \mathbf{P} \end{bmatrix} \equiv \beta h^2 \hat{\mathbf{H}} \tag{23}$$

Writing eq.(9) in partitioned from yields

$$\mathbf{\Phi}_{\mathbf{u}} \dot{\mathbf{u}} = -\mathbf{\Phi}_{\mathbf{v}} \dot{\mathbf{v}} \tag{24}$$

Since the coefficient matrix on the left is nonsingular,

$$\dot{\mathbf{u}} = -\mathbf{\Phi}_{\mathbf{u}}^{-1}\mathbf{\Phi}_{\mathbf{v}}\dot{\mathbf{v}} = \mathbf{H}\dot{\mathbf{v}} \tag{25}$$

Differentiating both sides of eq.(24) with respect to $\ddot{\mathbf{v}}$ and using eq.(22) yields

$$\mathbf{\Phi}_{\mathbf{u}}\dot{\mathbf{u}}_{\ddot{\mathbf{v}}} = -\mathbf{\Phi}_{\mathbf{v}}\dot{\mathbf{v}}_{\ddot{\mathbf{v}}} - \left(\mathbf{\Phi}_{\mathbf{q}}\dot{\mathbf{q}}\right)_{\mathbf{q}}\mathbf{q}_{\ddot{\mathbf{v}}} = -\gamma h\mathbf{\Phi}_{\mathbf{v}} - \beta h^2\left(\mathbf{\Phi}_{\mathbf{q}}\dot{\mathbf{q}}\right)_{\mathbf{q}}\begin{bmatrix}\mathbf{H}\\ \mathbf{I}\end{bmatrix} \tag{26}$$

Recalling that the coefficient matrix on the left is nonsingular,

$$\dot{\mathbf{u}}_{\ddot{\mathbf{v}}} = \gamma h\mathbf{H} - \beta h^2\mathbf{\Phi}_{\mathbf{u}}^{-1}\left(\mathbf{\Phi}_{\mathbf{q}}\dot{\mathbf{q}}\right)_{\mathbf{q}}\begin{bmatrix}\mathbf{H}\\ \mathbf{I}\end{bmatrix} \equiv \gamma h\mathbf{H} + \beta h^2\mathbf{J} \tag{27}$$

where matrix $\mathbf{J}$ is obtained by solving

$$\mathbf{\Phi}_{\mathbf{u}}\mathbf{J} = -\left(\mathbf{\Phi}_{\mathbf{q}}\dot{\mathbf{q}}\right)_{\mathbf{q}}\begin{bmatrix}\mathbf{H}\\ \mathbf{I}\end{bmatrix} \tag{28}$$

Finally, combining eqs.(15) and (27),

$$\dot{\mathbf{q}}_{\ddot{\mathbf{v}}} = \gamma h\begin{bmatrix}\mathbf{H}\\ \mathbf{I}\end{bmatrix} + \beta h^2\begin{bmatrix}\mathbf{J}\\ \mathbf{0}\end{bmatrix} \tag{29}$$

which may be extended, using the chain rule of differentiation and eq.(21), to obtain

$$\dot{\mathbf{q}}_{\ddot{\mathbf{q}}} = \dot{\mathbf{q}}_{\ddot{\mathbf{v}}}\ddot{\mathbf{v}}_{\ddot{\mathbf{q}}} = \gamma h\begin{bmatrix}\mathbf{HP}\\ \mathbf{P}\end{bmatrix} + \beta h^2\begin{bmatrix}\mathbf{JP}\\ \mathbf{0}\end{bmatrix} \equiv \gamma h\hat{\mathbf{H}} + \beta h^2\hat{\mathbf{J}} \tag{30}$$

It is important to recall that eqs.(15)-(30) apply only for variables defined at time step n + 1, the associated subscript having been suppressed for notational convenience.

3.2 Discretized Descriptor-Form Equations

Equation (5), at time step n + 1, may be written as

$$\mathbf{\Psi} \equiv \begin{bmatrix}\mathbf{M}(\mathbf{q}_{n+1})\ddot{\mathbf{q}}_{n+1} + \mathbf{\Phi}_{\mathbf{q}}^{T}(\mathbf{q}_{n+1})\boldsymbol{\lambda}_{n+1} - \mathbf{Q}^{A}(\mathbf{q}_{n+1}, \dot{\mathbf{q}}_{n+1}, t_{n+1})\\ \mathbf{\Phi}_{\mathbf{q}}(\mathbf{q}_{n+1})\ddot{\mathbf{q}}_{n+1} - \boldsymbol{\tau}(\mathbf{q}_{n+1}, \dot{\mathbf{q}}_{n+1})\end{bmatrix} = \mathbf{0} \tag{31}$$

where generalized coordinates and their first time-derivatives are functions of independent accelerations, as expressed by eqs.(10)-(12) and (25). Equation (31) is thus the discretization of the equations of motion in descriptor form, with $\ddot{\mathbf{q}}_{n+1}$ and $\boldsymbol{\lambda}$ as unknowns.

Since equations of eq.(31) are nonlinear, they are solved using Newton-Raphson iteration (Atkinson 1989). For this purpose, elements of the Jacobian matrix of the left side of eq.(31) with respect to the unknowns are computed, using the chain rule of differentiation and eqs.(23) and (30), as

$$\boldsymbol{\Psi}_{\ddot{\mathbf{q}}_{n+1}} = \begin{bmatrix} \mathbf{M}+\beta h^2\left[(\mathbf{M}\ddot{\mathbf{q}}_{n+1})_{\mathbf{q}}+\left(\boldsymbol{\Phi}_{\mathbf{q}}^T\boldsymbol{\lambda}_{n+1}\right)-\mathbf{Q}_{\mathbf{q}}^A\right]\hat{\mathbf{H}}-\mathbf{Q}_{\dot{\mathbf{q}}}^A\left[\gamma h\hat{\mathbf{H}}+\beta h^2\hat{\mathbf{J}}\right] \\ \boldsymbol{\Phi}_{\mathbf{q}}+\beta h^2\left[\left(\boldsymbol{\Phi}_{\mathbf{q}}\ddot{\mathbf{q}}_{n+1}\right)_{\mathbf{q}}-\boldsymbol{\tau}_{\mathbf{q}}\right]\hat{\mathbf{H}}-\boldsymbol{\tau}_{\dot{\mathbf{q}}}\left[\gamma h\hat{\mathbf{H}}+\beta h^2\hat{\mathbf{J}}\right] \end{bmatrix} \tag{32}$$

$$\boldsymbol{\Psi}_{\boldsymbol{\lambda}_{n+1}} = \begin{bmatrix} \boldsymbol{\Phi}_{\mathbf{q}}^T \\ \mathbf{0} \end{bmatrix} \tag{33}$$

Newton-Raphson iteration to solve eq.(31) is thus carried out as follows:

$$\begin{bmatrix} \mathbf{M}+\beta h^2\left[\left\{\mathbf{M}\ddot{\mathbf{q}}_{n+1}+\left(\boldsymbol{\Phi}_{\mathbf{q}}^T\boldsymbol{\lambda}_{n+1}\right)_{\mathbf{q}}-\mathbf{Q}_{\mathbf{q}}^A\right\}\hat{\mathbf{H}}-\mathbf{Q}_{\dot{\mathbf{q}}}^A\hat{\mathbf{J}}\right]-\gamma h\mathbf{Q}_{\dot{\mathbf{q}}}^A\hat{\mathbf{H}} & \boldsymbol{\Phi}_{\mathbf{q}}^T \\ \boldsymbol{\Phi}_{\mathbf{q}}+\beta h^2\left[\left\{\left(\boldsymbol{\Phi}_{\mathbf{q}}\ddot{\mathbf{q}}_{n+1}\right)_{\mathbf{q}}-\boldsymbol{\tau}_{\mathbf{q}}\right\}\hat{\mathbf{H}}-\boldsymbol{\tau}_{\dot{\mathbf{q}}}\hat{\mathbf{J}}\right]-\gamma h\boldsymbol{\tau}_{\dot{\mathbf{q}}}\hat{\mathrm{H}} & \mathbf{0} \end{bmatrix}^{(j-1)} \begin{bmatrix} \Delta\ddot{\mathbf{q}}_{n+1} \\ \Delta\boldsymbol{\lambda}_{n+1} \end{bmatrix}^{(j)} = -\boldsymbol{\Psi}^{(j-1)} \tag{34}$$

$$\begin{bmatrix} \ddot{\mathbf{q}}_{n+1} \\ \boldsymbol{\lambda}_{n+1} \end{bmatrix}^{(j)} = \begin{bmatrix} \ddot{\mathbf{q}}_{n+1} \\ \boldsymbol{\lambda}_{n+1} \end{bmatrix}^{(j-1)} + \begin{bmatrix} \Delta\ddot{\mathbf{q}}_{n+1} \\ \Delta\boldsymbol{\lambda}_{n+1} \end{bmatrix}^{(j)} \tag{35}$$

where j is the iteration counter. At each iteration, eqs.(11) and (12) determine $\dot{\mathbf{v}}_{n+1}^{(j)}$ and $\mathbf{v}_{n+1}^{(j)}$, and eqs.(8) and (9) are solved numerically for $\mathbf{u}_{n+1}^{(j)}$ and $\dot{\mathbf{u}}_{n+1}^{(j)}$. Iteration is continued until error criteria on $\Delta\ddot{\mathbf{q}}_{n+1}^{(j)}$, $\Delta\boldsymbol{\lambda}_{n+1}^{(j)}$, and $\boldsymbol{\Psi}^{(j)}$ are satisfied.

4. Newmark Integration: State-Space Form

The development in this section is organized as follows. In Subsection 4.1, the DAE are reduced to a set of state-space ODE in independent generalized coordinates. In Subsection 4.2, the Newmark implicit integration formula is applied to the resulting set of ODE. A description of the framework in which integration is carried out is followed by presentation of a method for generating the required derivative information and a procedure for solving the discretized equations of motion.

4.1 Reduction of DAE to State-Space ODE

A partitioning of the vector $\mathbf{q}$ of generalized coordinates into dependent and independent vectors $\mathbf{u}$ and $\mathbf{v}$ is first carried out, as in Section 2. Having partitioned the generalized coordinates, eqs.(1)-(4) can be rewritten in partitioned form (Haug 1989) as

$$\mathbf{M}^{vv}(\mathbf{u},\mathbf{v})\ddot{\mathbf{v}}+\mathbf{M}^{vu}(\mathbf{u},\mathbf{v})\ddot{\mathbf{u}}+\mathbf{\Phi}_{v}^{T}(\mathbf{u},\mathbf{v})\boldsymbol{\lambda}=\mathbf{Q}^{v}(\mathbf{u},\mathbf{v},\dot{\mathbf{u}},\dot{\mathbf{v}}) \tag{36}$$

$$\mathbf{M}^{uv}(\mathbf{u},\mathbf{v})\ddot{\mathbf{v}}+\mathbf{M}^{uu}(\mathbf{u},\mathbf{v})\ddot{\mathbf{u}}+\mathbf{\Phi}_{u}^{T}(\mathbf{u},\mathbf{v})\boldsymbol{\lambda}=\mathbf{Q}^{u}(\mathbf{u},\mathbf{v},\dot{\mathbf{u}},\dot{\mathbf{v}}) \tag{37}$$

$$\mathbf{\Phi}(\mathbf{u},\mathbf{v})=\mathbf{0} \tag{38}$$

$$\mathbf{\Phi}_{u}(\mathbf{u},\mathbf{v})\dot{\mathbf{u}}+\mathbf{\Phi}_{v}(\mathbf{u},\mathbf{v})\dot{\mathbf{v}}=\mathbf{0} \tag{39}$$

$$\mathbf{\Phi}_{u}(\mathbf{u},\mathbf{v})\ddot{\mathbf{u}}+\mathbf{\Phi}_{v}(\mathbf{u},\mathbf{v})\ddot{\mathbf{v}}=\boldsymbol{\tau}(\mathbf{u},\mathbf{v},\dot{\mathbf{u}},\dot{\mathbf{v}}) \tag{40}$$

With the partition of the generalized coordinate vector induced by eq.(6), eq.(36) is reduced to a set of ODE through a succession of steps, using information provided by eqs.(37-40). Since the coefficient matrix of $\dot{\mathbf{u}}$ in eq.(39) is nonsingular, $\dot{\mathbf{u}}$ can be determined as a function of $\mathbf{v}$ and $\dot{\mathbf{v}}$, where eq.(10) is used to eliminate explicit dependence on u. Next, eq.(40) uniquely determines $\ddot{\mathbf{u}}$ as a function of $\mathbf{v}$, $\dot{\mathbf{v}}$, and $\ddot{\mathbf{v}}$, where results of eqs.(10) and (25) are substituted. Since the coefficient matrix of $\boldsymbol{\lambda}$ in eq.(37) is nonsingular, $\boldsymbol{\lambda}$ can be determined uniquely as a function of $\mathbf{v}$, $\dot{\mathbf{v}}$, and $\ddot{\mathbf{v}}$, using previously derived results. Finally, each of the preceding results is substituted into eq.(36) to obtain a set of state-space ODE in only the independent generalized coordinates (Haug 1989),

$$\hat{\mathbf{M}}\ddot{\mathbf{v}}=\hat{\mathbf{Q}} \tag{41}$$

where

$$\begin{aligned}\hat{\mathbf{M}}&=\mathbf{M}^{vv}-\mathbf{M}^{vu}\mathbf{\Phi}_{u}^{-1}\mathbf{\Phi}_{v}-\mathbf{\Phi}_{v}^{T}\left(\mathbf{\Phi}_{u}^{-1}\right)\left[\mathbf{M}^{uv}-\mathbf{M}^{uu}\mathbf{\Phi}_{u}^{-1}\mathbf{\Phi}_{v}\right]\\ \hat{\mathbf{Q}}&=\mathbf{Q}^{v}-\mathbf{M}^{vu}\mathbf{\Phi}_{u}^{-1}\boldsymbol{\tau}-\mathbf{\Phi}_{v}^{T}\left(\mathbf{\Phi}_{u}^{-1}\right)^{T}\left[\mathbf{Q}^{u}-\mathbf{M}^{uu}\mathbf{\Phi}_{u}^{-1}\boldsymbol{\tau}\right]\end{aligned} \tag{42}$$

Using an implicit numerical integration formula to discretize the state-space ODE in eq.(41) is impractical, because solving the resulting set of nonlinear equations requires derivative information whenever a Newton-like method is applied. For the ODE in eq.(41), these derivatives would be difficult to generate. Using eq.(36), from which eq.(41) originates, leads to a consistent and tractable way to obtain the needed derivative information. Thus, rather than discretizing eq.(41), an implicit integration formula is used to discretize eq.(36). The dependent coordinates appearing in this equation, namely $\mathbf{u}$, $\dot{\mathbf{u}}$, $\ddot{\mathbf{u}}$, and $\boldsymbol{\lambda}$, are regarded as functions of the independent coordinates and treated accordingly when computing the Jacobian of the resulting nonlinear equations.

No explicit relationship between the dependent and independent coordinates is known. However, for a given set of independent states (positions and velocities), all dependent coordinates can be computed as follows: (1) $\mathbf{u}$ is determined from eq.(38), (2) $\dot{\mathbf{u}}$ is determined from eq.(39), and (3) $\ddot{\mathbf{u}}$, $\ddot{\mathbf{v}}$, and $\boldsymbol{\lambda}$ are simultaneously determined from eqs.(36), (37), and (40). The procedure for obtaining the derivative information needed is presented in Subsection 4.2, after introducing the implicit integrator used to discretize eq.(36).

4.2 Discretized State-Space Form Equations

The Newmark method of eqs.(11)-(14) is used here to discretize the state-space ODE. Regarding Eq. (36) as a set of second order ODE in the independent coordinates $\mathbf{v}$, and substituting $\mathbf{v}_{n+1}$ and $\dot{\mathbf{v}}_{n+1}$ from eqs.(38) and (39) yields

$$\boldsymbol{\Psi}(\ddot{\mathbf{v}}_{n+1}) \equiv \mathbf{M}^{vv}_{n+1}\ddot{\mathbf{v}}_{n+1} + \mathbf{M}^{vu}_{n+1}\ddot{\mathbf{u}}_{n+1} + \left[\boldsymbol{\Phi}_{v}^{T}\right]_{n+1}\boldsymbol{\lambda}_{n+1} - \mathbf{Q}^{v}_{n+1} = \mathbf{0} \tag{43}$$

This set of nonlinear equations must be solved for the independent accelerations $\ddot{\mathbf{v}}_{n+1}$. Newton's method is used for this purpose, requiring the Jacobian $\boldsymbol{\Psi}_{\ddot{\mathbf{v}}}$, where the time-step subscript is suppressed for notational convenience.

Taking the derivative of eq.(38) with respect to $\ddot{\mathbf{v}}$ results in $\boldsymbol{\Phi}_{u}\mathbf{u}_{\ddot{v}} = -\boldsymbol{\Phi}_{v}\mathbf{u}_{\ddot{v}}$. Using eq.(18) , this equation reduces to

$$\mathbf{u}_{\ddot{v}} = -\beta h^{2}\boldsymbol{\Phi}_{u}^{-1}\boldsymbol{\Phi}_{v} \equiv \beta h^{2}\mathbf{H} \tag{44}$$

where matrix $\mathbf{H}$ is obtained by solving eq.(19). Differentiating eq.(39) with respect to $\ddot{\mathbf{v}}$ and using the chain rule of differentiation yields

$$\begin{aligned}\boldsymbol{\Phi}_{u}\dot{\mathbf{u}}_{\ddot{v}} &= -\left\{(\boldsymbol{\Phi}_{u}\dot{\mathbf{u}})_{u}\mathbf{u}_{\ddot{v}} + (\boldsymbol{\Phi}_{u}\dot{\mathbf{u}})_{v}\mathbf{v}_{\ddot{v}} + \boldsymbol{\Phi}_{v}\dot{\mathbf{v}}_{\ddot{v}} + (\boldsymbol{\Phi}_{v}\dot{\mathbf{v}})_{u}\mathbf{u}_{\ddot{v}} + (\boldsymbol{\Phi}_{v}\dot{\mathbf{v}})_{v}\mathbf{v}_{\ddot{v}}\right\} \\ &= -\beta h^{2}\left[(\boldsymbol{\Phi}_{u}\dot{\mathbf{u}})_{u}\mathbf{H} + (\boldsymbol{\Phi}_{u}\dot{\mathbf{u}})_{v} + (\boldsymbol{\Phi}_{v}\dot{\mathbf{v}})_{u}\mathbf{H} + (\boldsymbol{\Phi}_{v}\dot{\mathbf{v}})_{v} - \gamma h\boldsymbol{\Phi}_{v}\right]\end{aligned} \tag{45}$$

Using eqs.(15), (16), and (44), $\dot{\mathbf{u}}_{\ddot{v}}$ reduces to

$$\begin{aligned}\dot{\mathbf{u}}_{\ddot{v}} = &-\beta h^{2}\left\{\boldsymbol{\Phi}_{u}^{-1}\left[(\boldsymbol{\Phi}_{u}\dot{\mathbf{u}})_{u} + (\boldsymbol{\Phi}_{v}\dot{\mathbf{v}})_{u}\right]\mathbf{H} + \boldsymbol{\Phi}_{u}^{-1}\left[(\boldsymbol{\Phi}_{u}\dot{\mathbf{u}})_{v} + (\boldsymbol{\Phi}_{v}\dot{\mathbf{v}})_{v}\right]\right\} \\ &-\gamma h\boldsymbol{\Phi}_{u}^{-1}\boldsymbol{\Phi}_{v} \equiv \beta h^{2}\mathbf{J} + \gamma h\mathbf{H}\end{aligned} \tag{46}$$

Differentiating eq.(40) with respect to $\ddot{\mathbf{v}}$ and reducing yields

$$\begin{aligned}\boldsymbol{\Phi}_{u}\ddot{\mathbf{u}}_{\ddot{v}} &= \boldsymbol{\tau}_{v}\mathbf{v}_{\ddot{v}} + \boldsymbol{\tau}_{\dot{u}}\dot{\mathbf{u}}_{\ddot{v}} + \boldsymbol{\tau}_{\dot{v}}\dot{\mathbf{v}}_{\ddot{v}} - \left[(\boldsymbol{\Phi}_{u}\ddot{\mathbf{u}})_{u}\mathbf{u}_{\ddot{v}} + (\boldsymbol{\Phi}_{u}\ddot{\mathbf{u}})_{v}\mathbf{v}_{\ddot{v}} + (\boldsymbol{\Phi}_{v}\ddot{\mathbf{v}})_{u}\mathbf{u}_{\ddot{v}}\right. \\ &\left.+(\boldsymbol{\Phi}_{u}\ddot{\mathbf{v}})_{v}\mathbf{v}_{\ddot{v}} + \boldsymbol{\Phi}_{v}\right] = \beta h^{2}\left[\boldsymbol{\tau}_{u}\mathbf{H} + \boldsymbol{\tau}_{v} + \boldsymbol{\tau}_{u}\mathbf{J} - (\boldsymbol{\Phi}_{u}\ddot{\mathbf{u}})_{u}\mathbf{H} - (\boldsymbol{\Phi}_{u}\ddot{\mathbf{u}})_{v}\right. \\ &\left.-(\boldsymbol{\Phi}_{u}\ddot{\mathbf{v}})_{u}\mathbf{H} - (\boldsymbol{\Phi}_{v}\ddot{\mathbf{v}})_{v}\right] + \gamma h\left(\boldsymbol{\tau}_{\dot{u}}\mathbf{H} + \boldsymbol{\tau}_{\dot{v}}\right) - \boldsymbol{\Phi}_{v}\end{aligned} \tag{47}$$

$$\ddot{\mathbf{u}}_{\ddot{\mathbf{v}}} = \beta h^2 \boldsymbol{\Phi}_{\mathbf{u}}^{-1}\left\{\left[\tau_{\mathbf{u}}\left(\boldsymbol{\Phi}_{\mathbf{u}}\ddot{\mathbf{u}}\right)_{\mathbf{u}} - \left(\boldsymbol{\Phi}_{\mathbf{v}}\ddot{\mathbf{v}}\right)_{\mathbf{u}}\right]\mathbf{H} + \tau_{\mathbf{v}} + \tau_{\mathbf{u}}\mathbf{J} - \left(\boldsymbol{\Phi}_{\mathbf{v}}\ddot{\mathbf{v}}\right)_{\mathbf{v}}\right\}\gamma h \boldsymbol{\Phi}_{\mathbf{u}}^{-1}\left(\tau_{\dot{\mathbf{u}}}\mathbf{H} + \tau_{\dot{\mathbf{v}}} + \mathbf{H}\right)$$
$$\equiv \beta h^2 \mathbf{L} + \gamma h \mathbf{N} + \mathbf{H} \qquad (48)$$

Finally, differentiating eq.(37) with respect to $\ddot{\mathbf{v}}$ and simplifying yields

$$\begin{aligned}
\boldsymbol{\Phi}_{\mathbf{u}}^{T}\boldsymbol{\lambda}_{\ddot{\mathbf{v}}} &= \mathbf{Q}_{\mathbf{u}}^{\mathbf{u}}\mathbf{u}_{\ddot{\mathbf{v}}} + \mathbf{Q}_{\mathbf{v}}^{\mathbf{u}}\mathbf{v}_{\ddot{\mathbf{v}}} + \mathbf{Q}_{\dot{\mathbf{u}}}^{\mathbf{u}}\dot{\mathbf{u}}_{\ddot{\mathbf{v}}} + \mathbf{Q}_{\dot{\mathbf{v}}}^{\mathbf{u}}\dot{\mathbf{v}}_{\ddot{\mathbf{v}}} - \Big\{\left(\boldsymbol{\Phi}_{\mathbf{u}}^{T}\boldsymbol{\lambda}\right)_{\mathbf{u}}\mathbf{u}_{\ddot{\mathbf{v}}} + \left(\boldsymbol{\Phi}_{\mathbf{u}}^{T}\boldsymbol{\lambda}\right)_{\mathbf{v}}\mathbf{v}_{\ddot{\mathbf{v}}} + \mathbf{M}^{\mathbf{uv}} \\
&\quad + \left(\mathbf{M}^{\mathbf{uv}}\ddot{\mathbf{v}}\right)_{\mathbf{u}}\mathbf{u}_{\ddot{\mathbf{v}}} + \left(\mathbf{M}^{\mathbf{uv}}\ddot{\mathbf{v}}\right)_{\mathbf{v}}\mathbf{v}_{\ddot{\mathbf{v}}} + \mathbf{M}^{\mathbf{uu}}\ddot{\mathbf{u}}_{\ddot{\mathbf{v}}} + \left(\mathbf{M}^{\mathbf{uu}}\ddot{\mathbf{u}}\right)_{\mathbf{u}}\mathbf{u}_{\ddot{\mathbf{v}}} + \left(\mathbf{M}^{\mathbf{uu}}\ddot{\mathbf{u}}\right)_{\mathbf{v}}\mathbf{v}_{\ddot{\mathbf{v}}}\Big\} \\
&= \beta h^2\Big[\mathbf{Q}_{\mathbf{u}}^{\mathbf{u}}\mathbf{H} + \mathbf{Q}_{\mathbf{v}}^{\mathbf{u}} + \mathbf{Q}_{\dot{\mathbf{u}}}^{\mathbf{u}}\mathbf{J} - \left(\boldsymbol{\Phi}_{\mathbf{u}}^{T}\boldsymbol{\lambda}\right)_{\mathbf{u}}\mathbf{H} - \left(\boldsymbol{\Phi}_{\mathbf{u}}^{T}\boldsymbol{\lambda}\right)_{\mathbf{v}} - \left(\mathbf{M}^{\mathbf{uv}}\ddot{\mathbf{v}}\right)_{\mathbf{v}}\mathbf{H} - \left(\mathbf{M}^{\mathbf{uv}}\ddot{\mathbf{v}}\right)_{\mathbf{v}} \\
&\quad - \mathbf{M}^{\mathbf{uu}}\mathbf{L} - \left(\mathbf{M}^{\mathbf{uu}}\ddot{\mathbf{u}}\right)_{\mathbf{u}}\mathbf{H} - \left(\mathbf{M}^{\mathbf{uu}}\ddot{\mathbf{u}}\right)_{\mathbf{v}}\Big] + \gamma h\left(\mathbf{Q}_{\dot{\mathbf{u}}}^{\mathbf{u}}\mathbf{H} + \mathbf{Q}_{\dot{\mathbf{v}}}^{\mathbf{u}} - \mathbf{M}^{\mathbf{uu}}\mathbf{N}\right) \\
&\quad - \mathbf{M}^{\mathbf{uv}} - \mathbf{M}^{\mathbf{uu}}\mathbf{H}
\end{aligned} \qquad (49)$$

$$\begin{aligned}
\boldsymbol{\lambda}_{\ddot{\mathbf{v}}} &= \beta h^2 \boldsymbol{\Phi}_{\mathbf{u}}^{-T}\Big\{\Big[\mathbf{Q}_{\mathbf{u}}^{\mathbf{u}} - \left(\boldsymbol{\Phi}_{\mathbf{u}}^{T}\boldsymbol{\lambda}\right)_{\mathbf{u}} - \left(\mathbf{M}^{\mathbf{uv}}\ddot{\mathbf{v}}\right)_{\mathbf{u}} - \left(\mathbf{M}^{\mathbf{uu}}\ddot{\mathbf{u}}\right)_{\mathbf{u}}\Big]\mathbf{H} + \mathbf{Q}_{\mathbf{v}}^{\mathbf{u}} + \mathbf{Q}_{\dot{\mathbf{u}}}^{\mathbf{u}}\mathbf{J} - \left(\boldsymbol{\Phi}_{\mathbf{u}}^{T}\boldsymbol{\lambda}\right)_{\mathbf{v}} \\
&\quad - \left(\mathbf{M}^{\mathbf{uv}}\ddot{\mathbf{v}}\right)_{\mathbf{v}} - \left(\mathbf{M}^{\mathbf{uu}}\ddot{\mathbf{u}}\right)_{\mathbf{v}} - \mathbf{M}^{\mathbf{uu}}\mathbf{L}\Big\} + \gamma h \boldsymbol{\Phi}_{\mathbf{u}}^{-T}\left(\mathbf{Q}_{\dot{\mathbf{u}}}^{\mathbf{u}}\mathbf{H} + \mathbf{Q}_{\dot{\mathbf{v}}}^{\mathbf{u}} - \mathbf{M}^{\mathbf{uu}}\mathbf{N}\right) \\
&\quad - \boldsymbol{\Phi}_{\mathbf{u}}^{-T}\left(\mathbf{M}^{\mathbf{uv}} + \mathbf{M}^{\mathbf{uu}}\mathbf{H}\right) \equiv -\boldsymbol{\Phi}_{\mathbf{u}}^{-T}\left[\beta h^2\mathbf{R} + \gamma h\mathbf{S} + \mathbf{M}^{\mathbf{uv}} + \mathbf{M}^{\mathbf{uu}}\mathbf{H}\right]
\end{aligned} \qquad (50)$$

The integration process progresses by determining the solution $\ddot{\mathbf{v}}_{n+1}$ of eq.(43) at each time-step, using Newton-Raphson iteration. The subscript n+1 is suppressed here, for notational convenience. Using results of the preceding derivation, the Jacobian of $\boldsymbol{\Psi}$ in eq.(43) with respect to $\ddot{\mathbf{v}}$ is

$$\begin{aligned}
\boldsymbol{\Psi}_{\ddot{\mathbf{v}}} &= \mathbf{M}^{\mathbf{vv}} + \left(\mathbf{M}^{\mathbf{vv}}\ddot{\mathbf{v}}\right)_{\mathbf{u}}\mathbf{u}_{\ddot{\mathbf{v}}} + \left(\mathbf{M}^{\mathbf{vv}}\ddot{\mathbf{v}}\right)_{\mathbf{v}}\mathbf{v}_{\ddot{\mathbf{v}}} + \mathbf{M}^{\mathbf{vu}}\ddot{\mathbf{u}}_{\ddot{\mathbf{v}}} + \left(\mathbf{M}^{\mathbf{vu}}\ddot{\mathbf{u}}\right)_{\mathbf{u}}\mathbf{u}_{\ddot{\mathbf{v}}} + \left(\mathbf{M}^{\mathbf{vu}}\ddot{\mathbf{u}}\right)_{\mathbf{v}}\mathbf{v}_{\ddot{\mathbf{v}}} \\
&\quad + \boldsymbol{\Phi}_{\mathbf{v}}^{T}\boldsymbol{\lambda}_{\ddot{\mathbf{v}}} + \left(\boldsymbol{\Phi}_{\mathbf{v}}^{T}\boldsymbol{\lambda}\right)_{\mathbf{u}}\mathbf{u}_{\ddot{\mathbf{v}}} + \left(\boldsymbol{\Phi}_{\mathbf{v}}^{T}\boldsymbol{\lambda}\right)_{\mathbf{v}}\mathbf{v}_{\ddot{\mathbf{v}}} - \mathbf{Q}_{\mathbf{u}}^{\mathbf{v}}\mathbf{u}_{\ddot{\mathbf{v}}} - \mathbf{Q}_{\mathbf{u}}^{\mathbf{v}}\mathbf{v}_{\ddot{\mathbf{v}}} - \mathbf{Q}_{\dot{\mathbf{u}}}^{\mathbf{v}}\dot{\mathbf{u}}_{\ddot{\mathbf{v}}} - \mathbf{Q}_{\dot{\mathbf{v}}}^{\mathbf{v}}\dot{\mathbf{v}}_{\ddot{\mathbf{v}}} \\
&= \mathbf{M}^{\mathbf{vv}} + \left(\mathbf{M}^{\mathbf{vv}}\ddot{\mathbf{v}}\right)_{\mathbf{u}}\beta h^2\mathbf{H} + \left(\mathbf{M}^{\mathbf{vv}}\ddot{\mathbf{v}}\right)_{\mathbf{v}}\beta h^2\mathbf{H} + \mathbf{M}^{\mathbf{vu}}\left[\beta h^2\mathbf{L} + \gamma h\mathbf{S} + \mathbf{M}^{\mathbf{uv}} + \mathbf{M}^{\mathbf{uu}}\mathbf{H}\right] \\
&\quad + \left(\mathbf{M}^{\mathbf{vu}}\ddot{\mathbf{u}}\right)_{\mathbf{u}}\beta h^2\mathbf{H} + \left(\mathbf{M}^{\mathbf{vu}}\ddot{\mathbf{u}}\right)_{\mathbf{v}}\beta h^2 + \mathbf{H}^{T}\left[\beta h^2\mathbf{R} + \gamma h\mathbf{S} + \mathbf{M}^{\mathbf{uv}} + \mathbf{M}^{\mathbf{uu}}\mathbf{H}\right] \\
&\quad + \left(\boldsymbol{\Phi}_{\mathbf{v}}^{T}\boldsymbol{\lambda}\right)_{\mathbf{u}}\beta h^2\mathbf{H} + \left(\boldsymbol{\Phi}_{\mathbf{v}}^{T}\boldsymbol{\lambda}\right)_{\mathbf{v}}\beta h^2 - \mathbf{Q}_{\mathbf{u}}^{\mathbf{v}}\beta h^2\mathbf{H} - \mathbf{Q}_{\mathbf{v}}^{\mathbf{v}}\beta h^2 - \mathbf{Q}_{\dot{\mathbf{v}}}^{\mathbf{v}}\gamma h - \mathbf{Q}_{\dot{\mathbf{u}}}^{\mathbf{v}}\left[\beta h^2\mathbf{J} + \gamma h\mathbf{H}\right]
\end{aligned} \qquad (51)$$

Collecting terms, this Jacobian is rewritten as

$$\begin{aligned}
\boldsymbol{\Psi}_{\ddot{\mathbf{v}}} &= \mathbf{M}^{\mathbf{vv}} + \mathbf{M}^{\mathbf{vu}}\mathbf{H} + \mathbf{H}^{T}\left(\mathbf{M}^{\mathbf{uv}} + \mathbf{M}^{\mathbf{uu}}\mathbf{H}\right) + \gamma h\left(\mathbf{M}^{\mathbf{vu}}\mathbf{N} + \mathbf{H}^{T}\mathbf{S} - \mathbf{Q}_{\dot{\mathbf{v}}}^{\mathbf{v}} - \mathbf{Q}_{\dot{\mathbf{u}}}^{\mathbf{v}}\mathbf{H}\right) \\
&\quad + \beta h^2\Big\{\Big[\left(\mathbf{M}^{\mathbf{vu}}\ddot{\mathbf{v}}\right)_{\mathbf{u}} + \left(\mathbf{M}^{\mathbf{vu}}\ddot{\mathbf{u}}\right)_{\mathbf{u}}\Big]\mathbf{H} + \Big[\left(\mathbf{M}^{\mathbf{vv}}\ddot{\mathbf{v}}\right)_{\mathbf{v}} + \left(\mathbf{M}^{\mathbf{vu}}\ddot{\mathbf{u}}\right)_{\mathbf{v}}\Big] + \mathbf{M}^{\mathbf{vu}}\mathbf{L} + \mathbf{H}^{T}\mathbf{R} \\
&\quad + \left(\boldsymbol{\Phi}_{\mathbf{v}}^{T}\boldsymbol{\lambda}\right)_{\mathbf{u}}\mathbf{H} + \left(\boldsymbol{\Phi}_{\mathbf{v}}^{T}\boldsymbol{\lambda}\right)_{\mathbf{v}} - \mathbf{Q}_{\mathbf{u}}^{\mathbf{v}}\mathbf{H} - \mathbf{Q}_{\mathbf{v}}^{\mathbf{v}} - \mathbf{Q}_{\dot{\mathbf{u}}}^{\mathbf{v}}\mathbf{J}\Big\}
\end{aligned} \qquad (52)$$

Newton-Raphson iteration is carried out in the form

$$\mathbf{\Psi}_{\ddot{\mathbf{v}}}^{(i)}\Delta\ddot{\mathbf{v}}^{(i)} = -\mathbf{\Psi}^{(i)}$$
$$\ddot{\mathbf{v}}^{(i+1)} = \ddot{\mathbf{v}}^{(i)} + \Delta\ddot{\mathbf{v}}^{(i)} \tag{53}$$

to iteratively solve for $\ddot{\mathbf{v}}$, until convergence criteria are satisfied. Within each iteration, eqs.(11) and (12) are used to obtain $\mathbf{v}^{i+1}$ and $\dot{\mathbf{v}}^{i+1}$, and eqs.(37)-(40) are solved for $\mathbf{u}^{i+1}$, $\dot{\mathbf{u}}^{i+1}$, $\ddot{\mathbf{u}}^{i+1}$, and $\boldsymbol{\lambda}^{i+1}$. The residual in satisfying eq.(43) is then evaluated and the process is terminated if error criteria are met. Otherwise, the Jacobian in eq.(52) is again evaluated and eqs.(53) are used to obtain a new estimate of the solution $\ddot{\mathbf{v}}$.

5. Runge-Kutta Integration

Runge-Kutta numerical integration methods have been well developed (Hairer, Norrett, Wanner 1993) for solution of first ordinary differential equations of the form

$$\dot{y} = f(t, y) \tag{54}$$

A broad range of Runge-Kutta integrators for this problem can be written in the form

$$k_i = f\left(t_n + c_i h,\, y_n + h\sum_{j=1}^{i} a_{ij}k_j\right), \quad i = 1, \ldots, s \tag{55}$$

$$y_{n+1} = y_n + h\sum_{i=1}^{s} b_i k_i \tag{56}$$

where t_n is the current time-step; y_n is the approximate solution at t_n; b_i, c_i, and a_{ij} are constants; s is the number of stages in integrating from t_n to t_{n+1}; and the constants satisfy the following conditions:

$$c_i = \sum_{j=1}^{i} a_{ij} \tag{57}$$

$$\sum_{i=1}^{s} b_i = 1 \tag{58}$$

In all cases treated here, $a_{ij} = 0$ for $j > i$; i.e., only Diagonally Implicit Runge-Kutta (DIRK) methods (Hairer, Norsett, Wanner 1993) with $a_{ii} \neq 0$ or explicit Runge-Kutta methods with $a_{ii} = 0$ are considered. For the explicit case, the upper limit of summation on the right of eq.(55) is $i-1$. Thus, successive steps using eq.(55) are carried out with only a single function evaluation to obtain k_i in the explicit case. In contrast, k_i appears on both sides of the equation if a_{ii} is not zero, the implicit case, and an iterative method of solving eq.(55) is required. For the remainder of this section, the upper limit of summation in eq.(55) is retained as i and the explicit case is obtained by setting $a_{ii} = 0$.

In order to make Runge-Kutta methods suitable for integration of the second order differential-algebraic equations of multibody dynamics, the second argument on the right of eq.(55) is interpreted as an approximate solution at time $t_i = t_n + c_i h$; i.e.,

$$z_i = y_n + h\sum_{j=1}^{i} a_{ij}\dot{z}_j \tag{59}$$

Substituting eq.(59) into eq.(54) and regarding the left side as $\dot{z}_i$, the same equation as eq.(55) is obtained, with $\dot{z}_i$ as the unknown. Once all stages in eq.(55) are solved for the associated $\dot{z}_i$, the results are substituted into eq.(56) to obtain

$$y_{n+1} = y_n + h\sum_{i=1}^{s} b_i\dot{z}_i \tag{60}$$

Applying eq.(59) again to integrate acceleration yields

$$\dot{z}_i = \dot{y}_n + h\sum_{j=1}^{i} a_{ij}\ddot{z}_j \tag{61}$$

This integration formula may be substituted into a second-order differential equation, with eq.(59), and the result solved at each of the s stages for the unknown $\ddot{z}_i$. Applying the integration formula of eq.(60) to second derivatives,

$$\dot{y}_{n+1} = \dot{y}_n + h\sum_{i=1}^{s} b_i\ddot{z}_i \tag{62}$$

In order to enable solution of second-order differential equations using eqs.(60) and (62), with $\ddot{z}_i$ as the solution variable in the discretized equations of motion, it is helpful to substitute eq.(61) into eq.(60), to obtain

$$\begin{aligned} y_{n+1} &= y_n + h\sum_{i=1}^{s} b_i\left(\dot{y}_n + h\sum_{j=1}^{i} a_{ij}\ddot{z}_j\right) \\ &= y_n + h\left(\sum_{i=1}^{s} b_i\right)\dot{y}_n + h^2\sum_{j=1}^{s}\left(\sum_{i=1}^{s} b_i a_{ij}\right)\ddot{z}_j \end{aligned}$$

$$= y_n + h\dot{y}_n + h^2 \sum_{j=1}^{s} \bar{b}_j \ddot{z}_j \tag{63}$$

where it is recalled that $a_{ij} = 0$ for $j > i$ and

$$\bar{b}_j = \sum_{i=1}^{s} b_i a_{ij} \tag{64}$$

This approach is commonly used with Newmark methods in structural dynamics (Hughes 1987) and is applicable for multibody dynamics (Haug, Iancu, Negrut 1997). Likewise, eq.(61) may be substituted into eq.(59) to obtain

$$z_i = y_n + h \sum_{j=1}^{i} a_{ij} \left(\dot{y}_n + h \sum_{\ell=1}^{j} a_{j\ell} \ddot{z}_\ell \right) \qquad j = 1, \ldots, i \quad \ell = 1, \ldots, j$$

$$= y_n + h \left(\sum_{j=1}^{i} a_{ij} \right) \dot{y}_n + h^2 \sum_{\ell=1}^{s} \left(\sum_{j=1}^{s} a_{ij} a_{j\ell} \right) \ddot{z}_\ell \qquad \ell < j < i$$

$$= y_n + h c_i \dot{y}_n + h^2 \sum_{\ell=1}^{i} \bar{a}_{i\ell} \ddot{z}_\ell \tag{65}$$

where the upper limits of the double summation in eq.(65) are changed from i and j to s, since $\ell \le j \le i$ and $a_{j\ell} = 0$ for $\ell > j$, and

$$\bar{a}_{i\ell} = \sum_{j=1}^{s} a_{ij} a_{j\ell} \tag{66}$$

Note that if $\ell > i$, each term in the sum on the right side of eq.(66) is zero and $a_{i\ell} = 0$.

Defining constants γ and β appropriately, and modifying the predictor definitions of eqs.(13) and (14), eqs.(59) and (65) can be put in precisely the same form as eqs.(11) and (12). Thus, implicit Runge-Kutta methods can be implemented in both the descriptor and state-space forms developed in Section 3 using the Newmark formulas. The approach taken in this section to integrate the differential-algebraic equations of motion for multibody dynamic systems using Runge-Kutta formulas is to insert eqs.(61) and (65) for independent coordinates into the underlying state-space ordinary differential equation of eq.(41). The resulting equations are equivalent to making the same substitution into the descriptor form of eq.(5), using eqs.(8) and (9) to determine dependent coordinates and their first time-derivatives. The resulting discretized equations of motion involve both independent and dependent accelerations and Lagrange multipliers as solution variables. They are solved numerically and the Runge-Kutta algorithm of eqs.(65), and (61)-(63) is used.

A variable step size algorithm with error control is presented by Negrut, Haug, and Iancu (1997), using embedded formulas for error estimation (Hairer and Wanner 1996).

6. Numerical Examples

6.1 Constant-Step-Size Newmark Method

The algorithms of Sections 3 and 4 are implemented with a constant-time-step Newmark integration formula obtained from eqs.(11) and (12) by taking $\gamma = 0.5$ and $\beta = 0.25$ (Hughes 1987). The focus of numerical experiments is on robustness of the algorithms, rather than efficiency. Three test problems are considered, based on models of a double pendulum, a slider-crank, and a seven-link mechanism that is used as a challenging test problem by Schiehlen (1990) and Hairer and Wanner (1996). Numerical results obtained with the implicit Newmark integrator are compared with results obtained by an explicit Adams Bashforth second-order integrator (Atkinson 1989). Numerical results for variants of these examples, using the descriptor form approach, are presented by Haug, Iancu, and Negrut (1997).

With nstep integration steps, such that for a given h, $(n\,step - 1)h < T \leq n\,step\,h$, the supremum norm is $\text{SupNorm} = \max e_i$, $1 \leq i \leq n\,step$, where, for a step size h, the error in variable p at time $t_i = ih$ is defined as $e_i = \left|p_i^{ref} - p_i\right|$. The reference solution p_i^{ref} is generated using a fourth-order explicit Runge-Kutta method (Hairer et al. 1993) with a time step of $h_{ref} = 10^{-6}$. Finally, if k is such that $e_k \geq e_i$ for $1 \leq i \leq n\,step$, then if $p_k^{ref} \neq 0$, the relative error is defined as $\text{Rel Err} \equiv \text{SupNorm} / \left|p_k^{ref}\right|$.

A Double Pendulum

The double pendulum shown in Fig. 1 is a two-body, two-degree-of-freedom, planar mechanical system. The mechanism is modeled using Cartesian coordinates, i.e., the x and y coordinates of the center of mass and angle θ that defines the orientation of the local centroidal reference frame of a body with respect to the global reference frame. A large amount of stiffness is incorporated in the system by means of two rotational spring-damper-actuators (RSDA). The parameters of the problem are provided in Table 1 (all data in this paper are in SI units; m, kg, sec). Initial conditions for the problem are given in Table 2. The first row contains position information and the second contains velocity information. The simulation is carried out for two seconds.

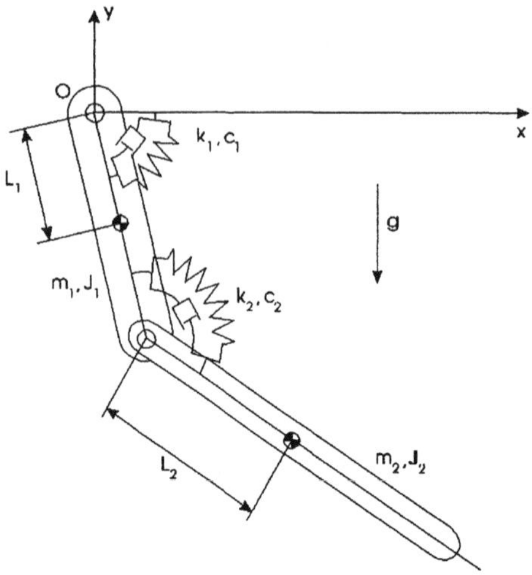

Fig. 1. Double pendulum

Table 1. Parameters for the double pendulum

L_1	m_1	k_1	c_1	L_2	m_2	k_2	c_2
1.0	3.0	400.0	15.0	1.5	0.3	300.0	10,000.0

Table 2. Initial conditions for the double pendulum

Body 1			**Body 2**		
x coordinate	y coordinate	θ coordinate	x coordinate	y coordinate	θ coordinate
1.0	0.0	2π	3.449	-0.388	23π/12
0.0	0.0	0.0	0.0	0.0	0.0

Plots of position and velocity error versus step size, for both the Newmark implicit and Adams Bashforth explicit integrators, are shown in Fig. 2. As can be seen, the implicit integrator is able to successfully function with step sizes approximately 100 times larger than for the explicit integrator. The straight line plot of error versus step size for the position variable, on a log-log scale, confirms that the implicit integrator is functioning properly, up to the point at which step size is large enough that Newton iteration for solution of independent accelerations fails. This performance of the implicit integrator for the DAE is characteristic of what would be expected if the implicit integrator were applied to the underlying stiff ODE.

Identical results were obtained using the Newmark integrator with both the descriptor and state-space forms of the governing equations. This behavior was observed in each of examples studied in this paper. For this example, both forms required comparable computing time.

A Slider-Crank

The slider-crank mechanism shown in Fig. 3 is a two-link, one-degree-of-freedom planar mechanical system, modeled using Cartesian coordinates. At the tip of body 2, a translational spring-damper-actuator (TSDA) element is attached. The parameters of the problem are provided in Table 3. They are selected to make the equations of motion stiff. Initial conditions for the problem are given in Table 4. The first row contains position information and the second contains velocity information. The length of the simulation is two seconds.

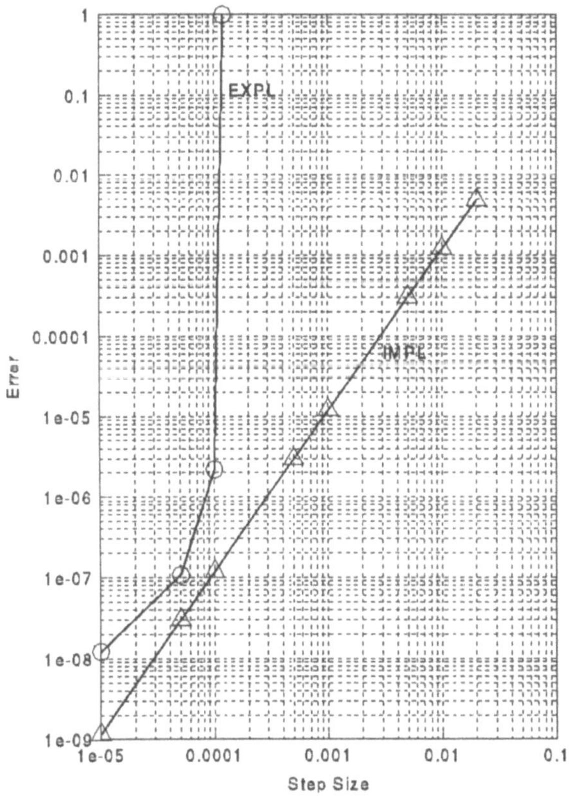

(a) Position

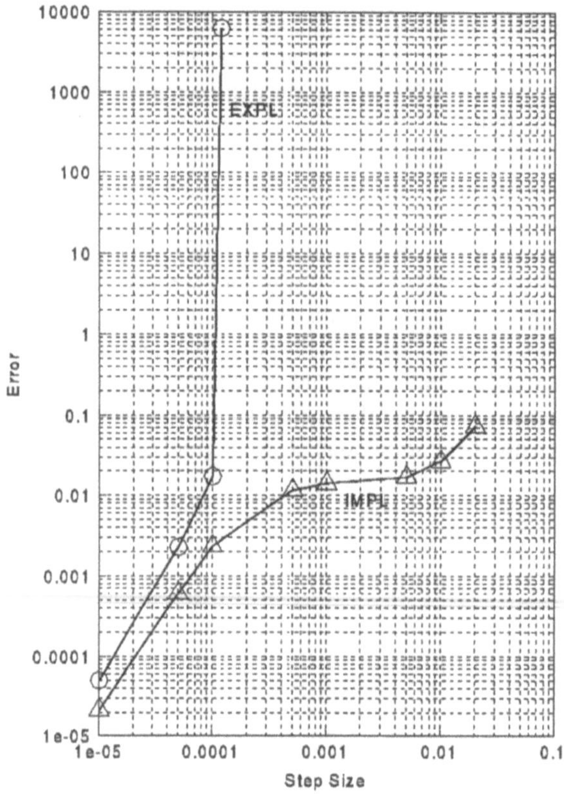

(b) Velocity

Fig. 2. Double pendulum results; constant step implicit & explicit

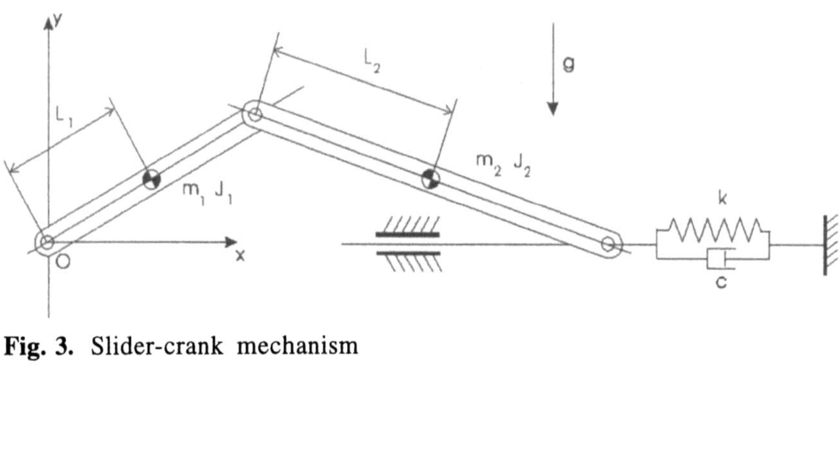

Fig. 3. Slider-crank mechanism

Table 3. Parameters for the slider-crank

L_1	m_1	L_2	m_2	k	c
0.3	3.0	0.6	9.0	100,000.0	300.0

Table 4. Initial conditions for the slider-crank

Body 1			**Body 2**		
x coordinate	y coordinate	θ coordinate	x coordinate	y coordinate	θ coordinate
1.0	- 0.3	3π/2	0.520	- 0.3	π/6
0.0	0.0	0.0	0.0	0.0	0.0

Plots of position and velocity error versus step size for this example are displayed in Fig. 4. As in the previous example, the implicit integrator is successful for step sizes approximately 100 times larger than for the explicit integrator. Note that, for this example, the error in implicit integration is not linear with step size at the largest step sizes tested. Nevertheless it is linear for small step size, as expected.

Again, essentially identical results were obtained with both the descriptor and state-space forms of the DAE, with comparable computing times.

A Seven-Link Mechanism

Numerical experiments for a larger-scale example are performed using "Andrew's squeezing mechanism" shown in Fig. 5. This is a classical test problem used in the literature, through the work of Schiehlen (1990) and Hairer and Wanner (1996). It consists of 7 rigid links connected by frictionless joints and undergoes planar motion. A large driving torque is applied to the crank (body 1) in Fig 5, while a stiff spring is connected to body 3. The mechanism experiences very large accelerations, limiting integrators to small step sizes to stably simulate the motion.

Cartesian coordinates are used to model the mechanism of Fig. 5. For this problem, the vector of generalized coordinates is of dimension 21, and 20 constraint equations must be satisfied. The mechanism has one degree of freedom. Parameters defining the problem (inertia properties, initial conditions, spring and damper characteristics, and driving forces) are as given by Hairer and Wanner (1996).

Plots of position and velocity error versus step size in Fig. 6 demonstrate that, for the stiff system being simulated, the implicit integrator is successful for step sizes approximately 100 times larger than for the explicit integrator.

While identical results were obtained with descriptor and state-space forms of the equations, in this somewhat larger-scale application, the state-space form required less computing time. This suggests that the state-space form may be preferable for large scale applications.

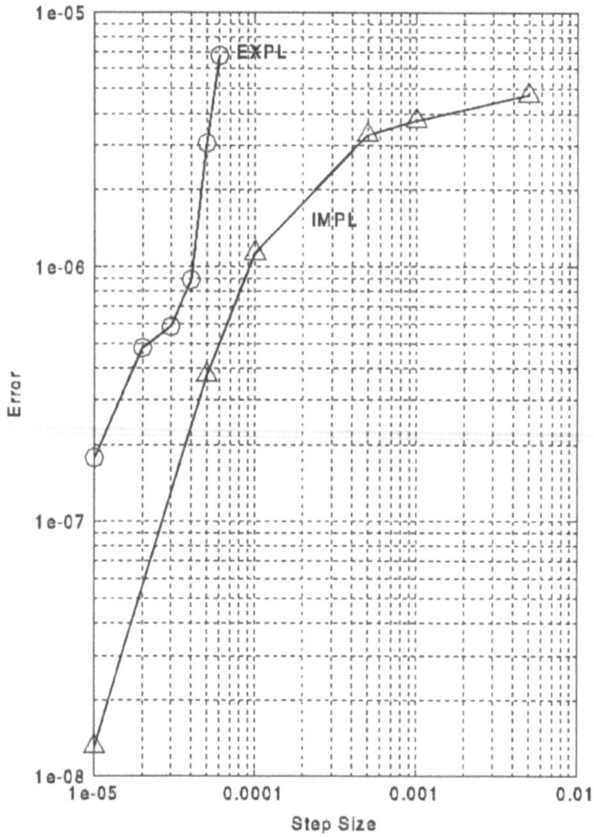

(a) Position

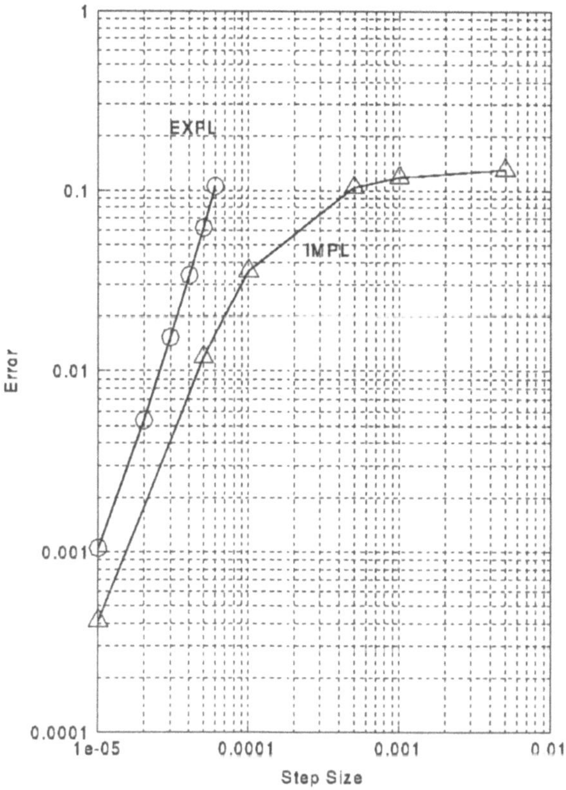

(b) Velocity

Fig. 4. Slider-crank results; constant step implicit & explicit

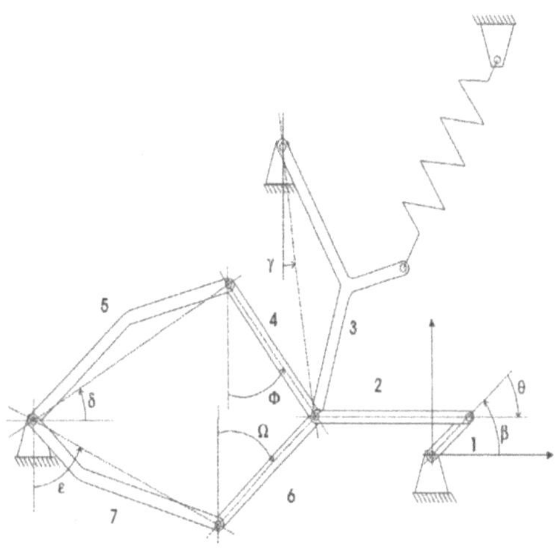

Fig. 5. Seven-link mechanism

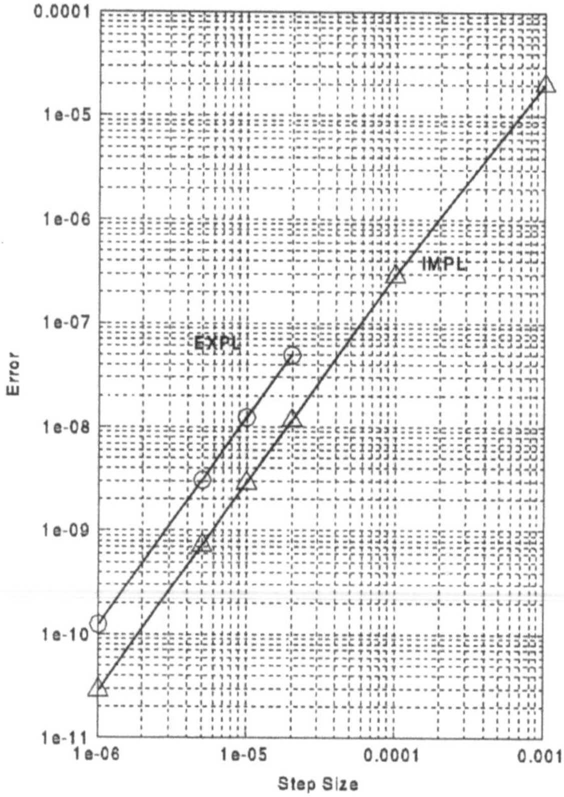
0.0001
1e-05
1e-06
1e-07
1e-08
1e-09
1e-10
1e-11
Error
IMPL
EXPL
1e-06
1e-05
0.0001
0.001
Step Size

(a) Position

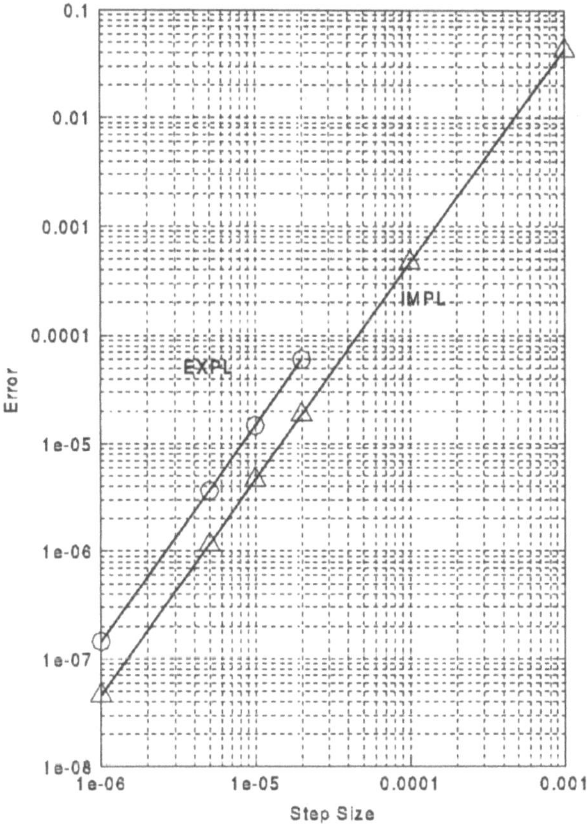

(b) Velocity

Fig. 6. Seven Body Results; Constant Step Implicit & Explicit

6.2 Variable Step Size Runge-Kutta Method

The SDIRK method of Section 5, with order $s = 4$ and with a third order error control algorithm to determine optimal step size (Negrut, Haug, and Iancu 1997) was used to simulate each of the examples treated above. Identical numerical results were obtained, but with significantly improved computational efficiency. To see why this superior performance occurs, Figs. 7 and 8 show step size selected by the algorithm and the number of Newton iterations required for solution of the double pendulum and slider-crank examples, respectively.

For the double pendulum, significantly larger step sizes were achieved than would be possible for the constant step size algorithm. In this example, low frequency motion of the first body continues to excite the highly damped stiff behavior of the second body; hence, the step size must occasionally be reduced to maintain accuracy.

For the slider-crank, after the initial stiff behavior, motion is damped and the algorithm is able to take a large step size. It is interesting to note that the aggressively large step size selected leads to the requirement for a nearly constant number of Newton iterations for the solution of the discretized equations of motion.

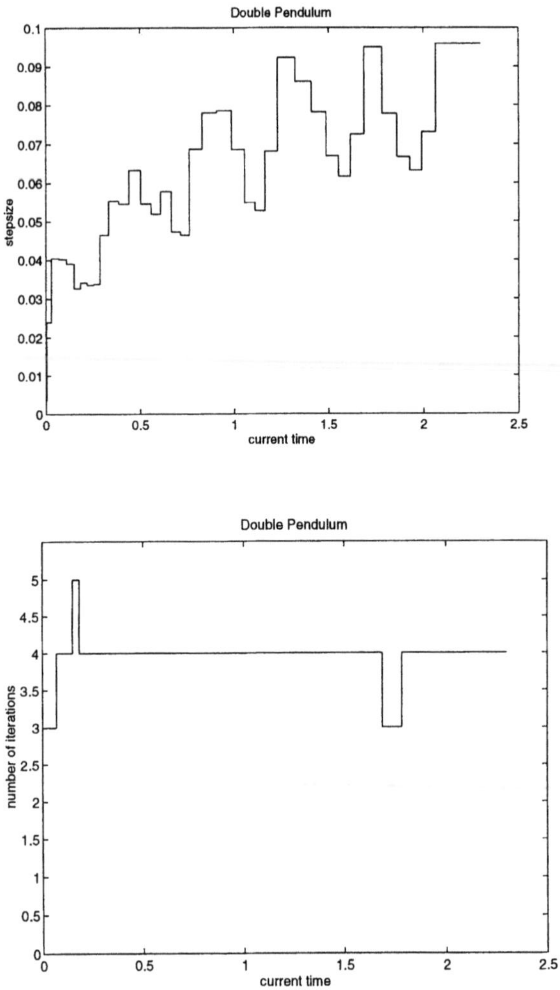

Fig. 7. Double pendulum results; SDIRK 4(3), descriptor form

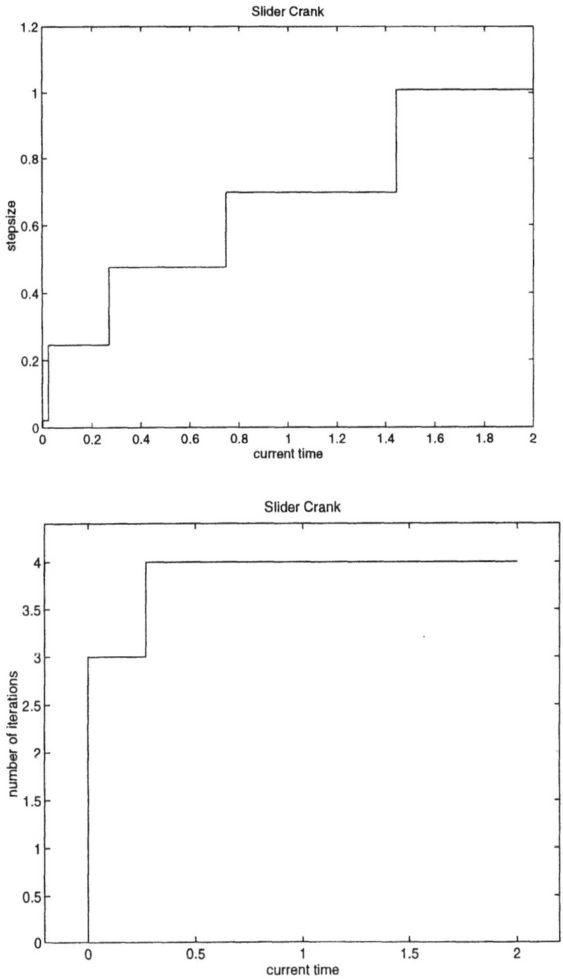

Fig. 8. Slider-crank results; SDIRK 4(3), descriptor form

7. Conclusions

As expected on theoretical grounds, application of implicit ODE methods for integration of the independent generalized coordinates in both the descriptor and state-space forms of the DAE algorithms yield reliable solutions for even stiff systems. While no definitive conclusion can be drawn regarding which of the two forms of the algorithm is more efficient, numerical results suggest that the state-space form may be preferred for large-scale applications.

It is important to note that each form of implicit integrator for DAE satisfies all the governing equations to within prescribed tolerance; i.e., the equations of motion, as well as position, velocity, and acceleration constraint equations. These integrators simultaneously yield accurate solutions and the excellent level of stability expected of implicit ODE solvers, even when applied to stiff DAE.

The variable-step-size Runge-Kutta algorithm with embedded formulas for step size and error control performed exceptionally well on each of the test problems studied. This is extremely important in enabling efficient solution of large-scale mechanical system dynamics problems with stiff components. Such problems arise often in automotive and aerospace applications involving flexible components and bushings.

Acknowledgment

This research was funded by the US Army Tank Automotive Command Center (TACOM) through the Automotive Research Center (Department of Defense contract number DAAE07-94-C-R094).

References

Atkinson, K.E. (1989): An Introduction to Numerical Analysis, Second Edition. Wiley, New York

Corwin, L.J. and Szczarba, R.H. (1982): Multivariable Calculus. Marcel Dekker, New York

Hairer, E., Norsett, S. P., and Wanner, G. (1993): Solving Ordinary Differential Equations I, Nonstiff Problems. Springer-Verlag, Berlin

Hairer, E. and Wanner, G. (1996): Solving Ordinary Differential Equations II: Stiff and Differential-Algebraic Problems. Springer-Verlag, Berlin

Haug, E.J. (1989): Computer-Aided Kinematics and Dynamics of Mechanical Systems Volume I: Basic Methods. Prentice-Hall, Englewood Cliffs, New Jersey

Haug, E.J., Negrut, D., and Iancu, M. (1997): A State-Space Based Implicit Integration Algorithm for Differential-Algebraic Equations of Multibody Dynamics. Mechanics of Structures and Machines, **25(3)**.

Haug, E. J. and Yen, J. (1992): Implicit Numerical Integration of Constrained Equations of Motion via Generalized Coordinate Partitioning. Journal of Mechanical Design, **114**: 296-304

Hughes, T.J. (1987): The Finite Element Method. Prentice-Hall, Englewood Cliffs, New Jersey

Haug, E.J., Iancu, M., and Negrut, D. (1997): Implicit Integration of the Equations of Multibody Dynamics in Descriptor Form. Proc. 1997 ASME Design Engineering Technical Conference. September 14-17, Sacramento, California. DETC97/DAC-3852

Negrut, D., Haug, E.J., and Iancu, M. (1997): Variable Step Implicit Numerical Integration of Stiff Multibody Systems. Technical Report No. 1, NADS and Simulation Center, University of Iowa, Iowa City, Iowa

Petzold, L.R. (1982): Differential/Algebraic Equations Are Not ODE's. SIAM Journal of Scientific and Statistical Computing, **3(3):** 367-384

Potra, F. A. (1993): Implementation of Linear Multistep Methods for Solving Constrained Equations of Motions. SIAM J. Numer. Anal., **30(3):** 774-789

Schiehlen, W. (1990): Multibody Handbook. Springer-Verlag, New York

Models for the Simulation of Power-Steering Systems

Rainer Dürr and Werner Schiehlen

Institute B of Mechanics, D-70550 Stuttgart, Germany,
rd@mechb.uni-stuttgart.de, wos@mechb.uni-stuttgart.de

Summary. This chapter deals with the modeling of power-steering systems in passenger vehicles. Based on simple models found in the literature a new model for power-steering systems is introduced. This model consists of two subsystems: a mechanical system and a hydraulic system. Different stages of model complexity for the hydraulic system are shown before it is assembled with the mechanical system to form the complex model of a power-steering system. Extensive simulation analysis is performed in different stages of model complexity in order to develop a model of the entire power-steering system in optimal quality. The feedback of tire forces is introduced for simulation purposes by a simple tire model consisting of a spring damper combination with Coulomb friction. The model is verified by time simulations.

Keywords. Models of power-steering system, simulation of power-steering system, multibody systems, hydraulics.

1. Introduction

High quality passenger cars are characterized by quite a number of accessory devices improving ride quality and ride safety. Most of the accessory devices interact with the vehicle by sensors, controllers and actuators resulting in an overall mechatronic system.

Common accessory devices are servobrakes, antilocking systems and anti-skid devices. Power-steering systems are other examples for useful accessory devices. For years, they support steering of vehicles, improve driving comfort of passengers and influence driving safety in a positive way in many motor vehicles.

The support of a power-steering system is normally realized with a hydraulic support force produced by a power cylinder. The entire power-steering system consists of:

- mechanical components, e.g. steering wheel, steering shaft and steering gear,
- hydraulic components of the steering system, e.g. oil, pump, orifices, pipes, and pressure chambers,
- electric components, e.g. in the Servotronic of ZF (1994).

Because subsystems out of different fields of engineering interact and form the overall system, this system is also called a mechatronic system.

As part of the entire motor vehicle the power-steering system has great impact on the overall dynamics of the system. The influence of the steering system cannot be neglected. When analyzing the dynamics and simulating the behavior of motor vehicles, the characteristics of the power-steering system are more and more of interest.

In order to estimate the influence of the power-steering system on the overall behavior of a motor vehicle, its thorough examination is necessary. For a first step, a model of the power-steering system is designed. The modeling of the power-steering system is done using a modular approach. There are several reasons to do so: it allows the exchange of model components in a simple way; therefore, different stages of model complexity can be realized quite easily. Each module can be treated as an autonomous part of the overall system. Furthermore, the tracing back of the influencing components can be done faster. The different modules can be assembled to the entire system. A division into a mechanical and a hydraulic module is chosen. The examination of the electric components is neglected due to the high-frequency dynamics.

The multibody system (MBS) approach is most suitable for modeling the mechanical parts of the power-steering system. Here the software package ADAMS is used. A short overview on the theory of multibody system dynamics and ADAMS is given, before the mechanical system is modeled. The system is composed of is eight rigid bodies with four degrees of freedom. For modeling the hydraulics of the power-steering system, several pressure chambers have to be used. From the hydraulic module the differential equations for the pressure in the pressure chambers and the algebraic equations for the volume flow through the orifices can be derived. The equations are implemented in the ADAMS model of the mechanical module. The hydraulic module is modeled with three, five, and seven pressure chambers. Linking the modules of mechanics and hydaulics forms the complete model of the power-steering system.

In order to excite and examine the model in simulations a tire model is needed. This tire model is linked to the power-steering system to provide feedback of tire forces. A simple tire model consisting of a spring damper in combination with Coulomb friction is used.

Extensive simulations are performed to investigate the different modules of the power-steering system and to find a model of optimal quality. A model of the power-steering system with a hydraulic module of five pressure chambers fits requirements the best.

2. Mechanical Module of the Power-Steering System

Figure 1 shows the mechanical steering system incorporated in a motor vehicle of Porsche. The mechanics of the steering system consist of a steering wheel, which transforms the steering torque of the driver into the rotation of the steering column and the steering screw. The rotation of the steering screw moves the steering rod, which steers the tires.

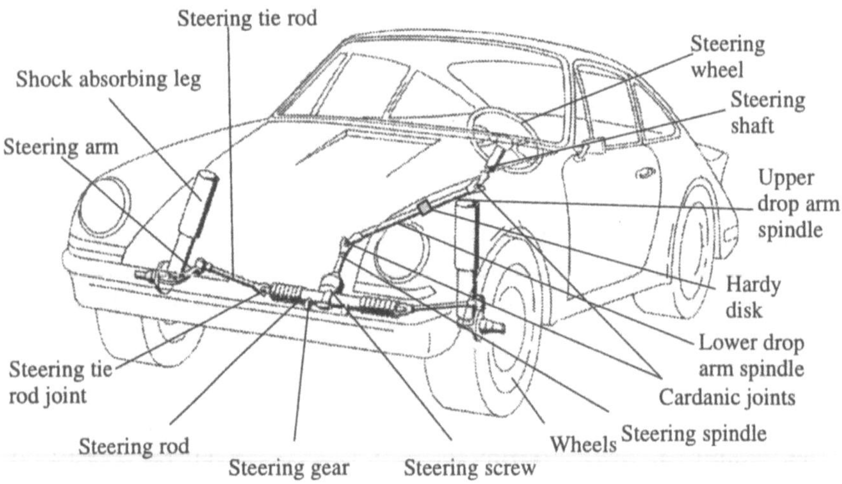

Fig. 1. Mechanical steering system, incorporated in a Porsche 911, in Reimpell (1984)

The steering column is built up from different parts: The steering shaft is connected to the steering wheel and is linked by a universal joint to the upper drop arm spindle. A Hardy disk is located between the upper and the lower drop arm spindle, to restrict the motion between these two bodies. The lower drop arm spindle is connected by another universal joint to the steering spindle. In a steering system without any hydraulic support force, the steering spindle is connected directly to the steering screw, while, in a power-steering system, a torsion bar with predefined stiffness connects the steering spindle with the steering screw.

The steering screw is kept with a bearing in the steering gear. If the steering screw is rotating, the steering rod moves sidewards. The steering rod is connected to the steering tie rods by the steering tie rod joints. The steering tie rods move the steering arms, which move the wheels. For the power-steering system considered, the steering screw is located in the center of the steering rod and the steering tie rods are attached to the sides of the steering rod.

Modeling the mechanics of the power-steering system, the multibody system approach is most suitable, see e.g. (Schiehlen 1990). This approach is based on a discrete description of individual mass-supporting bodies, which are linked to one another by massless joints and bearings as well as massless springs and damper elements. Many different program formalisms are available. The ADAMS program of Mechanical Dynamics (1987) is chosen for modeling and simulating the power-steering system. ADAMS is widely used for many mechanical problems in vehicle engineering and proved to be worthwhile, as shown by Zamow and Witte (1988). A description of the dynamics of motor vehicles and their components is found e.g. in (Popp and Schiehlen 1993).

Before modeling the mechanics of the power-steering system a short overview on the theory of multibody systems is given. For more detail see also (Schiehlen 1986) and (Schiehlen 1997). The approach of multibody systems utilizes a finite set of elements such as rigid bodies and particles, bearings, joints and supports, springs and dampers, acitve forces, and position actuators. To investigate the dynamics of a multibody system the system has to be described mathematically by equations and inequalities. In general, a multibody system can be described either by ordinary differential equations or by differential-algebraic equations.

To form the equations of a multibody system either in ordinary differential form or in differential-algebraic form, the kinematics of a multibody system has to be investigated first. Therefore, all constraints are omitted and the multibody system of p bodies has $6p$ degrees of freedom. The position of the system is given for each body relative to the inertial frame by the 3×1–translation vector

$$\boldsymbol{r}_i = \left[\begin{array}{ccc} r_{i1} & r_{i2} & r_{i3} \end{array} \right]^T \quad , \qquad i = 1(1)p \; , \tag{1}$$

of the center of mass C_i and the 3×3–rotation tensor

$$\boldsymbol{S}_i = \boldsymbol{S}_i(\alpha_i, \beta_i, \gamma_i) \; . \tag{2}$$

The rotation tensor $\boldsymbol{S}_i$ depends on three angles α_i, β_i, γ_i and corresponds with the direction cosine matrix relating the inertial frame I and the body-fixed frame i to each other. The $3p$ translational coordinates and the $3p$ rotational coordinates (angles) can be summarized in a $6p \times 1$–position vector

$$\boldsymbol{x} = [r_{11}\; r_{12}\; r_{13}\; r_{21} \;\; \cdots \;\; \alpha_p \;\; \beta_p \gamma_p]^T \quad . \tag{3}$$

Equations (1) and (2) read now

$$\boldsymbol{r}_i = \boldsymbol{r}_i(\boldsymbol{x}) \;, \quad \boldsymbol{S}_i = \boldsymbol{S}_i(\boldsymbol{x}) \; . \tag{4}$$

Secondly, the q holonomic, rheonomic constraints are added to the mechanical system in the form

$$\boldsymbol{\Phi}(\boldsymbol{x}, t) = \boldsymbol{0} \; , \tag{5}$$

where $\boldsymbol{\Phi}$ denotes a $q \times 1$–vector function.

The velocities of a multibody system with $6p$ degrees of freedom are found by differentiation of equation (4) with respect to the inertial frame:

$$\boldsymbol{v}_i = \dot{\boldsymbol{r}}_i \; , \tag{6}$$

$$\boldsymbol{\omega}_i = \dot{\boldsymbol{s}}_i \; . \tag{7}$$

The infinitesimal 3 × 1–rotation vector $\boldsymbol{s}_i$ used in (7) follows analytically from the corresponding infinitesimal skew-symmetric 3 × 3–rotation tensor.

The accelerations are obtained by a second differentiation with respect to the inertial frame:

$$\boldsymbol{a}_i = \dot{\boldsymbol{v}}_i \; , \tag{8}$$

$$\boldsymbol{\alpha}_i = \dot{\boldsymbol{\omega}}_i \; . \tag{9}$$

Thirdly, the r nonholonomic, rheonomic constraints, e.g. due to rigid wheels, are introduced implicitly by

$$\boldsymbol{\Psi}(\boldsymbol{x}, \dot{\boldsymbol{x}}, t) = \boldsymbol{0} \; , \tag{10}$$

where the symbol $\boldsymbol{\Psi}$ is denoting an $r \times 1$–vector function.

The accelerations are found again by differentiation with respect to the inertial frame I.

After the description of the kinematics of a multibody system the Newton and Euler equations can be applied using a free-body diagram. Now the rigid bearings and supports are replaced by adequate constraint forces and torques, as discussed below.

Newtons and Eulers equations are, for each body in the inertial frame,

$$m_i \dot{\boldsymbol{v}}_i = \boldsymbol{f}_i^e + \boldsymbol{f}_i^r \; , \quad i = 1(1)p \; , \tag{11}$$

$$\boldsymbol{I}_i \dot{\boldsymbol{\omega}}_i + \tilde{\boldsymbol{\omega}}_i \boldsymbol{I}_i \boldsymbol{\omega}_i = \boldsymbol{l}_i^e + \boldsymbol{l}_i^r \; , \quad i = 1(1)p \; . \tag{12}$$

The inertia is represented by the mass m_i and the 3 × 3–inertia tensor $\boldsymbol{I}_i$ with respect to the center of mass C_i of each body. The external forces and torques in eqs. (11) and (12) are composed by the 3×1–applied force vector $\boldsymbol{f}_i^e$ and torque vector $\boldsymbol{l}_i^e$ due to springs, dampers, actuators, weight, etc., and by the 3 × 1–constraint force vector $\boldsymbol{f}_i^r$ and torque vector $\boldsymbol{l}_i^r$. All torques are taken with respect to the center of mass C_i. The applied forces and torques, respectively, depend on the motion by appropriate laws and they may be coupled to the constraint forces and torques in the case of friction.

The constraint forces and torques originate from the reactions in joints, bearings, supports or wheels. They can be reduced by distribution matrices to the generalized constraint forces. The number of the generalized constraint forces is equal to the total number of constraints $(q + r)$ in the system. Introducing the $(q + r) \times 1$–vector of generalized constraint forces

$$\boldsymbol{\lambda} = [\lambda_1 \; \lambda_2 \; \lambda_3 \; \cdots \; \lambda_{q+r}]^T \tag{13}$$

and the 3 × $(q + r)$–distribution matrices

$$\boldsymbol{F}_i = \boldsymbol{F}_i(\boldsymbol{y}, \boldsymbol{z}, t) \; , \quad \boldsymbol{L}_i = \boldsymbol{L}_i(\boldsymbol{y}, \boldsymbol{z}, t) \tag{14}$$

it turns out

$$f_i^r = F_i\,\lambda\ ,\quad l_i^r = L_i\,\lambda\ ,\quad i = 1(1)p\ , \tag{15}$$

for each body. The constraint forces or the distribution matrices, respectively, are derived either by geometrical relations or by analysis.

The ideal applied forces and torques depend only on the kinematical variables of the system; they are independent of the constraint forces. Ideal applied forces are due to the elements of multibody systems, and further actions on the system, e.g. gravity. The forces may be characterized by any combination of proportional, differential and integral behavior.

Proportional forces are characterized by functions of the system's position and time, namely,

$$f_i^e = f_i^e(x, t)\ . \tag{16}$$

For example, conservative spring and weight forces as well as purely time-varying forces are proportional forces.

Proportional-differential forces depend on the position and the velocity:

$$f_i^e = f_i^e(x, \dot{x}, t)\ . \tag{17}$$

A parallel spring-damper configuration is a typical example for this kind of forces. Proportional-integral forces are, in turn, functions of the position and integrals thereof:

$$f_i^e = f_i^e(x, w, t)\ ,\quad \dot{w} = \dot{w}(x, w, t)\ , \tag{18}$$

where the $p \times 1$–vector w describes the position integrals. For example, serial spring-damper configurations and the eigendynamics of actuators result in proportional-integral forces. In vehicle systems proportional-integral forces appear, e.g., with modern engine mounts for simultaneous noise and vibration reduction. The same laws hold also for ideal applied torques.

An example of an ideal applied force in a mechanical multibody system would be the power-steering force, described later in this chapter.

In the case of nonideal constraints with sliding friction or contact forces, respectively, the applied forces are coupled with the constraint forces.

The Newton-Euler equations of a complete holonomic system can be cast in matrix notation. Therefore, the $f \times 1$–position vector

$$y = [y_1\ y_2\ y_3\ \cdots\ y_f]^T \tag{19}$$

has to be used, which represents the f generalized coordinates of a holonomic system. The Newton-Euler equations of a holonomic system in matrix notation reads then as

$$\bar{\bar{M}}\bar{J}\ddot{y} + \bar{q}^c(y, \dot{y}, t) = \bar{q}^e(y, \dot{y}, t) + \bar{Q}\lambda\ . \tag{20}$$

Here the symbol $\bar{\bar{M}}$ is a $6p \times 6p$–diagonal matrix, $\bar{Q}$ is a $6p \times (q + r)$–distribution matrix, $\bar{J}$ is a $6p \times f$–global matrix, $\bar{q}^c$ is the vector of Coriolis forces and $\bar{q}^e$ that of ideal applied forces, whereas $\bar{Q}\boldsymbol{\lambda}$ stands for the generalized constraint forces.

For nonholonomic systems the $g \times 1$–velocity vector

$$z(t) = [z_1\ z_2\ z_3\ \cdots\ z_g]^T \tag{21}$$

summarizing the g generalized velocities of the system is introduced. Hence, the Newton-Euler equations can be transformed to

$$\bar{\bar{M}}\bar{L}\dot{z} + \bar{q}^c(y, z, t) = \bar{q}^e(y, z, t) + \bar{Q}\boldsymbol{\lambda}\ , \tag{22}$$

where $\bar{L}$ is a $6p \times g$–matrix.

The Newton-Euler eqs. (11) and (12), as well as their compact forms of eqs. (20) and (22) are combined algebraical and differential equations.

To derive ordinary differential equations, the Newton-Euler eqs. (20) for holonomic systems can be transformed by applying d'Alembert's principle to the equation of motion

$$M(y,t)\ \ddot{y} + k(y, \dot{y}, t) = q(y, \dot{y}, t)\ , \tag{23}$$

where $M(y,t)$ is a completely symmetrized matrix and the constraint forces and torques are eliminated. The remaining $f \times 1$–vector k describes the generalized Coriolis forces and the $f \times 1$–vector q includes the generalized applied forces.

For nonholonomic systems the equation of motion can be obtained from the Newton-Euler equations in the form of eq. (22) by using Jourdain's principle. However, the equations of motion are not sufficient; they have to be completed by the explicit nonholonomic constraint eqs. (10). Thus, the complete equations read

$$\begin{aligned} &M(y,z,t)\ \dot{z} + k(y,z,t) = q(y,z,w,t)\ , \\ &\dot{y} = \dot{y}(y,z,t)\ , \quad \dot{w} = \dot{w}(y,z,t)\ . \end{aligned} \tag{24}$$

Here, $M(y,z,t)$ is the $g \times g$–symmetric inertia matrix, while k and q are $g \times 1$–vectors of generalized Coriolis and applied forces.

The Newton-Euler equations can also be transformed into differential-algebraic equations. The ADAMS program, used in this work, is based on differential-algebraic equations. ADAMS uses a variational form of the classical Newton-Euler eqs. (11) and (12), which lends to Lagrange's equation of the first kind

$$M(x)\ \ddot{x} + k(x, \dot{x}, t) = q(x, \dot{x}, t) - \Phi_x^T\ \boldsymbol{\lambda}\ , \tag{25}$$

where $\boldsymbol{\lambda}$ is the $q \times 1$–vector of Lagrangian multipliers.

The spatial generalized coordinates employed in ADAMS consist of three Cartesian coordinates of the mass center of each body relative to a global origin and three orientation angles (Euler Angles) defining the attitude of the centroidal principal axes of inertia of each body relative to mutually-perpendicular, inertially-fixed axes. In addition to these coordinates, ADAMS also employs body angular velocities and center-of-mass translational velocities, as well as applied force variables, and constraint forces, also known as Lagrange multipliers, as additional state variables.

The kinematical and dynamical equations of motion for all bodies have been derived a priori and precoded into the algorithm to be recalled whenever needed. This provides for a straightforward, systematic equation-assembly process and a standard set of equations, regardless of the application. As the solution of linear algebraic equations nearly always consumes the majority of CPU time in a multibody simualtion, ADAMS has been specifically designed to maximize the efficiency of the decomposition and back subsitution processes by taking full advantage of sparse matrix-processing techniques, and by symbolically encoding the decomposition path within the code prior to run-time. These efficiency considerations have continuously guided the development of ADAMS.

When constraints act between various parts of the system, ADAMS does not reduce the number of coordinates down to a minimal independent set, but rather appends all of the configuration and motion constraint equations, given by joints and actuators, onto the set of exisiting first-order differential equations defining the kinematical and dynamical equations for each body. The algebraic constraint equations have been prederived for each available ADAMS joint i.e., revolute, spherical, etc., and are recorded in the algorithm for use as needed.

Additionally, instead of embedding the complete form of complex external forces into the dynamical differential equations, ADAMS identifies these forces as state variables in the dynamical equations. Then, ADAMS adds further algebraic equations defining the full mathematical model of these forces. Any user-defined differential equations are appended last.

Thus, the system equations governing behavior of any mechanism in ADAMS consist of:

- six first-order dynamical equations of each body, relating forces to accelerations;
- six first-order kinematical equations of each body, relating positions to velocities;
- a single algebraic constraint for each scalar, holonomic, configuration constraint;
- a single differential equation for each scalar, nonholonomic, motion constraint;
- a single algebraic equation for each complicated, scalar force; and
- any number of user-defined algebraic or first-order differential equations.

The algebraic or first-order differential equations used by nonmechanical modules as the hydraulic module described later on would be an example of the user-defined equations. Therefore, ADAMS allows the implementation of equations of any number of subsystems.

The equations of ADAMS may be rearranged into the general form shown below:

$$\boldsymbol{M}(\boldsymbol{x}, \boldsymbol{u}, \dot{\boldsymbol{u}}, \boldsymbol{f}, t) = \boldsymbol{0} \ , \tag{26}$$

$$\boldsymbol{N}(\boldsymbol{u}, \boldsymbol{x}) = \boldsymbol{u} - \dot{\boldsymbol{x}} = \boldsymbol{0} \ , \tag{27}$$

$$\boldsymbol{\Phi}(\boldsymbol{x}, \boldsymbol{f}, t) = 0 \ . \tag{28}$$

The symbol $\boldsymbol{x}$ denotes the column matrix of generalized coordinates, $\boldsymbol{u}$ represents the column matrix of generalized speeds, $\boldsymbol{f}$ is the column matrix of constraint forces and applied forces, t represents time, $\boldsymbol{M}$ is the column matrix of all differential dynamical equations and any user-defined differential equations, $\boldsymbol{N}$ is the column matrix of differential kinematic equations and differential motion constraints, and $\boldsymbol{\Phi}$ is the column matrix of algebraic equations describing joint constraints and forces.

Equations (26-28) represent a very sparse, highly-uncoupled set of mixed, nonlinear differential-algebraic equations (DAE's) governing the behavior of an ADAMS model. These equations are solved in various combinations during kinematic, static, and dynamic analyses. The solution of sets of DAE's involves techniques that are quite different from those employed in solving standard ordinary differential equations (ODE's) found in most minimal-coordinate multibody formalism.

To solve the differential-algebraic equations a column matrix $\boldsymbol{v}$ of state variables has to be defined with

$$\boldsymbol{v} = [\boldsymbol{x}^T, \boldsymbol{u}^T, \boldsymbol{f}^T]^T \ . \tag{29}$$

Then, eqs. (26-28) can be written as the following single matrix equation:

$$\boldsymbol{G}(\boldsymbol{v}, \dot{\boldsymbol{v}}, t) = \boldsymbol{0} \ . \tag{30}$$

This equation represents a set of nonlinear equations which must, in general, be solved in an iterative fashion to obtain a solution at any instant. ADAMS solves these equations using a modified Newton-Raphson procedure. The equations are expanded in a Taylor series about an initial set of guesses for the state variables. At time zero, the initial guesses for the iteration procedure are obtained form the initial conditions; afterwards, they are set equal to past values of the state variables scaled in an appropriate manner.

More details about ADAMS and the computational methods employed in ADAMS are published in Ryan (1990), Ryan and Steigerwald (1988) and Wielenga (1987).

Using the multibody system approach implemented in ADAMS for modeling the mechanics of the power-steering system, the system is decomposed into eight rigid bodies with four degrees of freedom, see Fig. 2. The eight rigid bodies are the steering wheel, the steering shaft, the upper and lower drop arm spindle, the steering spindle, the steering screw, the steering gear and the steering rod. The body of the motor vehicle is considered to be fixed to an inertial frame. The mechanical steering system is fixed with bearings inside the body of the vehicle and is described by four degrees of freedom: the rotation of the steering wheel in the vehicle body; the rotation of the steering shaft caused by internal backlash between the steering wheel and the steering shaft; the rotation of the steering spindle; and the movement of the steering rod with respect to the vehicle body.

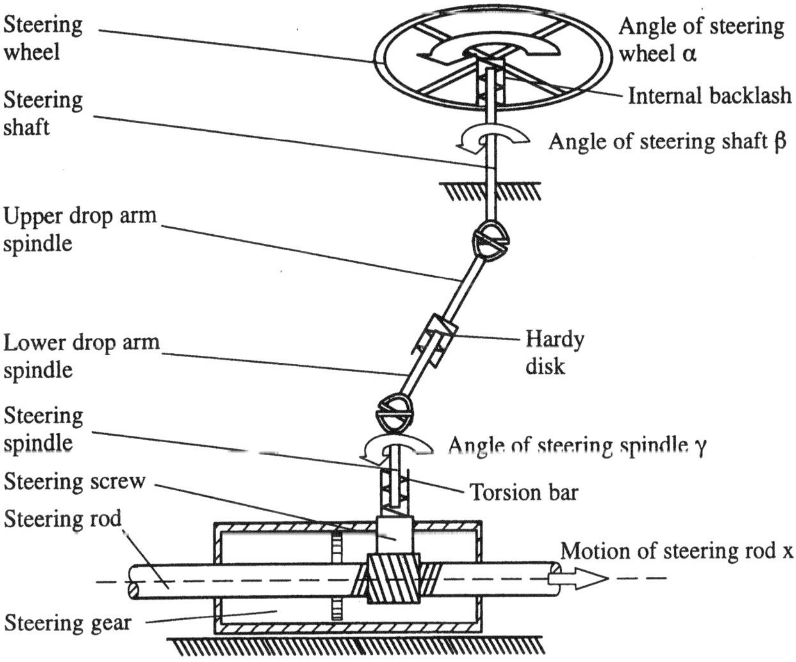

Fig. 2. Model of the mechanical module

The relative rotation between the steering wheel and the steering spindle is influenced by a rotational spring. In order to model the backlash in the steering column, a free rotation in a small range is possible. A Hardy disk restricts the motion between the upper and the lower drop arm spindle. The stiffness of the Hardy disk is supposed to be 100 Nm/rad. The torsion bar between the steering spindle and the steering screw is supposed to have a stiffness of 86 Nm/rad. The twisting angle of the torsion bar controls the servovalve of the hydraulics. To obtain proper hydraulic support of the power-steering system, its stiffness is adjusted by vehicle manufacturers.

Within the power-steering system two universal joints are modeled. The first one transforms the rotation of the steering shaft into the rotation of the upper drop arm spindle. The second universal joint is located between the lower drop arm spindle and the steering spindle and transforms the rotations of these bodies.

Inside the steering gear, the rotation of the steering screw is transformed into the motion of the steering rod. The radius of the steering screw is 8 mm. Therefore, a rotation of the steering screw by 1 rad leads to a translation of the steering rod by 8 mm. The steering tie rods and the steering arms are not included in the model.

Damping elements in the steering column are supposed to damp vibrations in the power-steering system. There is damping between the bodies of the mechanical steering system as well as between the mechanical steering system and the body of the surrounding vehicle. The Hardy disk is a damping element. It damps vibrations between the upper and the lower drop arm spindle. Another damper is located between the steering wheel and the vehicle body, which is assumed fixed to an inertial frame. A damper between the steering spindle and the vehicle body influences the rotation of the steering spindle in the vehicle body.

Damping elements are used to model viscous friction. There can also be Coulomb friction within the power-steering system. Coulomb friction is supposed to act between the steering spindle and the vehicle body and is modeled, too.

The input to the mechanics of the power-steering system is a steering torque on the steering wheel. The output is the motion of the steering rod and, therefore, a rotation of the wheels. The wheels react with a motion and a force acts on the steering rod.

In order to model the reaction of the wheels, a simple tire model is introduced. It consist of a spring-damper combination with Coulomb friction. Both wheels are combined into one reaction force. The reaction force acts on the steering rod, see Fig. 3.

3. Hydraulic Module of the Power-Steering System

The hydraulic module of a power-steering system consists of different components. There is a hydraulic supply pump, including a pressure relief valve, supply pipes, a servovalve, a hydraulic power cylinder, a reservoir and some pressure chambers, Figs. 4 and 5. The hydraulics of the power-steering system is controled by the twisting angle of the torsion bar within the mechanics of the system. The rotation of the torsion bar leads to an opening and closing of the servovalve orifices.

The torsion bar of the transmission connects the steering spindle with the steering screw, Fig. 4. The bar controls the servovalve of the hydraulics. The servovalve consists of two parts, A and B. Valve part A is linked to the steering spindle by a pin. Another pin connects valve part B with the steering screw. If the torsion bar is twisted, valve parts A and B rotate relative to each other.

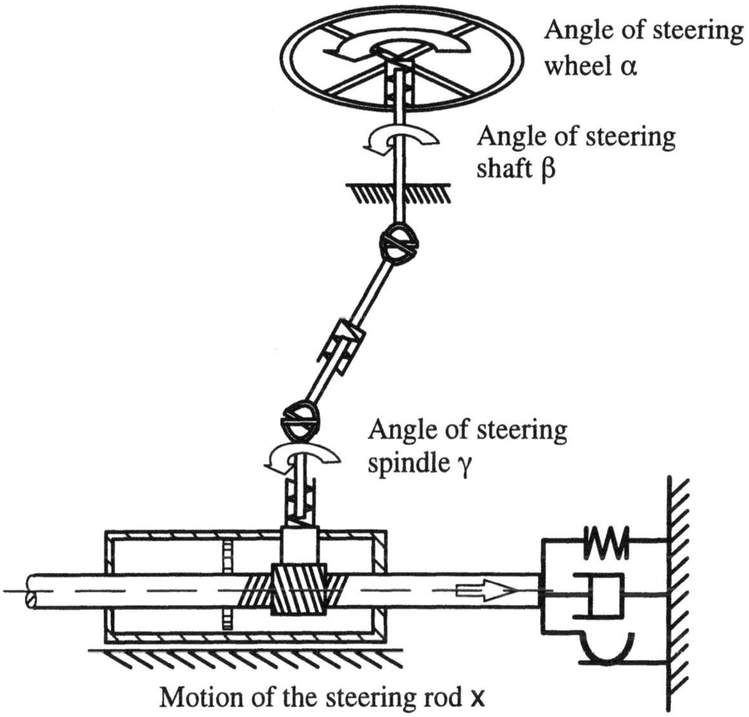

Fig. 3. Simplified tire model

The opening areas of the servovalve depend on the twisting angle of the torsion bar. The twisting angle is defined by the relative rotation between the steering spindle and the steering screw of the mechanical steering system. Since the torsion bar has a known rotational stiffness, it twists in a predictable manner if a torque acts on the power-steering system. Therefore, the servovalve parts A and B rotate relative to each other.

Figure 5 gives the C-C view of the servovalve in equilibrium. As can be seen, the servovalve has three parallel working sections around the perimeter of the torsion bar. The flows in the servovalve will be explained only for one out of the three sections.

The hydraulic supply pump supports the hydraulic system with oil. The oil is pumped through the supply pipes into the supply chamber p_{sup}. Under equilibrium conditions the oil inside the supply chamber is split into two flows. One flow goes into the valve chamber p_{val1}, while a second goes into the valve chamber p_{val2}. Both valve chambers are connected with a working chamber of the hydraulic power cylinder. Under equilibrium conditions, the pressures in the valve chambers p_{val1} and p_{val2} are equal to the pressure in the connected working chambers p_{work1} and p_{work2}. Therefore, there is no exchange of oil

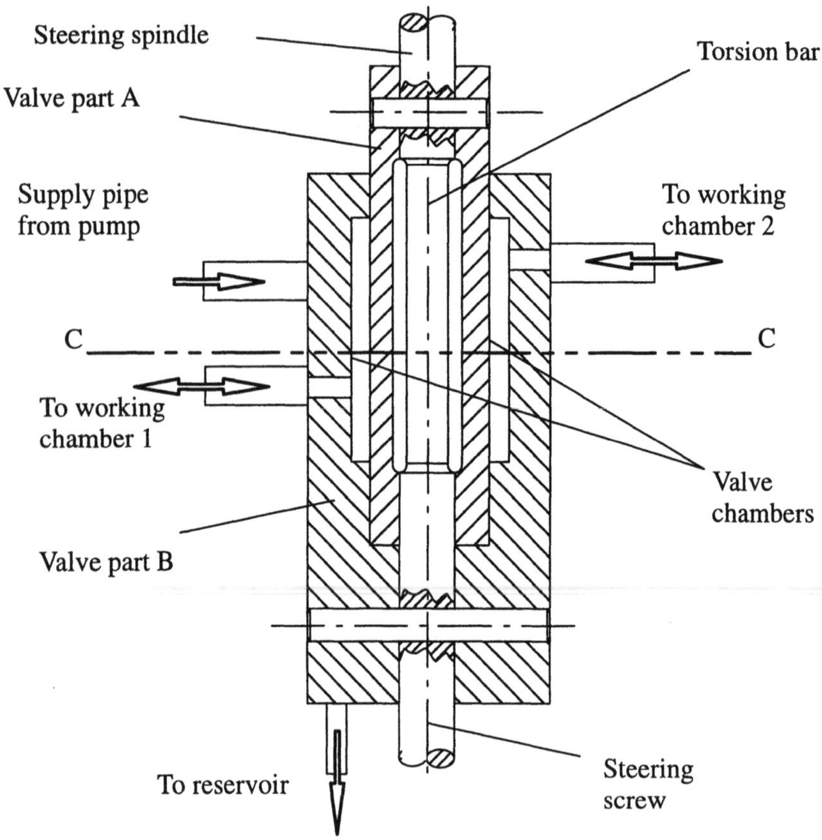

Fig. 4. Servovalve with torsion bar

between the chambers. The oil flows directly into the return chambers and back to the reservoir.

If a steering torque acts on the steering wheel and the torsion bar is twisted, the two valve parts A and B rotate relative to each other. If the steering wheel is, for example, turned to the right, the valve part A rotates in the clockwise direction relative to the valve part B. The opening area of the orifice between the supply chamber p_{sup} and the valve chamber p_{val1} opens, while the opening area of the orifice between the supply chamber p_{sup} and the valve chamber p_{val2} closes. Additionally, the opening area between the valve chamber p_{val1} and the return chamber closes, while the opening area between the valve chamber p_{val2} and the return chamber opens.

Opening the orifice area to valve chamber p_{val1}, while closing the orifice area to valve chamber p_{val2} increases the flow into valve chamber p_{val1}, while the flow into valve chamber p_{val2} decreases. On the other side the flow out

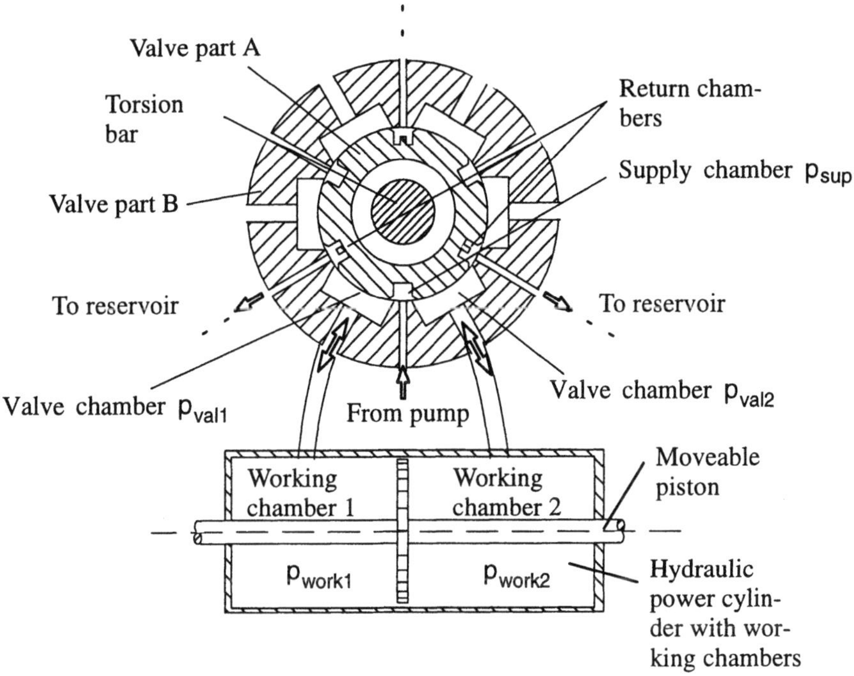

Fig. 5. Servovalve drawn in C-C section with working cylinder

of valve chamber p_{val1} to the return chamber decreases while the flow out of valve chamber p_{val2} to the return chamber increases. Therefore, the pressure in the valve chamber p_{val1} together with the pressure in the working chamber p_{work1} increases. On the other side the pressure in the valve chamber p_{val2} together with the pressure in the working chamber p_{work2} decreases. A pressure difference between the two working chambers is built up and produces a force on the piston to the right side.

Modeling the hydraulics of the power-steering system is done by splitting the system into different pressure chambers. Three approaches will be considered. The hydraulic system is split into three, five and seven pressure chambers. The results will be compared in simulations.

Dividing the hydraulic module into three pressure chambers is shown in Fig. 6. The volume of the supply chamber is built up with the volume of the supply pipe form the hydraulic pump to the servovalve. The volume of the working chambers 1 and 2 is formed by the volume of the pipes from the servovalve to the hydraulic cylinder and the volumes of the working areas in the hydraulic cylinder, respectively.

The three pressure chambers are supplied and emptied by six volume flows in and out of the chambers. The supply flow from the pump into the supply chamber is defined by

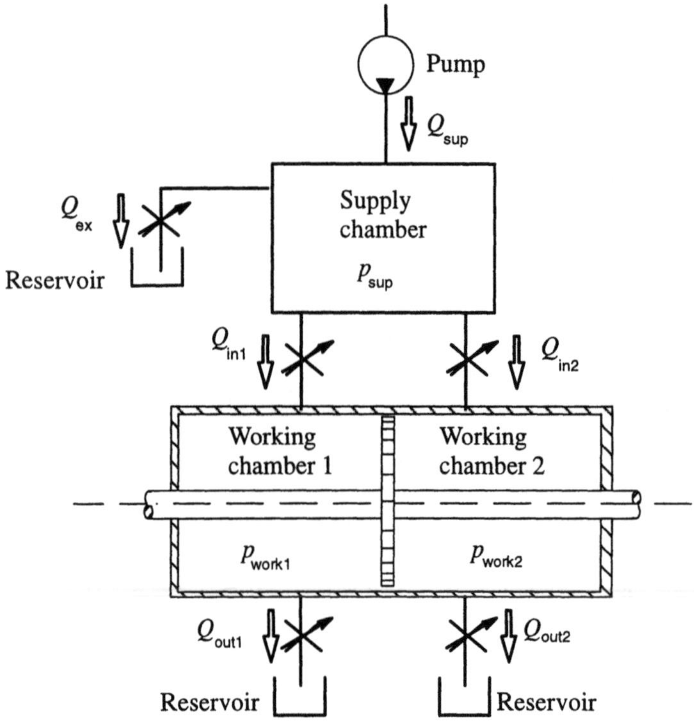

Fig. 6. Dividing the hydraulic module into three pressure chambers

$$Q_{\text{sup}} = Q_{\text{pump}} \frac{p_{\max} - 0.5 p_{\text{sup}}}{p_{\max} - p_0}. \tag{31}$$

In this equation the maximum of a possible volume flow from the pump Q_{pump} is given by $100 \frac{\text{cm}^3}{\text{s}}$, while the maximum of pressure $p_{\max}$, supported by the pump, is given with 10^7Pa. The pressure of the surrounding is supposed to be 10^5Pa and the pressure in the supply chambers is variable.

In order to limit the pressure in the hydraulic system a pressure relief valve is included. The relief valve controlls a pipe from the supply chamber to the reservoir. If the pressure in the supply chamber exceeds a value of $9 * 10^6$Pa, it opens and releases the system with a volume flow Q_{ex} out of the system. Therefore, a maximum pressure of $9 * 10^6$Pa is possible within the hydraulic system.

The working chambers 1 and 2 are supplied with fluid by the two volume flows Q_{in1} and Q_{in2} out of the supply chamber. The volume flows Q_{in1} and Q_{in2} depend on the opening areas of the valve. The opening areas of the valve are controled by the twisting angle of the torsion bar. The volume flows Q_{in1} and Q_{in2} are given by

$$Q_{\text{in}1} = A_{\text{in}1}\ \phi_{\text{loss}} \sqrt{\frac{2}{\rho}\left|p_{\text{sup}} - p_{\text{work}1}\right|}\ \text{sgn}\left(p_{\text{sup}} - p_{\text{work}1}\right) \tag{32}$$

and

$$Q_{\text{in}2} = A_{\text{in}2}\ \phi_{\text{loss}} \sqrt{\frac{2}{\rho}\left|p_{\text{sup}} - p_{\text{work}2}\right|}\ \text{sgn}\left(p_{\text{sup}} - p_{\text{work}2}\right), \tag{33}$$

with ϕ_{loss} as a factor for the loss in the orifice, the density ρ of the hydraulic oil, and the opening areas $A_{\text{in}1}$ and $A_{\text{in}2}$ of the considered orifice. A sign test is done with the signum function *sgn* (.).

The opening area $A_{\text{in}1}$ of the orifice depends on the twisting angle ψ of the torsion bar. The area controls the volume flow from the supply chamber into the working chamber 1 and is given by

$$A_{\text{in}} = A_{\text{norm}} + A_{\text{norm}}\ \sin\left(\frac{\frac{\pi}{2}\Delta\psi}{\Delta\psi_{\text{max}}}\right). \tag{34}$$

This equation is valid for a twisting angle ψ, which satisfies the condition

$$-\Delta\psi_{\text{max}} \leq \Delta\psi \leq \Delta\psi_{\text{max}}. \tag{35}$$

For ψ smaller than -ψ_{max} or greater than ψ_{max}, the opening area is either 0 or 2 A_{norm}, while A_{norm} is defined to be the negative valve lap. Figure 7 shows the graph of the opening area depending on the twisting angle of the torsion bar.

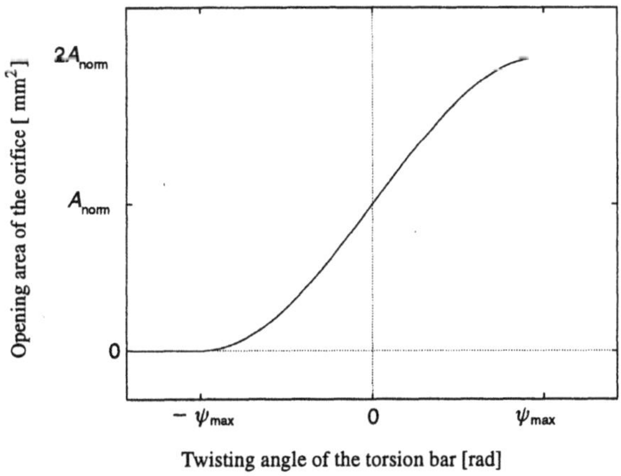

Fig. 7. Graph of the orifice opening area

The equation for the opening area of the orifice to the working chamber 2 correlates to the equation for the opening area to working chamber 1, but with the negative twisting angle of the torsion bar.

The volume flows Q_{out1} and Q_{out2} out of the working chambers 1 and 2 back to the reservoir are defined by

$$Q_{\text{out1}} = A_{\text{out1}}\, \phi_{\text{loss}} \sqrt{\frac{2}{\rho}\, |p_{\text{work1}} - p_0|}\ \operatorname{sgn}\left(p_{\text{work1}} - p_0\right) \text{ and} \tag{36}$$

$$Q_{\text{out2}} = A_{\text{out2}}\, \phi_{\text{loss}} \sqrt{\frac{2}{\rho}\, |p_{\text{work2}} - p_0|}\ \operatorname{sgn}\left(p_{\text{work2}} - p_0\right). \tag{37}$$

The foregoing volumes are controled by the opening areas A_{out1} and A_{out2} of the orifices to the reservoir. These opening areas of the return orifices are defined by a graph similar to the graph of the opening areas of the supply orifices and eq. (34). The negative valve lap is smaller for the return orifices and so is the maximum of the opening area.

The volume flows depend on the pressure in the pressure chambers. Oil is compressible and, therefore, the differential equation for the pressure in the supply chamber can be written, according to Backé (1992) or Merrit (1967), as

$$\dot{p}_{\text{sup}} = \frac{E}{V_{\text{sup}}} \left(Q_{\text{sup}} - Q_{\text{in1}} - Q_{\text{in2}} - Q_{\text{ex}}\right). \tag{38}$$

In this equation, E is the bulk modulus of the hydraulic oil, whereas V_{sup} is the volume of the supply chamber, where the volume of the supply pipe from the hydraulic pump is added. The volume flows come from the pump and go to the working cylinder and back to the reservoir, if the pressure relief valve is opened.

Considering the differential equations for the pressure in the working chambers 1 and 2, the volume change, due to the motion of the piston have to be taken into account. Therefore, the differential equations for the pressure in the working chambers are given by

$$\dot{p}_{\text{work1}} = \frac{E}{V_{\text{work}} + x A_{\text{piston}}} \left(Q_{\text{in1}} - Q_{\text{out1}} - \dot{x} A_{\text{piston}}\right) \text{ and} \tag{39}$$

$$\dot{p}_{\text{work2}} = \frac{E}{V_{\text{work}} - x A_{\text{piston}}} \left(Q_{\text{in2}} - Q_{\text{out2}} + \dot{x} A_{\text{piston}}\right). \tag{40}$$

with A_{piston} denoting the area of the piston in the hydraulic cylinder, while x is the translation of the piston inside the hydraulic cylinder and $\dot{x}$ is the velocity of the piston. The piston is part of the steering rod. Therefore, the translation and the velocity of the piston are the same for the steering rod. The volume V_{work} of one side of the hydraulic power cylinder in equilibrium is calculated by the volume of the working chamber and the volume of the pipes to the power cylinder.

If there is a pressure difference between the two working chambers in the hydraulic power cylinder, a force F_{servo} acts on the piston, due to the pressure difference. This force depends on the piston area A_{piston} and can be calculated in the form

$$F_{\text{servo}} = A_{\text{piston}} \left(p_{\text{work1}} - p_{\text{work2}} \right) . \tag{41}$$

In order to do more research on the hydraulic module of the power-steering system a more detailed model of the hydraulics is investigated in a next step. Two more chambers inside the servovalve, called valve chamber 1 and 2, are taken into consideration, as shown in Fig. 8. Due to the new pressure chambers, the volume flows in and out of the chambers change. Two additional volume flows Q_{work1} and Q_{work2} are necessary. These flows supply the working chambers 1 and 2 with oil and empty them. The supply of the valve chambers is done by the supply flows Q_{in1} and Q_{in2} while the return flows Q_{out1} and Q_{out2} empty the valve chambers. The working chambers are not connected to the reservoir anymore.

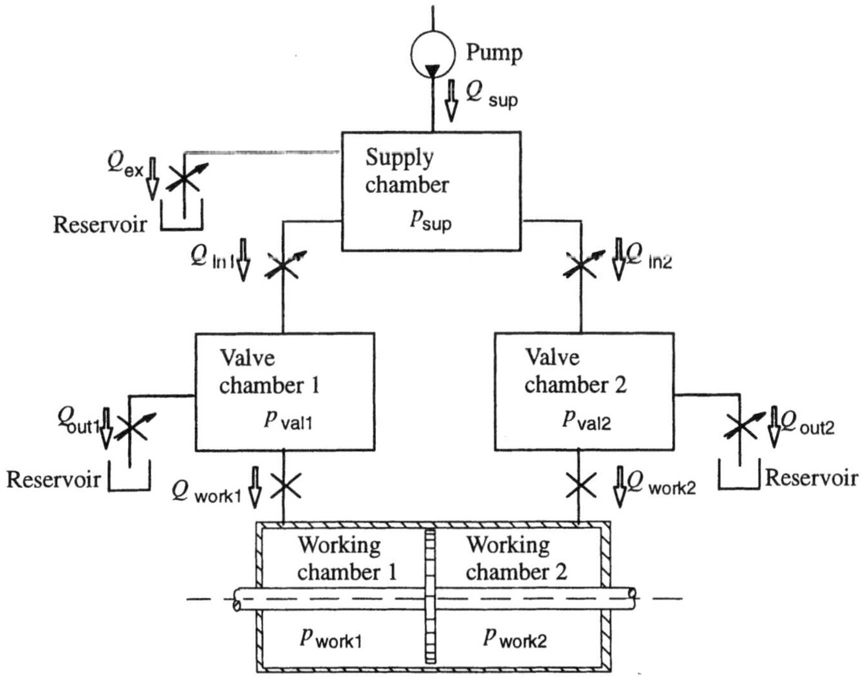

Fig. 8. Model of the hydraulic system with five pressure chambers

For the two valve chambers the differential equations of the pressure can be written as

$$\dot{p}_{\text{val1}} = \frac{E}{V_{\text{val}}} \left(Q_{\text{in1}} - Q_{\text{out1}} - Q_{\text{work1}} \right) \text{ and} \tag{42}$$

$$\dot{p}_{\text{val2}} = \frac{E}{V_{\text{val}}} \left(Q_{\text{in2}} - Q_{\text{out2}} - Q_{\text{work2}}\right) \tag{43}$$

with V_{val} indicating the volume of the valve. This volume equals the volume inside the servovalve. The differential equations for the pressure in the working chambers 1 and 2 change. There is only one volume flow going in and out of them, called volume flow Q_{work1} and Q_{work2}, respectively.

The volume flows Q_{work1} and Q_{work2} in and out of the working chambers are defined below:

$$Q_{\text{work1}} = A_{\text{norm}}\, \phi_{\text{loss}} \sqrt{\frac{2}{\rho} \left|p_{\text{val1}} - p_{\text{work1}}\right|}\, \operatorname{sgn}\left(p_{\text{val1}} - p_{\text{work1}}\right) \tag{44}$$

and

$$Q_{\text{work2}} = A_{\text{norm}}\, \phi_{\text{loss}} \sqrt{\frac{2}{\rho} \left|p_{\text{val2}} - p_{\text{work2}}\right|}\, \operatorname{sgn}\left(p_{\text{val2}} - p_{\text{work2}}\right). \tag{45}$$

The flows into the working chambers 1 and 2 are controled by an orifice of constant cross section. The cross section is considered in the term A_{norm} within eqs. (44) and (45).

Extending the hydraulic module by two more pressure chambers gives the model of the hydraulic system shown in Fig. 9. This model consists of seven pressure chambers. With the aid of this model we can decide whether a more detailed model improves the results of the simulations significantly.

In the model with seven pressure chambers, the return flows to the reservoir do not go back directly, but through two additional chambers called return chambers 1 and 2. The rest of the model is equal to the model with five pressure chambers. The volume flows to the return chambers are controled by variable opening areas, while the flows from the return chambers back to the reservoir go through an orifice of constant cross section. The equations for the volume flows to and out of the return chambers as well as the differential equations for the pressure in the return chambers are similar to eqs. (36), (37), (42) and (43).

4. Assembly of the Entire Power-Steering System Model

A power-steering system is a mechatronic system. Therefore, functions of the different fields of engineering interact as part of the overall systems. Here, the mechanical steering system interacts with the hydraulics of the system. In the model, the interaction between the different fields is realized by input and output variables. The connection of the modules with input and output variables can be seen in Fig. 10.

In order to simulate the system, an excitation of the system has to be defined. A user-defined excitation is chosen, which acts on the mechanical

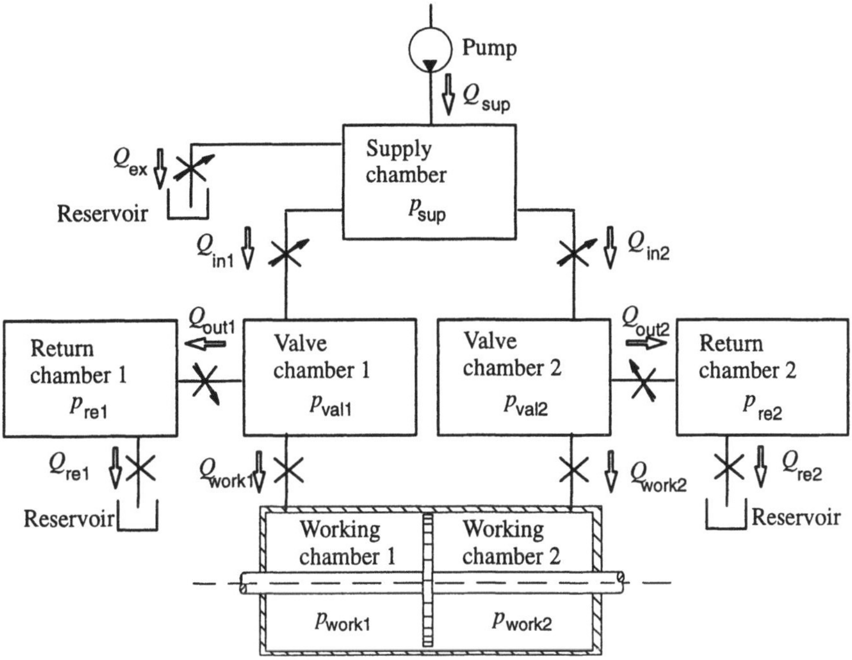

Fig. 9. Model of the hydraulic system with seven pressure chambers

module of the system. Either a steering torque T_{steer} acts on the steering wheel or a defined steering angle α of the steering wheel has to be reached.

The defined excitation is the input of the mechanical module. The mechanical module reacts dynamically to the excitation. Other input variables to the mechanical module are the reaction force of the tires F_{tire} and the hydraulic support force F_{servo}. The output variables of the mechanical module are the translation and the velocity of the steering rod, x and $\dot{x}$. Another output variable is the twisting angle of the torsion bar ψ, which is calculated by

$$\Delta\psi_{\max} = \gamma - \delta, \tag{46}$$

where γ is the rotation of the steering spindle and δ is the rotation of the steering screw. The rotation of the steering screw can be calculated with the translation of the steering rod and the radius r of the steering screw.

The input variable of the hydraulic module is the twisting angle of the torsion bar, which is identical with the relative angle of rotation ψ between the steering spindle and the steering screw of the mechanical module. Other input variables are the translation and the velocity of the steering rod, x and $\dot{x}$. With the input variables the hydraulic module calculates the output variables of the system using the volume flow equations and the differential equations for the pressure.

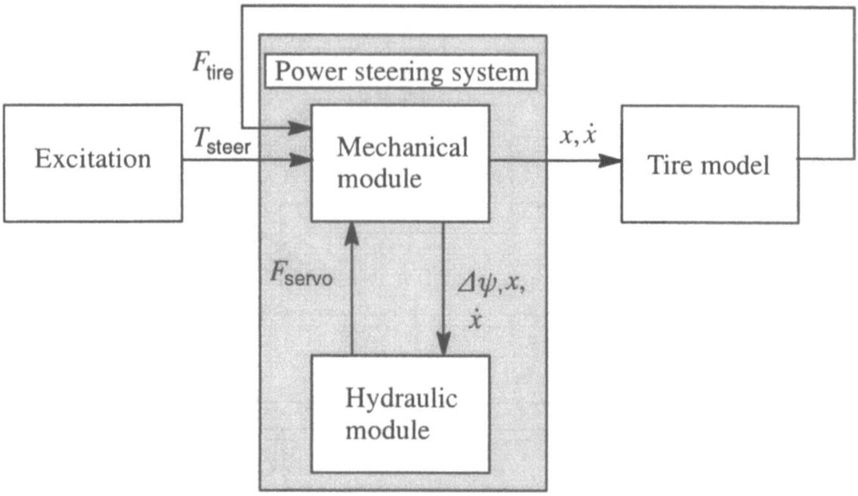

Fig. 10. Assembly of the modules of the entire power-steering system

During simulation, reaction forces of the tires have to act on the power-steering system. The tire force is calculated using the translation and the velocity of the steering rod, x and $\dot{x}$. This force acts on the steering rod of the mechanical system and influences its dynamic behavior.

5. Simulation

Three hydraulic modules of the power-steering system have been developed. To find the module of optimum quality, which corresponds to the behavior found in reality, the model behavior of the power-steering system has to be investigated in simulations. For excitation, a steering torque, which increases linearly over time, a steering angle, and a disturbance force on the tires are examined.

5.1 Excitation with a Steering Torque

In a first simulation the power-steering system is excited with a steering torque. The steering torque increases linearly within 500 seconds from 0 to 5 Nm. This excitation is also called quasistatic excitation. The motion of the tires, and therefore of the steering rod, is locked. This can be realized by a tire model with very high spring stiffness. Figure 11 shows the graph of the hydraulic support force over the steering torque when using the hydraulic module with five pressure chambers.

Up to a steering torque of about 2 Nm the hydraulic support force can be neglected. If the steering torque is greater than 2 Nm, the hydraulic support

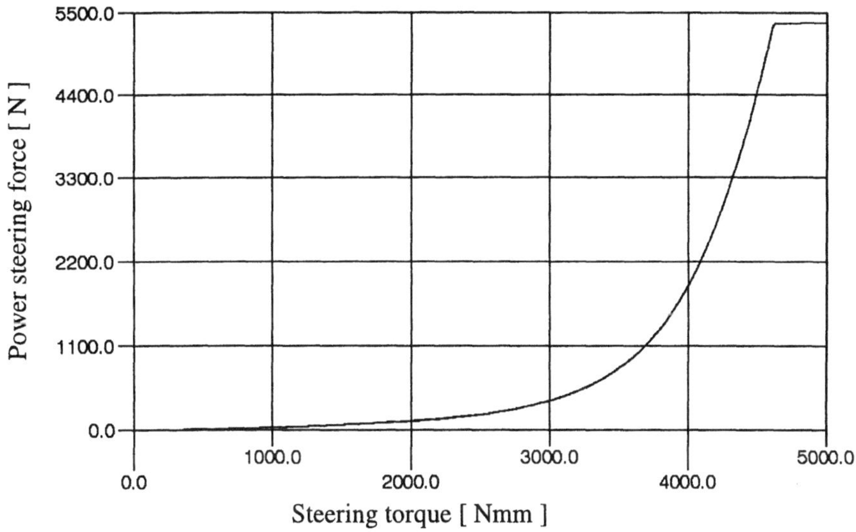

Fig. 11. Supporting force plotted over steering torque for quasistatic excitation

force starts to act evidently on the system and increases up to a maximum value of 5360 N, when the steering torque is about 4.6 Nm. Using a radius of 8 mm for the steering screw, a supporting steering torque of 43 Nm can be calculated. Simulations with the hydraulic modules of three and seven pressure chambers have not shown any significant difference in the results. Therefore, their graphs are not shown here. The graph of the hydraulic support force is identical to the requirements given by car manufacturers for all three hydraulic modules Raith (1993). Furthermore, the simulations showed, that by adjusting the stiffness of the torsion bar the beginning of the hydraulic support force can be varied over a wide range.

When examining the volume flows in the hydraulic module with five pressure chambers, the graphs of Fig. 12 are found. They show the volume flow into the supply chamber, the valve chamber 1 and 2 and the flow through the pressure relief valve.

The volume flow to the valve chamber 1 decreases. Only a small amount of oil is necessary to built up the pressure in the valve chamber 1 and the working chamber 1 to obtain a hydraulic support force to the power-steering system. A greater part of oil flows into valve chamber 2 and back to the reservoir. Increasing the steering torque leads to a reduction of the flow from the valve chamber 1 back to the reservoir. When the pressure in working chamber 1 and also in the valve chamber 1 and the supply chamber reaches a level of $9 * 10^6$Pa after 460 seconds, the pressure relief valve opens and releases the system. The flow through the pressure relief valve increases for the rest of the simulation time.

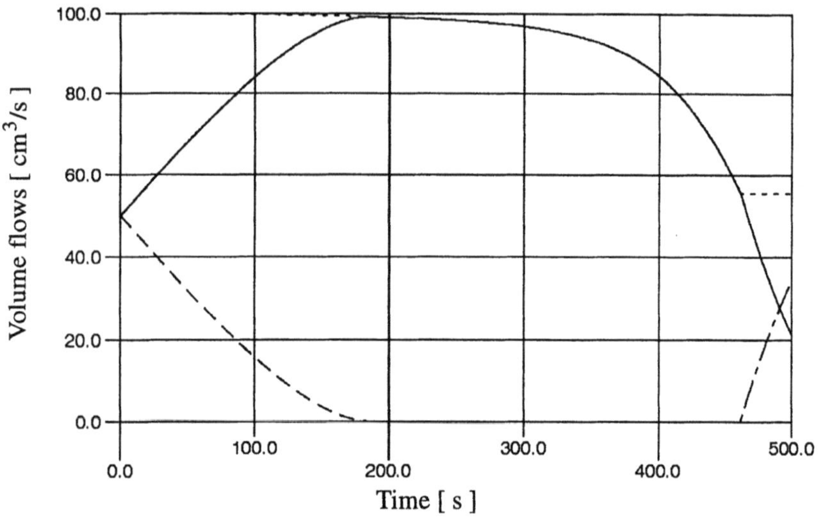

------- Volume flow from the pump to the supply chamber

----- · Volume flow to valve chamber 1

——— Volume flow to valve chamber 2

— - — Volume flow through the pressure relief valve to the reservoir

Fig. 12. Volume flows in the hydraulic module with five pressure chambers for quasistatic excitation

5.2 Excitation with a Steering Angle

Exciting the power-steering system with a quasistatic steering torque does not show any differences in the results for the different modules. Therefore, in a second simulation series, dynamic excitations are chosen. A sinusoidal steering angle of the steering wheel is given. The results for the hydraulic modules with three and five pressure chambers will be shown. The results of the simulation for the module with seven pressure chambers are similar to those of the module with five pressure chambers. Therefore, they will not be shown here.

The sinusoidal steering angle has a frequency of 2 Hz and an amplitude of 90 degrees. Simulations with the hydraulic module of three and five pressure chambers both without any disturbances and with a reduced supply volume flow result in the four graphs of Fig. 13 for the hydraulic support force over the steering torque. The steering torque is necessary to reach the desired steering angle of the steering wheel. The reduction of the supply flow is done by decreasing the flow from $100\frac{\mathrm{cm}^3}{\mathrm{s}}$ to $80\frac{\mathrm{cm}^3}{\mathrm{s}}$.

Reducing the supply flow for the hydraulic module with three pressure chambers has only a quantitative effect on the results of the simulation. The reduction of the supply flow leads only to a reduced hydraulic support force by

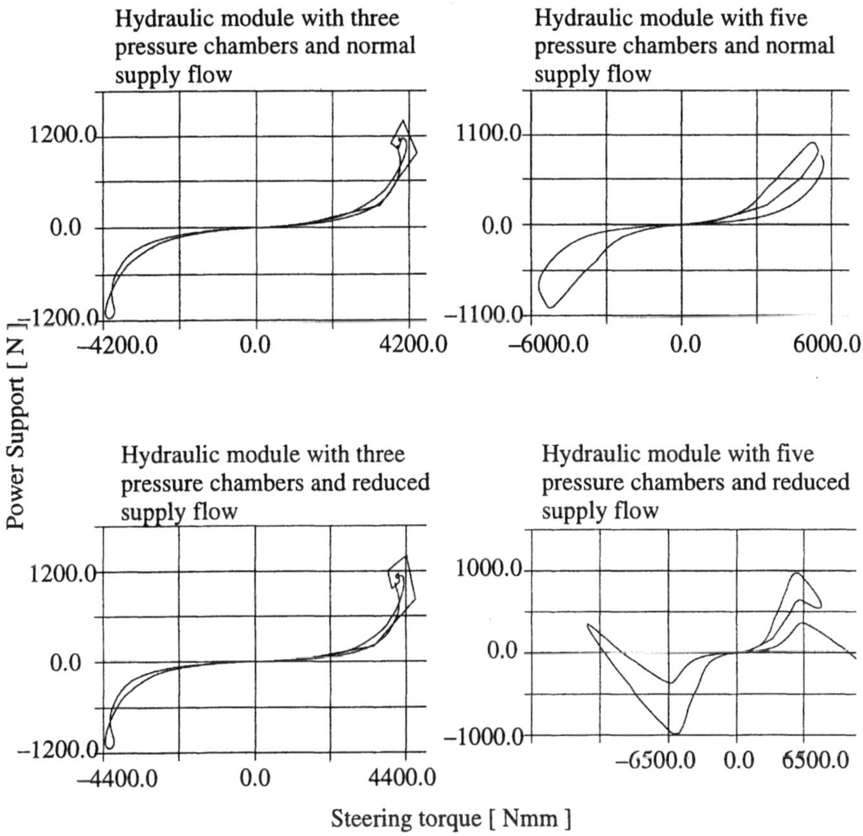

Fig. 13. Graphs of the hydraulic support for different hydraulic modules

the power-steering system after overcoming the disturbances in the beginning. Reducing the supply flow for the hydraulic module with five pressure chambers leads to a quantitative and especially a qualitative change of the results.

When using the hydraulic module with five pressure chambers under normal conditions the hydraulic support force starts later compared with the module with three pressure chambers. The driver has to steer with a higher torque to gain the desired steering angle. This can be seen in a graph going more to the side. The maximum hydraulic support force is smaller than when using the module with three pressure chambers; therefore, the driver has to drive with a higher steering torque. The pressure difference is built up more slowly and with a delay, due to more pressure chambers and more orifices, which have to be traversed by the oil.

If the supply flow from the pump is reduced for the module with five pressure chambers, this delay can be seen more clearly. To reach the desired steering angle, the driver not only has to overcome the reaction forces of the tires but also has to act against the pressure forces of the power cylinder. Not enough oil is supplied in a short period by the hydaulic pump. The empty volume caused by the motion of the piston is not filled fast enough, and, therefore, a pressure acts against the steering movement.

In reality, vehicle manufacturers try to have this decreasing support by the power-steering system. Therefore, they implement hydraulic pressure pumps with a decreasing scale function for the supported volume flow. If the motor speed is high, less oil is pumped into the system. One reason for this is that when driving at high speeds large steering actions can cause dangerous driving situations. Only small driving actions should be possible. If the power-steering system were to support small steering actions while driving with high velocities, dangerous driving situations could occur. If the power-steering system hinders the steering actions, the driver's feeling of the road improves.

The hydraulic module with five pressure chambers as well as the module with seven pressure chambers has a decreasing support function when driving at high velocities. The module with three pressure chambers does not show this behavior. Although the hydraulic modules with five and seven pressure chambers have shown the same results in simulation, the calculation time for the module with seven pressure chambers is much longer, because more equations have to be evaluated. This concludes the verification of the hydraulic module.

5.3 Excitation with a Tire Disturbance Force

In the last simulations the behavior of the power-steering system with three and five pressure chambers under a disturbing tire force is investigated. Because the hydraulic module with seven pressure chambers shows the same results as the one with five chambers, these results are not shown here. In simulations all steering wheel motions are prohibited. The wheels act with a disturbing sinusoidal force on the steering rod. The frequency is 25 Hz and the amplitude is supposed to be 400 N.

Considering the hydraulic module with three pressure chambers, Fig. 14 is found for the graphs of the forces and torques. The hydraulic support force and the steering torque increase over time. This is due to an undamped degree of freedom of the steering rod. The torsion bar is modeled as a spring without damping. Because the oil in the three chamber module flows back to the reservoir directly hindered only by one orifice, the hydraulic cylinder does not damp the motion of the steering rod significantly. Simulating the same system with an additional damper for the torsion bar confirms this statement. No increase of the hydraulic support force or steering force can be seen.

Simulation of the power-steering system with the five chamber hydraulic module leads to the force and torque graphs shown in Fig. 15. There is no increase of the steering torque or the hydraulic support force, although the torsion bar is undamped, too. Because the oil flow from the working chambers back to the reservoir leads through the valve chambers and more orifices, it

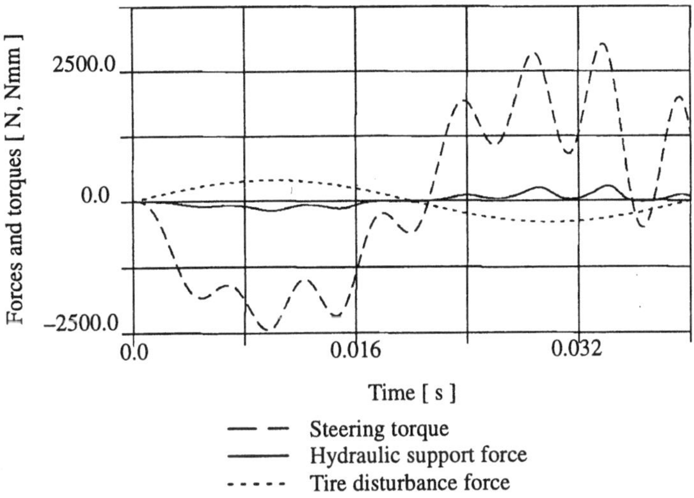

Fig. 14. Excitation of the three chambered steering system with a sinusoidal tire force

is slower. A hydraulic support force is produced by the hydraulic cylinder, which hinders the motion of the steering rod. Additionaly, a steering torque of the driver has to act on the steering wheel, which improves the feeling of the driver for the road.

6. Conclusions

This chapter discussed different models of power-steering systems. The hydraulic modules of the power-steering systems considered here differ in the number of pressure chambers. Modules with three, five, and seven pressure chambers are developed. In quasistatic simulations with a steering torque of increasing value, no significant differences between the results of the three models were found. When more dynamic simulations are conducted differences between the model with three and five pressure chambers occur.

In simulations with a given steering angle of high frequency the model with three pressure chambers shows differences with the behavior found in reality. The behavior of the models with five and seven pressure chambers is closer to the behavior seen in reality.

Simulations with a tire disturbance force support this statement. In the case of a three-chamber model, instabilities can be seen. No instabilities are noticed for the models with five and seven pressure chambers. The hydraulic cylinder damps the oscillations for the five- and seven-chamber models.

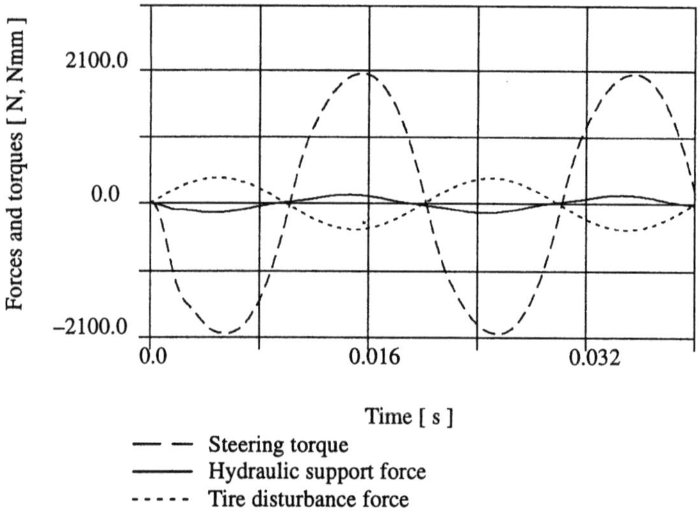

Fig. 15. Excitation of the five chambered steering system with a sinusoidal tire force

The results of the simulations for the models with five and seven pressure chambers are similar in all cases. Therefore, it can be said that a model with more than five pressure chambers does not improve the results but increases the simulation time. The more pressure chambers are included in the model, the more equations have to be solved and therefore, the calculation time increases. The model of the power-steering system with five pressure chambers turns out to fit reality best.

For future work the model of the power-steering system should be implemented in a complete vehicle model. More realistic tire forces can be gained. The behavior and influence of the power-steering system on the overall vehicle dynamics can be investigated in more detail and be judged more extensively.

References

Backé, W. (1992): Servohydraulik. Umdruck zur Vorlesung. Rheinisch-Westfälische Technische Hochschule, Institute for Hydraulic and Pneumatic Engines and Control, Aachen.

Mechanical Dynamics (1987): ADAMS User's Guide. 3055 Plymouth Rd, Ann Arbor, Michigan.

Merrit, H.E. (1967): Hydraulic Control Systems. John Wiley & Sons, New York.

Popp, U. and Schiehlen, W. (1993): Fahrzeugdynamik. Teubner-Verlag, Stuttgart.

Raith, A. (1993): Rechnerische Simulation von Lenkgetrieben mit Servounterstützung. Haus der Technik e.V., Documents for Presentations, Simulation in Development of Vehicles, Essen.

Reimpell, J. (1984): Fahrwerktechnik: Lenkung. Vogel-Fachbuch, Technik, Kraftfahrzeugwesen, Würzburg.

Ryan, R.R. and Steigerwald, M.F. (1988): ADAMS Theory and Application. Proceedings of the 1988 ADAMS Software Theory Seminar, Mechanical Dynamics, Inc., Ann Arbor, Michigan.

Ryan, R.R. (1990): ADAMS – Multibody System Analysis Software. Multibody Systems Handbook. Edited by W. Schiehlen, Springer-Verlag, Berlin.

Schiehlen, W. (1986): Technische Dynamik. Teubner, Stuttgart.

Schiehlen, W. (Editor) (1990): Multibody Systems Handbook. Springer-Verlag, Berlin.

Schiehlen, W. (1997): Multibody System Dynamics – Roots and Perspectives. Multibody System Dynamics, Vol. 1. Kluwer Academic Publishers, Dordrecht.

Wielenga, T.W. (1987): Analysis Methods and Model Representation in ADAMS. MDI Technical Paper No. 41.

Zamow, J. and Witte, L. (1988): Fahrzeugsimulation unter Verwendung des Starrkörperprogramms ADAMS. VDI-Berichte 699, Reihe 14, Nr. 71. VDI-Verlag, Düsseldorf.

ZF (1994): ZF-Kurzmitteilungen. Edited by Zahnradfabrik Friedrichshafen AG, Section Steeringsystem, Friedrichshafen.

Control of Robotic Systems by Exact Linearization

Christoph Woernle

Institute of Mechanical Engineering, University of Rostock
D–18059 Rostock, Germany
woernle@mbst.uni-rostock.de

Summary. Dynamic control of motions and reaction forces by exact input-output linearization is examined for robotic systems that are constrained at one or several points. Similar to exact linearization of unconstrained systems, linear decoupled input-output channels for position and force control variables are obtained that can be separately controlled. In particular, hybrid position-and-force control of manipulators with redundant degrees of freedom and redundant contact forces is regarded. Redundant variables are used to achieve repeatability, obstacle avoidance and minimization of driving forces. Further, motion control of a nonholonomic mobile platform is investigated.

Keywords. Exact input-output linearization, mobile robot, multibody system, nonlinear control, position-and-force control, redundant manipulator, robotics, nonholonomic system.

1. Introduction

Industrial robots are widely used in factory automation, while 'service robots' are being developed in various other fields. Examples are robots in civil engineering, robots for inspection of inaccessible areas, surgery robots, or mobile working platforms. Coming along with these applications are increased demands on control, for example collision avoidance, docking with targets, hybrid position-and-force control, or sensor-based motion planning.

Controller design is based on a dynamic model of the plant. Robotic systems are typically modeled as mechanical multibody systems that describe well effects of large spatial motions. Since mechanical systems generally show a highly nonlinear behavior, linear control strategies are often unsatisfactory, and tools of nonlinear control theory have to be applied. An important method for nonlinear control design is exact input-output linearization of the system by a nonlinear state feedback.

The basic idea of this approach is to compensate the nonlinearities of the plant by its 'inverse dynamics'. Thus, a fully or at least partly linear system is obtained to which linear control design methods can be applied. While early developments of this method were described and applied to the position control of robot manipulators, for example by Freund (1975), the underlying differential-geometric theory has been considerably extended during the last twenty years. A deep overview is given by the textbooks of Isidori (1989) and Nijmeijer and van der Schaft (1990).

The objective of this chapter is a unified formulation of the feedback linearization method for control of position and velocity variables and, in the case of constrained robotic systems, for hybrid position and force control. With regard to the control of mobile wheeled robots, both holonomic and nonholonomic mechanical models are taken into account.

Among other applications, the method covers the hybrid position-and-force control of manipulators with a single contact with the environment (Khatib 1987, Yoshikawa 1987), or multiple contacts with the environment (Zimmermann and West 1990), and the control of two manipulators commonly grasping an object (Yun and Kumar 1991, Koivo and Unseren 1991). An additional overview about existing methods is given by Kreutz (1989). Control of nonholonomic systems refers to d'Andréa-Novel et al. (1991), Campion et al. (1991), d'Andréa-Novel et al. (1992), and Sarkar et al. (1994).

Based on the dynamic multibody models developed in Section 2, exact input-output linearization is formulated in Section 3. Hybrid position-and-force control of 'redundant' robots contacting their environment or other robots is treated in Section 4. Here, redundancy includes both redundant degrees of freedom and redundant contact forces. Another application is motion control of a mobile wheeled platform that is treated in Section 5.

2. Dynamics of a Multibody System

To formulate the equations of motion of a controlled robotic system, the mechanical constraints are subdivided into two categories: system constraints and contact constraints. *System constraints* do not interest during the controlled motion of the system. Examples are the reaction forces in the joints of robots or mechanisms. Reaction forces of *contact constraints* are, however, of interest during motion. These constraints typically occur in combination with particular control tasks and may change during motion. An example is the reaction force on a robot gripper being in contact with the environment.

A multibody system without contact constraints is here described in terms of independent coordinates leading to a system of ordinary differential equations. The contact constraints are then introduced as algebraic conditions in terms of these coordinates in combination with reaction forces in the kinetic equations. Altogether, a differential-algebraic set of equations is obtained for the constrained system.

2.1 Dynamics Without Contact Constraints

The motion of a multibody system can be described by n_P independent position coordinates $\boldsymbol{q}_P = [q_{P1} \dots q_{P,n_P}]^{\mathrm{T}}$ – refer, for example, to Shabana (1989). Their total time derivatives $\dot{\boldsymbol{q}}_P$ are related to n_V independent generalized velocities $\boldsymbol{q}_V = [q_{V1} \dots q_{V,n_V}]^{\mathrm{T}}$ by the kinematic differential equation

$$\dot{\boldsymbol{q}}_P = \boldsymbol{H}(\boldsymbol{q}_P)\,\boldsymbol{q}_V \qquad \text{with} \qquad \boldsymbol{H} \in \mathbb{R}^{n_P, n_V}\,. \tag{1}$$

For $n_P = n_V$ the system is holonomic, for $n_P > n_V$ it is nonholonomic.

It is assumed that the system is controlled by generalized forces generated by $p \leq n_V$ mutually independent control inputs $\boldsymbol{u} = [u_1 \ldots u_p]^T$, and that the actuator dynamics is neglected. The dynamic behavior of the system is then described by

$$\boldsymbol{M}(\boldsymbol{q}_P, t)\,\dot{\boldsymbol{q}}_V = \boldsymbol{g}(\boldsymbol{q}_P, \boldsymbol{q}_V, t) + \boldsymbol{G}(\boldsymbol{q}_P, t)\,\boldsymbol{u}\,. \tag{2}$$

Here, $\boldsymbol{M} \in \mathbb{R}^{n_V, n_V}$ is the symmetric, positive-definite inertia-matrix, $\boldsymbol{g} \in \mathbb{R}^{n_V}$ is the vector of the generalized centrifugal and Coriolis forces as well as all applied forces except for the control forces, and $\boldsymbol{G} \in \mathbb{R}^{n_V, p}$ is a projection matrix for the control inputs $\boldsymbol{u}$. Explicit time dependence of the matrices only occurs for rheonomic systems.

2.2 Contact Constraints

It is assumed that the motion of the system (1) and (2) is constrained by d holonomic contact constraints that are independent from the system constraints. In the following, the contact constraints are described.

2.2.1 Contact Constraints in Cartesian Coordinates. The i-th holonomic contact constraint occurs between a pair of coordinate systems $\mathcal{K}_{ai}$, $\mathcal{K}_{bi}$ attached to bodies of the system (Fig. 1). Their position relative to a reference system $\mathcal{K}_0$ is described by position vectors $\boldsymbol{r}_{ai}$, $\boldsymbol{r}_{bi}$, and rotation tensors $\boldsymbol{R}_{ai}$, $\boldsymbol{R}_{bi}$, respectively. The i-th holonomic contact constraint then is an algebraic equation in terms of these position coordinates:

$$\varphi_{Fi}(\boldsymbol{R}_{ai}, \boldsymbol{r}_{ai}, \boldsymbol{R}_{bi}, \boldsymbol{r}_{bi}) = 0\,, \qquad i = 1, \ldots, d\,. \tag{3}$$

To express the time derivative of (3) needed in the sequel, the angular velocities of $\mathcal{K}_{ai}$, $\mathcal{K}_{bi}$ are introduced,

$$\boldsymbol{\omega}_{\beta i} = \frac{\mathrm{d}\boldsymbol{\vartheta}_{\beta i}}{\mathrm{d}t} \qquad \text{with} \qquad \mathrm{d}\widetilde{\boldsymbol{\vartheta}}_{\beta i} = \mathrm{d}\boldsymbol{R}_{\beta i}\,\boldsymbol{R}_{\beta i}^{\mathrm{T}}\,, \qquad \beta = a, b\,. \tag{4}$$

Here, $\boldsymbol{\vartheta}_{\beta i} \in \mathbb{R}^3$ are quasi-coordinates of rotation. The tilde is the operator to convert a Cartesian vector $\boldsymbol{r}$ into a skew-symmetric-tensor $\widetilde{\boldsymbol{r}}$ in such a way that holds $\widetilde{\boldsymbol{r}}\boldsymbol{e} = \boldsymbol{r} \times \boldsymbol{e}$.

Differentiation of eq. (3) with respect to time then leads to the i-th contact constraint at the velocity level,

$$\underbrace{\begin{bmatrix} \dfrac{\partial \varphi_{Fi}}{\partial \boldsymbol{\vartheta}_{ai}} & \dfrac{\partial \varphi_{Fi}}{\partial \boldsymbol{r}_{ai}} & \dfrac{\partial \varphi_{Fi}}{\partial \boldsymbol{\vartheta}_{bi}} & \dfrac{\partial \varphi_{Fi}}{\partial \boldsymbol{r}_{bi}} \end{bmatrix}}_{\boldsymbol{n}_i^{\mathrm{T}}(\boldsymbol{R}_{ai}, \boldsymbol{r}_{ai}, \boldsymbol{R}_{bi}, \boldsymbol{r}_{bi})} \begin{bmatrix} \boldsymbol{\omega}_a \\ \boldsymbol{v}_a \\ \boldsymbol{\omega}_b \\ \boldsymbol{v}_b \end{bmatrix}_i = 0\,, \quad i = 1, \ldots, d\,. \tag{5}$$

Here, $\boldsymbol{n}_i \in \mathbb{R}^{12}$ can be regarded as the normal vector of the generalized constraint surface (3) in the space of the altogether 12 absolute position coordinates of frames $\mathcal{K}_{ai}$ and $\mathcal{K}_{bi}$.

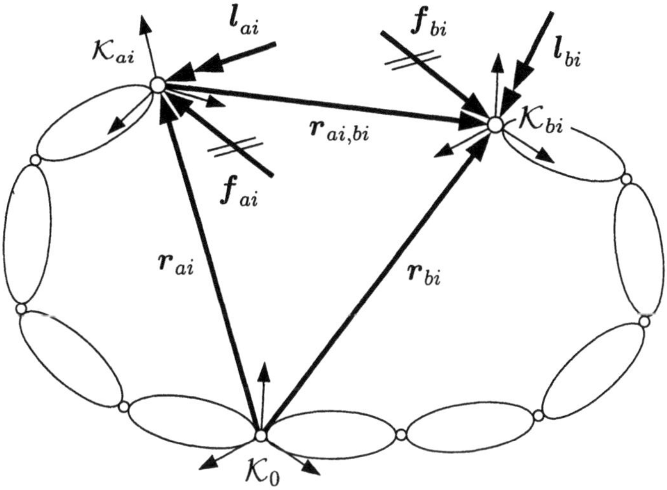

Fig. 1. Contact Forces

2.3 Kinematics of the Constraint Coordinate Systems

The position coordinates of the constraint systems $\mathcal{K}_{ai}$, $\mathcal{K}_{bi}$ depend on the position coordinates $\boldsymbol{q}_P$ of the multibody system (1) and (2),

$$\boldsymbol{R}_{ai}(\boldsymbol{q}_P, t)\,, \quad \boldsymbol{R}_{bi}(\boldsymbol{q}_P, t)\,, \quad \boldsymbol{r}_{ai}(\boldsymbol{q}_P, t)\,, \quad \boldsymbol{r}_{bi}(\boldsymbol{q}_P, t)\,. \tag{6}$$

Accordingly, the rotational and translational velocities $\boldsymbol{\omega}_{ai}$, $\boldsymbol{\omega}_{bi}$ and $\boldsymbol{v}_{ai}$, $\boldsymbol{v}_{bi}$ of $\mathcal{K}_{ai}$ and $\mathcal{K}_{bi}$, respectively, are expressed in terms of the system coordinates $\boldsymbol{q}_P$ and $\boldsymbol{q}_V$:

$$\begin{bmatrix} \boldsymbol{\omega}_a \\ \boldsymbol{v}_a \\ \boldsymbol{\omega}_b \\ \boldsymbol{v}_b \end{bmatrix}_i = \begin{bmatrix} \boldsymbol{J}_{Ra}(\boldsymbol{q}_P, t) \\ \boldsymbol{J}_{Ta}(\boldsymbol{q}_P, t) \\ \boldsymbol{J}_{Rb}(\boldsymbol{q}_P, t) \\ \boldsymbol{J}_{Tb}(\boldsymbol{q}_P, t) \end{bmatrix}_i \boldsymbol{q}_V + \begin{bmatrix} \bar{\boldsymbol{\omega}}_a(\boldsymbol{q}_P, t) \\ \bar{\boldsymbol{v}}_a(\boldsymbol{q}_P, t) \\ \bar{\boldsymbol{\omega}}_b(\boldsymbol{q}_P, t) \\ \bar{\boldsymbol{v}}_b(\boldsymbol{q}_P, t) \end{bmatrix}_i . \tag{7}$$

Here, $\boldsymbol{J}_{Ra}$, $\boldsymbol{J}_{Ta}$, $\boldsymbol{J}_{Rb}$, $\boldsymbol{J}_{Tb}$ are Jacobian matrices and $\bar{\boldsymbol{\omega}}_a$, $\bar{\boldsymbol{v}}_a$, $\bar{\boldsymbol{\omega}}_b$, $\bar{\boldsymbol{v}}_b$ are local velocities (only for rheonomic systems) of $\mathcal{K}_{ai}$ and $\mathcal{K}_{bi}$, respectively.

2.4 Contact Constraints in System Coordinates

By means of the kinematic functions (6), the contact constraints (3) are expressed in terms of the position coordinates $\boldsymbol{q}_P$,

$$\begin{bmatrix} \varphi_{F1}(\boldsymbol{q}_P, t) \\ \vdots \\ \varphi_{Fd}(\boldsymbol{q}_P, t) \end{bmatrix} = \begin{bmatrix} 0 \\ \vdots \\ 0 \end{bmatrix} \tag{8}$$

$$\boldsymbol{\varphi}_F(\boldsymbol{q}_P, t) = \boldsymbol{0}\,.$$

Inserting eq. (7) into eq. (5) yields the total time derivative of eq. (8),

$$\begin{bmatrix}\boldsymbol{\Phi}_{F1}(\boldsymbol{q}_P,t)\\ \vdots \\ \boldsymbol{\Phi}_{Fd}(\boldsymbol{q}_P,t)\end{bmatrix}\boldsymbol{q}_V+\begin{bmatrix}\zeta_{F1}(\boldsymbol{q}_P,t)\\ \vdots \\ \zeta_{Fd}(\boldsymbol{q}_P,t)\end{bmatrix}=\begin{bmatrix}0\\ \vdots \\ 0\end{bmatrix} \tag{9}$$

$$\boldsymbol{\Phi}_F(\boldsymbol{q}_P,t)\quad \boldsymbol{q}_V+\quad \boldsymbol{\zeta}_F(\boldsymbol{q}_P,t)\quad=\quad \mathbf{0}$$

with

$$\boldsymbol{\Phi}_{Fi}=\boldsymbol{n}_i^{\mathrm{T}}\begin{bmatrix}\boldsymbol{J}_{Ra}\\ \boldsymbol{J}_{Ta}\\ \boldsymbol{J}_{Rb}\\ \boldsymbol{J}_{Tb}\end{bmatrix}_i\in\mathbb{R}^{1,nv}\ ,\qquad \zeta_{Fi}=\boldsymbol{n}_i^{\mathrm{T}}\begin{bmatrix}\bar{\boldsymbol{\omega}}_a\\ \bar{\boldsymbol{v}}_a\\ \bar{\boldsymbol{\omega}}_b\\ \bar{\boldsymbol{v}}_b\end{bmatrix}_i\in\mathbb{R}\,.$$

Differentiation of eq. (9) with respect to time yields the contact constraints at the acceleration level,

$$\boldsymbol{\Phi}_F(\boldsymbol{q}_P,t)\,\dot{\boldsymbol{q}}_V+\boldsymbol{\xi}_F(\boldsymbol{q}_P,\boldsymbol{q}_V,t)=\mathbf{0}\quad\text{with}\quad \boldsymbol{\xi}_F=\frac{\mathrm{d}\boldsymbol{\Phi}_F}{\mathrm{d}t}\boldsymbol{q}_V+\frac{\mathrm{d}\boldsymbol{\zeta}_F}{\mathrm{d}t}\,. \tag{10}$$

2.5 Contact Forces

Each contact constraint (3) involves contact forces $\boldsymbol{f}_{ai}$, $\boldsymbol{f}_{bi}$ and contact torques $\boldsymbol{l}_{ai}$, $\boldsymbol{l}_{bi}$ in the corresponding constraint frames $\mathcal{K}_{ai}$, $\mathcal{K}_{bi}$, as per Fig. 1. With the principle of action and reaction holds

$$\begin{bmatrix}\boldsymbol{l}_{bi}\\ \boldsymbol{f}_{bi}\end{bmatrix}=-\begin{bmatrix}\boldsymbol{l}_{ai}+\widetilde{\boldsymbol{r}}_{ai,bi}\,\boldsymbol{f}_{bi}\\ \boldsymbol{f}_{ai}\end{bmatrix}. \tag{11}$$

The contact forces due to the i-th contact constraint are a scalar multiple λ_i of the normal vector $\boldsymbol{n}_i$ from eq. (5), namely,

$$\begin{bmatrix}\boldsymbol{l}_a\\ \boldsymbol{f}_a\\ \boldsymbol{l}_b\\ \boldsymbol{f}_b\end{bmatrix}_i=\boldsymbol{n}_i\,\lambda_i\,,\qquad i=1,\ldots,d\,. \tag{12}$$

The generalized contact forces belonging to the system coordinates are obtained by means of the transposed Jacobian matrices from eq. (7), i. e.,

$$\boldsymbol{g}_{Fi}\;=\;\underbrace{\left[\boldsymbol{J}_{Ra}^{\mathrm{T}}\,\boldsymbol{J}_{Ta}^{\mathrm{T}}\,\boldsymbol{J}_{Rb}^{\mathrm{T}}\,\boldsymbol{J}_{Tb}^{\mathrm{T}}\right]_i\boldsymbol{n}_i}_{\boldsymbol{\Phi}_{Fi}^{\mathrm{T}}}\,\lambda_i\quad\in\mathbb{R}^{nv}\,,\qquad i=1,\ldots,d\,. \tag{13}$$

The resulting generalized reaction forces $\boldsymbol{g}_F$ of all d contact constraints then read

$$\boldsymbol{g}_F\;=\;\sum_{i=1}^{d}\boldsymbol{g}_{Fi}\;\overset{(13)}{=}\;\sum_{i=1}^{d}\boldsymbol{\Phi}_{Fi}^{\mathrm{T}}\lambda_i\;=\;\underbrace{\left[\boldsymbol{\Phi}_{F1}^{\mathrm{T}}\ \ldots\ \boldsymbol{\Phi}_{Fd}^{\mathrm{T}}\right]}_{\boldsymbol{\Phi}_F^{\mathrm{T}}}\underbrace{\begin{bmatrix}\lambda_1\\ \vdots\\ \lambda_d\end{bmatrix}}_{\boldsymbol{\lambda}}. \tag{14}$$

2.6 Example: Contact Constraint Point – Plane

The end-effector of a robot (frame $\mathcal{K}_a$) contacts puntually a planar surface (frame $\mathcal{K}_b$), as depicted in Fig. 2.

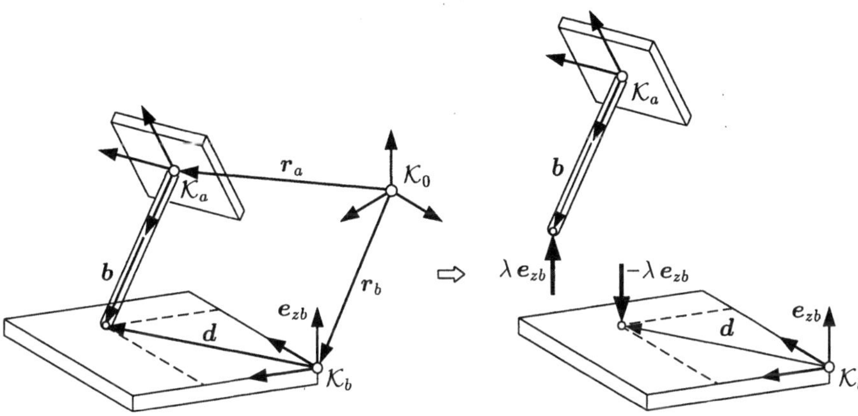

Fig. 2. Contact constraint Point – Plane

a) **Position**, with reference to eq. (3):

$$\varphi_F \equiv \boldsymbol{d}^{\mathrm{T}} \boldsymbol{e}_{zb} = 0 \qquad \text{with} \qquad \boldsymbol{d} = \boldsymbol{r}_a + \boldsymbol{b} - \boldsymbol{r}_b\,. \tag{15}$$

b) **Velocity**, with reference to eq. (5):

$$\boldsymbol{d}^{\mathrm{T}}\dot{\boldsymbol{e}}_{zb} + \dot{\boldsymbol{d}}^{\mathrm{T}}\boldsymbol{e}_{zb} = 0$$

$$\text{with} \qquad \dot{\boldsymbol{e}}_{zb} = \widetilde{\boldsymbol{\omega}}_b\,\boldsymbol{e}_{zb}\,, \qquad \dot{\boldsymbol{d}} = \dot{\boldsymbol{r}}_a + \overbrace{\widetilde{\boldsymbol{\omega}}_a\,\boldsymbol{b}}^{\dot{\boldsymbol{b}}} - \dot{\boldsymbol{r}}_b$$

$$\rightarrow \qquad \underbrace{\begin{bmatrix} -\boldsymbol{e}_{zb}^{\mathrm{T}}\widetilde{\boldsymbol{b}} & \boldsymbol{e}_{zb}^{\mathrm{T}} & -\boldsymbol{d}^{\mathrm{T}}\widetilde{\boldsymbol{e}}_{zb} & -\boldsymbol{e}_{zb}^{\mathrm{T}} \end{bmatrix}}_{\boldsymbol{n}^{\mathrm{T}}} \begin{bmatrix} \boldsymbol{\omega}_a \\ \boldsymbol{v}_a \\ \boldsymbol{\omega}_b \\ \boldsymbol{v}_b \end{bmatrix} = 0\,. \tag{16}$$

c) **Contact forces** in $\mathcal{K}_a$ and $\mathcal{K}_b$, with reference to eq. (12):

$$\begin{bmatrix} \boldsymbol{l}_a \\ \boldsymbol{f}_a \\ \boldsymbol{l}_b \\ \boldsymbol{f}_b \end{bmatrix} = \underbrace{\begin{bmatrix} \widetilde{\boldsymbol{b}}\,\boldsymbol{e}_{zb} \\ \boldsymbol{e}_{zb} \\ \widetilde{\boldsymbol{e}}_{zb}\,\boldsymbol{d} \\ -\boldsymbol{e}_{zb} \end{bmatrix}}_{\boldsymbol{n}} \lambda\,. \tag{17}$$

2.7 Dynamics with Contact Constraints

The motion of the multibody system (1) and (2) is now constrained by the d holonomic contact constraints (8). The numbers of independent position and velocity coordinates are reduced to

$$f_P = n_P - d\,, \qquad f_V = n_V - d\,, \tag{18}$$

respectively, as depicted in Fig. 3.

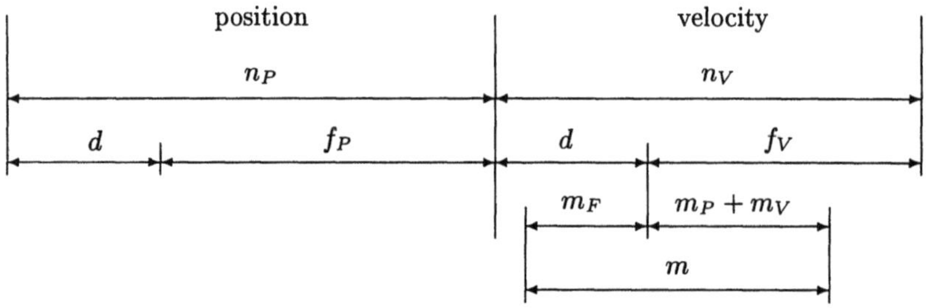

Fig. 3. Numbers of variables for a controlled multibody system

The equations of motion of the multibody system under the contact constraints (8) are now formulated in terms of the coordinates $\boldsymbol{q}_P$, $\boldsymbol{q}_V$ of system (1), (2). For that purpose, the generalized contact forces $\boldsymbol{g}_F$ from eq. (14) are added to the right-hand side of eq. (2), namely,

$$\boldsymbol{M}(\boldsymbol{q}_P,t)\,\dot{\boldsymbol{q}}_V = \boldsymbol{g}(\boldsymbol{q}_P,\boldsymbol{q}_V,t) + \underbrace{\boldsymbol{\Phi}_F^{\mathrm{T}}(\boldsymbol{q}_P,t)\,\boldsymbol{\lambda}}_{\boldsymbol{g}_F} + \boldsymbol{G}(\boldsymbol{q}_P,t)\,\boldsymbol{u}\,. \tag{19}$$

The differential-algebraic system of equations (1), (19), and (8) then describe the dynamics of the constrained multibody system. Hereby, the contact constraints (8) have also to be satisfied at the velocity level (9) and at the acceleration level (10). If the positions $\boldsymbol{q}_P$ and the velocities $\boldsymbol{q}_V$, which have to be consistent with the constraints (8) and their first-order time derivatives (9), as well as the control inputs $\boldsymbol{u}$ are given, eqs. (19) and (10) together represent a set of linear equations to determine uniquely the unknown accelerations $\dot{\boldsymbol{q}}_V$ and the unknown contact force coordinates $\boldsymbol{\lambda}$ (numbers of equations in $\langle\,\rangle$ braces):

$$\begin{array}{l}(19):\\(10):\end{array}\begin{bmatrix}\boldsymbol{M} & \boldsymbol{\Phi}_F^{\mathrm{T}}\\ \boldsymbol{\Phi}_F & \boldsymbol{0}\end{bmatrix}\begin{bmatrix}\dot{\boldsymbol{q}}_V\\ -\boldsymbol{\lambda}\end{bmatrix} = \begin{bmatrix}\boldsymbol{g}+\boldsymbol{G}\boldsymbol{u}\\ -\boldsymbol{\xi}_F\end{bmatrix} \quad \begin{array}{l}\langle n_V\rangle\\ \langle d\rangle\,.\end{array} \tag{20}$$

With $\boldsymbol{\lambda}$ from eq. (20), the contact forces and torques in the contact frames $\mathcal{K}_{ai}$, $\mathcal{K}_{bi}$, $i = 1,\ldots,d$, are obtained by means of eq. (12).

3. Exact Input-Output Linearization

For the multibody system with contact constraints described by eqs. (1), (19), and (8), exact input-output linearization (IO linearization) is investigated. First, the output variables (control variables) that shall be made track prescribed time functions, are defined. IO linearization for these outputs is then obtained by means of a nonlinear feedback of the system coordinates $\boldsymbol{q}_P$, $\boldsymbol{q}_V$.

3.1 Output Variables (Control Variables)

The objective of the considered control is to make altogether m output variables $\boldsymbol{y}(t)$ of the constrained multibody system track prescribed functions $\hat{\boldsymbol{y}}(t)$. In particular, m_P position outputs $\boldsymbol{y}_P$, m_V velocity outputs $\boldsymbol{y}_V$, and m_F contact force outputs $\boldsymbol{y}_F$ with $m_P+m_V+m_F = m$ are distinguished.

3.1.1 Position Outputs. The m_P position outputs $\boldsymbol{y}_P$ are defined as nonlinear time-varying functions of the position coordinates $\boldsymbol{q}_P$. Thus, these functions have the form of the holonomic constraints (8):

$$\begin{aligned} \begin{bmatrix} y_{P1} \\ \vdots \\ y_{P,m_P} \end{bmatrix} &= \begin{bmatrix} \varphi_{P1}(\boldsymbol{q}_P,t) \\ \vdots \\ \varphi_{P,m_P}(\boldsymbol{q}_P,t) \end{bmatrix} \\ \boldsymbol{y}_P &= \boldsymbol{\varphi}_P(\boldsymbol{q}_P,t) \quad . \end{aligned} \tag{21}$$

The functions $\boldsymbol{\varphi}_P(\boldsymbol{q}_P,t)$ are assumed to be at least two times differentiable with respect to t. According to eqs. (9) and (10), their time-derivatives are

$$\dot{\boldsymbol{y}}_P = \boldsymbol{\Phi}_P(\boldsymbol{q}_P,t)\,\boldsymbol{q}_V + \boldsymbol{\zeta}_P(\boldsymbol{q}_P,t)\ , \quad \boldsymbol{\Phi}_P = \frac{\partial \boldsymbol{\varphi}_P}{\partial \boldsymbol{q}_P}\boldsymbol{H}\ , \quad \boldsymbol{\zeta}_P = \frac{\partial \boldsymbol{\varphi}_P}{\partial t}\ , \tag{22}$$

$$\ddot{\boldsymbol{y}}_P = \boldsymbol{\Phi}_P(\boldsymbol{q}_P,t)\,\dot{\boldsymbol{q}}_V + \boldsymbol{\xi}_P(\boldsymbol{q}_P,\boldsymbol{q}_V,t)\ , \quad \boldsymbol{\xi}_P = \frac{\mathrm{d}\boldsymbol{\Phi}_P}{\mathrm{d}t}\boldsymbol{q}_V + \frac{\mathrm{d}\boldsymbol{\zeta}_P}{\mathrm{d}t}\ . \tag{23}$$

3.1.2 Velocity Output. The m_V velocity outputs $\boldsymbol{y}_V$ are defined by 'nonholonomic' functions

$$\begin{aligned} \begin{bmatrix} y_{V1} \\ \vdots \\ y_{V,m_V} \end{bmatrix} &= \begin{bmatrix} \boldsymbol{\Phi}_{V1}(\boldsymbol{q}_P,t) \\ \vdots \\ \boldsymbol{\Phi}_{V,m_V}(\boldsymbol{q}_P,t) \end{bmatrix} \boldsymbol{q}_V + \begin{bmatrix} \zeta_{V1}(\boldsymbol{q}_P,t) \\ \vdots \\ \zeta_{V,d_V}(\boldsymbol{q}_P,t) \end{bmatrix} \\ \boldsymbol{y}_V &= \boldsymbol{\Phi}_V(\boldsymbol{q}_P,t)\quad \boldsymbol{q}_V + \quad \boldsymbol{\zeta}_V(\boldsymbol{q}_P,t) \quad . \end{aligned} \tag{24}$$

The velocity outputs need not be integrable, i. e. the position variables whose time-derivative yields $\boldsymbol{y}_V$ do not exist in general. An example is treated in Section 5.

The velocity outputs are assumed to be at least once differentiable with respect to t,

$$\dot{y}_V = \Phi_V(q_P,t)\,\dot{q}_V + \xi_V(q_P,q_V,t)\,, \quad \xi_V = \frac{\mathrm{d}\Phi_V}{\mathrm{d}t}\,q_V + \frac{\mathrm{d}\zeta_V}{\mathrm{d}t}\,. \tag{25}$$

The position and velocity outputs have to be consistent with the contact constraints (8). This requires that the row vectors of the functional matrix (numbers of rows in $\langle\,\rangle$ braces)

$$\hat{\Phi} = \begin{bmatrix} \Phi_F \\ \Phi_P \\ \Phi_V \end{bmatrix} \qquad \begin{matrix} \langle d \rangle \\ \langle m_P \rangle \\ \langle m_V \rangle \end{matrix} \tag{26}$$

with n_V columns have to be linearly independent, i. e. the condition

$$\mathrm{rank}(\hat{\Phi}) = d + m_P + m_V \tag{27}$$

must be satisfied, with reference to Fig. 3. A necessary condition is that the number of rows must not exceed the number of columns, i. e.,

$$d + m_P + m_V \le n_V \qquad \Rightarrow \qquad m_P + m_V \le \underbrace{n_V - d}_{(18)\,:\; f_V}\,. \tag{28}$$

Thus, the number of position and velocity outputs $m_P + m_V$ must not exceed the number of independent velocities f_V of the constrained system.

3.1.3 Contact Force Outputs. The m_F contact force outputs y_F are chosen from the natural coordinates λ of the contact forces:

$$\begin{array}{ccccc} \begin{bmatrix} y_{F1} \\ \vdots \\ y_{F,m_F} \end{bmatrix} & = & \begin{bmatrix} L_{1,1} & \dots & L_{1,d} \\ \vdots & & \vdots \\ L_{m_F,1} & \dots & L_{m_F,d} \end{bmatrix} & \begin{bmatrix} \lambda_1 \\ \vdots \\ \lambda_d \end{bmatrix} \\ y_F & = & L & \lambda\,. \end{array} \tag{29}$$

Here, $L \in \mathbb{R}^{m_F,d}$ is an assignment matrix whose rows contain zeros except for one element with a '1'. With d mutually independent contact force coordinates, a necessary condition holds for the number m_F of contact force coordinates (Fig. 3):

$$m_F \le d\,. \tag{30}$$

It should be noted that conditions (28) and (30) only give an upper boundary for the number of output variables that is theoretically admissible by the constraints. The actual number of outputs, for which an IO linearization can be achieved further depends on the structure of the input-output equation introduced in the next subsection.

The nonlinear state feedback for exact IO linearization is now derived on the basis of the differential-algebraic equations of motion (1), (19), and (8) in combination with outputs defined by eqs. (21), (24), and (29).

3.2 Input-Output Equation

For the derivation of the IO linearization, the relative degree δ_i of the i-th output y_i is of importance. This is exactly the number of times one has to differentiate y_i with respect to t in order to have at least one component of input vector $\boldsymbol{u}$ explicitly appearing (Isidori 1989). By this, for each output y_i an equation is obtained which is linear in the δ_i-th-order derivative of y_i and the inputs $\boldsymbol{u}$. The set of m equations of that kind represents a linear relation between the m-dimensional vector of the δ_i-th-order derivatives of y_i and the p-dimensional input vector $\boldsymbol{u}$. Its solution with respect to $\boldsymbol{u}$ yields the desired 'inverse system' needed for IO linearization.

A first overview is given by the particular structure of the equations of motion. The solution of eq. (20) with respect to $\dot{\boldsymbol{q}}_V$ and $\boldsymbol{\lambda}$ shows that the lowest achievable relative degrees of the velocities $\boldsymbol{q}_V$ and contact forces $\boldsymbol{\lambda}$ are one and zero, respectively. The relative degrees of the position outputs $\boldsymbol{y}_P$ from eq. (21), therefore, are $\delta_{Pi} \geq 2$, of the velocity outputs $\boldsymbol{y}_V$ from eq. (24), $\delta_{Vi} \geq 1$, and of the contact force outputs $\boldsymbol{y}_F$ from eq. (29), $\delta_{Fi} \geq 0$.

Inserting accelerations $\dot{\boldsymbol{q}}_V$ and contact forces $\boldsymbol{\lambda}$ from eq. (20) into the output equations

$$\begin{array}{l} (23): \\ (25): \\ (29): \end{array} \quad \begin{bmatrix} \ddot{\boldsymbol{y}}_P \\ \dot{\boldsymbol{y}}_V \\ \boldsymbol{y}_F \end{bmatrix} = \begin{bmatrix} \boldsymbol{\Phi}_P & \boldsymbol{0} \\ \boldsymbol{\Phi}_V & \boldsymbol{0} \\ \boldsymbol{0} & \boldsymbol{L} \end{bmatrix} \begin{bmatrix} \dot{\boldsymbol{q}}_V \\ \boldsymbol{\lambda} \end{bmatrix} + \begin{bmatrix} \boldsymbol{\xi}_P \\ \boldsymbol{\xi}_V \\ \boldsymbol{0} \end{bmatrix} \quad \begin{array}{l} \langle m_P \rangle \\ \langle m_V \rangle \\ \langle m_F \rangle \end{array} \tag{31}$$

leads to an *input-output equation* (IO equation) of the form

$$\begin{aligned} \begin{bmatrix} \ddot{\boldsymbol{y}}_P \\ \dot{\boldsymbol{y}}_V \\ \boldsymbol{y}_F \end{bmatrix} &= \begin{bmatrix} \boldsymbol{c}_P(\boldsymbol{q}_P, \boldsymbol{q}_V, t) \\ \boldsymbol{c}_V(\boldsymbol{q}_P, \boldsymbol{q}_V, t) \\ \boldsymbol{c}_F(\boldsymbol{q}_P, \boldsymbol{q}_V, t) \end{bmatrix} + \begin{bmatrix} \boldsymbol{D}_P(\boldsymbol{q}_P, t) \\ \boldsymbol{D}_V(\boldsymbol{q}_P, t) \\ \boldsymbol{D}_F(\boldsymbol{q}_P, t) \end{bmatrix} \boldsymbol{u} \\ &= \quad \boldsymbol{c}(\boldsymbol{q}_P, \boldsymbol{q}_V, t) \quad + \quad \boldsymbol{D}(\boldsymbol{q}_P, t) \quad \boldsymbol{u}. \end{aligned} \tag{32}$$

with the *input-output matrix* (IO matrix) $\boldsymbol{D} \in \mathbb{R}^{m,p}$. The further procedure now depends on the rank of $\boldsymbol{D}$, whereby two cases may be distinguished, with reference to Fig. 3:

1. The rank of the IO matrix $\boldsymbol{D} \in \mathbb{R}^{m,p}$ is m:

   All outputs $\boldsymbol{y} \in \mathbb{R}^m$ are independently controllable by the inputs $\boldsymbol{u} \in \mathbb{R}^p$. The relative degrees of the outputs $\boldsymbol{y}_P$, $\boldsymbol{y}_V$, and $\boldsymbol{y}_F$ are, respectively,

$$\delta_{Pi} = 2\,, \quad i = 1, \ldots, m_P\,, \tag{33}$$

$$\delta_{Vi} = 1\,, \quad i = 1, \ldots, m_V\,, \tag{34}$$

$$\delta_{Fi} = 0\,, \quad i = 1, \ldots, m_F\,. \tag{35}$$

   The IO matrix $\boldsymbol{D} \in \mathbb{R}^{m,p}$ from eq. (32), then, is the *decoupling matrix* of the system, and an exact IO linearization is possible by a *static state feedback* that is described by algebraic equations. With the number p of inputs $\boldsymbol{u}$ and the overall number m of outputs $\boldsymbol{y}$, two subcases may be distinguished:

a) $m = p$: Matrix $\boldsymbol{D}$ is uniquely invertible ('quadratic' decoupling problem).

b) $m < p$: Matrix $\boldsymbol{D}$ has no unique inverse. The decoupling feedback is not uniquely determined. The available inputs $\boldsymbol{u} \in \mathbb{R}^p$ could control $p - m$ additional outputs. The decoupling problem is *redundant.*

2. The rank of the IO matrix $\boldsymbol{D} \in \mathbb{R}^{m,p}$ is less than m:

   The time derivatives $\ddot{\boldsymbol{y}}_P$, $\dot{\boldsymbol{y}}_V$, and $\boldsymbol{y}_F$ of the outputs cannot be independently controlled by the inputs $\boldsymbol{u}$. To obtain an IO matrix with maximum rank, $\ddot{\boldsymbol{y}}_P$, $\dot{\boldsymbol{y}}_V$ and $\boldsymbol{y}_F$ are further differentiated with respect to t. In general, the *dynamic extension* procedure has to be applied leading to a *dynamic state feedback* that is described by differential equations (Isidori 1995). The relative degrees then are $\delta_{Pi} > 2$, $\delta_{Vi} > 1$ and/or $\delta_{Fi} > 0$. An example is treated in Section 5.

Below, the first case (static state feedback) is further investigated.

3.3 Static Decoupling Feedback

It is now assumed that the decoupling matrix $\boldsymbol{D} \in \mathbb{R}^{m,p}$ in the IO equation (32) has full rank m. A static decoupling feedback (for $m < p$ not uniquely determined) is obtained by solving the IO equation (32) with respect to $\boldsymbol{u}$, whereby the output variables $\ddot{\boldsymbol{y}}_P$, $\dot{\boldsymbol{y}}_V$, and $\boldsymbol{y}_F$ are replaced by new inputs $\boldsymbol{w}_P$, $\boldsymbol{w}_V$, and $\boldsymbol{w}_F$, respectively. The static decoupling feedback then is

$$\boldsymbol{u} = \underbrace{-\boldsymbol{D}^{\mathrm{R}}\boldsymbol{c}}_{\boldsymbol{k}(\boldsymbol{q}_P,\boldsymbol{q}_V,t)} + \underbrace{\boldsymbol{D}^{\mathrm{R}}}_{\boldsymbol{K}(\boldsymbol{q}_P,t)} \begin{bmatrix} \boldsymbol{w}_P \\ \boldsymbol{w}_V \\ \boldsymbol{w}_F \end{bmatrix} + \underbrace{(\boldsymbol{I}_p - \boldsymbol{D}^{\mathrm{R}}\boldsymbol{D})}_{\boldsymbol{K}^*(\boldsymbol{q}_P,t)}\,\boldsymbol{u}^* \,. \tag{36}$$

Here, $\boldsymbol{D}^{\mathrm{R}} \in \mathbb{R}^{p,m}$ is a generalized right-inverse of $\boldsymbol{D}$ with $\boldsymbol{D}\,\boldsymbol{D}^{\mathrm{R}} = \boldsymbol{I}_m$. Vector $\boldsymbol{u}^* \in \mathbb{R}^p$ is projected by matrix $\boldsymbol{K}^* \in \mathbb{R}^{p,p}$ into the nullspace of $\boldsymbol{D}$. The choice of a particular $\boldsymbol{D}^{\mathrm{R}}$ and $\boldsymbol{u}^*$ influences the behavior of the redundant coordinates. An application is the control of manipulators with redundant degrees of freedom in Section 4.

Cancellation of the nonlinearities of the mechanical system by feedback (36) is shown by inserting eq. (36) into IO equation (32). If the matrices $\boldsymbol{D}$ and $\boldsymbol{c}$ in eq. (32) and eq. (36) coincide, a linear and decoupled system is obtained consisting of m decoupled input-output channels:

$$\left.\begin{array}{llll} \ddot{\boldsymbol{y}}_P = \boldsymbol{w}_P & \text{or} & \ddot{y}_{Pi} = w_{Pi}\,, & i = 1,\dots,m_P \\ \dot{\boldsymbol{y}}_V = \boldsymbol{w}_V & \text{or} & \dot{y}_{Vi} = w_{Vi}\,, & i = 1,\dots,m_V \\ \boldsymbol{y}_F = \boldsymbol{w}_F & \text{or} & y_{Fi} = w_{Fi}\,, & i = 1,\dots,m_F \end{array}\right\} . \tag{37}$$

Remark. If feedback (36) exists, it can be directly obtained as a solution of a system of equations composed of eqs. (20) and (31). Hereby, outputs $\ddot{\boldsymbol{y}}_P$, $\dot{\boldsymbol{y}}_V$, and $\boldsymbol{y}_F$ are replaced by new inputs $\boldsymbol{w}_P$, $\boldsymbol{w}_V$, and $\boldsymbol{w}_F$. Further, accelerations $\dot{\boldsymbol{q}}_V$ and contact force coordinates $\boldsymbol{\lambda}$ of the mechanical system are replaced by intermediate variables $\dot{\boldsymbol{q}}_V^*$ and $\boldsymbol{\lambda}^*$ of the feedback, respectively:

$$\begin{array}{r} (19): \\ (10): \\ (23): \\ (25): \\ (29): \end{array} \begin{bmatrix} \boldsymbol{M} & \boldsymbol{\Phi}_F^{\mathrm{T}} & \boldsymbol{G} \\ \boldsymbol{\Phi}_F & \boldsymbol{0} & \boldsymbol{0} \\ \boldsymbol{\Phi}_P & \boldsymbol{0} & \boldsymbol{0} \\ \boldsymbol{\Phi}_V & \boldsymbol{0} & \boldsymbol{0} \\ \boldsymbol{0} & -\boldsymbol{L} & \boldsymbol{0} \end{bmatrix} \begin{bmatrix} \dot{\boldsymbol{q}}_V^* \\ -\boldsymbol{\lambda}^* \\ -\boldsymbol{u} \end{bmatrix} = \begin{bmatrix} \boldsymbol{g} \\ -\boldsymbol{\xi}_F \\ -\boldsymbol{\xi}_P + \boldsymbol{w}_P \\ -\boldsymbol{\xi}_V + \boldsymbol{w}_V \\ \boldsymbol{w}_F \end{bmatrix} \begin{array}{l} \langle n_V \rangle \\ \langle d \rangle \\ \langle m_P \rangle \\ \langle m_V \rangle \\ \langle m_F \rangle . \end{array} \tag{38}$$

These are $(n_V + d) + m$ equations for $(n_V + d) + p$ unknowns $\dot{\boldsymbol{q}}_V^*$, $\boldsymbol{\lambda}^*$, and $\boldsymbol{u}$ ($m \le p$). The solution with respect to $\boldsymbol{u}$ (not unique for $m < p$) is a decoupling feedback. Especially, if all degrees of freedom are independently controllable by control inputs (e.g. manipulators), this system leads more directly to a decoupling feedback (36) than the previously described explicit formulation of IO equation (32) and its solution with respect to $\boldsymbol{u}$, as discussed in Section 4.

3.4 Control of the Linearized Subsystems

The objective of the control is to make outputs $\boldsymbol{y}_P(t)$, $\boldsymbol{y}_V(t)$, and $\boldsymbol{y}_F(t)$ track prescribed functions $\hat{\boldsymbol{y}}_P(t)$, $\hat{\boldsymbol{y}}_V(t)$, and $\hat{\boldsymbol{y}}_F(t)$ in the form

$$\lim_{t\to\infty} \boldsymbol{e}_P(t) = \boldsymbol{0} \quad \text{with} \quad \boldsymbol{e}_P(t) = \hat{\boldsymbol{y}}_P(t) - \boldsymbol{y}_P(t) , \tag{39}$$

$$\lim_{t\to\infty} \boldsymbol{e}_V(t) = \boldsymbol{0} \quad \text{with} \quad \boldsymbol{e}_V(t) = \hat{\boldsymbol{y}}_V(t) - \boldsymbol{y}_V(t) , \tag{40}$$

$$\lim_{t\to\infty} \boldsymbol{e}_F(t) = \boldsymbol{0} \quad \text{with} \quad \boldsymbol{e}_F(t) = \hat{\boldsymbol{y}}_F(t) - \boldsymbol{y}_F(t) . \tag{41}$$

This asymptotically-stable tracking control of the outputs $\boldsymbol{y}$ can be achieved by outer feedbacks, namely,

$$w_{Pi}(t) = \ddot{\hat{y}}_{Pi}(t) + \alpha_{P1}^{[i]}\, \dot{e}_{Pi}(t) + \alpha_{P0}^{[i]}\, e_{Pi}(t) , \tag{42}$$

$$w_{Vi}(t) = \dot{\hat{y}}_{Vi}(t) + \alpha_{V0}^{[i]}\, e_{Vi}(t) , \tag{43}$$

$$w_{Fi}(t) = \hat{y}_{Fi}(t) + \alpha_{F,-1}^{[i]} \int_0^t e_{Fi}(\tau)\, \mathrm{d}\tau . \tag{44}$$

A block diagram of the multibody system with decoupling feedback (36) and the outer feedback loops (42) to (44) is shown in Fig. 4. Further analysis of the IO linearized system reveals that it includes in general besides the decoupled system (37) a still nonlinear internal dynamics that is non-observable from the outputs (Woernle 1992).

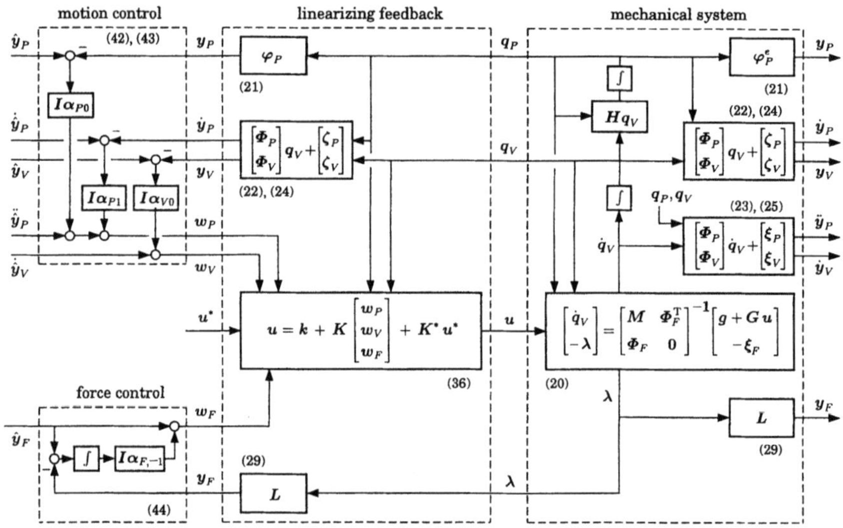

Fig. 4. Multibody system with feedback linearization and output error feedback

4. Control of Manipulators with Redundant Position Variables and Contact Forces

Complex manipulator tasks involve motions around obstacles in a complex workspace as well as material contacts between the end-effector or other links and the environment or other manipulators. These contacts are constraints for the motion of the manipulator reducing its number of degrees of freedom. Execution of a task requires simultaneous control of both position variables and contact forces.

Here, two kinds of redundant variables are defined (Woernle 1994): *redundant position variables* leading to redundant degrees of freedom and *redundant contact forces*. Redundant position variables occur, if a manipulator has more independently controllable position variables than position output variables specified by the particular task. Typical examples are large manipulators in civil engineering or for service applications, as shown in Fig. 5. In a similar manner, redundant contact forces occur, if there are more independently controllable contact forces than force output variables specified by the particular task. Redundant control variables provide increased flexibility for the execution of complex tasks, for example motions in an environment with obstacles or minimization of control forces.

4.1 Equations of Motion

In this subsection, holonomic open-loop manipulator systems with one or several arms and $n_P = n_V = n$ degrees of freedom are considered. A rigid-body model with $p = n$ control forces $\boldsymbol{u} \in \mathbb{R}^n$ acting directly along the n

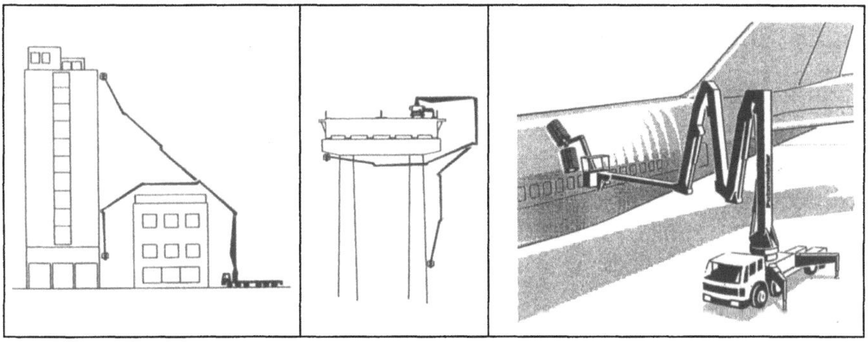

Fig. 5. Large manipulators with redundant degrees of freedom (Wanner 1993)

independent joint coordinates $q_P \in \mathbb{R}^n$ is assumed. During the execution of a task, the manipulator may be constrained by material contacts of the end-effector or other links with the environment. These contacts can be formulated in terms of d holonomic constraints, as per eqs. (8) and (10),

$$\varphi_F(q_P) = 0 \quad \in \mathbb{R}^d \qquad \rightarrow \qquad \Phi_F(q_P)\,\dot{q}_V + \xi_F(q_P, q_V) = 0\,. \quad (45)$$

If these contact constraints are independent, they reduce the number of degrees of freedom of the manipulator from n to $f_P = f_V = f$ with $f = n - d$, as depicted in Fig. 6.

With contact forces belonging to the contact constraints, the equations of motion of the holonomic system read, as per eqs. (1) and (19),

$$\dot{q}_P = q_V\,, \quad (46)$$

$$M(q_P)\,\dot{q}_V = g(q_P, q_V) + \Phi_F^{\mathrm{T}}(q_P)\,\lambda + u\,. \quad (47)$$

4.2 Output Variables

It was assumed that all n degrees of freedom of the unconstrained manipulator are independently controllable by the given control inputs u, and that the constraints (45) are independent. Then, $f = n - d$ position variables y_P and $m_F = d$ contact force variables y_F of the constrained manipulator described by eqs. (46), (47), and (45) can be independently controlled by inputs u (Fig. 6).

4.2.1 Position Outputs. In the considered case of redundant output variables, only $m_P < f$ position output variables according to eqs. (21) and (23), i. e.,

$$y_P = \varphi_P(q_P) \in \mathbb{R}^{m_P} \qquad \rightarrow \qquad \ddot{y}_P = \Phi_P(q_P)\,\dot{q}_V + \xi_P(q_P, q_V) \quad (48)$$

are specified by the particular manipulator task. Therefore, $f - m_P$ 'redundant' position variables remain (Fig. 6).

4.2.2 Force Outputs. The d controllable contact forces $\boldsymbol{\lambda}$ are subdivided into $m_F < d$ 'functional' forces $\boldsymbol{\lambda}_f$ specified by the particular manipulator task, and $d - m_F$ 'redundant' forces $\boldsymbol{\lambda}_r$, in the form

$$\boldsymbol{\lambda} = \begin{bmatrix} \boldsymbol{\lambda}_f \\ \boldsymbol{\lambda}_r \end{bmatrix} \quad \begin{matrix} \langle m_F \rangle \\ \langle d - m_F \rangle \, . \end{matrix} \tag{49}$$

The force outputs then are the functional contact forces, as per eq. (29),

$$\boldsymbol{y}_F = \underbrace{\begin{bmatrix} \boldsymbol{I} & \boldsymbol{0} \end{bmatrix}}_{\boldsymbol{L}} \begin{bmatrix} \boldsymbol{\lambda}_f \\ \boldsymbol{\lambda}_r \end{bmatrix} \, . \tag{50}$$

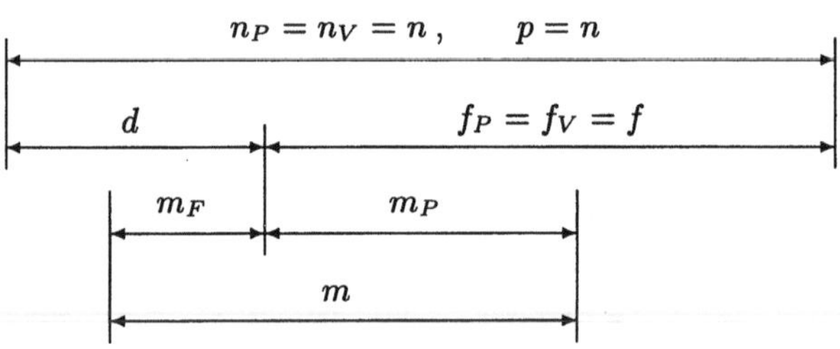

Fig. 6. Numbers of variables for redundant manipulators

4.2.3 Example. The motion of a planar manipulator with $n = 5$ joint angles $\boldsymbol{q}_P$ driven by joint torques $\boldsymbol{u} \in \mathbb{R}^5$ is constrained by $d = 2$ contact surfaces $\varphi_{F1} = 0$, $\varphi_{F2} = 0$ (Fig. 7). It has $f = n - d = 3$ degrees of freedom. The translation $y_P = s$ of point A along the surface $\varphi_{F1} = 0$ and the corresponding contact force $y_F = \lambda_1$ are defined as $m_P = 1$ position output and $m_F = 1$ force output, respectively. There remain $f - m_P = 2$ redundant position variables and $d - m_F = 1$ redundant contact force λ_2 at the surface $\varphi_{F2} = 0$.

By means of controllable redundant variables, it is possible to accomplish desirable secondary goals during tracking the desired values of the control outputs. Such goals are, for example,

- achieving repeatable motions in the coordinates $\boldsymbol{q}_P$ for cyclic trajectories of the position outputs $\boldsymbol{y}_P$,
- avoiding collisions of the manipulator with obstacles in the workspace,
- minimization of control inputs $\boldsymbol{u}$.

The practical problem is, however, that these secondary goals are not directly formulated by time functions of the redundant variables. The control outputs $\boldsymbol{y}_P$, $\boldsymbol{y}_F$, and the redundant variables are therefore controlled in a different manner. For the control outputs, decoupled input-output systems are obtained by IO linearization. They can be independently stabilized by linear control loops. Control of the redundant variables is achieved by defining a special linearization feedback that takes into account the secondary goals.

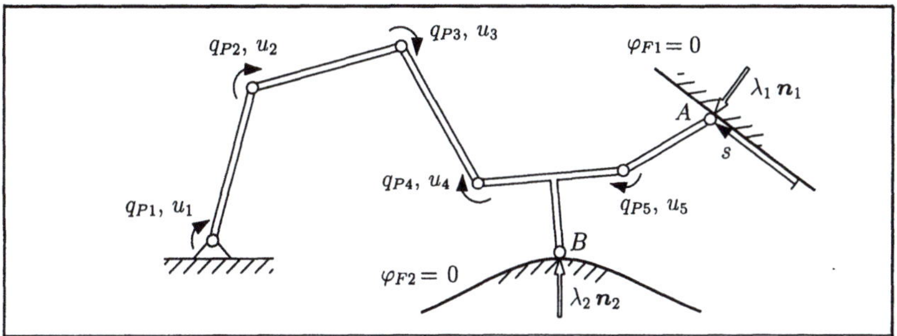

$n = 5$ position coordinates:

$$\boldsymbol{q}_P = \begin{bmatrix} q_{P1} \\ \vdots \\ q_{P5} \end{bmatrix}$$

$\rightarrow f = n - d = 3$ degrees of freedom

$d = 2$ contact constraints:

$$\begin{bmatrix} \varphi_{F1}(\boldsymbol{q}_P) \\ \varphi_{F2}(\boldsymbol{q}_P) \end{bmatrix} = \begin{bmatrix} 0 \\ 0 \end{bmatrix}$$

$m_P = 1$ position output:

$y_P = s(\boldsymbol{q}_P)$

$f - m_P = 2$ redundant position coordinates, e.g.:

q_{P1}, q_{P2}

$m_F = 1$ force output:

$y_F = \lambda_1$

$d - m_F = 1$ redundant contact force:

λ_2

Fig. 7. Constrained manipulator with redundant position variables and contact forces

4.3 IO Linearization

According to Section 3.2, the IO equation for the constrained manipulator is directly obtained by inserting the accelerations $\ddot{\boldsymbol{q}}$ and the contact forces $\boldsymbol{\lambda}$ resulting from eqs. (45) and (47) into output equations (48) and (50). Referring to eq. (32), the IO equation can be written as

$$\begin{aligned} \begin{bmatrix} \ddot{\boldsymbol{y}}_P \\ \boldsymbol{y}_F \end{bmatrix} &= \begin{bmatrix} \boldsymbol{c}_P(\boldsymbol{q}_P, \boldsymbol{q}_V) \\ \boldsymbol{c}_F(\boldsymbol{q}_P, \boldsymbol{q}_V) \end{bmatrix} + \begin{bmatrix} \boldsymbol{D}_P(\boldsymbol{q}_P) \\ \boldsymbol{D}_F(\boldsymbol{q}_P) \end{bmatrix} \boldsymbol{u} \\ &= \quad \boldsymbol{c}(\boldsymbol{q}_P, \boldsymbol{q}_V) \quad + \quad \boldsymbol{D}(\boldsymbol{q}_P) \quad \boldsymbol{u}\,. \end{aligned} \tag{51}$$

Under the assumption of controllable functional variables $\boldsymbol{y}_P$ and $\boldsymbol{y}_F$, the IO matrix $\boldsymbol{D} \in \mathbb{R}^{m,p}$ has maximal rank m (relative degrees are $\delta_{Pi} = 2$, $\delta_{Fi} = 0$), and it is the decoupling matrix of the system, as described in Section 3.2. Because $m < p$ in the redundant case, the linearizing feedback has the general form (36) and is not uniquely determined. Therefore, the next goal is to determine a particular feedback (36) in such a way that desirable secondary goals are achieved while outputs $\boldsymbol{y}_P$, $\boldsymbol{y}_F$ are tracked.

For that purpose, it is appropriate to proceed from eq. (38) that reads here (replacements $\ddot{\boldsymbol{y}}_P \rightarrow \boldsymbol{w}_P$, $\boldsymbol{y}_F \rightarrow \boldsymbol{w}_F$, $\boldsymbol{\lambda} \rightarrow \boldsymbol{\lambda}^*$, $\dot{\boldsymbol{q}}_V \rightarrow \dot{\boldsymbol{q}}_V^*$)

$$
\begin{array}{l}
(47): \\ (45): \\ (48): \\ (50):
\end{array}
\begin{bmatrix}
\boldsymbol{M} & \boldsymbol{\Phi}_F^{\mathrm{T}} & \boldsymbol{I} \\
\boldsymbol{\Phi}_F & \mathbf{0} & \mathbf{0} \\
\boldsymbol{\Phi}_P & \mathbf{0} & \mathbf{0} \\
\mathbf{0} & -\boldsymbol{L} & \mathbf{0}
\end{bmatrix}
\begin{bmatrix}
\dot{\boldsymbol{q}}_V^* \\ -\boldsymbol{\lambda}^* \\ -\boldsymbol{u}
\end{bmatrix}
=
\begin{bmatrix}
\boldsymbol{g} \\ -\boldsymbol{\xi}_F \\ -\boldsymbol{\xi}_P + \boldsymbol{w}_P \\ \boldsymbol{w}_F
\end{bmatrix}
\begin{array}{l}
\langle n \rangle \\ \langle d \rangle \\ \langle m_P \rangle \\ \langle m_F \rangle \,.
\end{array}
\tag{52}
$$

A physically illustrative solution with respect to $\boldsymbol{u}$ is obtained in two steps:

- The *inverse kinematics* determines desired accelerations $\dot{\boldsymbol{q}}_V^*$ taking into account the contact constraints (45) and the accelerations of the position outputs $\ddot{\boldsymbol{y}}_P$ prescribed by $\boldsymbol{w}_P$ in eq. (48). If there are redundant position variables, the inverse kinematics is not uniquely determined. A special solution is obtained by minimizing an objective function in terms of $\dot{\boldsymbol{q}}_V^*$ to fulfill desirable subtasks for the motion, for example repeatable cyclic motions or collision avoidance.
- By means of the *inverse dynamics*, control inputs $\boldsymbol{u}$ are computed to realize the accelerations $\dot{\boldsymbol{q}}_V^*$ obtained from the inverse kinematics as well as force outputs $\boldsymbol{y}_F$ prescribed by $\boldsymbol{w}_F$ in eq. (50). If there are redundant force variables, the inverse dynamics is not uniquely determined. A special solution is obtained by minimizing an objective function in terms of $\boldsymbol{u}$, for example minimizing a norm of $\boldsymbol{u}$.

4.3.1 Inverse Kinematics. The second and third row equation in eq. (52) determine accelerations $\dot{\boldsymbol{q}}_V^* \in \mathbb{R}^n$ that fulfill the contact constraints (45) and the prescribed accelerations $\boldsymbol{w}_P$ of the position outputs defined by eq. (48):

$$
\begin{array}{l} (45): \\ (48): \end{array}
\begin{bmatrix} \boldsymbol{\Phi}_F \\ \boldsymbol{\Phi}_P \end{bmatrix} \dot{\boldsymbol{q}}_V^* = -\begin{bmatrix} \boldsymbol{\xi}_F \\ \boldsymbol{\xi}_P \end{bmatrix} + \begin{bmatrix} \mathbf{0} \\ \boldsymbol{I} \end{bmatrix} \boldsymbol{w}_P \qquad \begin{array}{l} \langle d \rangle \\ \langle m_P \rangle \end{array}
$$

$$
\boldsymbol{\Phi} \quad \dot{\boldsymbol{q}}_V^* = \quad -\boldsymbol{\xi} \quad + \boldsymbol{W} \;\; \boldsymbol{w}_P \,. \tag{53}
$$

These are $d + m_P$ equations for n accelerations $\dot{\boldsymbol{q}}_V^*$. If there are $f - m_P$ redundant degrees of freedom, matrix $\boldsymbol{\Phi} \in \mathbb{R}^{d+m_P,n}$ has, according to Fig. 6, an $(f - m_P)$-dimensional nullspace, and the solution of eq. (53) is not uniquely determined.

A special solution is obtained by minimizing an objective function $Z(\dot{\boldsymbol{q}}_V^*)$ that is constrained by eq. (53). Here, a quadratic function

$$
Z(\dot{\boldsymbol{q}}_V^*) \;=\; \frac{1}{2}\dot{\boldsymbol{q}}_V^{*\mathrm{T}}\, \boldsymbol{P}\, \dot{\boldsymbol{q}}_V^* - \boldsymbol{p}^{\mathrm{T}}\, \dot{\boldsymbol{q}}_V^* \;\rightarrow\; \min_{\dot{\boldsymbol{q}}_V^*} \tag{54}
$$

with a symmetric, positive-definite matrix $\boldsymbol{P} \in \mathbb{R}^{n,n}$ and a vector $\boldsymbol{p} \in \mathbb{R}^n$ is defined (Woernle 1994).

To solve this constrained minimization problem, the extended objective function

$$
\bar{Z}(\dot{\boldsymbol{q}}_V^*) \;=\; Z(\dot{\boldsymbol{q}}_V^*) + \boldsymbol{\mu}^{\mathrm{T}}\,(\boldsymbol{\Phi}\, \dot{\boldsymbol{q}}_V^* + \boldsymbol{\xi} - \boldsymbol{W}\, \boldsymbol{w}_P) \;\rightarrow\; \min_{\dot{\boldsymbol{q}}_V^*} \tag{55}
$$

with Lagrange-multipliers $\boldsymbol{\mu} \in \mathbb{R}^{d+m_P}$ is minimized. The solution of this minimization problem reads

$$\dot{\boldsymbol{q}}_V^* = \boldsymbol{\Phi}^{\#}\,(\boldsymbol{W}\,\boldsymbol{w}_P - \boldsymbol{\xi}) + (\boldsymbol{I} - \boldsymbol{\Phi}^{\#}\,\boldsymbol{\Phi})\,\boldsymbol{P}^{-1}\,\boldsymbol{p} \tag{56}$$

with the pseudoinverse of $\boldsymbol{\Phi}$ weighted by the symmetric matrix $\boldsymbol{P}$, given as

$$\boldsymbol{\Phi}^{\#} = \boldsymbol{P}^{-1}\,\boldsymbol{\Phi}^{\mathrm{T}}\,(\boldsymbol{\Phi}\,\boldsymbol{P}^{-1}\,\boldsymbol{\Phi}^{\mathrm{T}})^{-1} \in \mathbb{R}^{n,d+m_P} \,. \tag{57}$$

which verifies $\boldsymbol{\Phi}\,\boldsymbol{\Phi}^{\#} = \boldsymbol{I}$.

4.3.2 Inverse Dynamics. By means of the inverse dynamics, the control inputs $\boldsymbol{u}$ are determined in such a way that both the accelerations $\dot{\boldsymbol{q}}_V^*$ resulting from eq. (56) and the prescribed contact forces $\boldsymbol{w}_F$ from eq. (50) are realized. With the contact constraints (45) partitioned according to the functional and redundant contact forces defined in eq. (49), i. e.,

$$\begin{bmatrix} \boldsymbol{\Phi}_{F_f}(\boldsymbol{q}_P) \\ \boldsymbol{\Phi}_{F_r}(\boldsymbol{q}_P) \end{bmatrix} \dot{\boldsymbol{q}}_V^* + \begin{bmatrix} \boldsymbol{\xi}_{F_f}(\boldsymbol{q}_P, \boldsymbol{q}_V) \\ \boldsymbol{\xi}_{F_r}(\boldsymbol{q}_P, \boldsymbol{q}_V) \end{bmatrix} = \begin{bmatrix} \boldsymbol{0} \\ \boldsymbol{0} \end{bmatrix} \quad \begin{matrix} \langle m_F \rangle \\ \langle d - m_F \rangle \end{matrix} \tag{58}$$

$$\boldsymbol{\Phi}_F(\boldsymbol{q}_P)\ \ \dot{\boldsymbol{q}}_V^* + \ \boldsymbol{\xi}_F(\boldsymbol{q}_P, \boldsymbol{q}_V) = \boldsymbol{0} \qquad \langle d \rangle \,,$$

the dynamic equations of the constrained manipulator are given by the first and fourth row equation of eq. (52), namely,

$$\boldsymbol{M}\,\dot{\boldsymbol{q}}_V^* = \boldsymbol{g} + \boldsymbol{\Phi}_{F_f}^{\mathrm{T}}\,\boldsymbol{\lambda}_f^* + \boldsymbol{\Phi}_{F_r}^{\mathrm{T}}\,\boldsymbol{\lambda}_r^* + \boldsymbol{u} \,, \tag{59}$$

$$\boldsymbol{\lambda}_f^* = \boldsymbol{w}_F \,. \tag{60}$$

With accelerations $\dot{\boldsymbol{q}}_V^*$ obtained from eq. (56), the control inputs $\boldsymbol{u}$ can be expressed in terms of unknown values $\boldsymbol{\lambda}_r^*$ of the redundant contact forces:

$$\boldsymbol{u} = \underbrace{\boldsymbol{M}\,\dot{\boldsymbol{q}}_V^* - \boldsymbol{g} - \boldsymbol{\Phi}_{F_f}^{\mathrm{T}}\,\boldsymbol{w}_F}_{\boldsymbol{u}_N} \ - \underbrace{\boldsymbol{\Phi}_{F_r}^{\mathrm{T}}}_{\boldsymbol{U}}\,\boldsymbol{\lambda}_r^* \,. \tag{61}$$

The control inputs $\boldsymbol{u}_N$ are necessary to realize both accelerations $\dot{\boldsymbol{q}}_V^*$ and functional contact forces $\boldsymbol{w}_F$ for vanishing redundant contact forces $\boldsymbol{\lambda}_r^* \equiv \boldsymbol{0}$. The second contribution generates redundant contact forces $\boldsymbol{\lambda}_r^*$. Since $\boldsymbol{\lambda}_r^*$ is not prescribed by the manipulator task, the control inputs $\boldsymbol{u}$ are not uniquely determined.

A special solution for $\boldsymbol{u}$ is obtained, for example, by minimizing an objective function $Z'(\boldsymbol{u})$ constrained by eq. (61). An appropriate objective function is a weighted norm of $\boldsymbol{u}$,

$$Z'(\boldsymbol{u}) = \boldsymbol{u}^{\mathrm{T}}\,\boldsymbol{S}\,\boldsymbol{u} \rightarrow \min_{u} \tag{62}$$

with a symmetric, positive-definite weighting matrix $\boldsymbol{S} \in \mathbb{R}^{p,p}$. Inserting $\boldsymbol{u} = \boldsymbol{u}(\lambda_r^*)$ from eq. (61) into eq. (62) gives the objective function in terms of the independent unknowns $\boldsymbol{\lambda}_r^*$:

$$Z'(\boldsymbol{\lambda}_r^*) = (\boldsymbol{u}_N - \boldsymbol{U}\,\boldsymbol{\lambda}_r^*)^{\mathrm{T}}\,\boldsymbol{S}\,(\boldsymbol{u}_N - \boldsymbol{U}\,\boldsymbol{\lambda}_r^*) \rightarrow \min_{\lambda_r^*} \,. \tag{63}$$

The necessary extremal condition for the optimization problem (63) yields

$$\frac{\partial Z'}{\partial \boldsymbol{\lambda}_r^{*\mathrm{T}}} = 0 : \qquad \boldsymbol{\lambda}_r^* = \underbrace{\left(\boldsymbol{U}^{\mathrm{T}} \boldsymbol{S} \boldsymbol{U}\right)^{-1} \boldsymbol{U}^{\mathrm{T}} \boldsymbol{S}}_{\boldsymbol{U}^{\#}} \, \boldsymbol{u}_N \,. \tag{64}$$

Here, $\boldsymbol{U}^{\#} \in \mathbb{R}^{d-m_F, p}$ ist a left-inverse of $\boldsymbol{U} \in \mathbb{R}^{p, d-m_F}$, i.e. $\boldsymbol{U}^{\#}\boldsymbol{U} = \boldsymbol{I}$. With $\boldsymbol{\lambda}_r^*$ obtained from eq. (64), the control inputs in eq. (61) are

$$\boldsymbol{u} = \boldsymbol{\Gamma} \left(\boldsymbol{M} \, \dot{\boldsymbol{q}}_V^* - \boldsymbol{g} - \boldsymbol{\Phi}_{F_f}^{\mathrm{T}} \, \boldsymbol{w}_F \right) \quad \text{with} \quad \boldsymbol{\Gamma} = \boldsymbol{I} - \boldsymbol{U}\,\boldsymbol{U}^{\#} \in \mathbb{R}^{p,p} \,. \tag{65}$$

4.3.3 Linearizing Feedback. The overall decoupling feedback is obtained by inserting eq. (56) into eq. (65):

$$\begin{aligned} \boldsymbol{u} = & \overbrace{-\boldsymbol{\Gamma}\left(\boldsymbol{M}\,\boldsymbol{\Phi}^{\#}\,\boldsymbol{\xi} + \boldsymbol{g}\right)}^{\boldsymbol{k}(\boldsymbol{q},\dot{\boldsymbol{q}})} + \overbrace{\boldsymbol{\Gamma}\left[\boldsymbol{M}\,\boldsymbol{\Phi}^{\#}\,\boldsymbol{W} \quad -\boldsymbol{\Phi}_{F_f}^{\mathrm{T}}\right]}^{\boldsymbol{K}(\boldsymbol{q})} \begin{bmatrix} \boldsymbol{w}_P \\ \boldsymbol{w}_F \end{bmatrix} \\ & + \underbrace{\boldsymbol{\Gamma}\,\boldsymbol{M}\,(\boldsymbol{I} - \boldsymbol{\Phi}^{\#}\,\boldsymbol{\Phi})\,\boldsymbol{P}^{-1}}_{\boldsymbol{K}^*(\boldsymbol{q})} \; \underbrace{\boldsymbol{p}}_{\boldsymbol{u}^*} \,. \end{aligned} \tag{66}$$

If the corresponding matrices of the mechanical system and feedback (66) coincide, decoupled IO channels

$$\ddot{\boldsymbol{y}}_P = \boldsymbol{w}_P \,, \qquad \boldsymbol{y}_F = \boldsymbol{w}_F \tag{67}$$

are obtained. A block diagram of the overall control structure is shown in Fig. 8.

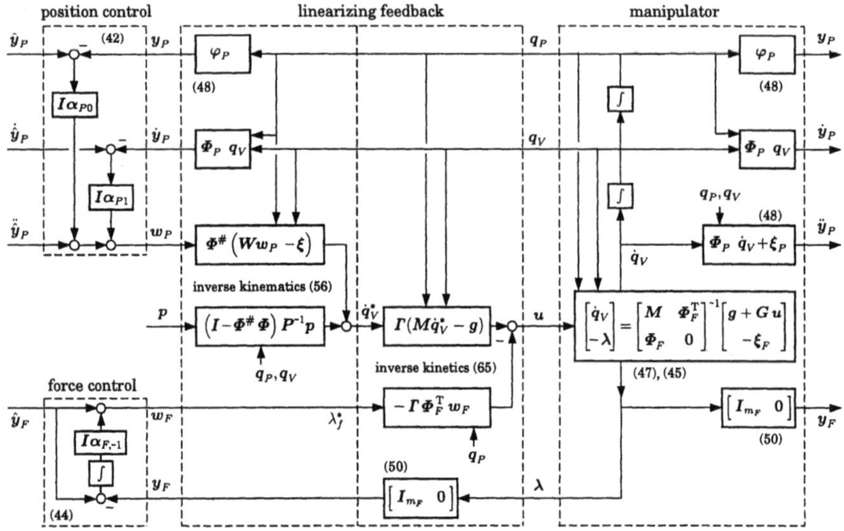

Fig. 8. Nonlinear control structure

4.4 Simulation Results

4.4.1 Control of Redundant Position Variables by Potential Forces. If the constrained manipulator has $f - m_P$ redundant degrees of freedom, the components of the accelerations $\dot{q}_V^*$ from eq. (56) in the $(f - m_P)$-dimensional nullspace of matrix $\boldsymbol{\Phi}$ from eq. (53) are not uniquely determined. They can be prescribed by choice of $\boldsymbol{P}$ and $\boldsymbol{p}$. Since the coordinates of $\boldsymbol{p}$ can be regarded as generalized forces belonging to coordinates $\boldsymbol{q}_P$, they are defined in such a way that they generate a motion of the system towards a desired configuration. This can be achieved by an artificial potential energy field $\Pi(\boldsymbol{q}_P)$ that is minimal in the desired configurations of the manipulator. The forces $\boldsymbol{p}$ then are conservative forces $\boldsymbol{p}_c(\boldsymbol{q}_P)$ belonging to $\Pi(\boldsymbol{q}_P)$ together with dissipative forces $\boldsymbol{p}_d(\boldsymbol{q}_P, \boldsymbol{q}_V)$ for asymptotic stabilization (Khatib 1986, Woernle 1994):

$$\boldsymbol{p}(\boldsymbol{q}_P, \boldsymbol{q}_V) = \underbrace{-\frac{\partial \Pi(\boldsymbol{q}_P)}{\partial \boldsymbol{q}_P^{\mathrm{T}}}}_{\boldsymbol{p}_c} \underbrace{- \begin{bmatrix} k_1\, q_{V1} \\ \vdots \\ k_n\, q_{Vn} \end{bmatrix}}_{\boldsymbol{p}_d}, \quad k_i = \text{const}\,. \tag{68}$$

The definition $\boldsymbol{P} = \boldsymbol{I}$ for the artificial inertia matrix turned out to be reasonable for manipulators with revolute joints. Compared with the definition $\boldsymbol{P} = \boldsymbol{M}(\boldsymbol{q}_P)$, these decoupled 'masses' avoid undesired dynamic couplings between the coordinates that make the dynamic transition to the equilibrium position less predictable, even leading to instability, as reported by Hollerbach and Suh (1987). Examples are discussed below.

a) Repeatable Motions in Joint Space. If there are no other restrictions, it is usually desired to reach the initial joint positions at the end of a closed-loop trajectory of position outputs $\boldsymbol{y}_P$ in order to execute repeated tasks always in the same manner. For kinematic position control (i. e. control at the velocity level), it was discussed, for example, by Klein and Huang (1983), Chung et al. (1992), and de Luca et al. (1992) that pure pseudoinverse feedback does not guarantee repeatable motions. For the dynamic control (i. e. control at the acceleration level) considered in this paper, this phenomenon is illustrated by an example of an unconstrained planar manipulator with $n = 5$ revolute joints (Fig. 9). Defining the coordinates of the tip point as $m_P = 2$ position outputs $\boldsymbol{y}_P$, the manipulator has $f - m_P = 3$ redundant position variables. Repeatable motions are obtained by means of artificial spring forces in the joints that vanish (or are in equilibrium) in the initial configuration $\boldsymbol{q}_P(t_0)$. A possible potential is

$$\Pi(\boldsymbol{q}_P) = \frac{1}{2} \sum_{i=1}^{n} c_i \left[q_{Pi}(t) - q_{Pi}(t_0) \right]^2 , \qquad c_i = \text{const}\,. \tag{69}$$

With constants $c_i = 5\,\mathrm{s}^{-2}$, $k_i = 20\,\mathrm{s}^{-1}$, $i = 1, \ldots, 5$, and $\boldsymbol{P} = \boldsymbol{I}$, repeatable joint motions $\boldsymbol{q}_P(t)$ are achieved while tracking a circular trajectory $\hat{\boldsymbol{y}}_P(t)$ (cycle time 20 s), as depicted in Fig. 9.

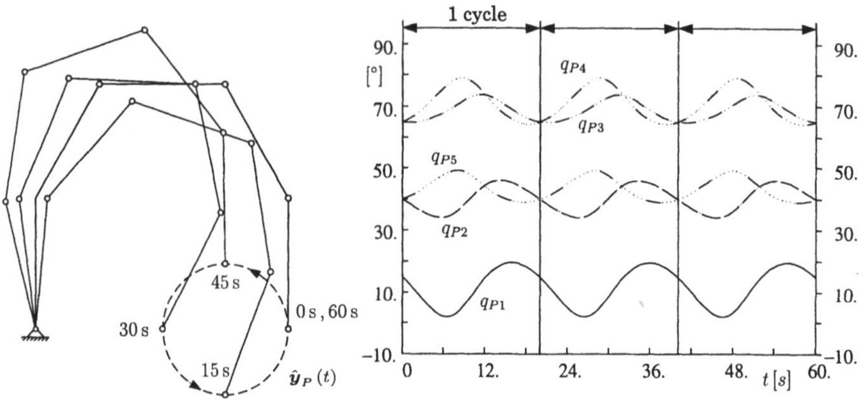

Fig. 9. Repeatable motion by potential forces

b) Collision avoidance. A main advantage of robots with redundant position variables is their ability to avoid collisions with obstacles in the workspace while tracking a desired position output trajectory. This can be achieved by artificial potential forces repulsing the manipulator arm off the obstacle. For that purpose, discrete points are defined on the manipulator. To each point with the absolute position vector $\boldsymbol{r}_i(\boldsymbol{q}_P)$ and the shortest distance $d(\boldsymbol{r}_i)$ from the obstacle a potential $\Pi_i(\boldsymbol{r}_i(\boldsymbol{q}_P))$ is assigned (Khatib 1986):

$$\Pi_i(\boldsymbol{r}_i(\boldsymbol{q}_P)) = \begin{cases} \frac{1}{2}\kappa\left(\frac{1}{d(\boldsymbol{r}_i)} - \frac{1}{d_0}\right)^2, & d(\boldsymbol{r}_i) \leq d_0\,, \\ 0 &, \quad d(\boldsymbol{r}_i) > d_0\,. \end{cases} \tag{70}$$

Here, κ is a scaling factor and d_0 denotes the distance of influence around the obstacle. The overall potential then is the sum of the contributions (70).

To illustrate the effect of this potential, a rectangular obstacle near the second joint is added in the previous example (Fig. 10). Compared with Fig. 9, the joint angles $\boldsymbol{q}_P(t)$ show a reduced amplitude of angle q_{P1} and increased amplitudes of the other angles. Due to the still active potential (69), repeatable joint motions are nevertheless obtained.

4.4.2 Hybrid Control of Position and Force Variables. As an example for hybrid position-and-force control with redundant position and force variables, a planar four-link manipulator with $n = 4$ joint angles $\boldsymbol{q}_P$ controllable by torques $\boldsymbol{u}$ is considered in Fig. 11. The points A and B are constrained by the $d = 2$ surfaces $\varphi_{F1} = 0$ and $\varphi_{F2} = 0$, respectively, leaving $f = 2$ degrees of freedom. The translation $y_P = x$ of point A and the contact force $y_F = \lambda_1$ are defined as $m_P = 1$ position output and $m_F = 1$ force output, respectively. Thus, there are $f - m_P = 1$ redundant position variable and $d - m_F = 1$ redundant contact force λ_2.

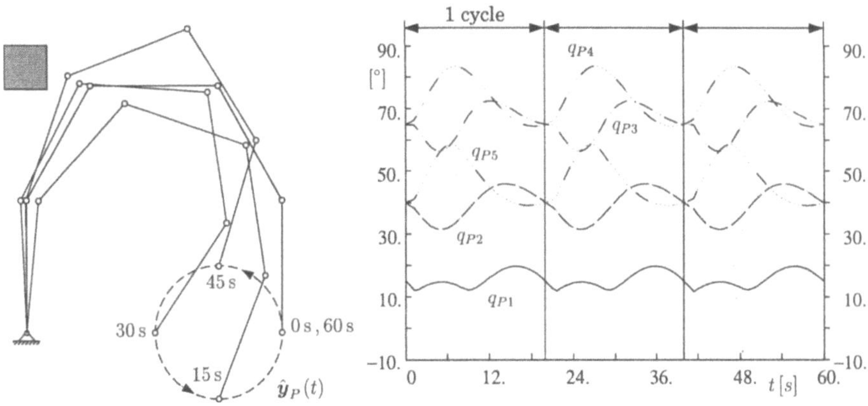

Fig. 10. Repeatable motion and collision avoidance by potential forces

Because of the $f - m_P = 1$ redundant position variable, a potential field according to eq. (69) is defined. The desired functions of the outputs are the motion of point A with constant velocity,

$$\hat{y}_P(t) = v_0\, t + x(t_0) \qquad \text{with} \qquad v_0 = 0.375\,\mathrm{m\,s^{-1}} \,,$$

and a constant contact force

$$\hat{y}_F(t) = 100\,\mathrm{N} \,.$$

The initial position shown in Fig. 11 is defined by

$$\boldsymbol{q}_P(t_0) = [\,30° \;\; 120° \;\; -60° \;\; -90°\,]^{\mathrm{T}} \,, \qquad \boldsymbol{q}_V(t_0) = \boldsymbol{0} \,.$$

With pole-placements $s_{1,2} = (-8.6 \pm 5.0\,\mathrm{i})\ \mathrm{s}^{-1}$ for the position controller and $s = -20.0\ \mathrm{s}^{-1}$ for the force controller, the feedback gains are:

$$\alpha_{P1} = 17.3\,\mathrm{s} \,, \;\; \alpha_{P0} = 100.0 \,, \;\; \alpha_{F,-1} = 10.0\,\mathrm{s}^{-1} \,.$$

The arm lenghts ℓ_i, the masses m_i and the moments of inertia Θ_{Si} with respect to the mass centers in the middle of the links are, for $i = 1, \ldots, 4$,

$$\ell_i = 1.0\,\mathrm{m} \,, \qquad m_i = 10.0\,\mathrm{kg} \,, \qquad \Theta_{Si} = 0.833\,\mathrm{kg\,m^2} \,.$$

In Fig. 11, the motion of the manipulator and the corresponding time histories of the error $e_P(t)$ of the position output, the force output $y_F(t)$, the joint angles $\boldsymbol{q}_P(t)$, and the redundant contact force λ_2 are shown. Further, the contribution $\boldsymbol{u}_N(t)$ of the driving torques necessary for execution of the motion with vanishing redundant contact force λ_2 according to eq. (61) and the actual torques $\boldsymbol{u}(t)$ according to eq. (65) for $\boldsymbol{S} = \boldsymbol{I}$ are represented. Support of the kinematic chain at point B by the contact force λ_2 reduces the torques u_1 and u_2, while u_3 and u_4 are unchanged.

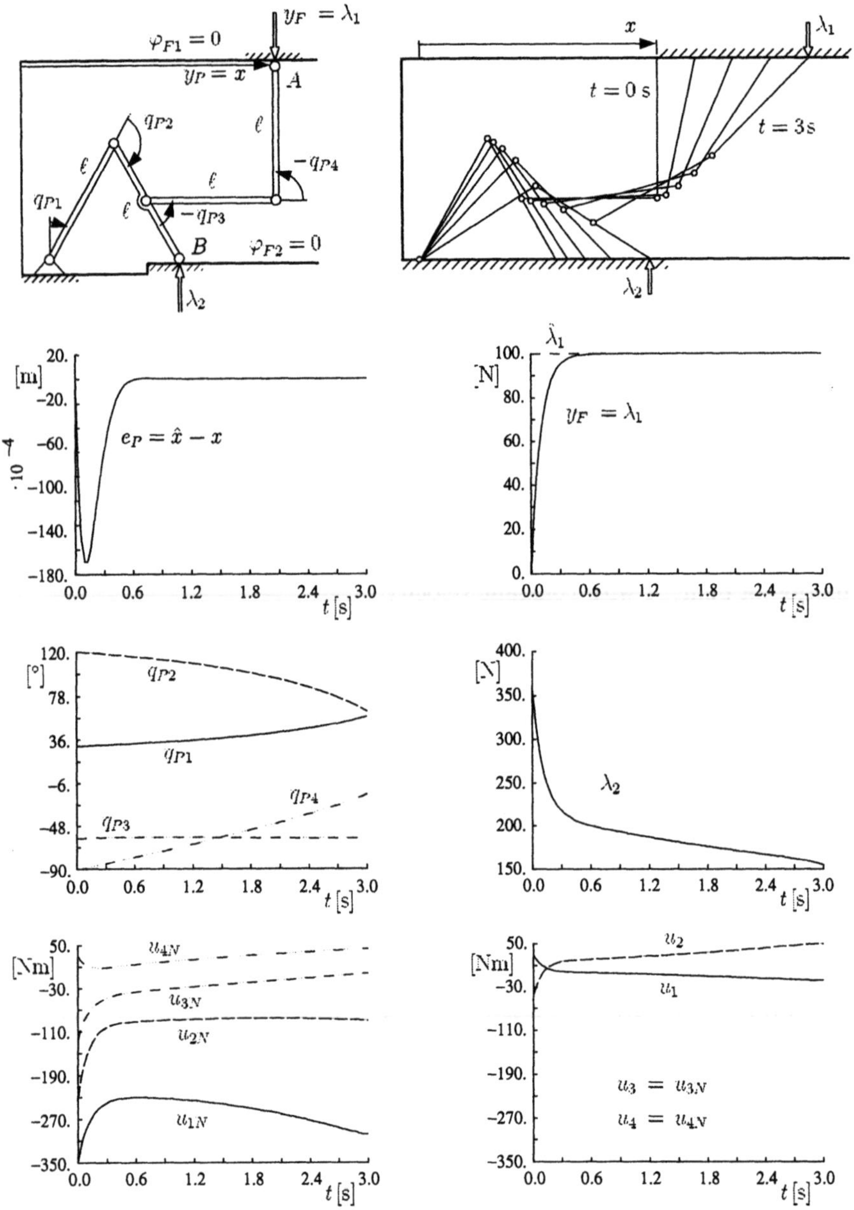

Fig. 11. Example for hybrid position-and-force control with redundant variables

5. Motion Control of a Nonholonomic Mobile Platform

Wheeled mobile robots can be modeled by nonholonomic constraints, if slipping between the ground and the wheels is neglectable because of slow motions. Position-and-velocity control of a mobile platform with two independently driven wheels is regarded in the section below (Woernle 1995).

5.1 Equations of Motion

The mobile platform is supported by two independently driven wheels with coinciding platform-fixed axes and a third passive castor wheel (Fig. 12). Ignoring the passive wheel, the positions of the platform and the driven wheels are described by five position coordinates

$$\boldsymbol{q}_P = \begin{bmatrix} r_{1x} & r_{1y} & \beta & \gamma_1 & \gamma_2 \end{bmatrix}^{\mathrm{T}} \tag{71}$$

with

r_{1x}, r_{1y}, β	-	position coordinates of the platform,
γ_1, γ_2	-	angles of rotation of the driven wheels.

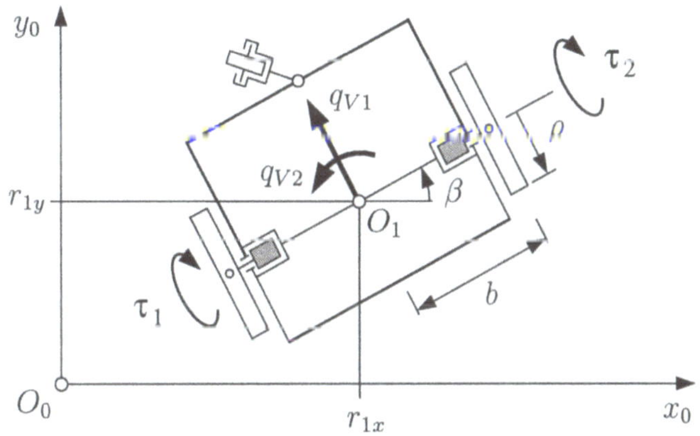

Fig. 12. Model of a nonholonomic mobile platform

The assumption of pure rolling of the wheels leads to a system of three independent constraints at the velocity level. It includes one integrable, holonomic constraint and two nonintegrable, nonholonomic constraints, as described, for example, by Campion et al. (1991) and d'Andréa-Novel et al. (1991). Thus, there are two independent velocities that are appropriately defined as q_{V1} (speed of reference point O_1) and $q_{V2} = \dot{\beta}$ (angular velocity around the vertical z-axis). The time derivatives of the five position coordinates are expressed by the explicit constraint equations (distance of the wheels $2b$, wheel radius ρ), as per (1),

$$\begin{bmatrix} \dot{r}_{1x} \\ \dot{r}_{1y} \\ \dot{\beta} \\ \dot{\gamma}_1 \\ \dot{\gamma}_2 \end{bmatrix} = \begin{bmatrix} -\sin\beta & 0 \\ \cos\beta & 0 \\ 0 & 1 \\ \frac{1}{\rho} & \frac{b}{\rho} \\ \frac{1}{\rho} & -\frac{b}{\rho} \end{bmatrix} \begin{bmatrix} q_{V1} \\ q_{V2} \end{bmatrix} \qquad (72)$$

$$\dot{\boldsymbol{q}}_P \quad = \quad \boldsymbol{H}(\boldsymbol{q}_P) \qquad \boldsymbol{q}_V \; .$$

With driving torques $\boldsymbol{\tau} = \begin{bmatrix} \tau_1 & \tau_2 \end{bmatrix}^{\mathrm{T}}$, the dynamic equations of motion are

$$\boldsymbol{M}(\boldsymbol{q}_P)\,\dot{\boldsymbol{q}}_V = \boldsymbol{g}(\boldsymbol{q}_P, \boldsymbol{q}_V) + \boldsymbol{G}(\boldsymbol{q}_P)\,\boldsymbol{\tau} \, . \qquad (73)$$

The equations for the angles of rotation of the wheels γ_1 and γ_2 in eq. (72) are not essential for the motion of the platform as long as the generalized forces in eq. (73) do not depend on γ_1 and γ_2. This holds in most applications.

5.2 Output Variables for Path Following

It can be shown that the nonholonomic system (72), (73), although it is completely controllable, cannot be asymptotically stabilized to a rest position in all state variables by means of a smooth feedback law (Brockett 1983). Nevertheless, motion control along prescribed paths with desired velocity is possible, whereby two output variables are independently controlled, according to the $p = 2$ available inputs, as discussed, for example, by d'Andréa-Novel et al. (1991), d'Andréa-Novel et al. (1992), Sarkar et al. (1994). Two formulations of that control problem are (Woernle 1995):

- *Trajectory tracking* (Fig. 13a). A platform-fixed point Q is made track a point $\hat{Q}$ with prescribed motion $\hat{\boldsymbol{r}}_Q(t)$. Thus, $m_P = 2$ position outputs $y_{P1} = r_{Qx}$, $y_{P2} = r_{Qy}$ are defined. If the distance h between Q and the wheel axis is nonzero, a static decoupling feedback exists; otherwise, a dynamic feedback has to be used, as described by d'Andréa-Novel et al. (1992).
- *Path following* (Fig. 13b). The control objective is to make the reference point $Q \equiv O_1$ track a given path with desired speed (Woernle 1995a). The path is defined as a planar curve in parameter form (parameter μ) independently from the desired speed. The output variables are the shortest distance d of O_1 from the path and the speed q_{V1} of O_1 (forward speed of the platform), as depicted in Fig. 14. These are $m_P = 1$ position output and $m_V = 1$ velocity output, namely,

$$y_P = d(\boldsymbol{q}_P)\,, \qquad (74)$$

$$y_V = q_{V1}\,. \qquad (75)$$

5.3 IO Linearization by Dynamic Feedback

5.3.1 Compensation of the Dynamic Nonlinearities. A feedback to compensate the dynamic nonlinearities is obtained by solving the dynamic

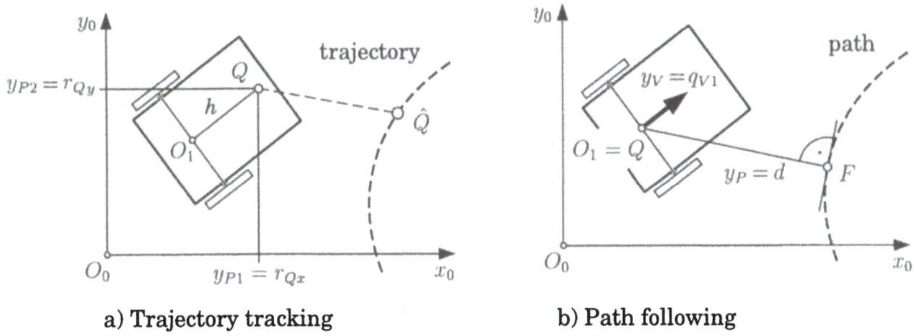

Fig. 13. Control tasks

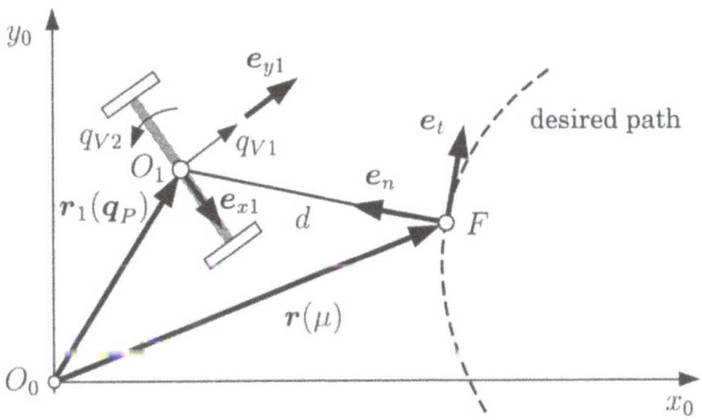

Fig. 14. Definitions of control outputs

equations of motion (73) with respect to the driving torques $\boldsymbol{\tau}$, whereby the accelerations $\dot{\boldsymbol{q}}_V$ are replaced by the new inputs $\boldsymbol{u}$:

$$\boldsymbol{\tau} = \boldsymbol{G}^{-1}(\boldsymbol{M}\,\boldsymbol{u} - \boldsymbol{g}) \; . \tag{76}$$

This state feedback transforms system (72), (73) into

$$\begin{bmatrix} \dot{\boldsymbol{q}}_P \\ \dot{\boldsymbol{q}}_V \end{bmatrix} = \begin{bmatrix} \boldsymbol{H}(\boldsymbol{q}_P)\,\boldsymbol{q}_V \\ \boldsymbol{0} \end{bmatrix} + \begin{bmatrix} \boldsymbol{0} \\ \boldsymbol{I} \end{bmatrix} \boldsymbol{u} \, . \tag{77}$$

5.3.2 IO Equation. The outputs y_P from eq. (74) and y_V from eq. (75) are differentiated with respect to time along the trajectories of eq. (77) until at least one component of the inputs $\boldsymbol{u}$ appears explicitly. This leads to the IO-equation, compare with eq. (32),

$$\begin{aligned} \begin{bmatrix} \ddot{y}_P \\ \dot{y}_V \end{bmatrix} &= \begin{bmatrix} c_P(\boldsymbol{q}_P, \boldsymbol{q}_V) \\ 0 \end{bmatrix} + \begin{bmatrix} \boldsymbol{e}_2^{\mathrm{T}} \boldsymbol{e}_n & 0 \\ 1 & 0 \end{bmatrix} \begin{bmatrix} u_1 \\ u_2 \end{bmatrix} \\ &= \quad \boldsymbol{c}(\boldsymbol{q}_P, \boldsymbol{q}_V) \quad + \quad \boldsymbol{D}(\boldsymbol{q}_P) \qquad \boldsymbol{u} \quad . \end{aligned} \tag{78}$$

Since the IO matrix $\boldsymbol{D}$ is singular, the solution with respect to $\boldsymbol{u}$ is not possible, and a static decoupling feedback does not exist – refer to Section 3.2.

5.3.3 Dynamic Extension. The idea of dynamic extension is to define new inputs that do not explicitly appear in the IO equation with singular IO matrix. The differentiation process can then be continued (Isidori 1995). For the given system (77), this is simply achieved by positioning an integrator before the input u_1 that appears explicitly in the IO equation (78). Thus, new inputs $\bar{u}_1$, $\bar{u}_2$ are defined by (additional state variable η), namely,

$$\begin{bmatrix} u_1 \\ u_2 \end{bmatrix} = \begin{bmatrix} \eta \\ \bar{u}_2 \end{bmatrix} \quad \text{with} \quad \dot{\eta} = \bar{u}_1 \,. \tag{79}$$

Further differentiation of eq. (78) with respect to time yields under consideration of system (77) extended by eq. (79) an IO equation for the new inputs $\bar{u}_1$, $\bar{u}_2$, i.e.,

$$\begin{aligned} \begin{bmatrix} y_P^{(3)} \\ \ddot{y}_V \end{bmatrix} &= \begin{bmatrix} \bar{c}_P(\boldsymbol{q}_P, \boldsymbol{q}_V, \eta) \\ 0 \end{bmatrix} + \begin{bmatrix} \boldsymbol{e}_2^{\mathrm{T}} \boldsymbol{e}_n & -\boldsymbol{e}_1^{\mathrm{T}} \boldsymbol{e}_n \, q_{V1} \\ 1 & 0 \end{bmatrix} \begin{bmatrix} \bar{u}_1 \\ \bar{u}_2 \end{bmatrix} \\ &= \quad \bar{\boldsymbol{c}}(\boldsymbol{q}_P, \boldsymbol{q}_V, \eta) \quad + \quad \bar{\boldsymbol{D}}(\boldsymbol{q}_P, q_{V1}) \quad \bar{\boldsymbol{u}} \,. \end{aligned} \tag{80}$$

The IO matrix $\bar{\boldsymbol{D}}$ is regular except for the following cases:

- $q_{V1} = 0$, vanishing speed of the platform. The path-following control for point O_1 cannot be applied for motions into rest positions. The same restriction is found for trajectory control of reference point O_1, as treated by d'Andréa-Novel et al. (1992).
- $\boldsymbol{e}_1^{\mathrm{T}} \boldsymbol{e}_n = 0$, the platform must not be positioned with its wheel axis $\boldsymbol{e}_1$ tangential to the path as depicted in Fig. 14. The same restriction holds for path-following control of a reference point not located on the wheel axis, as discussed by Sarkar et al. (1994).

Except for these cases, a decoupling and linearizing feedback is obtained. The partial relative degrees of the outputs y_P and y_V are $\delta_P = 3$ and $\delta_V = 2$, respectively.

5.3.4 Linearizing State Feedback. To obtain the linearizing state feedback, the IO equation (80) is solved with respect to the inputs $\bar{\boldsymbol{u}}$. The time derivatives $y_P^{(3)}$ and $\ddot{y}_V$ of the outputs are hereby replaced by new inputs $\boldsymbol{w} = \begin{bmatrix} w_P & w_V \end{bmatrix}^{\mathrm{T}}$, respectively:

$$\begin{aligned} \bar{\boldsymbol{u}} &= \bar{\boldsymbol{D}}^{-1} (\boldsymbol{w} - \bar{\boldsymbol{c}}) \\ &= \bar{\boldsymbol{u}}(\boldsymbol{q}_P, \boldsymbol{q}_V, \eta, \boldsymbol{w}) \,. \end{aligned} \tag{81}$$

If the matrices in eqs. (80) and (81) coincide, a linear and decoupled system is obtained that can be asymptotically stabilized by feedbacks (42) and (43), respectively,

$$y_P^{(3)} = w_P \,, \tag{82}$$

$$\ddot{y}_V = w_V \,. \tag{83}$$

5.4 Simulation Results

As an example, a simulation of the motion of the platform along a circular path (radius R) is considered. The initial position is not located on the path (Fig. 15). The desired time functions for the outputs are $\hat{y}_P(t) = \hat{d}(t) \equiv 0$ (vanishing distance) and $\hat{y}_V(t) = \hat{q}_{V1}(t) \equiv \omega_0 R$ with $\omega_0 = 1.0\,\mathrm{s}^{-1}$ (constant speed). The feedback gains are determined according to assumed pole-placements

$$s_1 = -2\,\mathrm{s}^{-1}\,, \qquad s_{2,3} = (-1.6 \pm 1.2\,\mathrm{i})\,\mathrm{s}^{-1}\,, \tag{84}$$

$$s_{1,2} = (-1.6 \pm 1.2\,\mathrm{i})\,\mathrm{s}^{-1} \tag{85}$$

for the control loops belonging to the IO channels (82) and (83), respectively. Assuming an initial speed $\hat{q}_{V1}(0) = 0.4\,\omega_0 R$, the trajectory of point O_1 in Fig. 15 shows that the vehicle moves asymptotically towards the circular path. Due to the nonholonomic constraint, the wheel axis is always orthogonal to the trajectory of O_1. Thus, the orientation of the platform is uniquely defined by the path. This is an advantage of the formulation introduced here, compared with path-following control for a reference point not located on the wheel axis, as discussed by Sarkar et al. (1994). However, it is necessary to switch over e.g. to that control law for motions into rest positions.

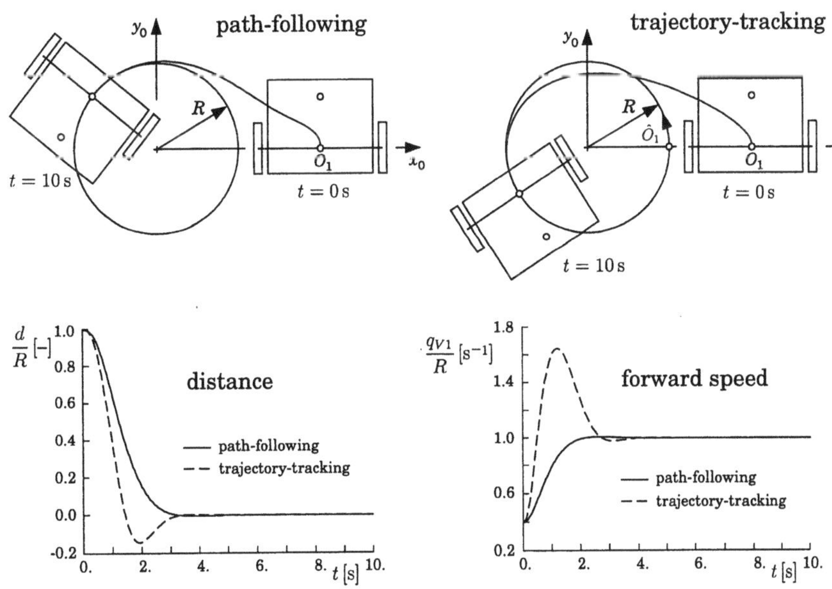

Fig. 15. Path following and trajectory tracking

The path-following control is compared with the trajectory-tracking control, where the coordinates r_{1x} and r_{1y} of point O_1 are made track desired time functions $\hat{r}_{1x}(t) = R\cos\omega_0 t$ and $\hat{r}_{1y}(t) = R\sin\omega_0 t$ with $\omega_0 = 1.0\,\mathrm{s}^{-1}$, as depicted in Fig. 13. IO linearization for this control problem leads to two triple-integrating IO channels, as investigated by d'Andréa-Novel et al. (1992). With assigned poles according to eq. (84) for the control loops belonging to outputs y_{P1} and y_{P2}, respectively, the platform moves asymptotically towards the circular trajectory too (Fig. 15). In order to catch up the moving target point $\hat{O}_1$, the forward velocity has to be further increased compared with the path-following control. In general, path-following control is more appropriate compared with trajectory-tracking control, if a given path is to be tracked with a desired speed, and if it it is not necessary to reach a point on the path at a particular instant.

6. Conclusion

Control of position, velocity, and contact forces in robotic systems by means of exact IO linearization has been discussed in this chapter. By means of a nonlinear state feedback, the nonlinearities of the mechanical system are compensated, and a linear input-output behavior for the task-specific control outputs is obtained. Here, the crucial step is to find an invertible input-output matrix. If it exists, static (algebraic) feedbacks described by algebraic equations or dynamic feedbacks described by differential equations are obtained for exact IO linearization.

For manipulators that are constrained by contacts with the environment, redundancy in both position coordinates and contact forces is defined. By means of suitable linearizing feedbacks, it is possible to achieve secondary goals such as obstacle avoidance or minimizing driving forces, while the control outputs are asymptotically-stable tracked. To make a nonholonomic mobile platform follow a desired path, a dynamic state feedback is necessary.

Ongoing work is devoted to the control of robotic systems with higher-order relative degrees that are caused, for example, by elastic transmission elements in the drives. Further, control strategies for underwater robots that have similar control properties to the nonholonomic platform discussed here are developed.

References

Brockett, R.W. (1983): Asymptotic Stability and Feedback Stabilization. In: R.W. Brockett et al. (Eds.): Differential Control Theory, Birkhäuser, Boston, 181–191

d'Andréa-Novel, B., Bastin, G., and Campion, G. (1991): Modeling and Control of Non-Holonomic Wheeled Mobile Robots. Proc. 1991 IEEE Int. Conference on Robotics and Automation, Sacramento, 1130–1135 d'Andréa-Novel, B., Bastin, G., and Campion, G. (1992): Dynamic Feedback Linearization of Non-Holonomic Wheeled Mobile Robots. Proc. 1992 IEEE Int. Conf. on Robotics and Automation, Nice, 2527–2532

Campion, G., d'Andréa-Novel, B., and Bastin, B. (1991): Controllability and State Feedback Stabilization of Nonholonomic Mechanical Systems. In: C. Canudas de Wit (Ed.): Advanced Robot Control. Springer-Verlag, Berlin

Chung, Y. S., Griffis, M., and Duffy, J. (1992): Unique Joint Displacement Generation for Redundant Robotic Systems. Proc. 22nd ASME Mechanisms Conference, Phoenix, 1992, DE-Vol. 45, 637–641

Freund, E. (1975): The Structure of Decoupled Nonlinear Systems. International Journal of Control **21**, 443–450

Hollerbach, M.J. and Suh, K.C. (1987): Redundancy Resolution of Manipulators through Torque Optimization. IEEE J. Rob. Autom. **3**, 308–316

Isidori, A. (1989): Nonlinear Control Systems. Springer-Verlag, Berlin

Khatib, O. (1986): Real-time Obstacle Avoidance for Manipulators and Mobile Robots. The International Journal of Robotics Research, **5**, 90–98

Khatib, O. (1987): A Unified Approach for Motion and Force Control of Robot Manipulators: The Operational Space Formulation. IEEE Journal of Robotics and Automation **3**, 43–53

Klein, C.A. and Huang, C.H. (1983): Review of Pseudoinverse Control for Use with Kinematically Redundant Manipulators. IEEE Transactions on Systems, Man, and Cybernetics **13**, 245–250

Kreutz, K. (1989): On Manipulator Control by Exact Linearization. IEEE Transactions on Automatic Control **34**, 763–767

de Luca, A., Lanari, L., and Oriolo, G. (1992): Control of Redundant Robots on Cyclic Trajectories. Proc. 1992 IEEE Int. Conf. on Robotics and Automation, Nice, 500–506

Nijmeijer, H., and v. d. Schaft, A.J. (1990) Nonlinear Dynamical Control Systems. Springer-Verlag, Berlin

Sarkar, N., Yun, X., and Kumar, V. (1994): Control of Mechanical Systems With Rolling Constraints: Application to Dynamic Control of Mobile Robots. The International Journal of Robotics Research **13**, 55–69.

Shabana, A. (1989): Dynamics of Multibody Systems. John Wiley, New York

Wanner, M.C. (1993): Mechatronic Components for the Aircraft Cleaning System 'SKYWASH'. In Hiller, M. and Fink, B.: Proc. 2nd conference on Mechatronics and Robotics in Moers (Germany), 217–224

Woernle, C. (1992): Position and Force Control of Multibody Systems by Feedback Linearization. Proc. 22nd ASME Mechanisms Conference, Phoenix, DE-Vol. 47, 11–18

Woernle, C. (1994): Control of Constrained Manipulators with Redundant Position Variables and Contact Forces. Proc. 23nd ASME Mechanisms Conference, Minneapolis, DE-Vol. 72, 427–436

Woernle, C. (1995a): Nonlinear Motion Control of a Nonholonomic Mobile Platform by Exact Linearization. Proc. 9th IFToMM World Congress on the Theory of Machines and Mechanisms, Milano, 1995, Vol. 2, 1416–1420

Woernle, C. (1995b): Regelung von Mehrkörpersystemen durch externe Liarisierung. VDI Fortschritt-Berichte Reihe 8, Nr. 517, VDI-Verlag, Düsseldorf

Yoshikawa, T. (1987): Dynamic Hybrid Position/Force Control of Robot Manipulators - Description of Hand Constraints and Calculation of Joint Driving Force. IEEE Journal of Robotics and Automation **3**, 386–392

Zimmermann, M. and West, H. (1990): Dynamics and Sensor-Based Control of Manipulators in Contact with the Environment. Proc. 8th CISM-IFToMM Symposium Ro.Man.Sy.'90, Cracow, 410–417

PART III
Dynamics of Flexible Multibody Systems

Flexible Bodies in Multibody Systems

Richard Schwertassek

Institute for Robotics and System Dynamics, DLR, D-82234 Oberpfaffenhofen, Germany
Richard.Schwertassek@dlr.de

Summary. The motion of flexible bodies in multibody systems can be represented in many applications as a superposition of a large reference motion and small deformations, which allows linearisation in the deformation variables. The linearisation procedure requires a careful analysis of geometric stiffening. The consideration of the effect is discussed for general flexible structures here including theoretical background, computer implementation and demonstration of the necessity of its consideration in specific examples.

Keywords. Flexible multibody systems, geometric stiffening.

1. Introduction

Current design and analysis of engineering systems from a wide variety of applications is based on digital computer simulation. An interdisciplinary effort of various disciplines such as dynamics, control system design, numerical analysis, optimisation, computer science, etc. is required for the development of the corresponding software tools. Dynamic system simulation usually is based on detailed system models resulting in complex system equations. By the middle of this century it was proposed for the first time to use the computer not only for the solution but also for the generation of the complex system equations. A computer-based generation of mechanical system equations requires a systematic methodology for their derivation, which means that the corresponding methods, proposed before, have to be modified and adapted to exploit the specific strength of the computer. The idea resulted in the appearance of a new discipline called computational mechanics.

Initially, computational mechanics had been a synonym for the Finite Element Method (FEM). The development of multibody system dynamics started in the mid-sixties, some ten years after the first publications on finite element methodologies. The abbreviation MBS is used for both, multibody system and multibody simulation. Often, the question arises about the relative advantages of the two methodologies, FEM and MBS, and about their typical areas of application. Multibody simulation is of particular value when dealing with system models, in which the *flexibility is concentrated* in interconnections between more or less rigid bodies. The early applications of the method have been to *dynamical* problems exclusively. By contrast, the Finite Element Method has its origins in the analysis of the *statics* of structures, where the *flexibility is distributed* over the entire system. Recent work in the area of computational mechanics is dealing increasingly with attempts to unify the two approaches and the corresponding software tools.

Considering the history of mechanics, the term computational mechanics arises quite naturally to describe the latest period of the development. The seventeenth century saw the rise of classical mechanics. The corresponding concepts and methodologies had matured by the end of the last century, but this does not imply that mechanics was going to turn into a dying science. Rather, the main stimulus began to shift by that time from theory to application from science to technology, as new problems needing analysis arose in engineering applications. Besides classical or analytical mechanics "applied mechanics" became a separate area of research. Special techniques for handling linear vibration problems were developed during the first decades of this century and they were in general use, although the computational burden of linear algebra limited their practical application to small-scale systems. Since the fourth decade of this century there was also growing interest in nonlinear systems but the mathematical difficulties of nonlinear mechanics focused mostly on the development of approximate literal or numerical methods and confined solutions largely to one-degree-of-freedom systems. Applications to engineering problems were rare. In the fifties the situation changed completely due to the rise of the digital computer, which broke down the computational barriers. Now the methods of engineering mechanics had to be adapted to the way computers work and the new discipline of computational mechanics was born.

The main steps of these developments in the two areas of multibody dynamics and finite element analysis have been described in the literature, e.g. by Knothe and Wessels (1991) and Roberson (1977). The foundations of multibody dynamics, the basic laws of translational and rotational motion, are due to Newton and Euler (Truesdell 1964); in 1788 Lagrange contributed the statement on the behaviour of the constraint forces – d'Alembert's principle in Lagrangian form – which is required to study general constrained motion (Szabo 1979). By the end of the last century, Resal formulated, probably for the first time, multibody system equations, but without going beyond writing them down. The first practical application of multibody dynamics is in biomechanics and it is due to Fischer (1906, 1907). From a modern point of view his motivation is strange – he received support from the army, which was interested in answering questions such as how a soldier might salute in a most comfortable way (Braune and Fischer 1889). Fischer had all the equations, redeveloped some sixty years later, but without a computer he could not capitalise on his work. Aerospace and vehicle dynamics problems motivated new interest in an analysis of multibody systems in the mid-sixties as described by Roberson (1977) and Schwertassek and Roberson (1986). The development of the finite element method followed similar routes stimulated by needs in civil engineering and aircraft design. An excellent description of the development of the foundations of the method, in the realm of the theory of elasticity, may be found in (Love 1944) and as well in (Szabo 1979). Information about the rise of the method itself is given in the introduction of (Knothe and Wessels 1991).

Multibody systems were the focus of a special issue of a journal dedicated to the 65th birthday of R. E. Roberson (Likins, 1989). An analysis of such systems requires the knowledge of the system motion and of the resulting forces, including constraint forces and internal forces due to body deformation. All of these systems can be mapped onto the general multibody system model shown in Fig. 1. The elements of the model are a global reference frame, rigid or flexible bodies and (massless) interconnections, which are subdivided into force elements and joints. The former are used to model the applied interaction forces between the bodies, whereas the latter represent the devices constraining relative motion. Multibody formalisms are computer oriented procedures for generating the equations of motion for such models, based on data, which describe the system elements and the system topology (García de Jalón and Bayo 1994, Haug 1989, Huston 1990, Nikravesh 1988, Roberson and Schwertassek 1988, Schiehlen 1986, Shabana 1989, and Wittenburg 1977).

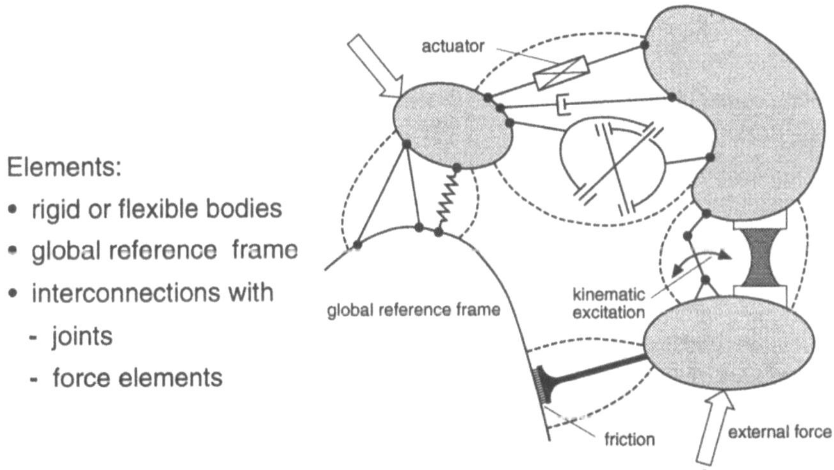

Fig. 1. General multibody system model.

The notation used here is as described in (Roberson and Schwertassek 1988) with minor modifications to distinguish matrices from scalars and single variables: Vectors considered as invariants are represented by underlined letters from any alphabet and matrices are denoted by boldface letters. Vector arrays, a convenient device to represent the resolution of vectors in a specific basis, are denoted by underlined boldface letters. Greek subscripts always range from 1 to 3.

Applying concepts detailed in (Roberson and Schwertassek 1988), the equations of motion of a multibody system with N rigid bodies and N_I interconnections are found to be

$$\left.\begin{aligned} &\dot{\mathbf{p}}_A = \mathbf{Y}_A\,\mathbf{g}_A && \text{(a)}\\ &\mathbf{J}_I\,\dot{\mathbf{g}}_A = \boldsymbol{\chi} + \boldsymbol{\lambda}_A + \mathbf{W}_I\,\boldsymbol{\lambda}_C + \mathbf{W}_{II}\,\bar{\boldsymbol{\lambda}}_C && \text{(b)}\\ &\mathbf{f}_C(\mathbf{p}_A, t) = \mathbf{0} \quad \text{and} \quad \mathbf{W}_{II}^T\,\mathbf{g}_A = \mathbf{Z} && \text{(c)} \end{aligned}\right\} \qquad (1)$$

Here

$$\left.\begin{array}{l}\mathbf{p}_A=\left[\mathbf{p}^a\right]\\ \mathbf{g}_A=\left[\mathbf{g}^a\right]\end{array}\right\}\quad a=1,2,\ldots N,\quad \text{where}\quad \left\{\begin{array}{l}\mathbf{p}^a=\left[p_j^a\right]\\ \mathbf{g}^a=\left[g_j^a\right]\end{array}\right.\quad j=1,2,\ldots n_f^a \tag{2}$$

are minimum sets of variables describing relative position and velocity of the system bodies across the N "primary" interconnections a, i.e. the interconnections in a spanning tree of the system graph, and n_f^a is the number of degrees of freedom in the relative motion of the two bodies interacting across the interconnection a. Matrices

$$\bar{\boldsymbol{\lambda}}_C=\left[\bar{\boldsymbol{\lambda}}^c\right],\quad c=N+1,N+2,\ldots N_I,\quad \text{where}\quad \bar{\boldsymbol{\lambda}}^c=\left[\lambda_j^c\right], j=1,2,\ldots n_c^c \tag{3}$$

contain constraint forces across the "secondary" interconnections, which close the loops of the system graph and n_c^c is the number of constrained modes across an interconnection c. Matrices

$$\boldsymbol{\lambda}_A=\left[\boldsymbol{\lambda}^a\right],\quad a=1,2,\ldots N,\quad \boldsymbol{\lambda}_C=\left[\boldsymbol{\lambda}^c\right],\quad c=N+1,N+2,\ldots N_I \tag{4}$$

contain generalised applied forces across both types of interconnections. Detailed expressions for the system matrices $\mathbf{Y}_A$, $\mathbf{J}_I$, $\boldsymbol{\chi}$, $\mathbf{W}_I$, $\mathbf{W}_{II}$, $\mathbf{f}_C$ and $\mathbf{Z}$, appearing in the kinematical and in the dynamical equations (1.a,b) and in the loop-closure-conditions (1.c), are given in Roberson and Schwertassek, 1988. Based on these expressions they can be computed, given the system-data and the system-state described by $\mathbf{p}_A$, $\mathbf{g}_A$ and $\bar{\boldsymbol{\lambda}}_C$. Interpretations of the various terms appearing in the system matrices have been detailed by Roberson and Schwertassek (1988) and by Wittenburg (1977) and in the literature cited there.

It was stated in 1989 that the "beautiful" equations (1) are not very useful for multibody analysis – see (Likins 1989). The reason behind this statement is the appearance of so called $O(N)$-formalisms. A numerical integration of the system equations (1) requires the inversion of the generalised mass matrix $\mathbf{J}_I$ resulting in a computational effort, which grows with the third power of the number N of system bodies. By contrast, the $O(N)$-formalisms generate the explicit form of the dynamical system equations (1.b), i.e. the equations

$$\dot{\mathbf{g}}_A=\mathbf{J}_I^{-1}\left(\boldsymbol{\chi}+\boldsymbol{\lambda}_A+\mathbf{W}_I\,\boldsymbol{\lambda}_C+\mathbf{W}_{II}\,\bar{\boldsymbol{\lambda}}_C\right) \tag{5}$$

with a number of operations growing linearly with N. The development of $O(N)$-formalisms as detailed in Schwertassek and Rulka, 1991, is not a very glorious part of the history of multibody dynamics. The method was published by Vereshchagin (1974) but the importance of this paper was not appreciated. In the mid eighties the formulation was reinvented by various authors and only then it made its way to contribute to the improvement of multibody simulation codes. An additional enhancement of the methodology may be obtained by tuning the generation and the numerical solution of multibody system equations properly, as proposed by Eichberger (1993, 1994) and Eichberger et al. (1994).

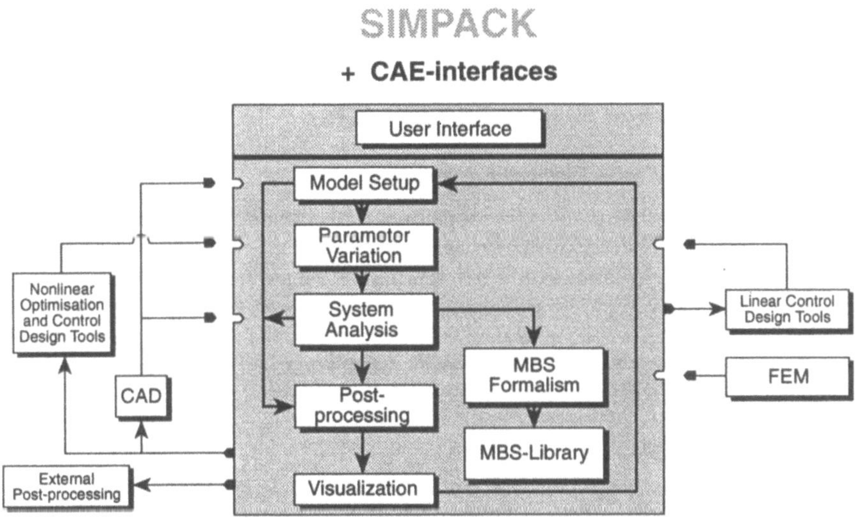

Fig. 2. General structure of the multibody code SIMPACK and its interfaces

The formalisms mentioned here are the kernel of the multibody simulation code SIMPACK – see (Rulka 1990, Rulka and Eichberger 1993 and Kortüm et al. 1995). Multibody simulation is not the only tool required for analysis and design of engineering systems. SIMPACK has thus been designed as an open system, which is able to communicate with other software such as control-system design, optimisation or CAD tools – see Fig. 2. Such interfaces are visualised in the figure by plugs. One of these plugs – the interface to FEM-software – will be discussed in Sect. 2.

2. Flexible Bodies in Multibody Systems

2.1 Background

The initial proposals for the modeling of flexible bodies in multibody systems date back to the early seventies, but the subject attracted not too much attention in those days. In Kane et al. (1987) it was pointed out that some of the most popular multibody codes did yield wrong results because of forgetting so called geometric stiffening terms, when modeling the deformation of flexible bodies superimposed on a large reference- or rigid-body motion. The publication stimulated a series of workshops on the subject (Man and Laskin 1988, Taylor 1988). A survey reveals three methodologies being pursued for the modeling of flexible bodies – see (Shabana 1996 a and b 1997, and Shabana and Schwertassek 1997):

2.1.1 Floating Frame of Reference Formulation

The flexible body motion is represented as a superposition of a large reference motion and small deformations, which allows the linearisation of the equations of motion in the deformation variables. This is the most common procedure applied in multibody simulation. An incorrect linearisation resulted in overlooking the geometric stiffening terms in some cases. The method yields a correct mass matrix – a necessary statement, as will be seen immediately.

2.1.2 Incremental Finite Element Formulation

This is the common finite element approach. The motion of a structure is represented in terms of nodal coordinates, which represent a rigid body motion in case of beam and plate elements to a linear approximation only. As a consequence one does not obtain a correct representation of generalised masses when using such a model for large-deformation dynamics – see (Shabana 1996 b, Shabana and Schwertassek 1997). This disadvantage is not relevant when dealing with static problems, for which the method has been developed initially.

2.1.3 Large Rotation Vector Approach

The problems of the incremental finite element formulation are avoided in a methodology, initially proposed by Simo (1985) and Simo and Vu-Quoc (1986). Models of beams and plates are introduced, in which the shape functions allow a representation of large rigid-body motion. But the method brings about a problem, too: The application of the models to beams without shear deformation seems to be complicated because the coordinates introduced to represent the rotation of the cross sections of the beam become redundant in such cases.

2.1.4 Summary

Flexible-body modeling is an area of current research interest. The analysis of *small* deformation problems can be done using the floating frame of reference formulation. The approach requires a careful analysis of geometric stiffening. *Large* deformation dynamic analysis is still a research area. In particular, there is a fundamental problem which remains to be solved: the establishment of the relationships between the different methodologies used in flexible multibody dynamics and a clear definition of the underlying assumptions as required to develop guidelines indicating which method is most appropriate for a specific application.

2.2 Floating Frame of Reference Formulation

In many applications of multibody simulation a small deformation of specific system bodies superimposed on a large reference motion needs to be considered. Linearising the equations of motion in the deformation variables yields computational advantages as will be demonstrated by examples. The following topics will be discussed now:

- representation of system motion;
- data describing a flexible body as required for multibody simulation codes and resulting interfaces and pre-processors;

- general methodology to include geometric stiffening and clarification of its importance by examples.

The details of how to derive the linearised flexible-body equations containing the geometric stiffening terms will be discussed in Section 3.

2.2.1 Motion Representation

The representation of the motion of a flexible body in a multibody system suitable for dealing with the above-mentioned problems will be introduced now for the example of an Euler-Bernoulli beam (with a straight reference configuration). Figure 3 shows the model and the notation used to represent the motion of the points P of the beam with respect to an inertial frame $\{O^I, \underline{\mathbf{e}}^I\}$.

Considering the beam as a body i of a multibody system the beam reference frame is denoted by $\{O^i, \underline{\mathbf{e}}^i\}$. Its translational motion is described by

$$\underline{\rho}^i = \underline{\mathbf{e}}^{i^T} \boldsymbol{\rho}^i, \quad \boldsymbol{\rho}^i = \left[\rho^i_\alpha\right] \text{ and } \underline{\mathbf{e}}^i = \left[\underline{e}^i_\alpha\right] \tag{6}$$

and the rotation of the beam reference frame is given in terms of the rotation matrix $\mathbf{A}^i$ representing the orientation of $\underline{\mathbf{e}}^i$ with respect to $\underline{\mathbf{e}}^I$ as

$$\underline{\mathbf{e}}^i = \mathbf{A}^i \underline{\mathbf{e}}^I, \quad \mathbf{A}^i = \left[A^i_{\alpha\beta}\right] \text{ and } \underline{\mathbf{e}}^I = \left[\underline{e}^I_\alpha\right]. \tag{7}$$

A representative point P of the beam is identified by the coordinates $\mathbf{R}$ of the vector

$$\underline{R} = \underline{\mathbf{e}}^{i^T} \mathbf{R}, \quad \mathbf{R} = \left[R_\alpha\right] \tag{8}$$

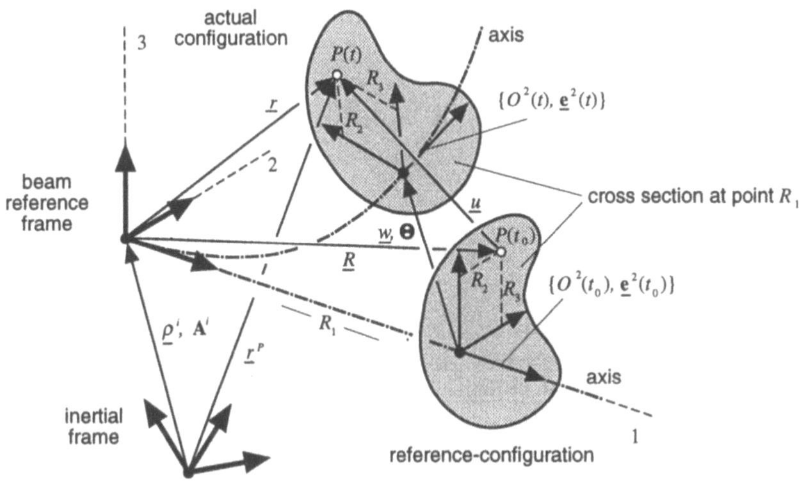

Fig. 3. A straight Euler-Bernoulli beam and notation representing its motion.

specifying the location of P in the reference configuration. In an actual configuration the location of P with respect to the inertial frame $\{O^I, \underline{\mathbf{e}}^I\}$ is given by

$$\underline{r}^P = \underline{\mathbf{e}}^{I^T}\ {}^I\mathbf{r}^P = \underline{\mathbf{e}}^{i^T}\ \mathbf{r}^P\ , \quad {}^I\mathbf{r}^P = \left[{}^I r_\alpha^P\right], \quad \mathbf{r}^P = \left[r_\alpha^P\right]\ . \tag{9}$$

Thus the absolute motion of the beam is described by the functions

$${}^I\mathbf{r}^P = {}^I\mathbf{r}^P(\mathbf{R}, t) \quad \text{or} \quad {}^I r_\alpha^P = {}^I r_\alpha^P(R_1, R_2, R_3, t)\ . \tag{10}$$

The displacements of the points P of the beam are

$$\underline{u} = \underline{\mathbf{e}}^{i^T}\mathbf{u}\ , \quad \mathbf{u} = [u_\alpha] \quad \text{where} \quad \mathbf{u} = \mathbf{u}(\mathbf{R}, t) \quad \text{or} \quad u_\alpha = u_\alpha(R_1, R_2, R_3, t)\ . \tag{11}$$

The cross sections of the beam being rigid allows one to represent the displacement field $\mathbf{u}(\mathbf{R}, t)$ by the motion of a frame $\{O^2, \underline{\mathbf{e}}^2\}$ fixed in a representative cross section at R_1, i.e. by the displacements

$$\underline{w} = \underline{\mathbf{e}}^{i^T}\mathbf{w}\ , \quad \mathbf{w} = [w_\alpha] \quad \text{where} \quad w_\alpha = w_\alpha(R_1, t) \quad \text{or} \quad \mathbf{w} = \mathbf{w}(R_1, t) \tag{12}$$

of the points on the axis of the beam with respect to its straight reference configuration, and by the rotation of the basis $\underline{\mathbf{e}}^2$ with respect to $\underline{\mathbf{e}}^i$, as described by the orientation matrix $\boldsymbol{\Theta}$ satisfying

$$\underline{\mathbf{e}}^2 = \boldsymbol{\Theta}\,\underline{\mathbf{e}}^i, \quad \boldsymbol{\Theta} = \left[\Theta_{\alpha\beta}\right] \quad \text{where} \quad \Theta_{\alpha\beta} = \Theta_{\alpha\beta}(R_1, t) \quad \text{or} \quad \boldsymbol{\Theta} = \boldsymbol{\Theta}(R_1, t)\ . \tag{13}$$

The relations provide the representation (10) of the motion of the beam with respect to inertial space as

$${}^I\mathbf{r}^P(\mathbf{R}, t) = \mathbf{A}^{1^T}(t)\ \boldsymbol{\rho}^1(t) + \mathbf{A}^{1^T}(t)\ \left(\mathbf{R}^A + \mathbf{w}(\mathbf{R}_1, t)\right) + \mathbf{A}^{1^T}(t)\ \boldsymbol{\Theta}^T(\mathbf{R}_1, t)\ \mathbf{R}^C \tag{14}$$

with the matrices

$$\mathbf{R}^A = \begin{bmatrix} R_1 & 0 & 0 \end{bmatrix}^T \quad \text{and} \quad \mathbf{R}^C = \begin{bmatrix} 0 & R_2 & R_3 \end{bmatrix}^T \tag{15}$$

specifying the location of the cross section on the axis (index A) and the location of P in the cross section (index C). As the cross sections remain perpendicular to the axis, the matrices $\mathbf{w}$ and $\boldsymbol{\Theta}$ may be represented by four deformation variables

$$\mathbf{y} = \mathbf{y}(R_1, t) = \left[y_i(R_1, t)\right], \quad i = 1, 2, 3, 4\ . \tag{16}$$

With eqs. (14) and (16) the absolute motion of the beam is given by

- four deformation variables $\mathbf{y}(\mathbf{R}_1, t)$ representing the motion of the points P of the beam with respect to the reference frame $\{O^i, \underline{e}^i\}$,
- six variables, e.g. the three coordinates $\rho^i_\alpha(t)$ of $\underline{\rho}^i(t)$ together with three parameters $p^i_\alpha(t)$ specifying the nine elements of $\mathbf{A}^i(t)$, which represent the reference motion of $\{O^i, \underline{e}^i\}$ superimposed on the deformation of the beam.

Selecting a set of shape-functions and introducing a Ritz-approximation

$$\mathbf{y}(R_1, t) = \mathbf{\Phi}(R_1)\,\mathbf{q}(t) \tag{17}$$

the deformation of the beam is given by the unknown functions $\mathbf{q}(t)$ solely. Such a representation of motion may be derived for any flexible body. If the deformations are small, equations linearised in $\mathbf{q}$ are sufficient to describe the motion.

2.2.2 *Flexible Body Data*

The next problem of flexible body modeling for multibody simulation is the definition of the data describing such a body. For that purpose we first consider the two contiguous *rigid* bodies of a multibody system, as shown in Fig. 4.

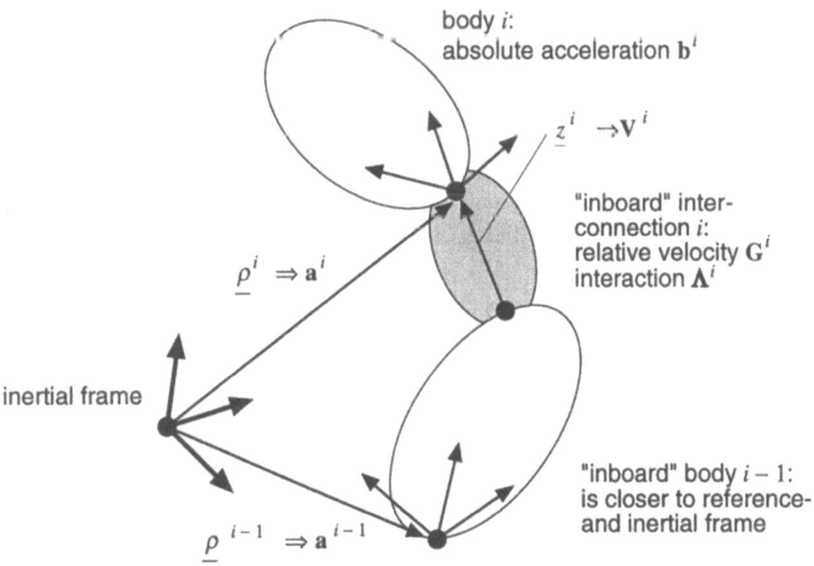

Fig. 4: Two contiguous bodies of a multibody system and associated notation.

The coordinates (in the i-th body frame) of the absolute acceleration of a reference point on body i – here the attachment point of the inboard joint – is $\mathbf{a}^i$ and the matrix $\dot{\boldsymbol{\omega}}^i$ contains the corresponding coordinates of the absolute angular acceleration. Considering primary interconnections only[1], there exists a unique inboard body $i-1$ and the two bodies i and $i-1$ interact across the inboard interconnection labelled i. The relative velocities (resolved in the body frame $i-1$) of the two bodies are $\mathbf{V}^i$ and $\boldsymbol{\Omega}^i$ and the interactions (also resolved in the body frame $i-1$) are $\mathbf{F}^i$ and $\mathbf{L}^i$. In these terms we introduce the 6×1-matrices of absolute accelerations $\mathbf{b}^i$, of relative velocities $\mathbf{G}^i$ and of interactions $\boldsymbol{\Lambda}^i$ as

$$\mathbf{b}^i = \begin{bmatrix} \mathbf{a}^i \\ \dot{\boldsymbol{\omega}}^i \end{bmatrix}, \quad \mathbf{G}^i = \begin{bmatrix} \mathbf{V}^i \\ \boldsymbol{\Omega}^i \end{bmatrix}, \quad \boldsymbol{\Lambda}^i = \begin{bmatrix} \mathbf{F}^i \\ \mathbf{L}^i \end{bmatrix}. \tag{18}$$

The development of an $O(N)$-formalism for the generation of the explicit form (5) of the dynamical system equations requires the knowledge of three equations in terms of these quantities – see Schwertassek and Rulka, 1991:

- the representation of the acceleration $\mathbf{b}^i$ in terms of the acceleration $\mathbf{b}^{i-1}$ and of the relative acceleration $\dot{\mathbf{G}}^i$ across the interconnection, i.e.

$$\mathbf{b}^i = \mathbf{C}^i \left(\mathbf{D}^{i-1T}\, \mathbf{b}^{i-1} + \boldsymbol{\alpha}^{i-1} + \dot{\mathbf{G}}^i \right); \tag{19}$$

- the description of the effects of the constraints across the interconnection, i.e. the representation of the relative velocity $\mathbf{G}^i$ in terms of n_f^i unknown velocities $\mathbf{g}^i$ – see eq. (2) – and of the velocities $\overline{\mathbf{g}}^i$ known from the constraint equations, as well as the separation of the interaction $\boldsymbol{\Lambda}^i$ into the known generalised applied forces $\boldsymbol{\lambda}^i$ and the unknown generalised constraint forces $\overline{\boldsymbol{\lambda}}^i$ corresponding to the velocities $\mathbf{g}^i$ and $\overline{\mathbf{g}}^i$, i.e.

$$\mathbf{G}^i = \boldsymbol{\varphi}^i \mathbf{g}^i + \overline{\boldsymbol{\varphi}}^i \overline{\mathbf{g}}^i, \quad \boldsymbol{\Lambda}^i = \boldsymbol{\psi}^i \boldsymbol{\lambda}^i + \overline{\boldsymbol{\psi}}^i \overline{\boldsymbol{\lambda}}^i \quad \text{where} \quad \left[\boldsymbol{\varphi}^i, \overline{\boldsymbol{\varphi}}^i\right]^T = \left[\boldsymbol{\psi}^i, \overline{\boldsymbol{\psi}}^i\right]^{-1}; \tag{20}$$

- the dynamical equations of motion in terms of $\mathbf{b}^i$ formulated without considering the constraints

$$\mathbf{J}^i \mathbf{b}^i = \mathbf{Q}^i + \mathbf{C}^i \boldsymbol{\Lambda}^i - \mathbf{D}^i \boldsymbol{\Lambda}^{i+1}. \tag{21}$$

The data of a multibody system are the parameters and functions required to compute the matrices $\mathbf{C}^i, \mathbf{D}^i, \boldsymbol{\alpha}^i, \mathbf{Q}^i$ and $\mathbf{J}^i$ appearing in the eqs. (19), (20) and (21). Matrices $\boldsymbol{\varphi}^i, \overline{\boldsymbol{\varphi}}^i$ are obtained from a library of joints, while the applied forces $\boldsymbol{\lambda}^i$ from a library of force-elements.

---

[1] For simplicity we consider systems having primary interconnections which form simple chains only. This allows the simple labeling scheme used here.

Generalising eqs. (19) - (21) for flexible bodies considering the representation of motion given by eq. (17) yields the flexible body data. Just the two equations (19) and (21) need to be reformulated, because the joints, as before, are assumed to be not deformed. It may be worth mentioning that interconnection elements between two bodies – see Fig. 1 – cannot be combined in one single interconnection anymore, because contrary to rigid bodies, a reduced system of interaction forces does not have the same physical effect as the original system when bodies are flexible.

Equations (19) and (21), generalised for flexible bodies, were developed in (Wallrapp 1994 b). As an example, consider the generalised equation (21), which has the general form

$$\mathbf{M}^i\,\mathbf{b}^i + \mathbf{k}^i_\omega + \mathbf{k}^i = \mathbf{f}^i \quad \text{with } \mathbf{b}^i = \begin{bmatrix} \mathbf{a}^i \\ \dot{\boldsymbol{\omega}}^i \\ \hline \ddot{\mathbf{q}}^i \end{bmatrix}, \mathbf{a}^i = \left[a^i_\alpha\right], \boldsymbol{\omega}^i = \left[\omega^i_\alpha\right], \mathbf{q}^i = \left[q^i_j\right]. \tag{22}$$

In eq. (22) the matrix $\mathbf{b}^i$ of accelerations as defined in eq. (18) has been augmented by the second derivatives $\ddot{\mathbf{q}}^i$ representing accelerations due to the deformation of body i. Matrix $\mathbf{M}^i$ contains the generalised masses. Suppressing body-label i for simplicity, its general form, as developed in Wallrapp, 1994.b, is

$$\mathbf{M}^i = \mathbf{M} = \left[\begin{array}{cc|c} m\mathbf{E} & m\tilde{\mathbf{d}}^T(\mathbf{q}) & \mathbf{C}_t^T(\mathbf{q}) \\ m\tilde{\mathbf{d}}(\mathbf{q}) & \mathbf{I}(\mathbf{q}) & \mathbf{C}_r^T(\mathbf{q}) \\ \hline \mathbf{C}_t(\mathbf{q}) & \mathbf{C}_r(\mathbf{q}) & \mathbf{M}_e(\mathbf{q}) \end{array}\right]. \tag{23}$$

The elements of the matrix depend on $\mathbf{q} \equiv \mathbf{q}^i$. The submatrix in the upper left corner, which is multiplied by the reference accelerations $\mathbf{a}^i$ and $\dot{\boldsymbol{\omega}}^i$, goes into $\mathbf{J}^i$ for $\mathbf{q} = \mathbf{0}$. The submatrix $\mathbf{M}_e(\mathbf{q})$ contains the generalised masses corresponding to the coordinates $\mathbf{q}$ describing body deformation. Matrices $\mathbf{C}_t$ and $\mathbf{C}_r$ contain the coupling terms between the reference motion and the deformation. The matrices $\mathbf{k}^i_\omega$, $\mathbf{k}^i$ and $\mathbf{f}^i$ have a similar structure and they contain, respectively, the Coriolis and centrifugal terms, the generalised internal forces due to body deformation, and the generalised external forces upon the body.

For small values of $\mathbf{q}$ the elements of all the matrices may be developed into Taylor series about the reference configuration $\mathbf{q} = \mathbf{0}$. The coefficients of these expansions contain integrals over the shape functions $\boldsymbol{\Phi}$ introduced in eq. (17) to approximate the body deformation. The entity of these integrals together with some additional data yields the set of data required to describe flexible bodies in a multibody system – see (Wallrapp 1994 b). The definition of these standard input data (SID) for flexible bodies is based on a proposal of standard input data for systems of rigid bodies[2] given in (Otter et al. 1993). Considering the first terms of the Taylor expansions only yields the multibody system equations linearised in the variables **q**. To compute such a specific set of the standard input data corresponding to these linearised equations for beams and finite element structures the two pre-processors BEAM and FEMBS have been developed – see Fig. 5 and (Wallrapp 1991, 1993 a, 1993 b, 1994 a).

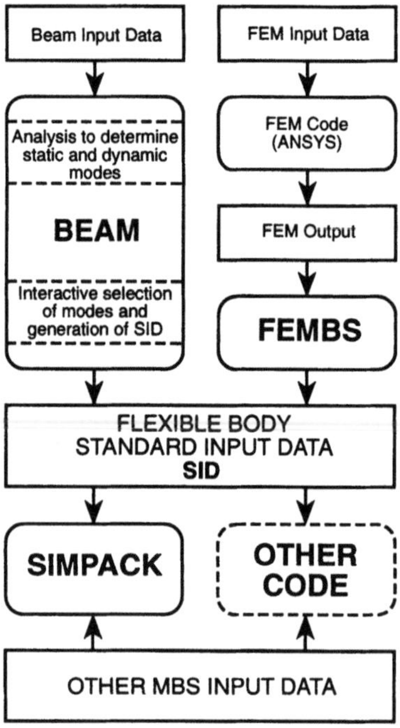

Fig. 5. Interfaces of standard input data (SID) and pre-processors BEAM and FEMBS for flexible multibody simulation.

2.2.3 Generalities on Geometric Stiffening

The system equations linearised in **q** may contain geometric stiffening terms in certain situations. Overlooking such terms resulted in the problems mentioned at the beginning of this section. Geometric stiffening is well known from investigations on the buckling of rods and plates – see e.g. (Love 1944) – as well as from an analysis of rotating machinery (Weidenhammer 1970) or from the design of motion control systems for flexible robots, (Bremer 1983, Truckenbrodt 1980). The effect needs to be considered in the linearised equations of motion assuming small deformation of flexible bodies if (Botz 1992, Bremer and Pfeiffer 1992, Schwertassek 1994, Sorge 1993)

---

[2] Based on such a standardisation of representations of multibody systems the comparison of simulation results of different multibody codes may be facilitated significantly.

- the stiffness of a structure against a specific deformation is much larger than against the other deformations;
- there are forces acting in the direction of the high stiffness, which are large enough to produce a deformation of the same order of magnitude as small forces do in the directions of lower stiffness.

In multibody system models large forces upon a flexible body often result from the reference motion of the body, i.e. from the motion of the body reference frame (in Fig. 3 the beam reference frame). A large reference motion yields large inertia forces and the deformations described by $\mathbf{q}$ remain small only if the body is sufficiently stiff in the direction of those forces. In most of the publications on the modelling of geometric stiffening in multibody simulation all (i.e. large and small) inertia forces in the direction of the high stiffness had been considered for flexible-body models in multibody systems. An analysis of geometric stiffening based on perturbation techniques – see (Schwertassek 1995) – demonstrates that:

- geometric-stiffening terms result from nonlinearities in the transformations between deformation quantities (required to formulate the strain-energy expressions) and displacement quantities (which are most convenient to formulate the kinetic energy expression);
- an analysis of motion in the direction of the high stiffness requires an evaluation of the equilibrium of large forces only, i.e. of large external forces and large internal forces due to small deformations in the direction of high stiffness.

Neglecting small inertial forces in the direction of high stiffness – and correspondingly suppressing the high frequency oscillations in those directions – yields considerable computational savings.

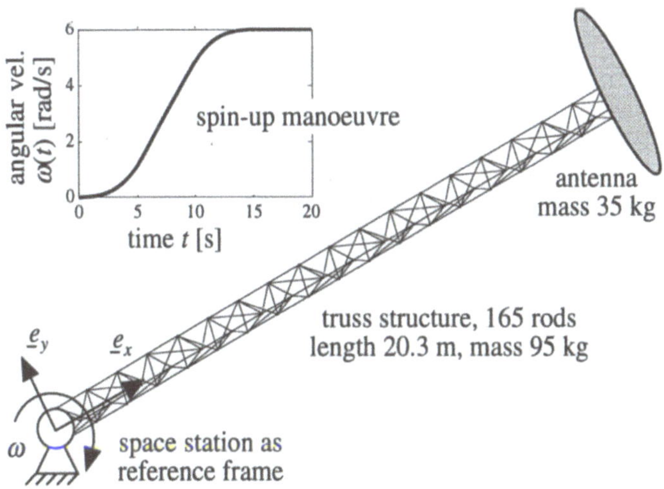

Fig. 6: Vibration analysis of an antenna, attached by a flexible truss (NASA-Minimast) to a space station undergoing a spin-up-manoeuvre as shown in the diagram $\omega(t)$.

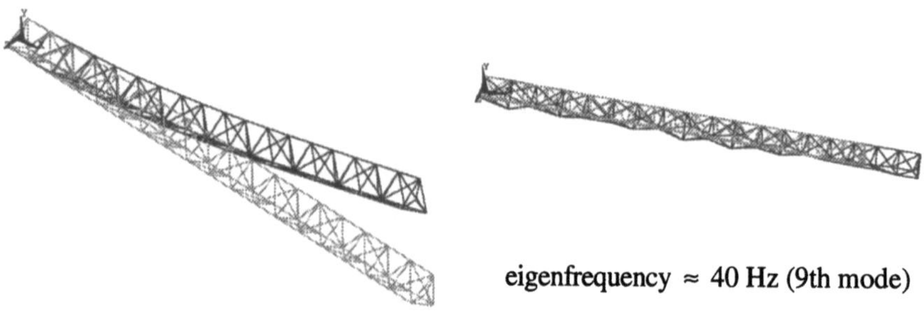

eigenfrequency ≈ 40 Hz (9th mode)

eigenfrequency ≈ 2 Hz (1st mode)

Fig. 7. First eigenfunctions of the truss structure with significant lateral and longitudinal motion and corresponding eigenfrequencies.

These results, developed for beams initially, have been generalised in (Sachau 1996) to finite element structures. An example taken from the latter reference demonstrates the advantage of the new methodology. Figure 6 shows an antenna attached by a truss structure to a space station undergoing a spin-up-manoeuvre. The shape functions obtained from a finite element analysis, and used in the flexible-body model of the truss for a multibody simulation, are shown in Fig. 7. The truss (NASA-Minimast) is very stiff against stretching (longitudinal direction) as compared to bending (lateral direction), but nonetheless the large centrifugal forces resulting from the spin-up manoeuvre yield longitudinal deformations of the same order of magnitude as small loads in the lateral direction.

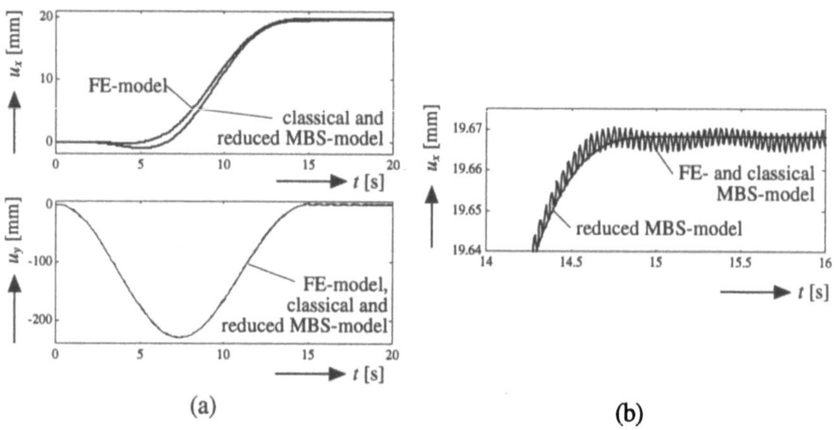

Fig. 8. Plots of (a) the longitudinal and lateral motion of the antenna and (b) a detail of the longitudinal motion.

Figure 8 shows simulation results representing the motion of the antenna due to the manoeuvre, which have been obtained using three models: a FE-model; a "classical" MBS model; and the new "reduced" MBS model. The three models yield approximately the same results but the ratio of simulation times is found to be 360:15:1. The FE analysis is costly because it allows an analysis of stress in addition to the analysis of motion. The reduction of computer time with the reduced MBS model results from neglecting the high frequency oscillations in the longitudinal motion. Such oscillations, as obtained by the FE and the classical MBS model, are shown in Fig. 8(b) together with the result obtained by the reduced model. The high-frequency oscillations cannot be represented correctly in a linear model as demonstrated in (Sachau 1996). In this reference it is shown as well that such a reduction of computer time is most valuable when designing active vibration control systems – see also (Gaul et al. 1997).

3. Geometric Stiffening

3.1 Introduction

A well known example of geometric stiffening is shown in Fig. 9. Starting from a discussion of this specific case we will clarify the modeling of the effect for more general problems. To prepare all of these discussions it is helpful to formulate the partial differential equations of motion and the corresponding boundary conditions as provided by Hamilton's principle for a three-dimensional motion of the Euler-Bernoulli beam of Fig. 3. A linearisation of the equations assuming small deformations in case of large external loads requires knowledge of an approximation of the nonlinear equations up to second-order terms. These relations will be developed first. Hamilton's principle states

$$\int_{t_0}^{t_1} \left(\delta T - \delta U + \delta' A_a\right) dt = 0 . \tag{24}$$

Here δT and δU are the variations of the kinetic and the strain energy of the beam and $\delta' A_a$ is the virtual work of the external forces.

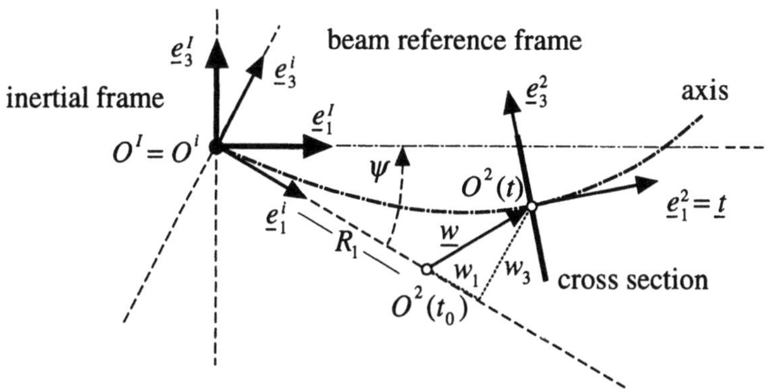

Fig. 9. Planar motion of a rotating beam and corresponding notation.

3.2 Expressions for the Kinetic and Potential Energies

The computation of δT is most convenient in terms of displacement quantities. They will be introduced followed by a discussion of deformation quantities, which are more suitable to formulate the strain energy expression and its variation δU.

3.2.1 Displacement Quantities

As stated in eq. (16), four variables are required to represent the deformation of the beam of Fig. 3. We now select

$$\mathbf{y} = \mathbf{y}(R_1, t) = \begin{bmatrix} w_1(R_1, t) & w_2(R_1, t) & w_3(R_1, t) & \vartheta_1(R_1, t) \end{bmatrix}^T \tag{25}$$

where ϑ_1 is one of the three Tait-Bryan or Cardan angles ϑ_α – see (Roberson and Schwertassek 1988) – parameterising $\mathbf{\Theta}$ from eq. (13) as

$$\mathbf{\Theta} = \begin{bmatrix} +c_2\,c_3 & +c_2\,s_3 & -s_2 \\ -c_1 s_3 + s_1 s_2\,c_3 & +c_1 c_3 + s_1 s_2 s_3 & +s_1 c_2 \\ +s_1 s_3 + c_1 s_2 c_3 & -s_1 c_3 + c_1 s_2 s_3 & +c_1 c_2 \end{bmatrix} \quad \text{where} \begin{cases} c_\alpha = \cos\vartheta_\alpha \\ s_\alpha = \sin\vartheta_\alpha \, . \end{cases} \tag{26}$$

Problems of how to select advantageous sets of angle representations of $\mathbf{\Theta}$ have been discussed, among others, in (Houbolt and Brooks 1957, Hodges and Ormiston 1980 and Hodges 1985). Angles ϑ_2 and ϑ_3 are obtained in terms of the elements of $\mathbf{y}$ from eq. (25) by the condition which postulates that the orientation of the cross sections remains perpendicular to the axis of the beam. For small values of w_α and ϑ_α a second-order approximation is found to be

$$\vartheta_2 = -\,w_3' + w_1'\,w_3' \,, \qquad \vartheta_3 = w_2' - w_1'\,w_2' \; . \tag{27}$$

Thus $\mathbf{\Theta}$ can be represented (to a second-order approximation) in terms of $\mathbf{y}$ as

$$\mathbf{\Theta} = \begin{bmatrix} 1 - \frac{1}{2}\left(w_2'^2 + w_3'^2\right) & w_2' - w_1'\,w_2' & w_3' - w_1'\,w_3' \\ -w_2' + w_1'\,w_2' - \vartheta_1\,w_3' & 1 - \frac{1}{2}\left(w_2'^2 + \vartheta_1^2\right) & \vartheta_1 \\ -w_3' + w_1'\,w_3' + \vartheta_1\,w_2' & -\vartheta_1 - w_2'\,w_3' & 1 - \frac{1}{2}\left(w_3'^2 + \vartheta_1^2\right) \end{bmatrix} . \tag{28}$$

In view of eq. (28) the displacement-field, as given by eq. (11), can be represented in terms of the displacement quantities $\mathbf{y}$ from eq. (25) in the form

$$\left.\begin{aligned} u_1 &= w_1 + R_2\left(-w_2' + w_1'\,w_2' - \vartheta_1\,w_3'\right) + R_3\left(-w_3' + w_1'\,w_3' + \vartheta_1\,w_2'\right) \\ u_2 &= w_2 - R_2\,\tfrac{1}{2}\left(w_2'^2 + \vartheta_1^2\right) - R_3\left(\vartheta_1 + w_2'\,w_3'\right) \\ u_3 &= w_3 + R_2\,\vartheta_1 - R_3\,\tfrac{1}{2}\left(w_3'^2 + \vartheta_1^2\right). \end{aligned}\right\} \tag{29}$$

Equations (29) show one of the possibilities to describe the motion of the beam with respect to its reference frame $\{O^i, \underline{\mathbf{e}}^i\}$ in terms of a minimum set of variables. Using eqs. (14) and (28) the variation δT of the kinetic energy is found in terms of the displacement quantities defined by eq. (25) from

$$\delta T = \int_{V_0} {}^I\dot{\mathbf{r}}^{PT}\, \boldsymbol{\delta}^I\dot{\mathbf{r}}^P \rho_0 \, dV_0 \,. \tag{30}$$

Here ρ_0 is the density of the material in the reference configuration of the beam and V_0 is the corresponding volume.

3.2.2 Deformation Quantities

We now define a set of deformation variables resulting in a simple strain energy expression U. In case of a beam the set of variables preferred by most authors is defined as follows:

$$\mathbf{y} = \mathbf{y}(R_1, t) = \begin{bmatrix} v_1(R_1, t) & v_2(R_1, t) & v_3(R_1, t) & \tau(R_1, t) \end{bmatrix}^T \tag{31}$$

where

$$v_1(R_1, t) = s - R_1 = \int_0^{R_1} \sqrt{\left(1 + w_1'(\xi, t)\right)^2 + \left(w_2'(\xi, t)\right)^2 + \left(w_3'(\xi, t)\right)^2}\, d\xi - R_1 \tag{32}$$

with the arc-length $s = s(R_1, t)$ of the axis of the beam – see Fig. 10 – and where

$$v_2(R_1,t) \;=\; + \int_0^{R_1}\int_0^{\xi} \kappa_3(\eta,t)\, d\eta\, d\xi \tag{33}$$

$$v_3(R_1,t) \;=\; - \int_0^{R_1}\int_0^{\xi} \kappa_2(\eta,t)\, d\eta\, d\xi \tag{34}$$

$$\tau(R_1,t) \;=\; + \int_0^{R_1} \kappa_1(\xi,t)\, d\xi \tag{35}$$

with the three "curvatures" κ_α of the axis. As detailed in Schwertassek, 1994, each κ_α is defined (as a function of $s(R_1, t)$, not of R_1) as follows:

$$\boldsymbol{\kappa}(s,t) = \left[\kappa_\alpha(s,t)\right] = \begin{bmatrix} k_1(s,t) + \partial\beta(s,t)/\partial s \\ k_3(s,t)\sin\beta(s,t) \\ k_3(s,t)\cos\beta(s,t) \end{bmatrix} \tag{36}$$

and verifies

$$\frac{d\mathbf{\Theta}}{ds} = -\tilde{\mathbf{\kappa}}\,\mathbf{\Theta}, \quad \mathbf{\Theta} = \mathbf{\Theta}(s,t) \tag{37}$$

where $s = s(R_1, t)$, as given in eq. (32), and with $\tilde{\mathbf{\kappa}}$ denoting the cross product matrix of $\mathbf{\kappa}$.

In eq. (36) k_1 and k_3 are the two curvatures of a spatial curve as known from the differential geometry of spatial curves (Frenet's equations) and β is an additional parameter representing the angle between the orientation of the 2- and 3-axis of the "natural basis" of a spatial curve and the principal axes of the area moments of inertia of the cross sections of the beam – see Fig. 10. In view of these definitions and their interpretation v_1 represents stretching of the the beam, while v_2 and v_3 bending, and τ denotes torsion. The latter is defined in terms of the twist k_1 of the axis and of the angle β. The need to introduce the additional parameter β becomes evident when considering torsional motions of a straight beam. In this case $k_1 \equiv k_3 \equiv 0$ and β alone captures the rotations of the cross sections about the 1-axis.

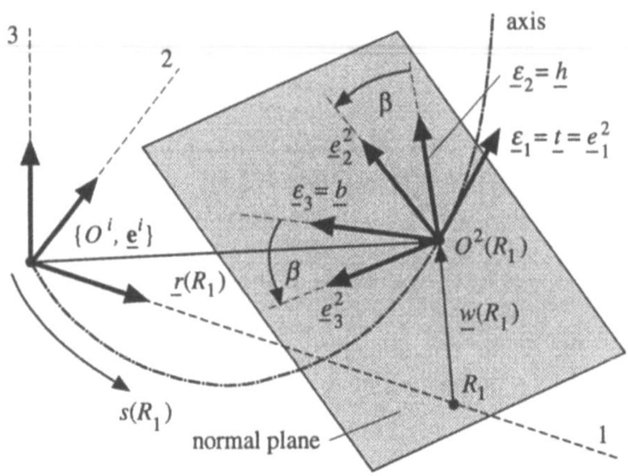

Fig. 10. The axis of an Euler-Bernoulli-beam interpreted as a spatial curve together with the corresponding "natural basis" $\underline{\boldsymbol{\varepsilon}} = [\underline{\varepsilon}_\alpha]$, formed by the tangent vector $\underline{t}$, the normal vector $\underline{h}$ and the binormal vector $\underline{b}$. The principal axes of the area moments of inertia of the cross sections of the beam are given by vectors $\underline{e}_2^2$ and $\underline{e}_3^2$. The angle between the foregoing axes and vectors $\underline{h}$ and $\underline{b}$ is β.

We now introduce the

- area of the cross sections

$$A = \int_{A_0} dR_2 \, dR_3 = \int_{A_0} dA \tag{38}$$

where A_0 is the cross section in the reference configuration (here $A_0 \equiv A$),

- area moments of inertia of the cross sections

$$J_{22} = \int\limits_{A_0} R_2^2 \, dA\,, \quad J_{33} = \int\limits_{A_0} R_3^2 \, dA\,, \quad J_p = J_{22} + J_{33} = \int\limits_{A_0} \left(R_2^2 + R_3^2\right) dA\,, \tag{39}$$

- length L of the beam in its reference configuration,
- modulus of elasticity $\mathcal{E}$ and the modulus of rigidity $\mathcal{G}$, together with the density ρ_0 already known from eq. (30).

With these definitions and with $J_T = J_p$ for circular cross sections the strain energy function can be expressed conveniently by the variables defined in eqs. (32) to (35) as

$$U = \tfrac{1}{2}\int\limits_0^L \left(\mathcal{E}\,A\,v_1'^2 + \mathcal{G}\,J_T\,\tau'^2 + \mathcal{E}\,J_{22}\,v_2''^2 + \mathcal{E}\,J_{33}\,v_3''^2\right) dR_1\,. \tag{40}$$

3.2.3 Transformations

To develop the dynamical equations of motion based on eq. (24) both T and U need to be represented in terms of the same set of variables $\mathbf{y}$, either the displacement quantities of eq. (25) or the deformation quantities of eq. (31). Both sets of variables have their advantages and disadvantages. The displacement quantities yield a simple description of the kinematics of the beam as given by eq. (14) and as a consequence, a representation of the kinetic energy, which is straightforward to develop. On the other hand, the deformation quantities provide a simple representation of the strain energy. The discussion demonstrates that the arguments available at this point give no hint whether to prefer one of the two possibilities, eq. (25) or eq. (31), to represent the motion of the beam. In any case the transformations between the two sets of variables are required to develop the dynamical equations of motion based on eq. (24).

A derivation of the dynamical equations in terms of the displacement quantities requires the representation of the deformation quantities by the displacement variables to obtain the potential energy expression in these terms. To a second order approximation one finds from a corresponding approximation of eqs. (37) and (32) (Schwertassek, 1994):

$$v_1 = w_1 + \tfrac{1}{2}\int\limits_0^{R_1} \left(w_2'^2 + w_3'^2\right) d\xi \tag{41.a}$$

$$v_2 = w_2 - \int\limits_0^{R_1} w_1'\, w_2'\, d\xi - \int\limits_0^{R_1}\!\int\limits_0^{\xi} w_1'\, w_2''\, d\eta\, d\xi + \int\limits_0^{R_1}\!\int\limits_0^{\xi} \vartheta_1\, w_3''\, d\eta\, d\xi \tag{41.b}$$

$$v_3 = w_3 - \int\limits_0^{R_1} w_1'\, w_3'\, d\xi - \int\limits_0^{R_1}\!\int\limits_0^{\xi} w_1'\, w_3''\, d\eta\, d\xi - \int\limits_0^{R_1}\!\int\limits_0^{\xi} \vartheta_1\, w_2''\, d\eta\, d\xi \tag{41.c}$$

$$\tau \; = \vartheta_1 \; - \int_0^{R_1} w_1' \, \vartheta_1' \, d\xi + \int_0^{R_1} w_3' \, w_2'' \, d\xi = \varphi_1 - \int_0^{R_1} w_1' \, \varphi_1' \, d\xi , \; \varphi_1' = w_3' \, w_2'' + \vartheta_1' . \text{(41.d)}$$

Correspondingly a comparison of a second-order approximation of $\mathbf{\Theta}$ in terms of the deformation quantities (31) with the expression for $\mathbf{\Theta}$ given in eq. (28) yields a representation of ϑ_1 in terms of these variables. The approximation of $\mathbf{\Theta}$ required here is found from eq. (37) as

$$\mathbf{\Theta} =$$

$$\begin{bmatrix} 1-\frac{1}{2}\left(v_2'^2+v_3'^2\right) & v_2' + \int_0^{R_1} v_1' v_2'' \, d\xi - \int_0^{R_1} \tau v_3'' d\xi & v_3' + \int_0^{R_1} v_1' v_3'' \, d\xi + \int_0^{R_1} \tau v_2'' \, d\xi \\ -v_2' - \int_0^{R_1} v_1' v_2'' \, d\xi - \int_0^{R_1} v_3' \tau' d\xi & 1-\frac{1}{2}\left(v_2'^2+\tau^2\right) & \tau + \int_0^{R_1} v_1' \tau' d\xi - \int_0^{R_1} v_3' v_2'' \, d\xi \\ -v_3' - \int_0^{R_1} v_1' v_3'' \, d\xi + \int_0^{R_1} v_2' \tau' d\xi & -\tau - \int_0^{R_1} v_1' \tau' d\xi - \int_0^{R_1} v_2' v_3'' \, d\xi & 1-\frac{1}{2}\left(v_3'^2+\tau^2\right) \end{bmatrix} \quad (42)$$

Besides ϑ_1 we also need to know the w_α in terms of the deformation quantities. They are found from a second order approximation of the constraint equation $\underline{e}_1^2 = \underline{t}$ – see Fig. 10 – with $\underline{e}_1^2$ given by the variables of eq. (25) and $\underline{t}$ given by the variables of eq. (31). Following these routes the entity of transformations is found to be

$$w_1 = v_1 \; - \frac{1}{2} \int_0^{R_1} \left(v_2'^2 + v_3'^2\right) d\xi \quad \text{(43.a)}$$

$$w_2 = v_2 \; + \int_0^{R_1} v_1' \, v_2' \, d\xi + \int_0^{R_1} \int_0^{\xi} v_1' \, v_2'' \, d\eta \, d\xi - \int_0^{R_1} \int_0^{\xi} \tau \, v_3'' \, d\eta \, d\xi \quad \text{(43.b)}$$

$$w_3 = v_3 \; + \int_0^{R_1} v_1' \, v_3' \, d\xi + \int_0^{R_1} \int_0^{\xi} v_1' \, v_3'' \, d\eta \, d\xi + \int_0^{R_1} \int_0^{\xi} \tau \, v_2'' \, d\eta \, d\xi \quad \text{(43.c)}$$

$$\vartheta_1 = \tau \; + \int_0^{R_1} v_1' \, \tau' \, d\xi - \int_0^{R_1} v_3' \, v_2'' \, d\xi \; . \quad \text{(43.d)}$$

Now the kinetic energy can be formulated in terms of the deformation quantities.

3.3 Geometric Stiffening due to Centrifugal Forces

Applying the representation of the motion of the beam given above we study the simple example shown in Fig. 9. The beam rotates about the 2-axis of the beam reference frame with a known angular velocity $\dot{\psi}(t)$ resulting in centrifugal forces in the longitudinal direction. The linearised equations of motion, assuming the displacements $w_1(R_1,t)$ and $w_3(R_1,t)$ to be small, will be developed now.

3.3.1 Modeling Considerations

The linearised equations of motion are found to be

$$\rho_0\, A\, \ddot{w}_1 - \left(\mathcal{E}\, A\, w_1'\right)' = Q_{w1} \text{ (a) and } \rho_0\, A\, \ddot{w}_3 + \left(\mathcal{E}\, J_{33}\, w_3''\right)'' = Q_{w3} \text{ (b)}, \quad (44)$$

where Q_{w1} and Q_{w3} are the generalised external forces acting upon the beam, corresponding to the coordinates $w_1(R_1,t)$ and $w_3(R_1,t)$. In multibody dynamics problems the reference motion, i.e. the motion of the beam reference frame is large in many applications. In the example of Fig. 9 the reference motion is given by the angular velocity $\dot{\psi}(t)$. If it is large, the rotation yields large centrifugal forces in the longitudinal direction of the beam. Neglecting dynamics in eq. (44.a), i.e. assuming $\ddot{w}_1 \equiv 0$, the relation reveals that large longitudinal forces yield large deformations w_1, which is a contradiction to the assumption formulated above. The equation demonstrates that the longitudinal stiffness $\mathcal{E}A$ of the beam has to be very large to keep w_1 small. In other words, the product $\mathcal{E}A\, w_1'$ has to be of the same order of magnitude as large centrifugal forces Q_{w1} due to $\dot{\psi}(t)$ – $\mathcal{E}A\, w_1'$ has to be large, despite the fact that w_1 is a small-order quantity.

The discussion demonstrates that the linearised equations (44) cannot be used to solve our problem. The linearised equations need to be redeveloped considering in addition to the order of magnitude of the variables w_1 and w_3 as well the order of magnitude of the system parameters. Accordingly, the linearised equations of motion of the beam of Fig. 9 are developed now assuming that

- the deformation variables w_1 and w_3 are small,
- the angular velocity $\dot{\psi}(t)$ of the reference frame $\{O^i, \underline{e}^i\}$ is large and results in large centrifugal forces,
- the longitudinal stiffness of the beam is very large such that $\mathcal{E}A\, w_1'$ is not a small term despite the first assumption.

Perturbation techniques – see e.g. (Coddington and Levinson 1955, Nayfeh and Mook 1979) – provide the book-keeping capabilities to monitor the order of magnitude of terms as required here. Moreover, they are a valuable tool to develop equations, which guarantee to describe the system behaviour in an approximation having a small and well defined error. Such techniques are used now to develop the desired linearised equations.

3.3.2 Equations of Motion in Terms of Displacement Quantities

Using eqs. (41) we reformulate the potential energy expression (40) in terms of the displacement quantities of eq. (25), here w_1 and w_3, and apply Hamilton's principle, eq. (24), to obtain the nonlinear, 2nd order approximations of the equations of motion

$$\left.\begin{aligned}&\rho_0 A\left(\ddot{w}_1 - w_3\,\ddot{\psi} - 2\,\dot{w}_3\,\dot{\psi} - (R_1 + w_1)\,\dot{\psi}^2\right)\\&\qquad\qquad - \left(\mathcal{E}A\left(w_1' + \tfrac{3}{2}w_1'^2 + \tfrac{1}{2}w_3'^2\right)\right)' = 0\end{aligned}\right\}\qquad(45.a)$$

$$\left.\begin{aligned}&\rho_0 A\left(\ddot{w}_3 + (R_1 + w_1)\,\ddot{\psi} + 2\,\dot{w}_1\,\dot{\psi} - w_3\,\dot{\psi}^2\right)\\&\quad - \left(\rho_0 J_{33}(1 - w_1')\,\ddot{\psi}\right)' - \left(\mathcal{E}A\,w_1'\,w_3'\right)' + \left(\mathcal{E}J_{33}\,w_3''\right)'' = 0\end{aligned}\right\}\qquad(45.b)$$

and the relations to be satisfied at the boundaries

$$\begin{aligned}&\Big(-\mathcal{E}A\left(w_1' + \tfrac{3}{2}w_1'^2 + \tfrac{1}{2}w_3'^2\right)\delta w_1\\&-\left(\rho_0 J_{33}(1 - w_1')\,\ddot{\psi} + \mathcal{E}A\,w_1'\,w_3' - \left(\mathcal{E}J_{33}\,w_3''\right)'\right)\delta w_3 - \mathcal{E}J_{33}\,w_3''\,\delta w_3'\Big)\Big|_{R_1=0}^{L} = 0.\end{aligned}\qquad(46)$$

Equations (45) and (46) still contain second-order terms, in particular those which are required to account for our third assumption, saying that $\mathcal{E}A\,w_1'$ is not small. The linearised equations are developed now introducing dimensionless variables. These variables allow one to compare orders of magnitude of various terms most conveniently (i.e. by comparing numbers without having to worry about physical dimensions), as suggested by techniques well known from the various forms of the perturbation method. Introducing appropriate dimensionless variables, a parameter

$$\varepsilon = \frac{\mathcal{E}J_{33}}{\rho_0 A L^4}\Big/\frac{\mathcal{E}}{\rho_0 L^2} = \frac{\mathcal{E}J_{33}}{L^3}\Big/\frac{\mathcal{E}A}{L} = \frac{J_{33}}{A L^2}\qquad(47)$$

appears automatically. It represents the ratio of longitudinal and lateral stiffnesses of the beam, a ratio that is small as compared to 1 because of the assumptions. Collecting all of the terms of the order $O(\varepsilon^0) = O(1)$ in the dimensionless form of eq. (45.a) and of the order $O(\varepsilon)$ in the dimensionless form of eq. (45.b) one arrives at equations yielding both w_1 and w_3 with an error of the order $O(\varepsilon^2)$, Schwertassek, 1994. The resulting equations are given here in the form obtained after switching back to the original variables. The equations of motion are

$$\left.\begin{aligned}&\mathcal{E}A\,w_1'' + \rho_0 A R_1\,\dot{\psi}^2 = 0 \quad \text{(a)}\\&\rho_0 A\left(\ddot{w}_3 + R_1\,\ddot{\psi} + 2\,\dot{w}_1\,\dot{\psi} - w_3\,\dot{\psi}^2\right) - \mathcal{E}A\left(w_1'\,w_3'\right)' + \mathcal{E}J_{33}\,w_3'''' = 0 \quad \text{(b)}.\end{aligned}\right\}\qquad(48)$$

Correspondingly, one derives from eq. (46) the boundary conditions, namely,

$$\left(-\mathcal{E} A\, w_1'\, \delta w_1 - \left(\rho_0\, J_{33}\, \ddot{\psi} + \mathcal{E} A\, w_1'\, w_3' - \mathcal{E} J_{33}\, w_3'''\right) \delta w_3 - \right.$$
$$\left. - \mathcal{E} J_{33}\, w_3''\, \delta w_3' \right)\Big|_{R_1=0}^{L} = 0\,. \tag{49}$$

The perturbation techniques used to develop eqs. (48) and (49) reveal moreover that the equations apply only if an additional condition is satisfied: The acceleration $\ddot{\psi}(t)$ has to remain small. At this point the well known advantage of the perturbation method of being "foolproof" becomes apparent: The technique gives a hint on whether the assumptions on which the derivation of the linearised equations has been based are inconsistent or not.

Equations (48) are the linearised equations of motion for a rotating beam, a result which led some confusion and misleading interpretations – see e.g. (Ryan 1987) – because eq. (48.b) contains an obviously nonlinear term $\mathcal{E} A\left(w_1'\, w_3'\right)'$. A careful interpretation of the equations readily demonstrates that they are linear indeed. From eq. (48.a) one concludes that $\mathcal{E} A\, w_1' = Z$, where

$$Z = -\tfrac{1}{2}\rho_0\, A\left(R_1^2 - L^2\right)\dot{\psi}^2 = -\,\rho_0\, A\left(R_1 - L\right)\left(R_1 + \frac{L - R_1}{2}\right)\dot{\psi}^2 \tag{50}$$

is the centrifugal force acting at the cross section at point R_1, as visualised in Fig. 11: The force is due to the mass outside of R_1 (grey part of the beam) rotating as a rigid body with angular velocity $\dot{\psi}$.

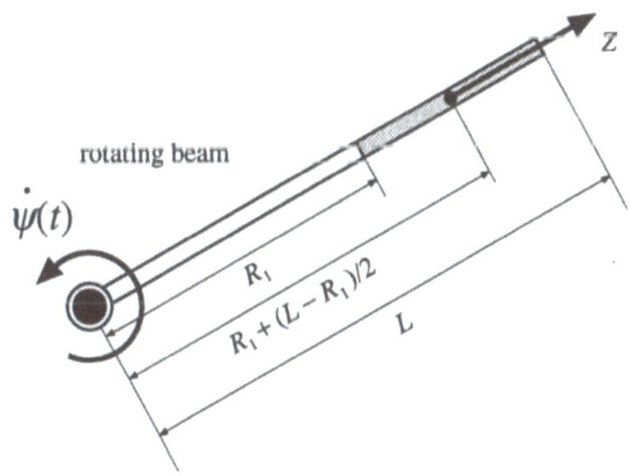

Fig. 11. Centrifugal force applied at a cross section of a rotating beam.

Introducing the condition $\mathcal{E} A\, w_1' = Z$ for equilibrium in longitudinal direction into eq. (48.b), one obtains the linear equation

$$\rho_0\, A\left(\ddot{w}_3 + R_1\, \ddot{\psi} + 2\, \dot{w}_1\, \dot{\psi} - \left(w_3 - R_1\, w_3' - \tfrac{1}{2}\left(R_1^2 - L^2\right) w_3''\right)\dot{\psi}^2\right)$$
$$+\, \mathcal{E} J_{33}\, w_3'''' = 0 \tag{51.a}$$

or, considering the definition (50) of Z,

$$\rho_0 A\left(\ddot{w}_3 + R_1 \ddot{\psi} + 2\dot{w}_1 \dot{\psi} - w_3 \dot{\psi}^2\right) + \left(Z w_3'\right)' + \mathcal{E} J_{33} w_3'''' = 0 . \tag{51.b}$$

Also, with eq. (50) the boundary conditions (49) take the linear form

$$\begin{aligned}\Big(\tfrac{1}{2}\rho_0 A\left(R_1^2 - L^2\right)\dot{\psi}^2\, \delta w_1 \\ - \left(\rho_0 J_{33} \ddot{\psi} - \tfrac{1}{2}\rho_0 A\left(R_1^2 - L^2\right)\dot{\psi}^2 w_3' - \mathcal{E} J_{33} w_3'''\right)\delta w_3 \\ - \mathcal{E} J_{33} w_3''\, \delta w_3' \Big)\Big|_{R_1=0}^{L} = 0 .\end{aligned} \tag{52}$$

Comparing eqs. (48.b) and (51.b) with eq. (44.b) one realises that a number of terms depending on $\dot{\psi}$ and $\ddot{\psi}$ have been added, together with an additional term $\mathcal{E} A (w_1' w_3')' = (Z w_3')'$, which is recognised as a linear term only after considering eq. (50). This term takes into account the influence of large longitudinal forces and results from nonlinear strain-displacement relations to be discussed later; it modifies the frequency of the bending motions. Accordingly, it is called a geometric or rotational stiffening term, depending on whether the nonlinearity of the (geometrical) strain-displacement relations or the centrifugal forces are to be stressed as its source.

3.3.3 Equations of Motion in Terms of Deformation Quantities

We now use eqs. (43) to represent the kinetic energy T in terms of the deformation quantities v_1 and v_3. Introducing the resulting expression together with U as given by eq. (40) into Hamilton's principle, eq. (24), and applying perturbation techniques to derive the linearised equations for the planar example shown in Fig. 9, one finds the linearised equations of motion

$$\left.\begin{aligned} &\mathcal{E} A v_1'' + \rho_0 A R_1 \dot{\psi}^2 = 0 \quad \text{(a)} \\ &\rho_0 A\left(\ddot{v}_3 + R_1 \ddot{\psi} + 2\dot{v}_1 \dot{\psi} - \left(v_3 - R_1 v_3' - \tfrac{1}{2}\left(R_1^2 - L^2\right) v_3''\right)\dot{\psi}^2\right) \\ &\qquad + \mathcal{E} J_{33} v_3'''' = 0 \quad \text{(b)} \end{aligned}\right\} \tag{53}$$

and the boundary conditions

$$\begin{aligned}\Big(\tfrac{1}{2}\rho_0 A\left(R_1^2 - L^2\right)\dot{\psi}^2\, \delta v_1 \\ - \left(\rho_0 J_{33} \ddot{\psi} - \tfrac{1}{2}\rho_0 A\left(R_1^2 - L^2\right)\dot{\psi}^2 v_3' - \mathcal{E} J_{33} v_3'''\right)\delta v_3 \\ - \mathcal{E} J_{33} v_3''\, \delta v_3' \Big)\Big|_{R_1=0}^{L} = 0\end{aligned} \tag{54}$$

i.e. the very same equations as obtained previously – see eqs. (48.a), (51.a) and (52). The two sets of linearised equations have to be identical, of course, a fact which can be concluded immediately from the transformations (43). The result reveals an advantage of the deformation quantities used here: the equations of motion (53.b) are obtained directly without having to solve the equation for the longitudinal motion first and to introduce the result into the equation for the lateral motion as in case of eqs. (48) in terms of the displacement quantities (31).

The derivation of equations of motion by perturbation techniques verifies the statement given in Subsection 2.2.3: Approximations of w_1 and w_3 or v_1 and v_3 having an error of the order $O(\varepsilon^2)$ require the evaluation of equilibrium between large external forces and large internal forces due to small deformations in the direction of high stiffness only – see eqs. (48.a) and (53.a). Additional linear terms in these equations do not increase the accuracy of the result, but rather decrease the efficiency of the simulation, as demonstrated by the discussion of Fig. 8.

3.3.4 Guidelines for Consideration of Geometric Stiffening Terms

The two sets of deformation variables defined in eqs. (25) and (31), respectively, entail advantages and disadvantages: Whereas the former result in a simple kinetic energy expression and its variation (30), the latter are more convenient to formulate the potential energy (40). Moreover, it was demonstrated in the preceding section that a linearisation of the equations of motion is much simpler when using variables (31): One just deletes nonlinear terms to obtain the linearised equations of motion without having to worry about orders of magnitude of stiffness parameters. For the present purpose another aspect is even more important: Variables (31) suggest a systematic way to develop equations of motion for straight beams as required in multibody system simulation, i.e. equations linearised about the straight reference configuration including the geometric stiffness terms due to large inertial loads.

The geometric stiffness term $\mathcal{E}A(w_1' w_3')' = (Z\, w_3')'$ in eq. (48.b) results from the nonlinearity in the transformation (41.a), i.e. from nonlinear coupling terms between the linear and angular displacements (25) of the cross sections of the beam and the deformations (31) representing elongation, bending and torsion. This can be seen as follows: The strain energy function (40) is obtained for the planar example shown in Fig. 9 as

$$U = \frac{1}{2}\int_0^L \left(\mathcal{E}A\, v_1'^2 + \mathcal{E}J_{33}\, v_3''^2\right) dR_1 \ . \tag{55}$$

Its variation, as required in Hamilton's principle, is found to be

$$\delta U = \int_0^L \mathcal{E}A\left(v_1'\, \delta v_1' \ + \mathcal{E}J_{33}\, v_3''\, \delta v_3''\right) dR_1 \ . \tag{56}$$

Using eq. (41.a) one finds in terms of the variables w_1 and w_3 to a second-order approximation considering the assumptions formulated in Subsection 3.3.1

$$\delta U = \int_0^L \mathcal{E}A\left(\left(\left(w_1' + \tfrac{1}{2}w_3'^2\right)\delta w_1' + w_1' w_3' \,\delta w_3'\right) + \mathcal{E}J_{33}\, w_3'' \,\delta w_3''\right) dR_1 \ . \tag{57}$$

Applying Hamilton's principle and perturbation techniques to generate the linearised equations one observes that the nonlinear term $w_1' w_3' \,\delta w_3'$ in the integral (57) results in the term $\mathcal{E}A\,(w_1' w_3')' = (Z\, w_3')'$ in eq. (48.b), i.e. in the geometric stiffening term. As demonstrated by the transformation of eq. (56) into eq. (57), this expression has its source in the nonlinear term $w_3'^2/2$ in the relation

$$v_1' = w_1' + \tfrac{1}{2}w_3'^2 \tag{58}$$

which is derived from eq. (41.a) that relates stretching v_1 (due to large longitudinal forces Z) to longitudinal and lateral displacements w_1 and w_3 ($w_2 \equiv 0$ for the example). In the linearised equations (44), applying for small external forces acting upon the beam, such coupling terms between deformations and displacements need not be considered, but, as the example demonstrates, the nonlinearity has to be taken into account in the linearised equations, when two conditions are satisfied:

1. the resistance against one of the deformations described by the variables v_1, v_2, v_3 and τ is much higher than against the other ones;
2. the forces acting in the direction with high resistance against deformation – "stiff directions" – are large enough to produce displacements, which are of the same order of magnitude as the displacements expected in the other "non-stiff directions".

Because of the high resistance against deformation, large parameters appear in the system equations. Their products with the second-order coupling terms in the transformations (41) yield the geometric stiffness expressions in the linearised equations. Based on such considerations the other cases in which nonlinear coupling terms need be considered in the linearised equations can be identified easily. They represent the well-known coupling effects appearing in the theory of buckling. One of the better known examples is the beam shown in Fig. 12. It has a low stiffness against torsion τ and bending v_2 about the 3-axis but a high stiffness against bending v_3 about the 2-axis. The coupling of the displacements w_2, w_3, and ϑ_1 due to bending v_3 in the stiff direction is obtained from eq. (41.c) for a beam whose elongation can be neglected (i.e. $w_1 \equiv 0$) as

$$v_3'' = w_3'' - \vartheta_1\, w_2'' . \tag{59}$$

The corresponding motions are visualised in Fig. 12. Such a nonlinear coupling needs to be considered – see e.g. (Truckenbrodt 1980) – if sufficiently large bending torques about the 2-axis – e.g. inertial torques resulting from an angular acceleration $\ddot{\psi}(t)$ about the 2-axis and from an endmass attached to the tip of the beam – are applied. Details on the modeling of the effect for multibody simulation may be found in (Sandlass 1996).

Another example is given in Fig. 13. It shows a long beam having high torsional stiffness and comparatively low longitudinal stiffness under the action of a torque L_1 producing a torsional displacement despite the high stiffness against this motion. Such situations appear in the modeling of the drill column of an oil derrick – see (Heisig 1993) – or of long lightweight space robotic arms moving large loads. In such cases eq. (41.d) applies, i.e.

$$\tau' = \varphi_1' - w_1' \varphi_1' \tag{60}$$

yields a geometric stiffness term, which needs to be considered in multibody simulation. The latter example was described in (Föppl 1900). A generalisation of these ideas of how to consider geometric stiffening in case of beams to finite element structures was given in (Wallrapp and Sachau 1994, Sachau 1996).

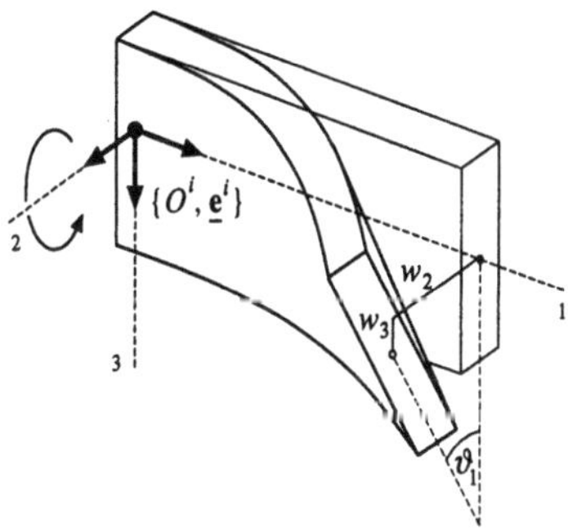

Fig. 12. Coupling of translational (lateral) and rotational displacements of cross sections of a beam due to bending about the 2-axis.

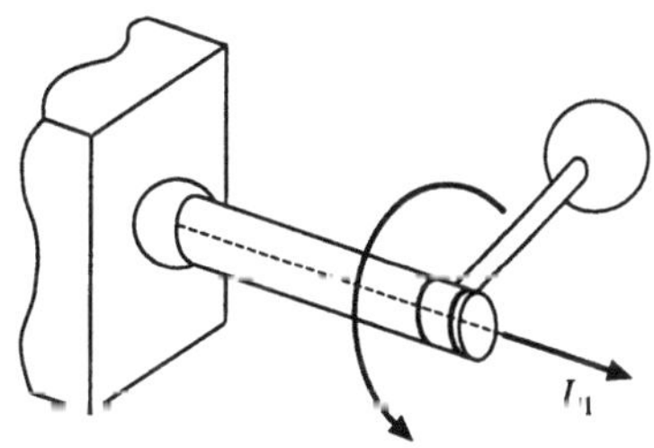

Fig. 13. An example requiring to consider geometric stiffening due to the coupling of (longitudinal) translational and rotational displacements of the cross sections resulting from torsion.

3.3.5 Simulation Results

The reference motion of the beam in Fig. 9 is now selected to be given by

$$\omega(t) = \left\{ \begin{array}{ll} 0 & \text{if } t \le 0 \\ \dfrac{\Omega}{T_0}\left(t - \dfrac{T_0}{2\pi}\sin\left(2\pi t / T_0\right)\right) & \text{if } 0 \le t \le T_0 \\ \Omega & \text{if } t \ge T_0 \end{array} \right\} \text{where} \left\{ \begin{array}{l} T_0 = 30\ \text{s}, \\ \Omega = 3\ \text{rad/s}. \end{array} \right. \tag{61}$$

The function represents a spin-up-manoeuvre. The data of the beam (approximating the NASA-minimast of Fig. 6) are

$$\left. \begin{array}{l} L = 20.3\ \text{m}, \quad \mathcal{E} = 1.17 \cdot 10^{11}\ \text{N/m}^2, \quad A = 4 \cdot 10^{-4}\ \text{m}^2, \\ \rho_0 = 1.2167 \cdot 10^4\ \text{kg/m}^3, \quad J_{33} = 9.83 \cdot 10^{-5}\ \text{m}^4. \end{array} \right\} \tag{62}$$

We compare the motions of the beam undergoing the spin-up-manoeuvre as obtained in three cases:

1. consideration of all the linear terms by including in the equation for the longitudinal motion in addition to the terms of the order $O(\varepsilon^0) = O(1)$ those $O(\varepsilon)$-terms, which are neglected in eq. (48.a);
2. equations (48), as provided by the perturbation techniques, discussed above;
3. equations (48) without the geometric stiffening terms $EA(w_1' w_3')' = (Z w_3')'$.

A solution of the partial differential equations (48) of motion is obtained by representing the displacement quantities w_1 and w_3 as proposed in eq. (17), i.e.

$$\left.\begin{aligned} & w_i(R_1,t) = \boldsymbol{\Phi}_i^T(R_1)\mathbf{q}_i(t) \\ & \Rightarrow \quad \boldsymbol{\Phi} = \begin{bmatrix} \boldsymbol{\Phi}_1^T & \mathbf{0} \\ \mathbf{0} & \boldsymbol{\Phi}_3^T \end{bmatrix}, \quad \boldsymbol{\Phi}_i^T = \left[\Phi_{ij}\right], \quad \mathbf{q} = \begin{bmatrix} \mathbf{q}_1 \\ \mathbf{q}_3 \end{bmatrix}, \quad \mathbf{q}_1 = \left[q_j\right], \quad \mathbf{q}_3 = \left[q_j\right] \\ & \text{with } \; j = 1, 2, \ldots n_{q1} \text{ for } i = 1 \text{ and } j = n_{q1}+1, n_{q1}+2, \ldots n_q \text{ for } i = 3. \end{aligned}\right\} \tag{63}$$

The Ritz procedure requires the shape functions to satisfy the boundary conditions

$$\Phi_{1j}(0) = 0, \quad \Phi_{1j}'(L) = 0, \quad \Phi_{3j}(0) = 0, \quad \Phi_{3j}'(0) = 0, \quad j = 1, 2, \ldots n_q. \tag{64}$$

Based on the representation of motion as suggested by eq. (63) the equations of motion[3] become

$$\mathbf{M}\ddot{\mathbf{q}} + \mathbf{D}\dot{\mathbf{q}} + \mathbf{K}\mathbf{q} = \mathbf{h} \quad \text{where} \quad \left\{\begin{aligned} & \mathbf{M} = \begin{bmatrix} \mathbf{M}_{11} & \mathbf{0} \\ \mathbf{0} & \mathbf{M}_{33} \end{bmatrix}, \quad \mathbf{D} = \begin{bmatrix} \mathbf{0} & \mathbf{D}_{13} \\ \mathbf{D}_{31} & \mathbf{0} \end{bmatrix}, \\ & \mathbf{K} = \begin{bmatrix} \mathbf{K}_{11} & \mathbf{K}_{13} \\ \mathbf{K}_{31} & \mathbf{K}_{33} \end{bmatrix}, \quad \mathbf{h} = \begin{bmatrix} \mathbf{h}_1 \\ \mathbf{h}_3 \end{bmatrix}. \end{aligned}\right. \tag{65}$$

The matrix-elements in eq. (65) are integrals over the shape functions $\boldsymbol{\Phi}_i(R_1)$, i.e.

$$\mathbf{M}_{11} = +\int_0^L \rho_0 A \boldsymbol{\Phi}_1 \boldsymbol{\Phi}_1^T \, dR_1 \qquad \mathbf{M}_{33} = +\int_0^L \rho_0 A \boldsymbol{\Phi}_3 \boldsymbol{\Phi}_3^T \, dR_1 \tag{66}$$

$$\mathbf{D}_{13} = +2\omega \int_0^L \rho_0 A \boldsymbol{\Phi}_1 \boldsymbol{\Phi}_3^T \, dR_1 \quad \mathbf{D}_{31} = -2\omega \int_0^L \rho_0 A \boldsymbol{\Phi}_3 \boldsymbol{\Phi}_1^T \, dR_1 = -\mathbf{D}_{13} \tag{67}$$

---

[3] The equations are derived conveniently when formulating Hamilton's principle (24) in terms of $\mathbf{q}$, while considering eq. (63) and not by introducing eq. (63) into the partial differential equations of motion, such as eq. (48), in terms of w_1 and w_3.

$$\left.\begin{aligned}
\mathbf{K}_{11} &= \mathbf{K}_{11l} - \omega^2 \mathbf{M}_{11} & \mathbf{K}_{33} &= \mathbf{K}_{33l} - \omega^2 \mathbf{M}_{33} + \omega^2 \mathbf{K}_{geo} \\
\mathbf{K}_{11l} &= \int_0^L \mathcal{E} A \, \mathbf{\Phi}_1' \mathbf{\Phi}_1'^T \, dR_1 & \mathbf{K}_{33l} &= \int_0^L \mathcal{E} J_{22} \, \mathbf{\Phi}_3'' \mathbf{\Phi}_3''^T \, dR_1 , \\
& & \mathbf{K}_{geo} &= \frac{1}{2} \int_0^L \varrho_0 A \left(L^2 - R_1^2 \right) \mathbf{\Phi}_3' \mathbf{\Phi}_3'^T \, dR_1 \\
\mathbf{K}_{13} &= + \dot{\omega} \int_0^L \varrho_0 A \, \mathbf{\Phi}_1 \mathbf{\Phi}_3^T \, dR_1 & \mathbf{K}_{31} &= - \dot{\omega} \int_0^L \varrho_0 A \, \mathbf{\Phi}_3 \mathbf{\Phi}_1^T \, dR_1 = - \mathbf{K}_{13}
\end{aligned}\right\} \quad (68)$$

$$\mathbf{h}_1 = + \omega^2 \int_0^L \varrho_0 A R_1 \mathbf{\Phi}_1 \, dR_1 \qquad \mathbf{h}_3 = + \dot{\omega} \int_0^L \varrho_0 A R_1 \mathbf{\Phi}_3 \, dR_1 \, . \quad (69)$$

In case 2 – the case corresponding to eqs. (48) – the equations are reduced into

$$\left.\begin{aligned}
\mathbf{K}_{11l} \quad \mathbf{q}_1 &= \mathbf{h}_1 , \\
\mathbf{M}_{33} \ddot{\mathbf{q}}_3 + \left(\mathbf{K}_{33l} - \omega^2 \mathbf{M}_{33} + \omega^2 \mathbf{K}_{geo} \right) \quad \mathbf{q}_3 &= \mathbf{h}_3 .
\end{aligned}\right\} \quad (70)$$

We use the first eigenfunctions for longitudinal and lateral motion of the beam, as shown in Fig. 14, as shape functions. A consideration of additional eigenfunctions does not change the following results significantly.

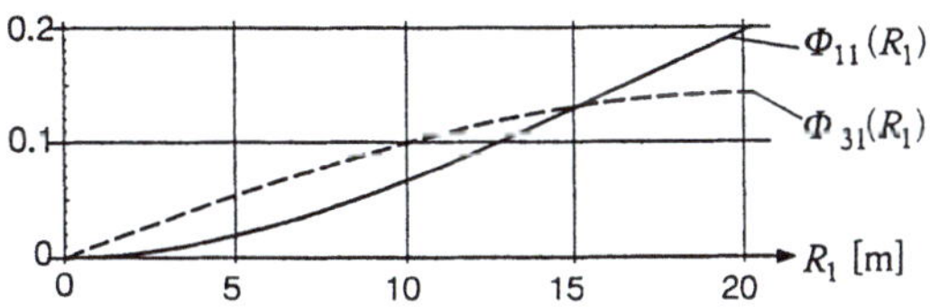

Fig. 14. Eigenfunctions for bending (Φ_{11}) and axial vibration (Φ_{31}) of the beam corresponding to natural frequencies of 2.0875 Hz and 38.19 Hz.

Evaluating the integrals (66) to (69) for this choice of shape functions yields

$$\left.\begin{aligned}
&\mathbf{M}_{11} = \mathbf{M}_{33} = M = 1.0 , && \mathbf{D}_{13} = - \mathbf{D}_{31} = D = 1.91728 \, \omega , \\
&\mathbf{K}_{13} = - \mathbf{K}_{31} = K = 0.95864 \, \dot{\omega} , && \\
&\mathbf{K}_{11l} = K_{11l} = 57577 , && \mathbf{K}_{33l} = K_{33l} = 172.034 , \\
&\mathbf{K}_{geo} = K_{geo} = 1.19334 \, \omega^2 , && \\
&\mathbf{h}_1 = h_1 = 115.649 \, \omega^2 , && \mathbf{h}_3 = h_3 = 114.774 \, \dot{\omega} .
\end{aligned}\right\} \quad (71)$$

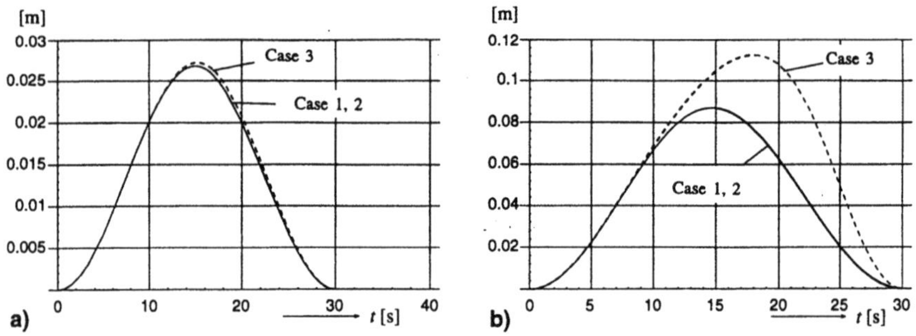

Fig. 15. Lateral displacement $w_3(L,t)$ of the tip of the beam for the three cases and for two stationary values Ω of $\omega(t)$: a) $\Omega = 3$ rad/s; b) $\Omega = 10$ rad/s.

The solutions of the equations of motion (65) for the three cases defined above are given in Figs. 15 and 16. Figure 15 shows the lateral displacement $w_3(L,t)$ of the tip of the beam for two angular velocities Ω – see eq. (61). Figure 15.a applies for $\Omega = 3$ rad/s and it demonstrates that there is no need to consider geometric stiffening in such a case because the centrifugal forces are too small. This may be concluded as well by evaluating $\mathbf{K}_{33}$ and in particular $\mathbf{K}_{geo}$ as given by eq. (68) using the values given in eq. (71). By contrast, geometric stiffening becomes important when the angular velocity Ω is increased to $\Omega = 10$ rad/s – see Fig. 15.b. Both diagrams demonstrate that there is no difference between the cases 1 and 2 as far as the accuracy of the result is concerned.

Figure 16 shows the longitudinal displacement $w_1(L,t)$ of the tip of the beam for the three cases and for $\Omega = 3$ rad/s – no difference appears within the accuracy of the diagram: the elongation of the beam follows the centrifugal force h_1 as given by eq. (71). This is also true for $\Omega = 10$ rad/s – just the final elongation is increased in the latter case.

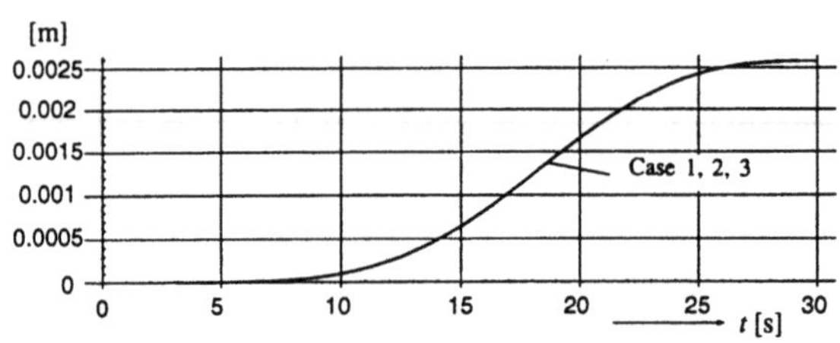

Fig. 16. Longitudinal displacement $w_1(L,t)$ of the tip of the beam for the three cases and for $\Omega = 3$ rad/s.

A major difference between the various models is observed in computational performance, however. Because of the presence of the high frequency of 38.19 Hz in case 1, there appear small high frequency oscillations due to the nonzero inertia term $\mathbf{M}_{11}\ddot{\mathbf{q}}_1$. These oscillations slow down the numerical integration of the equations of motion without enhancing accuracy of the result.

4. Large Deformation Analysis

The floating frame of reference formulation is used to deal with small deformation problems. Recent space applications – e.g. the analysis of the motion of an antenna mounted parallel to the axis of a rotating spacecraft or the generation of virtual reality worlds for surgeon-training – require a large deformation analysis. The two methods mentioned in Section 2.1 allow a solution of such problems, but they also result in the problems mentioned there. In particular, the incremental finite element formulation is difficult to apply to dynamical problems – see (Shabana 1996 a) – because the beam and plate elements as used in this method represent a rigid motion of the elements to a linear approximation only. To overcome the resulting problems a new methodology has been proposed, the "absolute nodal coordinate formulation" (Shabana 1996 b, Shabana and Schwertassek 1997). Its basic idea is to give up using nodal coordinates of finite elements, which have a simple geometric interpretation such as the angles describing the orientation of the cross sections at the nodes of a beam element. Rather, one introduces suitable combinations of material derivatives of the nodal coordinates to describe the orientation of the cross sections. As a consequence one arrives at nodal coordinates representing large rotations of the elements correctly.

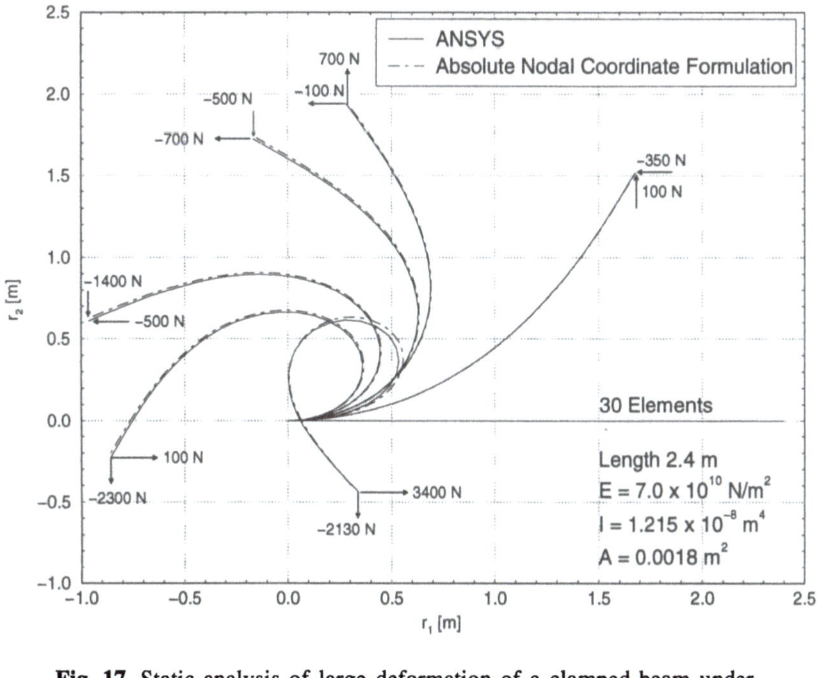

Fig. 17. Static analysis of large deformation of a clamped beam under the action of forces at its right endpoint.

Pursuing such ideas the diagram in Fig. 17 was generated (Dombrowski 1997). It shows the deformation of a clamped beam under the action of forces applied at its end, as obtained by the nodal coordinate formulation and by the incremental finite element formulation available in ANSYS. The results are in good agreement. In particular, the torque required to bend a beam into a circle is known to be $M = 2\pi \mathcal{E} I / L$. The parameters given in the figure yield $M = 2258\ Nm$ by comparison with $M = 2227\ Nm$ provided by the numerical procedure. Thus, for *static analysis* the new method yields the same results as well-established and tested procedures but it has a significant advantage for an analysis of *dynamical problems*. Contrary to the incremental finite element formulation and to the large rotation vector approach it yields correct generalised masses and inertias and a nonredundant representation of rigid body rotation. Exploring these concepts in more detail may help to unify the methodologies of multibody dynamics and finite element analysis.

References

Botz, M. (1992): Zur Dynamik von Mehrkörpersystemen mit elastischen Balken, Technische Hochschule Darmstadt, Fachbereich Mechanik, Dissertation.

Braune, W. and Fischer, O. (1889): Über den Schwerpunkt des menschlichen Körpers mit Rücksicht auf die Ausrüstung des deutschen Infanteristen. Abh. d. königl. sächs. Gesellschaft der Wissenschaften, 26, pp. 560-672.

Bremer, H. (1983): Kinetik starr-elastischer Mehrkörpersysteme. Fortschr. Ber. VDI-Z, Vol. 53(11), VDI-Verlag, Düsseldorf.

Bremer, H. and Pfeiffer, F. (1992): Elastische Mehrkörpersysteme. Teubner Studienbücher, Mechanik, B. G. Teubner, Stuttgart.

Coddington, E. A. and Levinson, N. (1955): Theory of Ordinary Differential Equations. McGraw-Hill Book Company, New York.

v. Dombrowski, S. (1997): Modellierung von Balken bei großen Verformungen für ein kraftreflektierendes Eingabegerät, Universität Stuttgart, Diplomarbeit.

Eichberger, A. (1993): Simulation von Mehrkörpersystemen auf parallelen Rechnerarchitekturen, Universität-Gesamthochschule Duisburg, Fachbereich Maschinenbau, Dissertation.

Eichberger, A. (1994): Transputer-Based Multibody System Dynamic Simulation, Part I: The Residual Algorithm - A Modified Inverse Dynamic Formulation, Part II: Parallel Implementation - Results. Mechanics of Structures and Machines, 22(2), pp. 211-261.

Eichberger, A., Führer, C., and Schwertassek, R. (1994): The Benefits of Parallel Multibody Simulation. Intl. Journal for Numerical Methods in Engineering, 37, pp. 1557-1572.

Fischer, O. (1906): Theoretische Grundlagen für eine Mechanik der lebenden Körper. B.G. Teubner, Leipzig.

Fischer, O. (1907): Kinematik organischer Gelenke. F. Vieweg, Braunschweig.

Föppl, A. (1900): Vorlesungen über Technische Mechanik, 3. Band, Festigkeitslehre. B. G. Teubner, Leipzig.

García de Jal6n, J. and Bayo, E. (1994): Kinematic and Dynamic Simulation of Multibody Systems, The Real-Time Challenge. Mechanical Engineering Series, ed. F. F. Ling, Springer-Verlag, New York.

Gaul, L., Lenz, J., and Sachau, D. (1997): Active Damping of Space Structures by Contact Pressure Control in Joints, 15th Intl. Modal Analysis Conf. (IMAC), Orlando, Florida.

Haug, E. J. (1989): Computer-Aided Kinematics and Dynamics of Mechanical Systems, Volume I: Basic Methods. Allyn and Bacon, Boston.

Heisig, G. (1993): Zum statischen und dynamischen Verhalten von Tiefbohrsträngen in räumlich gekrümmten Bohrlöchern, Technische Universität Braunschweig, Dissertation.

Hodges, D. H. and Ormiston, R. A. (1980): On the Nonlinear Deformation Geometry of Euler-Bernoulli Beams, NASA, TP 1566.

Hodges, D. H. (1985): Nonlinear Equations for Dynamics of Pretwisted Beams Undergoing Small Strains and Large Rotations, NASA, TP 2470.

Houbolt, J. C. and Brooks, G. W. (1957): Differential Equations of Motion for Combined Flapwise Bending, NASA, TN 3905.

Huston, R. L. (1990): Multibody Dynamics. Butterworth-Heinemann, Boston.

Kane, T. R., Ryan, R. R., and Banerjee, A. K. (1987): Dynamics of a Cantilever Beam Attached to a Moving Base. J. Guidance, Control, and Dynamics, 10(2), pp. 139-151.

Knothe, K. and Wessels, H. (1991): Finite Elemente. Springer-Verlag, Berlin.

Kortüm, W., Sachau, D., and Schwertassek, R. (1995): Analysis and Design of Flexible and Controlled Multibody Systems with SIMPACK, Paper IAF-95-A.4.04, 46th Intl. Astron. Congress, Oslo, Norway.

Likins, P. (1989): Robert E. Roberson: A Personal Tribute. Mech. Struct. & Mach., 17(2), pp. 131-133.

Love, A. E. H. (1944): A Treatise on the Mathematical Theory of Elasticity. First American Printing of the Fourth Edition 1927, Dover Publications, New York.

Man, G. K. and Laskin, R. A. (1988): Proc. of the Workshop on Multibody Simulation, JPL, Pasadena, Report D-5190.

Nayfeh, A. H. and Mook, D. T. (1979): Nonlinear Oscillations. John Wiley&Sons.

Nikravesh, P. E. (1988): Computer-Aided Analysis of Mechanical Systems. Prentice Hall, Englewood Cliffs, NJ.

Otter, M. et al. (1993): An Object Oriented Data Model for Multibody Systems, in Proc. Int. Symp. on Advanced Multibody System Dynamics, (W. Schiehlen, ed.), Kluwer Acad. Publ., pp. 19-48.

Roberson, R. E. (1977): Computer-oriented dynamic modeling of spacecraft: historical evolution of Eulerian multibody formalisms since 1750, IAF-Paper 77-A 11, 28th International Astronautical Congress, Prag.

Roberson, R. E. and Schwertassek, R. (1988): Dynamics of Multibody Systems. Springer-Verlag, Berlin.

Rulka, W. (1990): SIMPACK - A Computer Program for Simulations of Large-Motion Multibody Systems, in Multibody System Handbook, (W. Schiehlen, ed.), Springer-Verlag, Berlin, pp. 265-284.

Rulka, W. and Eichberger, A. (1993): SIMPACK: An Analysis and Design Tool for Mechanical Systems, in Multibody Computer Codes in Vehicle System Dynamics, (W. Kortüm and R. S. Sharp, ed.), Vol. 22, Supplement to Vehicle System Dynamics, Swets and Zeitlinger, Amsterdam, pp. 122-126.

Ryan, R. R. (1987): Flexible Multibody Dynamics: Problems and Solutions, in Proc. of Workshop on Multibody Simulation, (G. K. Man and R. A. Laskin, eds.), Vol. I, (3), JPL D-5190, Pasadena, CA, pp. 103-190.

Sachau, D. (1996): Berücksichtigung von Körperverformung und Fügestellen in der Mehrkörpersimulation mit Anwendung auf aktive Raumfahrtstrukturen, Universität Stuttgart, Dissertation.

Sandlass, T. (1996): Herleitung aller geometrischen Steifigkeitsterme eines Bernoulli-Balkens und Implementierung im SIMPACK-Präprozessor BEAM, Universität Stuttgart, Studienarbeit.
Schiehlen, W. (1986): Technische Dynamik. Teubner Studienbücher, Mechanik, B.G. Teubner, Stuttgart.
Schwertassek, R. and Roberson, R. E. (1986): A Perspective on Computer-Oriented Multibody Dynamical Formalisms and their Implementations, in Dynamics of Multibody Systems, IUTAM/ IFToMM Symposium Udine 1985, (G. Bianchi and W. Schiehlen, eds.), Springer-Verlag, Berlin, pp. 261-273.
Schwertassek, R. and Rulka, W. (1991): Aspects of Efficient and Reliable Multibody System Simulation, in Real-Time Integration Methods for Mechanical System Simulation, (E. J. Haug and R. C. Deyo, eds.), Springer-Verlag, Berlin, pp. 55-96.
Schwertassek, R. (1994): Modellierung deformierbarer Körper in Mehrkörpersystemen, Deutsche Forschungsanstalt für Luft- und Raumfahrt (DLR), Institut für Robotik und Systemdynamik, Oberpfaffenhofen, Bericht IB 515-94-19.
Schwertassek, R. (1995): On a Systematic Way to Compute Geometric Stiffness Terms in Multibody System Simulation, Paper AAS 95-391, AAS/AIAA Astrodynamics Specialist Conference, Halifax, Canada.
Shabana, A. A. (1989): Dynamics of Multibody Systems. J. Wiley & Sons, New York.
Shabana, A. A. (1996.a): An Absolute Nodal Coordinate Formulation for the Large Rotation and Deformation Analysis of Flexible Bodies, Department of Mechanical Engineering, University of Illinois at Chicago, Techn. Report MBS96-1-UIC.
Shabana, A. A. (1996.b): Incremental Finite Element Approach and Exact Rigid Body Inertia. ASME Journal of Mechanical Design, vol. 118(2), pp. 171-178.
Shabana, A. A. (1997): Flexible Multibody Dynamics: Review of Past and Recent Developments. to appear in Multibody Simulation.
Shabana, A. A. and Schwertassek, R. (1997): Floating Frame of Reference Formulation and Definition of Nodal Coordinates. accepted for publication in Intl. J. of Non-Linear Mechanics.
Simo, J. C. (1985): A Finite Strain Beam Formulation. Comp. Meth. Appl Mech. Eng., 49, pp. 55-70.
Simo, J. C. and Vu-Quoc, L. (1986): On the Dynamics of Flexible Beams under Large Overall Motions - The Plane Case, Parts I and II. J. Appl. Mech., 53, pp. 849-854 and 855-863.
Sorge, K. (1993): Mehrkörpersysteme mit starr-elastischen Subsystemen. Fortschr.-Ber. VDI, Reihe 11: Schwingungstechnik, Nr. 184, VDI-Verlag, Düsseldorf.
Szabo, I. (1979): Geschichte der mechanischen Prinzipien. Birkhäuser-Verlag, Basel.
Taylor, L. (1988): Proc. of the Workshop on Computational Aspects in the Control of Flexible Systems, NASA, TM-101578-Pt-1.
Truckenbrodt, A. (1980): Bewegungsverhalten und Regelung hybrider Mehrkörpersysteme mit Anwendung auf Industrieroboter. Fortschr.-Ber. der VDI-Zeitschr., Reihe 8: Meß-, Steuerungs- und Regelungstechnik, Nr. 33, VDI-Verlag, Düsseldorf.
Truesdell, C. (1964): Die Entwicklung des Drallsatzes. ZAMM, 44, pp. 149-158.
Vereshchagin, A. F. (1974): Computer Simulation of the Dynamics of Complicated Mechanisms of Robot-Manipulators. Engineering Cybernetics, (6), pp. 65-70.
Wallrapp, O. (1991): Simulation flexibler Mehrkörpersysteme bei Verwendung von FEM-Daten - Interface zwischen ANSYS und Mehrkörperprogrammen, ANSYS Users' Meeting, CAD-FEM, Ebersberg.
Wallrapp, O. (1993.a): Standard Input Data of Flexible Bodies for Multibody System Codes, DLR, German Aerospace Establishment, Institute for Robotics and System Dynamics, Oberpfaffenhofen, Report IB 515-93-04.

Wallrapp, O. (1993.b): Standard Input Data of Flexible Members for Multibody System Codes, in Advanced Multibody System Dynamics - Simulation and Software Tools, (W. Schiehlen, ed.), Kluwer Academic Publishers, Dordrecht, pp. 445-450.

Wallrapp, O. (1994.a): BEAM - A Pre-Processor for Mode Shape Analysis of Straight Beam Structures and Generation of the SID File for MBS Codes, User's Manual, Deutsche Forschungsanstalt für Luft- und Raumfahrt (DLR), Inst. Robotik und Systemdynamik, Oberpfaffenhofen, Report Version 3.0.

Wallrapp, O. (1994.b): Standardization of Flexible Body Modeling in Multibody System Codes, Part I: Definition of Standard Input Data. Mechanics of Structures and Machines, 22(3), pp. 283-304.

Wallrapp, O. and Sachau, D. (1994): Space Flight Dynamic Simulations Using Finite Element Analysis Results in Multibody System Codes, 2nd Intl. Conf. on Computational Structures Technology, Vol. Advances in Computational Mechanics, CIVIL-COMP Press, Athens, Greece.

Weidenhammer, F. (1970): Gekoppelte Biegeschwingungen von Laufschaufeln im Fliehkraftfeld. Ing. Archiv, 39, pp. 281.

Wittenburg, J. (1977): Dynamics of Systems of Rigid Bodies. Leitfäden der angewandten Mathematik und Mechanik, ed. H. Görtler, Vol. 33, B. G. Teubner, Stuttgart.

Parallelization of Large Mechanical Engineering Codes

A Case Study of the Multibody Simulation Package FEDEM

Arne Marthinsen[1] and Ole Ivar Sivertsen[2]

[1] SINTEF Applied Mathematics, N-7034 Trondheim, Norway.
Arne.Marthinsen@math.sintef.no
[2] Department of Machine Design and Materials Technology, Norwegian University of Science and Technology, N-7034 Trondheim, Norway.
ole.sivertsen@protek.unit.no

Summary. In this chapter we report some results from the EUROPORT project EUROPENGER–FEDEM and a porting project on a Cray T3E, and discuss the opportunities and problems involved in the parallelization of large industrial simulation codes in mechanical engineering. We give a brief presentation of the FEDEM code (Finite Element Dynamics in Elastic Mechanisms), which is a multidisciplinary, multibody simulation package for mechanical systems, and discuss the parallelization work. Real-world test problems have been provided by three end-users of the FEDEM code; we discuss some results from the benchmarking.

Keywords: Parallel computation, multibody simulation, mechanical engineering

1. Introduction

During the last few years, parallelization of simulation codes in many areas has become popular. Industry realizes that low efficiency in their codes is a problem, and assume that parallel implementations will speed up the computation. In a few years, if not already, industrial end-users will need more computing power than they can afford when buying a vector super computer, and parallel computation seems a natural solution.

The overall motion of a mechanical system is often simulated based on a nonlinear finite element formulation. Since large multibody simulations incorporating flexibility computations on the body elements are computer-intensive operations, a parallel implementation would allow for more accurate simulations at a reduced computational time and cost.

FEDEM (Finite Element Dynamics in Elastic Mechanisms) is a computer program for multibody simulation of mechanical systems. The program incorporates the flexibility of the different bodies of a system, using nonlinear finite element modeling. FEDEM provides a full dynamic simulation environment, including stress and displacement analysis at both mechanism and individual-link levels. One of the main reasons for using a simulation tool like FEDEM is that the costs for prototyping and testing are often high and sometimes even prohibitive due to the size, complexity and environment of the product.

FEDEM has been extensively used for analyzing industrial robots, automotive systems, marine, defense and space structures and mechatronic systems in general. The simulation package contains modules for finite element, mechanism, and control modeling. Simulation of the dynamic behavior is carried out for the multidisciplinary model by time integration. The FEDEM code has also an option for retracking of stresses at selected times for the different bodies of a mechanism. Eigenvalue analysis on the system model during simulation is also an option. Result presentation is performed as curve plots of the simulation variables, as animation of simulated motion of the mechanical structure and as animated motion simultaneously showing stresses and deformations.

Simulating a robot arm with FEDEM shows what effects motion will have on the different parts, thus giving hints as to whether the structures have to be enforced to grant reliable performance. This approach is also vital with satellite parts, bridges and car bodies. Development time and costs can be significantly reduced by allowing engineers to test design changes within short times. FEDEM provides a complete set of modeling tools: dynamic simulation and post processing modules including tools for analyzing stresses and deformations. While simulating, calculation of dynamic motion, stresses, and deformations may be done simultaneously. From an integrated software base for multidisciplinary simulation, optimization and design, the software partners of the porting consortium have extended the software capabilities by porting the code to a high performance computer environment.

FEDEM is a truly multidisciplinary simulation system that was developed further in the ESPRIT II #5524 MDS project and the ESPRIT III EUROPORT subproject #8586 EUROPENGER-FEDEM (see e.g. (Henriksson et al. 1995), (Colbrook et al. 1995), (Marthinsen and Sivertsen 1996), and (Marthinsen and Hansen 1995)). The parallel FEDEM was ported to a Cray T3E in a CUG-FEDEM project[1] (Marthinsen 1997). Information about FEDEM is available on the World Wide Web at URL

`http://www.fedem.no/`

2. The FEDEM Simulation Code

FEDEM is divided into three main parts: *prepro*, *simule*, and *stress*. While *prepro* is run only once at the start of a simulation, *simule* and *stress* are run every time-step and every n^{th} time-step, respectively, where n depends on the step size and the total time window. However, the time spent in *prepro* may significantly increase the total simulation time. Figure 1 shows the structure in the FEDEM code. The execution sequence proceeds from left to right.

What makes FEDEM special is that it integrates the flexibility of the links in the construction. In addition, FEDEM offers retracking of stresses at any

---

[1] A Cray University & Development Grant project

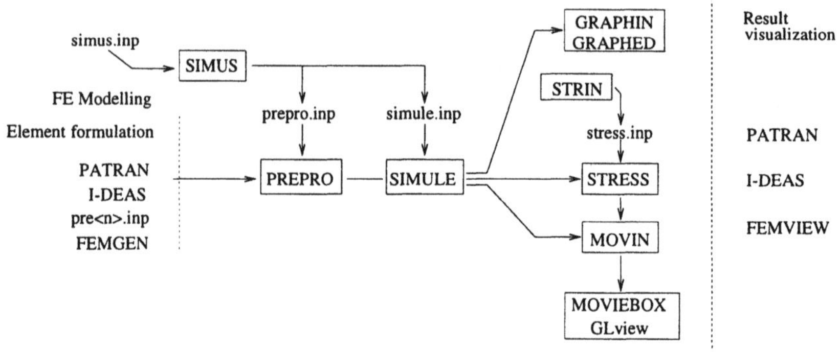

Fig. 1. Schematic overview of the structure in the FEDEM code.

time and at any location in global coordinates. Mechanical or mechatronic products may be modeled in much detail, for example, through modeling of flexible telescopic joints, through modeling of parallel joints between links, or through inclusion of distributed environmental forces on the links. The FEDEM system contains a comprehensive library of multibody simulation elements such as links, joints, springs, dampers, forces, drive elements, gears, Coulomb friction, and control elements. If no explicit finite element models has been made for the links, super elements are automatically generated from beam elements. The user has, however, a large library of beam, shell, and solid elements available for modeling. Some of the mathematical foundations for FEDEM are described in (Sivertsen 1994) and (Sivertsen and Waløen 1982). Below we describe the three main modules of FEDEM.

2.1 *prepro*

prepro is the preprocessor. It constructs the mass and stiffness matrices, and performs an eigenvalue analysis in order to reduce the internal degrees of freedom (DOF) by a Component Mode Synthesis (CMS) transformation. *prepro* has a library of elements that are used for building substructures for the different links in a simulation model. For each substructure, with external nodes fixed, a number of eigenvalues and corresponding eigenvectors may be calculated for the internal degrees of freedom, followed by transformation of super element mass and stiffness matrices.

The dynamic equation of motion for a substructure may be written as

$$\begin{bmatrix} M_{ee} & M_{ei} \\ M_{ie} & M_{ii} \end{bmatrix} \begin{bmatrix} \ddot{v}_e \\ \ddot{v}_i \end{bmatrix} + \begin{bmatrix} C_{ee} & C_{ei} \\ C_{ie} & C_{ii} \end{bmatrix} \begin{bmatrix} \dot{v}_e \\ \dot{v}_i \end{bmatrix} + \begin{bmatrix} K_{ee} & K_{ei} \\ K_{ie} & K_{ii} \end{bmatrix} \begin{bmatrix} v_e \\ v_i \end{bmatrix} = \begin{bmatrix} Q_e \\ Q_i \end{bmatrix},$$

where subscript e denotes external and i denotes internal degrees of freedom, or, in substructure matrix form,

$$M_s\ddot{v} + C_s\dot{v} + K_s v = Q_s,$$

where subscript s refers to substructure. The basis for the CMS transformation is to express the super element displacements by the external DOF, v_e, and by the generalized DOF, y, namely,

$$v = \begin{bmatrix} v_e \\ v_i \end{bmatrix} = \begin{bmatrix} I & 0 \\ B & \Phi \end{bmatrix} \begin{bmatrix} v_e \\ y \end{bmatrix} = H \begin{bmatrix} v_e \\ y \end{bmatrix},$$

where $B = -K_{ii}^{-1} K_{ie}$. Here, the eigenvector matrix $\Phi = [\phi_1, \phi_2, \ldots, \phi_S]$ is built from S eigenvectors defined by the eigenvalue problem

$$\left(K_{ii} - \omega^2 M_{ii}\right) \phi = 0,$$

typically those corresponding to the S smallest eigenvalues. This reduction leads to the equations

$$m \begin{bmatrix} \ddot{v}_e \\ \ddot{y} \end{bmatrix} + c \begin{bmatrix} \dot{v}_e \\ \dot{y} \end{bmatrix} + k \begin{bmatrix} v_e \\ y \end{bmatrix} + P = 0$$

where

$$m = H^T M_s H, \quad c = H^T C_s H, \quad k = H^T K_s H, \quad \text{and} \quad P = -H^T Q_s.$$

The substructures may be built from beam, shell or solid elements.

2.2 *simule*

simule is the mechanism simulation module that solves the general dynamic equation of motion, $F(t, r, \dot{r}, \ddot{r}) = 0$. The incremental form is iterated and solved:

$$M_k \Delta \ddot{r}_k + C_k \Delta \dot{r}_k + K_k \Delta r_k = Q_{k+1} - \left[F_k^I + F_k^D + F_k^S\right], \tag{1}$$

where F^I represents inertia forces, F^D represents damping forces and F^S represents elastic forces at the beginning of the time-step. The inertia forces at the beginning of the time-step also include Coriolis and gyroscopic forces. Moreover, Q contains the input loads and k denotes the time-step. The above matrices are banded, with bandwidth possibly quite large, and symmetric, but they are not necessarily positive definite.

Equation (1) may typically be solved with the well-known Newmark β-family of algorithms. The Newmark solution algorithm and the iteration process result in a set of linear equations to be solved, i.e.,

$$N_k \Delta r_k = \Delta \widehat{Q}_k,$$

where N_k is the Newton matrix at the system level and $\widehat{Q}_k$ represents either the external loading or the residual forces, at the kth iteration.

simule includes an integration algorithm with an automatic time-step control option, dynamic simulation in the specified time interval, a static equilibrium simulation option, an option to do eigenvalue analysis at specified time-steps, a restart option, and options for optimization and sensitivity analyses.

2.3 *stress*

stress calculates displacements and stresses in elastic links subjected to static or dynamic load in a process that may be thought of as the inverse of the CMS transformation. We essentially have that

$$v_i = v_i^i + v_i^e = \Phi y + B v_e$$

and the expanded substructure displacement vector may be written as $v = [v_i, v_e]^T$. The stress, according to Hooke's law, may then be written as

$$\sigma_j = E\widetilde{B}N b_j v,$$

where E is the elasticity matrix, $\widetilde{B}$ is the strain function matrix, N is the matrix of shape functions and b_j is the incidence matrix that represents the topology for primary element j in the substructure.

3. Parallel Programming: Languages and Communication

Most of the large industrial simulation codes today are written in FORTRAN 77 with some extensions possibly written in C. Many large hardware and software vendors have, however, constructed modern languages that are more suited for high performance computing.

Fortran 90 was accepted as a standard in 1990. It includes FORTRAN 77 as a subset but has extensions as, e.g., array operations, user-defined data types, pointers, and, perhaps most importantly, dynamic memory allocation. Many vendors offer Fortran 90 compilers, and Cray has in fact accepted Fortran 90 as their main compiler.

High Performance Fortran (HPF) is a specification of extensions to Fortran 90 developed by the High Performance Fortran Forum, which is a collaboration between vendors, universities, and research organizations. HPF supports data-parallel programming through automatic distribution of data, parallel statements, intrinsic functions, and standard libraries.

C and C++ are frequently used in the development of general applications, but they are not as widely used in high-performance computing. One of the reasons is the extensive use of pointers, which makes automatic vectorization or parallelization of the code difficult. There are, however, committees working on these problems, but no standard is currently available.

Automatic parallelization is normally taking the advantage of data-parallel sections in the code. This style is often used when the problem in nature is suited for parallel solution. The data are distributed onto the available processors, and all processors perform the same operations on their own data. Parallelization of data-parallel sections is normally easy and effective, and the use of optimized compilers may further simplify and improve the work. Automatic parallelization may be done by the compiler in the case of simple loops, but if the whole problem or large parts of it in nature is data-parallel, human interaction is needed.

The alternative parallelization approach is based on functional parallelization with message-passing communication. While automatic parallelization is most common on shared memory computers, message-passing is the basic communication system for distributed memory computers. Message-passing communication is more low level than automatic shared memory communication, since the user actually specifies how data are exchanged between the processors. Message-passing may be used on heterogeneous clusters as well as in homogeneous massively parallel computers. The most common message-passing library today is MPI (Message Passing Interface), but also PVM (Parallel Virtual Machine) and PARMACS (PARallel MACros) are still in use. Several hardware vendors have optimized versions of these libraries. In addition, most hardware vendors have their own message-passing library, e.g. NX/2 from Intel, MPL from IBM, and SHMEM from Cray.

4. Parallelization Strategies

Before porting industrial applications one should consider the suitability of the code for parallel implementation, whether functional parallelism or data-parallelism should be exploited, the potential scalability of the code, and how much effort to spend in parallelizing the code. Finite element computations are usually well suited for parallelization. Parallel linear algebra routines are in many cases very efficient, and construction of system matrices (stiffness and mass matrices, together with right-hand side vectors) may be done in a distributed fashion. In addition, evaluation of displacements and retracking of stresses may be done in parallel for each element in the model.

The FEDEM code is written in FORTRAN 77 and consists of several hundred-thousand lines of code. The data-structure is quite typical for programs written in this language. The program first allocates a huge data array and divides it into two main parts. Almost all data used by the program are stored in this memory. The first part is an array of indices in the second part, which is the data storage. Based on a certain data format, both integers and floating-point numbers are stored in the data part. As a consequence of this data format, all information about the problem is property oriented, i.e., stored as long vectors that contain the same property for all elements in the model. This data format is thus not suited for efficient partition of the computational work without much duplication of information, and therefore, efficient distributed manipulation of the data may be difficult without a complete rewriting of the data-structure.

FEDEM is based on substructuring of the physical problem, and this opens for a natural high level parallelization suitable for implementation in both the *prepro* and *stress* modules. Because of this, the steps in porting

FEDEM from serial to a full distributed program have differed much in complexity and time consumption. Because of the size and the complexity of the code, we have not performed any thorough analyses of how well the different parts are suited for parallelization. However, along with the description of the parallelization strategies for each main component of FEDEM in Section 4.2, we have discussed briefly what gains we would expect. The parallelization is, roughly speaking, done using a combination of three different steps:

substructure decomposition: the first step is to split the problem based on the natural substructure decomposition of the construction. This decomposition, however, is not a scalable approach, and the substructures may vary in size and computational requirements. On the other hand, this is a natural way of splitting the problem;

parallel linear algebra: the second step includes parallelization of the linear algebra routines used in the computation at the substructure level;

parallel updating of system matrices: the third step involves the distributed storage of system matrices and parallel updating on each processor

Note that FEDEM is a very large and complex code, and that it is, therefore, difficult to parallelize in a scalable way. Not only the most time-consuming parts should be rewritten – all parts must use the distributed data-structures in order to reduce the serial execution time within the parallel code. The parallelization of FEDEM is done using message-passing communication.

4.1 Data Distribution

When doing parallel computations, the data distribution strategy will often have a large impact on the performance of the parallel code. For an algorithm to perform efficiently, data must be available to each processor when needed, and the need for communication should be minimized. If parallelization is considered when writing the program from scratch, the data-structures may be designed to optimize efficiency on distributed-memory platforms. If, on the other hand, data-structures have been designed to suit the needs of a vector machine, much effort is needed to efficiently parallelize the code. In the industrial codes available today, the last approach is dominating. FEDEM also conforms to this picture.

In order to minimize the communication in FEDEM, we perform all assembly and computation with data distributed on all the processors. Still we have some communication during the assembly process, but afterwards we can apply the matrix manipulation routines directly to the data on each processor. An alternative approach would be to assemble all the system matrices on one processor, and whenever matrix manipulation is to be done, distribute the data onto the processors and collect them after the calculation has been completed. This would introduce a vast amount of communication activity in the interprocessor network, and would require a large memory on one single processor.

There exist a number of third-party libraries for parallel linear algebra, and several hardware vendors have supplied optimized parallel software routines. These libraries should be used (preferably public domain solvers to ensure portability) if data-structures can be effectively rewritten to conform to the needs of the libraries. It is, however, not very likely that the data-structures in a large industrial code correspond directly to those of the third-party libraries. The efficiency of the code drops rapidly if intensive communication activity is needed to adjust data before calling linear algebra routines. In large projects, it may therefore be decided to write the necessary routines from scratch to suit the actual data-structure.

4.2 Parallelization Strategies for the Modules in FEDEM

In this subsection we briefly explain the parallelization strategies that are implemented in the parallel version of FEDEM.

Parallelization of *prepro*: The algorithm implemented in *prepro* is:

for substructures smaller than a given size
  do in parallel
    perform eigenvalue analysis on each substructure
    reduce internal degrees of freedom
  enddo
end
for the rest of the substructures do
  perform distributed eigenvalue analysis
  perform distributed reduction of the system
enddo

Depending on the size of each substructure relative to machine-dependent parameters, it is decided whether the substructure should be computed on a single processor. If a small substructure is distributed onto several processors, the communication in the simulation process will be significant compared to the computation time, and increased simulation time may in fact be the result.

We may expect significant speed-up when the small substructures are simulated in parallel, since the work is totally decoupled and the need for communication is minimal. For large substructures there is more communication activity involved and a more thorough analysis is required to predict the speed-up.

Parallelization of *simule*: The algorithm implemented in *simule* is:

do in parallel (in a distributed fashion)
  perform distributed eigenvalue analysis if requested
  integrate next time-step:

update system matrices (in a distributed fashion)
solve linear equations (in a distributed fashion)
enddo

Note that the time integration itself is a typical serial application.
In the work on parallelizing FEDEM, we divided the work into two steps. The first step consisted of distributing data before and collect data after matrix manipulation was performed. All assembly of system matrices was done on one processor. This approach introduced a vast amount of communication and the need for much memory on a single processor. The second step was to perform all assembly and computation with data distributed on a number of processors. Still we have some communication, but now we can apply the matrix manipulation routines directly to the data on each processor. The data-structures initialized at the start of a simulation are never altered during the time-stepping (but the numerical data certainly changes).
simule is the most demanding part to parallelize efficiently. Without implementation of modified data-structures, we may expect the speed-up to be very limited; but with a suitable data-structure, *simule* should be able to take full advantage of effective parallel linear algebra routines.

Parallelization of *stress*: *stress* and *simule* should be run concurrently. When the time integrator has finished a time-step, the results are sent to *stress*. The two modules may thus, at the same time, work on time-step t_{n+1} and t_n, respectively. Results may be visualized as soon as *stress* has finished a time-step. This feature is not yet implemented, but it may be of future interest. The computation of the stresses by means of matrix multiplication can be done in parallel for all elements in the model. The following algorithm performs the inverse CMS transformation and computes the stress and displacement distributions in every element of the mechanism:

at the substructure level:
collect displacement vectors for each substructure from simule (global coordinates, including external DOF, and generalized coordinates)
do in parallel (in a distributed fashion)
compute displacement vectors for each substructure in local coordinates, including internal DOF
enddo
at the element level:
do in parallel for each element
compute stresses
enddo
end

If the whole simulation is performed without running *stress* concurrently, we may also parallelize over each time-step:

do in parallel for each time-step
perform the algorithm above
enddo

The algorithm is not completely implemented yet. So far, only the parallelization at the substructure level is implemented and tested.
The operations in *stress* are highly decoupled. Substructuring leads to the same expectations to speed-up as in *prepro*, but, in addition, we may ensure load-balancing by distributing work at the element level.

4.3 Practical Questions

Because of the special distributed data-structure in FEDEM, we decided to write linear algebra routines that effectively use this data-structure, instead of implementing existing parallel routines. The condition number of the system matrix in FEDEM may be very large. One has not succeeded in finding effective and cheap preconditioners, so a direct solution technique is implemented. The solver is based on an LDL^T-factorization of the system matrix, since it is not necessarily positive definite.

We have programmed message-passing using some communication primitives called *GC – Generalized Communication*, written at SINTEF Applied Mathematics,[2] that allows for using other communication libraries directly (Amundsen and Skålin 1995). *GC* is a package of subroutines that support efficient and portable communication. It is **not** a new communication library, but rather a consistent interface to many existing communication systems. *GC* supports both message-passing and shared memory communication, and has both a C and a FORTRAN interface. The primitives are provided free of charge.

Supported libraries as of Spring 1997 are MPI, PVM, NX/2, SHMEM, MPL and THD. The package allows the user to choose, at compile time, which underlying communication system to use, by using preprocessor flags. Information about *GC* is available on the World Wide Web at URL

`http://www.sima.sintef.no/GC`

5. Evaluation of the Parallelized Code

The purpose of benchmarking is to show the benefit of using parallel architectures for specific applications. The way the benefit is measured can vary. An important approach for the benchmarks in the EUROPORT 2 projects was to choose test cases which were relevant and commonly used in daily business. The "benefit" should not only incorporate the actual speed-up obtained, but also factors as, e.g., the gain in costs relative to the turnover,

[2] URL: `http://www.sima.sintef.no`

Fig. 2. Test Case 1: industrial robot from ABB Robotics Products AB

rationalization improvements, and the ability of growing as a company. The sum of these benefits should be sufficient such that the end-user will actually purchase a parallel computer or take parallel advantage of existing computers in the company.

5.1 Description of the Test Cases

The industrial partners have provided three real-world problems, pictured in Figures 2 to 4. These problems differ in size and complexity.

Test Case 1: *Robot Simulation Model "IRB 6000".* This test case is a medium-sized problem that models an industrial robot from ABB Robotics Products AB in Sweden. The robot model consists of 16 different parts, and the simulation model at the system level is composed of the super elements generated from the substructure finite element models and a number of system components. The model represents the actual robot design running through a typical robot cycle. FEDEM was a key tool during the development of the optimized IRB 6000 robot that is a commercially successful industrial robot.

Test Case 2: *Satellite Deployment Model.* This test case is a quite small problem modeling a satellite deployment for Dornier GmbH in Germany. The FEDEM simulations make it possible to study the dynamics of boom deployment and the flexible bodies undergoing latching of the booms at end-stop of deployment. Effects of different friction conditions in the joints are also studied.

Test Case 3: *Bridge Between Two North Sea Oil Platforms.* This test case is a simulation model of a personnel bridge between North sea oil platforms Veslefrikk A and B. The length of the bridge is 49 meters, and the model was

Fig. 3. Test Case 2: a satellite deployment model from Dornier GmbH

developed by SINTEF in a verification project for Statoil in Norway. The FE-DEM simulations make it possible to study the dynamics of the flexible bodies combined with telescopic motion of the bridge. Effects of different friction conditions in the sliding joints are also studied.

Fig. 4. Test Case 3: telescopic bridge between oil platforms Veslefrikk A and B in the North Sea from FEDEM AS

We have benchmarked the code on the following four platforms:

- Cray T3E with 64 nodes,[3].

[3] The Cray T3E has 64 DEC Alpha EV5 RISC processors with 128 MB memory each. It is a distributed memory MIMD machine with peak performance of 600 Mflops per processor, communication speed being 2×450 Mb/s.

- Intel Paragon XP/S with 56 nodes,[4]
- Parsytec GC/PowerPlus with 32 nodes,[5] and
- IBM SP2 with 8 nodes.[6]

Some selected results from the benchmarking of FEDEM are shown in Figures 5 to 8.

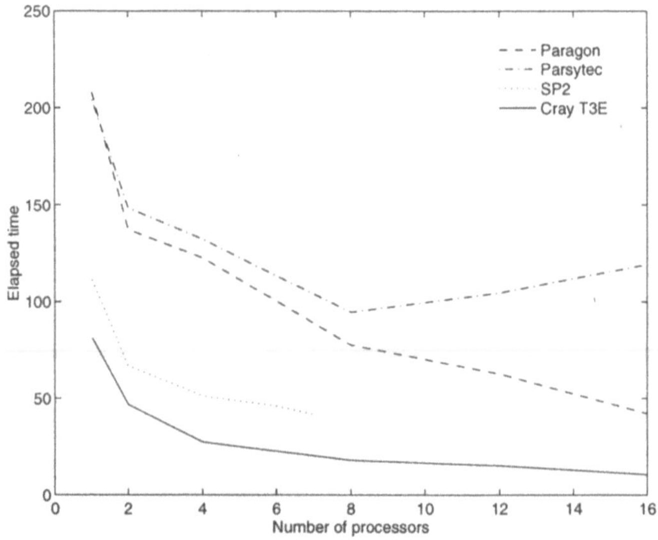

Fig. 5. Some results from Test Case 1, `prepro`

The benchmarking of the parallel FEDEM code has verified the theoretical differences between the benchmarking platforms. Cray T3E is by far the most powerful computer among the benchmarking platforms, both with regards to peak processor performance and communication speed. On the Parsytec GC/PP, the processors are quite fast, but the communication network is not scaled to match the computational performance. The sustained processor performance on the Intel Paragon is approximately as on the Parsytec, but the communication speed is significantly larger. The IBM SP2 has fast nodes, and the communication speed is approximately half the speed obtained on the Intel Paragon.

[4] The Intel Paragon has 32 MB memory on each node. Peak performance is 75 Mflops and communication speed is 80 Mb/s.

[5] Each node on the Parsytec consists of two PowerPC-601 and four T805 (for communication) with 64 MB memory. Peak performance is 160 Mflops and communication speed is 3.2 Mb/s.

[6] The nodes in the IBM SP2 are 590 RS/6000 with a total of 10.5 GB memory. Total peak performance is 15.4 Gflops and communication speed is 35 Mb/s.

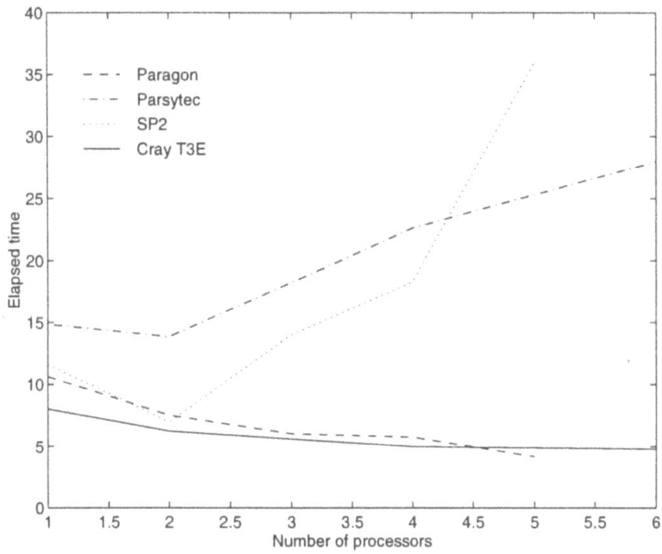

Fig. 6. Some results from Test Case 2, `prepro`

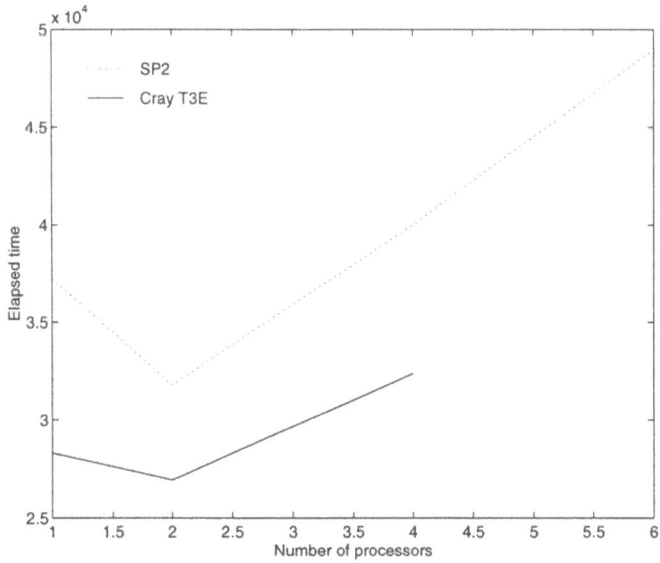

Fig. 7. Some results from Test Case 3, `simule`

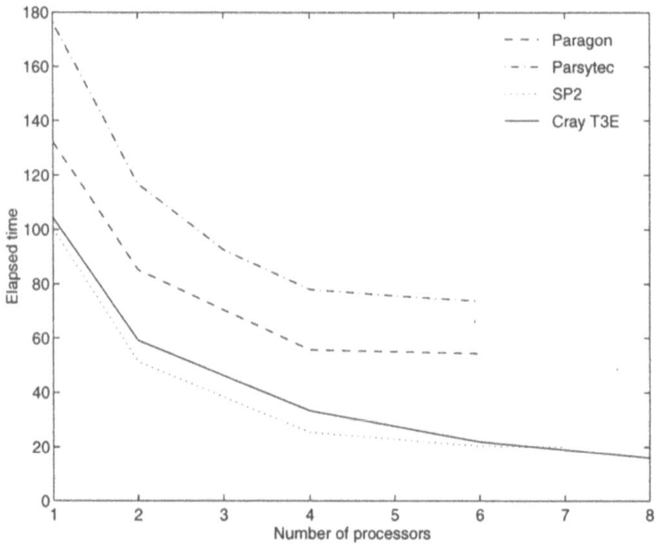

Fig. 8. Some results from Test Case 2, **stress**

Having said this, it should be stressed that if we did not know the theoretical performance of each of the benchmarking computers, the only information we could read from the graphs is the performance of the current parallel version of the code on the different computers. Optimization of the code with respect to communication or load balance may drastically alter the picture. Presentation of benchmarking results is very difficult, and one should be careful when drawing conclusions from the graphs. We cannot say anything about the general relative performance of the computers used in the benchmarking; we can conclude that this version of the parallel code runs best (to be understood as *uses least elapsed time on the simultations*) on a Cray T3E. We may also conclude that the code is scalable to a certain number of processors on a given computer. This does, however, not imply that the code exploits the computers in an optimal way.

Figure 5 shows the results from the benchmarking of **prepro** on Test Case 1 and Figure 6 shows the corresponding results on Test Case 2. The graphs show the elapsed time as a function of number of processors for each of the benchmarking computers. The execution time is significantly reduced in some parts of the parallel code, but the scalability is quite limited, as shown by the graphs. The early stage parallelization of **simule** is clearly reflected in Figure 7. The code includes much communication and when using more than two processors it dominates the picture. Results from **stress** on Test Case 2 are shown in Figure 8.

6. Concluding Remarks

FEDEM is a typical example of a computationally intensive industrial simulation code. The parallelization process has shown that it is difficult to rewrite such a system to a scalable parallel code. Ability to make efficient use of parallel machines with more than a modest number of processors (e.g. four to eight) in large industrial codes, seems far away. Although some parts of the code may scale very well, the overall process always includes a relatively large percentage that must be run serially. Despite this fact, the end-users support the parallelization work. They are interested in solving larger problems than those they are solving on existing vector computers, and not only in reducing the execution time of today's simulations (Gustafson 1988).

The speed-up of parallelized industrial codes is seldom as good as expected, and optimization must be considered. This is an iterative process, since removal of some bottlenecks will introduce or reveal some others, especially when parallelizing large industrial codes that involve significant amounts of sequential code. The parallelization process in FEDEM is quite young, and we have therefore not yet considered special optimization of, e.g., the I/O process, communication, and data storage. This will, however, be topics of future work.

The industrial relevance and some of the benefits from the parallelization of FEDEM may be summarized as follows:

- The design or redesign of an industrial product requires a number of design changes. For each design change, a dynamic simulation is required to verify the change in performance and reliability. To speed up this design process, it is important that the simulations run as fast as possible, because the effect of one design change often is the input for other design changes;
- The simulation times often set limits on how large simulation models it is practical to simulate. A parallel implementation will increase this limit significantly, especially because the parallelization effect is better the larger the model is;
- More detailed simulation models for increasing the accuracy in the simulations will increase the model size and, consequently, a parallel implementation will increase the reliability of the simulation results.

Industrial simulation codes are often used iteratively during the whole design process. For each proposed design modification, the simulation is run over again as a parameter study. It is thus essential that the elapsed time needed to do the simulation is as short as possible. Concurrent visualization of the results will allow the user to abort a simulation if the results do not look satisfactory, without having to wait for the whole simulation to come to an end. It is also important that the mental process of being creative (e.g. suggesting improved design solutions) not be interrupted unnecessarily by time-consuming simulations. Besides, being able to run many simulations efficiently will make it possible to study more concepts and design alternatives that, in the end, should give a better design. Also, in a competitive situation, being able to speed up the design process and achieve improved performance is of vital importance.

Having said this, note that speed-up of existing models is not the only requirement of the end-users. As important is that a parallel version of the code enables the simulation of larger problems, which cannot be run on sequential computers within reasonable time. The capability of running larger and more accurate simulations should be considered as very important.

Parallelization of general mechanical engineering codes is often a comprehensive work. Even though many of the computationally intensive operations are very well suited for parallelization (e.g. assembly of system matrices and solution of systems of linear equations), the resulting speed-up is quite limited. This is particularly true when parallelizing existing simulation codes. When developing new codes, however, the problems may be solved by carefully designed parallel algorithms using suitable data-structures. This will reduce development costs and increase performance of the code.

Because of more efficient modeling and more complex products, combined with increased performance of the computer hardware and utilization of parallel computing, larger simulation models are foreseen. Shorter product development times will increase the trend from prototyping and testing to more detailed modeling and simulation. More efficient simulation tools are therefore in strong demand from the designers.

Acknowledgement. We would like to thank Jørn Amundsen and Roar Skålin at SINTEF Applied Mathematics for enlightening discussions and valuable comments and ideas during the work with this chapter. This work was partly performed on the SINTEF Intel Paragon system, and was supported in part by Intel Corporation through the SINTEF/Intel Paragon Supercomputer Partnership Agreement. This work was in part supported by a Cray University Research & Development Grant. The authors would also like to thank the European Commission for supporting the porting of FEDEM through the EUROPENGER-FEDEM consortium in EUROPORT 2.

References

Amundsen, J. and Skålin, R. (1995): GC User's Guide. Technical report, SINTEF Report STF10 A95002

Colbrook, A., Elliott, J., Lemke, M., and Wray, F. (1995): EUROPORT2 - ESPRIT European Porting Action No 2. In B. Hertzberger and G. Serazzi, editors, *High-Performance Computing and Networking*, Lecture Notes in Computer Science, Vol. 919, Springer-Verlag, Berlin, 869–874

Gustafson, J. (1988): Reevaluating Amdahl's Law. *Communications of the ACM*, 31(5):532–533

Henriksson, T., Specht, B., Marthinsen, A., and Sivertsen, O. I. (1995): Final Benchmarking Report. Technical report, SINTEF Industrial Mathematics. EUROPORT project: *EUROPENGER-FEDEM*, Deliverable D6 (Confidential)

Marthinsen, A. (1997): Parallelization of the Multibody Simulation Package FEDEM on a Cray T3E. To appear. SINTEF Applied Mathematics, Norway

Marthinsen, A. and Hansen, T. H. (1995): Parallelization of the Multibody Simulation Package FEDEM. In B. Hertzberger and G. Serazzi, editors, *High-Performance Computing and Networking*, Lecture Notes in Computer Science, Vol. 919, Springer-Verlag, Berlin, 916–921. Also available as SINTEF Report STF10 S95001.

Marthinsen, A. and Sivertsen, O. I. (1996): A Parallel Version of the Combined Multibody, Control and Structural Dynamic Simulation Code FEDEM. In T. Iversen, editor, *Applied Modelling and Simulation – Proceedings of the 38th SIMS Simulation Conference*, 317–327. Tapir trykkeri, Norway. Also available as SINTEF Report STF42 S96501.

Sivertsen, O. I. (1994): Computed-Aided Modelling, Dynamic Simulation and Dimensioning of Mechanisms. Volume I: Fundamentals. Technical report, Department of Machine Design and Materials Technology, Norwegian Institute of Technology

Sivertsen, O. I. and Å. Ø. Waløen (1982): Non-linear Finite Element Formulation for Dynamic Analysis of Mechanisms with Elastic Components. *ASME Paper No. 82-DET-102*, 1982. Presented at the Mechanisms Conference, Washington DC, Oct. 12–15

Flexible Multibody Dynamics with Nonlinear Deformations: Vehicle Dynamics and Crashworthiness Applications

Jorge A.C. Ambrósio and Manuel F.O.S. Pereira

IDMEC Polo Instituto Superior Técnico, Av. Rovisco Pais, P-1096 Lisboa, Portugal
jorge@lemac.ist.utl.pt and d1023@beta.ist.utl.pt

Summary: Practical formulations to describe both linear and nonlinear deformations of flexible multibody systems are presented, discussed and applied to complex industrial systems. Initially the flexibility of the multibody system is described with rigid bodies connected by highly deformable elements using the finite segment and plastic hinge approaches. In turn, these formulations are used in the optimization of multibody systems. A Lagrangian formulation is used to describe the nonlinear geometric and material deformations of flexible multibody components. The coupling between the deformation and the rigid body motion is preserved though the method proposed only uses invariant diagonal mass matrices. Applications to the simulation and design for crashworthiness of transportation vehicles are presented to illustrate the use of these techniques. The formulation is also specialized to model flexible multibody systems with linear elastic deformations. It is shown that with the proposed approach, the body elastodynamics effects are all evaluated using standard commercial finite element codes. In this manner, the analyst can use any type of linear finite element in the multibody model and the same finite element model can also be used to evaluate the structural integrity of the system components. The methodology is applied to model a vehicle with a complex flexible chassis, which is simulated for different riding scenarios.

Keywords: Elastodynamics, plastic hinges, component mode synthesis, finite elements, impact, impact biomechanics, design sensitivity, optimization

1. Introduction

Multibody systems are generally complex arrangements of structural and mechanical subsystems with different design purposes and mechanical behavior. Depending on the type of applications, operating speeds, external or internal loading of the components, the multibody system may experience small or large deformations that lead to a change of the system performance. This is particularly well known in aerospace mechanisms where slender elements are present. Other cases include vehicle components under extreme conditions or machine rods operating at high speeds. The structures, on the other hand, may behave as multibody systems due to their large rotations or because they develop well defined mechanisms of deformation, as in crashworthiness applications.

Based on rigid multibody dynamics the system deformations can be described using lumped deformations modeled by spring-damper elements. Due to its simplicity, this approach has found its way into multibody systems where the system components are made of slender elements such as beams. Methodologies to describe flexibility effects such as the finite segment approach (Huston and Wang 1994) or the plastic hinge technique (Kamal 1970; Nikravesh, Chung and Benedict 1983; Ambrósio, Pereira and Dias 1996; Kindervater 1997) represent these methods. The advantage of these procedures lies in the simple mechanics associated with them and the small number of parameters required to describe the structural behavior that characterizes the lumped stiffness. This fact makes the lumped stiffness methods well suited to the optimal design of flexible multibody systems in vehicle dynamics and crashworthiness (Haug and Arora 1979; Dias and Pereira 1994). However, both the finite segment approach and the plastic hinge technique are unable to describe the coupling resulting from loading in different directions, such as beam-columns, or to capture other nonlinear flexible effects such as centrifugal stiffening. Moreover these methods assume that the mechanisms of deformation are known beforehand.

Finite element based methodologies have emerged in the past years to model the flexible effects of multibody systems efficiently (Erdman and Sandor 1972; Thompson and Sung 1986). Initially the description of the flexible effects have been superimposed on the gross rigid body motion with the structural deformations being calculated from the inertial loading of the mechanical components. However, the coupling between the gross motion and the system elastodynamics plays a important role in mechanisms operating at high speeds with slender elements. This problem has been addressed using finite element methods where the mass matrix includes the inertia coupling between the effects mentioned above (Song and Haug 1980; Shabana and Wehage 1983; Pereira and Proença 1991). Even multibody systems experiencing only elastic flexibility can have highly nonlinear deformations such as in the case of large aerospace structures on spin-up maneuvers. This problem has been approached by different researchers that proposed methodologies able to deal with such nonlinear geometrical effects (Kane, Ryan and Banerjee 1987; Cardona and Geradin 1988; Geradin, Cardona, Doan and Duysens 1994).

Based on an updated Lagrangian formulation the equations of motion for a flexible body experiencing general geometric and material nonlinear deformations have been obtained (Ambrósio and Nikravesh 1992). This is an integrated methodology which is able to capture the full nonlinear structural behavior of the system components and maintaining the body mass matrix constant while the inertial coupling is preserved. This methodology has been applied to a wide range of vehicle crashworthiness problems (Pereira, Ambrósio and Dias 1997; Ambrósio, Silva, and Gonçalves 1996; Ambrósio, Nikravesh and Pereira 1990).

The methodology is also specialized for multibody systems experiencing only linear elastic deformations. For this purpose the mode component synthesis is used to reduce the number of generalized elastic coordinates of the system (Gonçalves and Ambrósio 1996). It is shown that all information required in the methodology described to account for the system flexibility can be obtained directly from any commercial finite element code.

This work summarizes developments on flexible multibody dynamics methodologies, applicable to vehicle dynamics, flexible aerospace structures, crashworthiness analysis and design and general contact-impact problems which have been under development at the IDMEC - Instituto Superior Técnico.

2. Rigid Multibody Dynamics

Many novel methodologies for the description of rigid multibody systems have been proposed in the past two decades. In this work, Cartesian coordinates are used to describe the rigid multibody dynamics not because of their efficiency but because they are well understood. The objective is to describe the local nonlinear deformations that appear in the rigid multibody system as a result of contact or impact between the system components and external or internal obstacles and to incorporate the elastodynamics in the description of flexible multibody systems. Other formulations for rigid multibody systems have relative advantages and drawbacks but they will not be discussed here. These are summarized in recent books (Nikravesh 1988; Roberson and Schwertassek 1988; Haug 1989; Shabana 1989; García de Jalón and Bayo 1993; Pereira and Ambrósio 1994).

2.1. Rigid Multibody System Equations of Motion

A multibody system is a collection of rigid and flexible bodies joined together by kinematic joints and force elements as depicted in Fig. 2.1. The presence of the kinematic joints is the main reason why it is normally necessary to use an extended set of coordinates to describe the system. These are called the dependent coordinates and define uniquely the position and orientation of the body in the system.

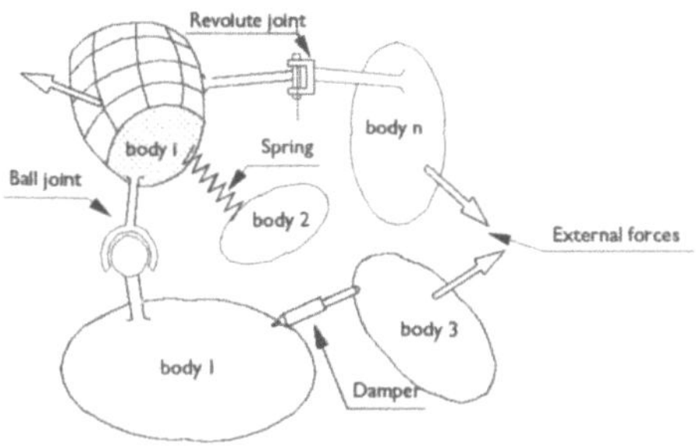

Fig. 2.1. Schematic representation of a multibody system

Reference Cartesian coordinates identify the position of each body through the position of a point and a set of parameters to define its angular orientation. For the ith body of the system $\mathbf{q}_i$ denotes a vector containing the translational coordinates $\mathbf{r}_i$, and a set of rotational coordinates $\mathbf{p}_i$. A vector of velocities for a rigid body i is defined as $\mathbf{v}_i$ contains a 3-vector of translational velocities $\dot{\mathbf{r}}_i$ and a 3-vector of angular velocities $\boldsymbol{\omega}_i$. The vector of accelerations for the body is denoted by $\dot{\mathbf{v}}_i$, the time derivative of $\mathbf{v}_i$. For a multibody system containing nb bodies, the vectors of coordinates, velocities, and accelerations are $\mathbf{q}$, $\mathbf{v}$ and $\dot{\mathbf{v}}$ that contain the elements of $\mathbf{q}_i$, $\mathbf{v}_i$ and $\dot{\mathbf{v}}_i$, respectively, for i=1,...,nb.

The constraint equations arise from the kinematic joints describing the relative motion between contiguous bodies. Let the kinematic constraints representing joints between rigid bodies be described by mr independent equations as

$$\boldsymbol{\Phi}(\mathbf{q}) - \mathbf{0} \tag{1}$$

The first and second derivatives of the constraints yield the kinematic velocity and acceleration equations, respectively

$$\dot{\boldsymbol{\Phi}}(\mathbf{q},t) = \mathbf{0} \quad \equiv \quad \mathbf{D}\,\mathbf{v} = \boldsymbol{\nu} \tag{2}$$

$$\ddot{\boldsymbol{\Phi}}(\mathbf{q},\dot{\mathbf{q}},t) = \mathbf{0} \quad \equiv \quad \mathbf{D}\,\dot{\mathbf{v}} = \boldsymbol{\gamma} \tag{3}$$

where $\mathbf{D}$ is the Jacobian matrix of the constraints. The equations of motion for the system of rigid bodies are written (Nikravesh 1988)

$$\mathbf{M}\ddot{\mathbf{q}} + \mathbf{D}^T\boldsymbol{\lambda} = \mathbf{g} \tag{4}$$

where $\mathbf{M}$ is the inertia matrix, $\boldsymbol{\lambda}$ is a vector of unknown Lagrange multipliers, and $\mathbf{g} = \mathbf{g}(\mathbf{q},\mathbf{v})$ contains the forces, moments and gyroscopic terms.

Eqs.(3) and (4) are a system of differential and algebraic equations that must be solved together to obtain the system accelerations. For the numerical solution of the system equations of motion a predictor-corrector, variable order algorithm is used (Shampine and Gordon 1975) together with a constraint violation stabilization scheme (Baumgarte 1972) or with an Augmented Lagrangian methodology (Garcia de Jalón and Bayo 1993). This algorithm also includes an error estimation and a step-size control feature which is particularly suitable for impact simulations where sudden forces may appear as a result of contact between different parts of the impacting structures.

2.2. Continuous Impact Force Including Local Deformations

During the contact between the components of the multibody and external or internal objects some level of local deformations are developed. When using a rigid multibody model to describe the system behavior extra care must be paid to the description of the impact between contacting surfaces. Even when the model accounts for the flexibility of the system components the local deformations are responsible for some level of energy dissipation. A model for the contact force must use the material and geometric properties of the surfaces in contact and contribute to a stable integration of the equations of motion. This is achieved with a continuous contact force model (Lankarani and Nikravesh 1990).

Let the contact force between two bodies of the system or between a component and an external object be a function of the pseudo-penetration and pseudo-velocity of penetration given by

$$\mathbf{f}_{s,i} = \left(K\delta^n + D\dot{\delta}\right)\mathbf{u} \tag{5}$$

where D is a damping coefficient and K is a generalized stiffness coefficient which depends on the geometry of the surfaces in contact and their material properties. The pseudo-penetration velocity $\dot{\delta}$ represents the actual relative velocity between the two contacting surfaces in the point of contact.

The damping coefficient D introduces the hysteresis form of the energy dissipation leading to the energy loss pictured by Fig. 2.2. This coefficient can be shown to be a function of the impact velocity $\dot{\delta}^{(-)}$, relative stiffness of the contacting surfaces and restitution coefficient e, and it is written as

$$D = \frac{3K\left(1-e^2\right)}{4\dot{\delta}^{(-)}}\delta^n \tag{6}$$

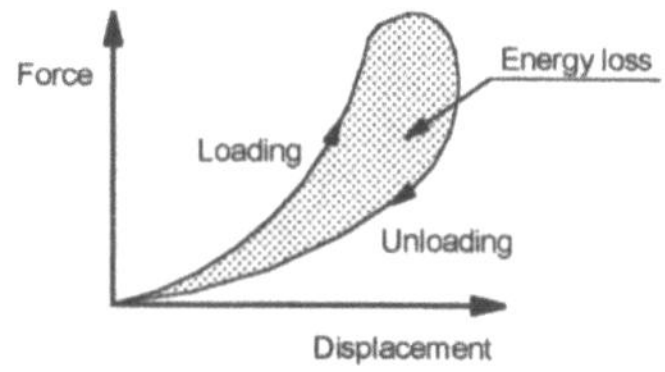

Fig. 2.2. Force displacement relationship accounting for histeresis

where the restitution coefficient e reflects the type of impact; for instance, for a fully elastic contact e is unity while for a fully plastic contact e is zero. The nonlinear contact force is obtained by substituting eq.(6) into eq.(5):

$$\mathbf{f}_{s,i} = K\ \delta^n \left[1 + \frac{3\left(1-e^2\right)}{4}\frac{\dot{\delta}}{\dot{\delta}^{(-)}}\right]\mathbf{u} \tag{7}$$

This equation represents impact conditions for contacting velocities lower than the propagation speed of elastic waves, i.e., $\dot{\delta}^{(-)} \leq 10^{-5}\sqrt{E/\rho}$.

2.3. Application to Biomechanical Impact Modeling

Biomechanical models of the human body for crashworthiness applications, with the exception of some aerospace applications, are commonly constructed with rigid bodies only. Local flexibility effects of the biomechanical segments are described by the contact and impact force model presented before. A three-dimensional, whole body response, biomechanical model of the human body is presented in Fig 2.3. Using a proper database, models with different sizes and masses can be easily obtained for different application cases.

The model is described using twelve rigid bodies interconnected by eleven kinematic joints, as represented in Fig. 2.3a. A set of contact surfaces is defined for the calculation of the external forces exerted on the model by the seat cushions, the floor or any other object coming in contact with a biomechanical component, including other model members. These surfaces are ellipsoids, as illustrated in Fig. 2.3b. Contact forces between seat and occupant components are described in Fig. 2.4. Joint resistance torques that characterize passive muscle action are modeled by viscous torsional dampers located in each kinematic joint. When two contiguous bodies interconnected by a joint reach physically unacceptable positions, penalty torques are applied to each body to limit such relative motion.

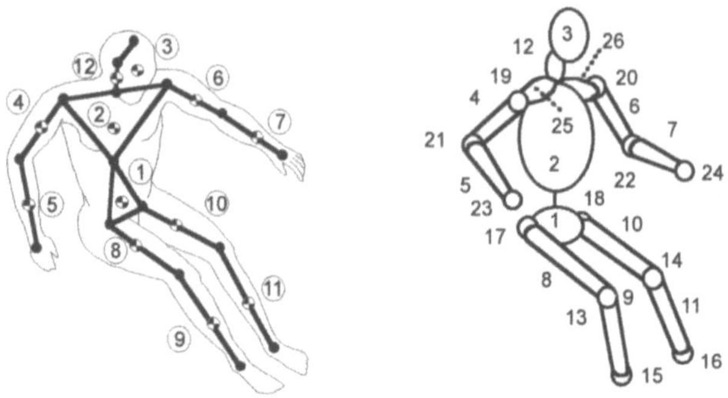

a) Bodies and joints

b) Contact surfaces

Fig. 2.3. Three-dimensional biomechanical model

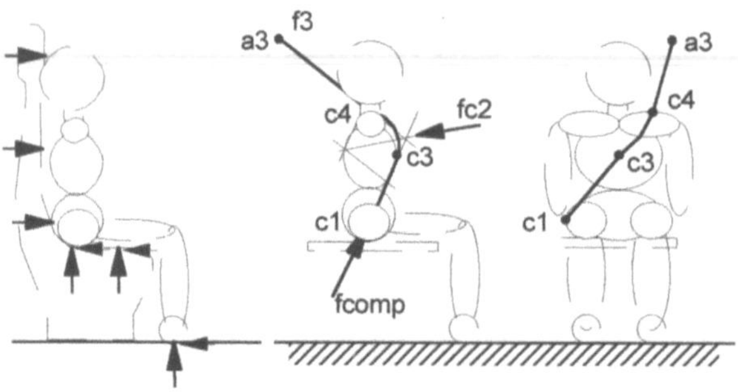

Fig. 2.4. Three-dimensional biomechanical model and restraint systems

A fully coupled vehicle-occupant model and the interaction between the compartment and occupant is described in the framework of a rigid multibody dynamics formulation in order to illustrate the use of the contact impact force model to describe local deformations during complex contact scenarios (Silva, Ambrósio and Pereira 1997). The vehicle model includes a detailed suspension system with a tire-terrain description, steering capabilities and a description of the bodywork with appropriate force-deflection characteristics for the potential points of contact between the chassis and the ground. The different body segments are treated as ellipsoids in order to handle the contact problem between the occupant and the geometry of the passenger compartment. A shoulder and lap safety belt system is also included in the model. The results of the application of this complex model are presented in Fig. 2.5 for a rollover simulation.

Fig. 2.5. Occupant and vehicle integrated models in a rollover simulation.

This type of models offer results with a good quality, provided that the large deformations of the system components are not relevant. However, many of the real life cases are characterized by elastic or plastic deformations of the system components that contribute decisively towards the vehicle response. For these cases it is important that the system elastodynamics is accounted for.

3. Flexible Multibody Dynamics by Lumped Deformations

Among the aspects of multibody modeling the problems concerned with the flexibility of its components are among the most challenging. Application cases where the system elastodynamics play important roles are, for instance, aerospace satellite deployment, vehicle dynamics, robotics, vehicle crashworthiness or high-speed machinery. A natural way to describe the flexibility effects is to assume specific patterns of deformation for the flexible components and to model the deformation itself by linear or nonlinear spring elements. In the framework of linear deformations, the finite segment approach uses this idea to model slender multibody components made of beams and bars (Huston and Wang 1994). The plastic hinge approach also uses the same principles but it is applied to models of systems experiencing nonlinear deformations (Ambrósio, Pereira and Dias 1996). Both formulations are reviewed here.

3.1. Finite Segment Approach to Flexible Multibody Systems

Let the flexible components of the multibody system be made of slender components such as the connecting rods of high-speed machinery or the structural frame of buses and trucks. In this case each slender component can be modeled as a collection of rigid bodies connected by linear springs, as shown in Fig. 3.1. These springs, representing the axial, bending and torsion properties of the beams, capture the flexibility of the whole component.

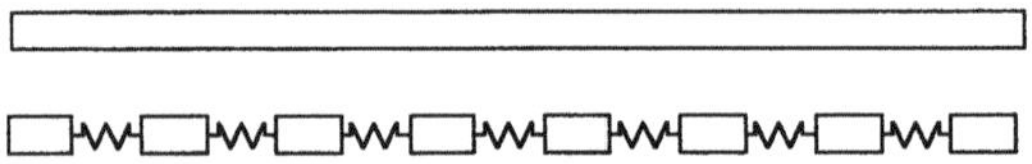

Fig. 3.1. Slender component and its finite segment model

Twelve generalized displacements are associated to each finite segment, this is, three translations and three rotations at each end. When the beams deform the reference frames attached to the rigid bodies used for their model rotate and translate with respect to each other. Then the relative displacements between the ends of adjacent segments can be easily calculated.

Forces and moments applied to the rigid bodies can be calculated from the relative end displacements and rotations assuming that each two adjacent bodies are connected by springs and eventually dampers. For this purpose to each rigid body there are deformation elements attached to each end as depicted by Fig. 3.2. The characteristics of these springs are related with the material and geometric properties of the system components.

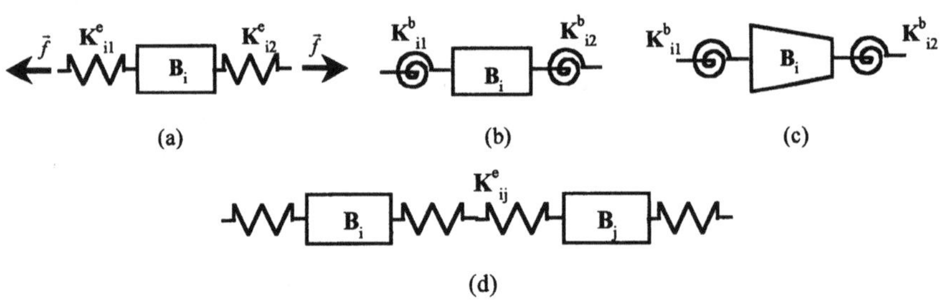

Fig. 3.2. Finite segments and their combinations: (a) extensional straight; (b) bending straight; (c) tapered bending; (d) combination extensional straight

Using the principles of structural analysis the stiffness coefficients for these springs are calculated (Huston 1990; Huston and Wang 1994). For instance, the straight extensional segment and the straight bending, respectively represented in Fig. 3.2a and 3.2b have their stiffness coefficients, respectively, given by:

$$k_i^e = k_j^e = 2\frac{E_i A_i}{l_i} \tag{8}$$

$$k_i^e = k_j^e = 2\frac{E_i I_i}{l_i} \tag{9}$$

$$k_{ij}^e = 2\frac{E_i E_j A_i A_j}{E_i A_j l_j + E_j A_i l_i} \tag{10}$$

where E_I is the Young modulus, A_I is the cross-section area, I_I the cross-section moment of inertia and l_I the length of the finite segment.

3.2. Plastic Hinges in Multibody Nonlinear Deformations

In many impact situations, the individual structural members are overloaded, principally in bending, giving rise to plastic deformations in highly localized regions, called plastic hinges. These deformations, presented in Fig. 3.3, develop at points where maximum bending moments occur, load application points, joints or locally weak areas (Murray 1983) and, therefore, for most practical situations, their location is predicted well in advance. Multibody models obtained with this method are relatively simple, which makes the procedure adequate for the early phases of vehicle design. The methodology described herein is known in the automotive, naval and aerospace industries as conceptual modeling (Ambrósio and Pereira 1997; Matolczy 1997; Kindervater 1997).

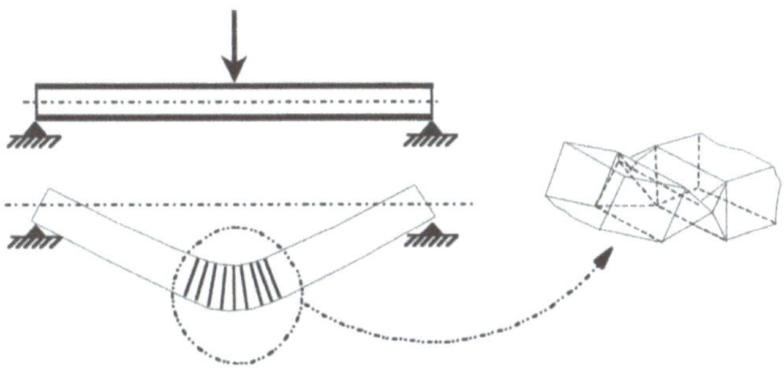

Fig. 3.3. Localized deformations on a beam and a plastic hinge

The plastic hinge concept has been developed by using generalized spring elements to represent constitutive characteristics of localized plastic deformation of beams and kinematic joints to control de kinematics of the deformation, as illustrated in Fig. 3.4. (Pereira, Ambrósio and Dias 1997). The characteristics of the spring-damper that describes the properties of the plastic hinge are obtained by experimental component testing, finite element nonlinear analysis or simplified analytical methods. For a flexural plastic hinge the spring stiffness is expressed as a function of the change of the relative angle between two adjacent bodies connected by the plastic hinge, as shown in Fig. 3.5.

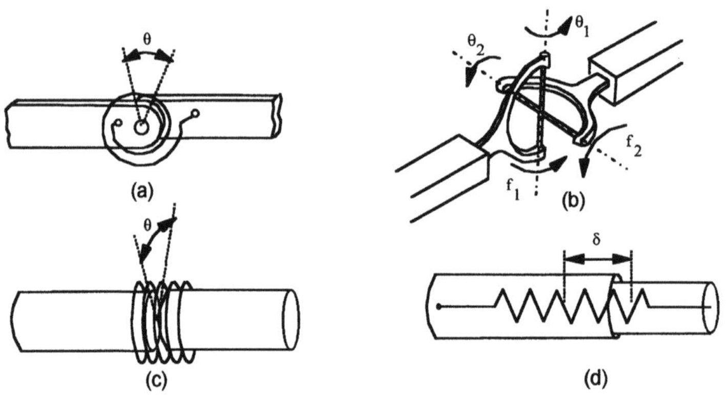

Fig. 3.4. Plastic hinge models for different loading conditions: a) bending with one axis; b) bending with two axis; c) torsion; d) axial

For a bending plastic hinge the revolute joint axis must be perpendicular to the neutral axis of the beam and to the plastic hinge bending plane simultaneously. The relative angle between the adjacent bodies measured in the bending plane is:

$$\theta_{ij} = \theta_i - \theta_j - \theta_{ij}^0 \tag{11}$$

where θ_{ij}^{0} is the initial relative angle between the adjacent bodies. Note that for the case of flexible adjacent bodies the relative angular values also include information on the nodal rotational displacements. For some other type of rigid body coordinates, such as joint coordinates, the joint angles are directly obtained as relative coordinates from the integration process (Nikravesh 1990).

The typical torque-angle constitutive relationship, as in Fig. 3.5, is found based on a kinematic folding model (Kecman 1983) by (Anceau, Drazetic and Ravalard 1992) for the case of a steel tubular cross section. This model is modified accounting for elastic-plastic material properties including strain hardening and strain rate sensitivity of some materials. A dynamic correction factor is used to account for the strain rate sensitivity (Winmer 1977)

$$P_d / P_s = 1 + 0.07 V_0^{0.82} \tag{12}$$

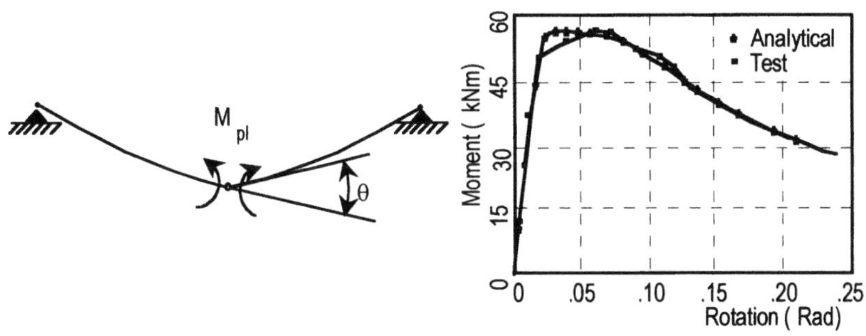

Fig. 3.5. Bending moment in a plastic hinge and its constitutive relationship

Here P_d and P_s are the dynamic and static forces, respectively, and V_0 is the relative velocity between the adjacent bodies. The coefficients appearing in eq.(12) are dependent on the type of cross section and material.

3.3. Validation of the Plastic Hinge Approach to Flexibility

The formulations describing the flexibility of multibody systems using a lumped description of the deformation, and in particular the plastic hinge approach, are used in the conceptual design of new systems. In order to validate the numerical technique proposed, comparisons with experimental results and analytical simulations provided in the literature are carried out (Ambrósio, Pereira and Dias 1996; Pereira, Drazetic and Ravalard 1993). The experimental procedure consists in a beam, with a length of 1 meter, a rectangular cross section of 50x25 mm and a thickness of 2.65 mm, made of E36 steel, articulated at an end and with a mass of 5.95 Kg in the other end. The beam is accelerated in such a way that the velocity of the tip is 11.85 m/s when it collides with a stop located 0.5 meter from the beam support, as shown in Fig. 3.6. The accelerations of the ballasting mass are measured with accelerometers mounted on it.

The observation of the impacting beam behavior clearly indicates that a plastic hinge develops in the region of contact with the stop, as pictured in Fig. 3.7. The permanent bending angle observed for the plastic hinge is 22°. Computer simulations are carried out with different models, shown in Fig. 3.6, where the contact between the beam and the rigid block is modeled with the continuous contact force described by eq.(7). Fig. 3.7 illustrates the displacement of the multibody model made of two rigid bodies and a plastic hinge. It is observed that the gross motion predicted is similar to the experimental behavior with a permanent bending angle of 21.25° between the rigid bodies.

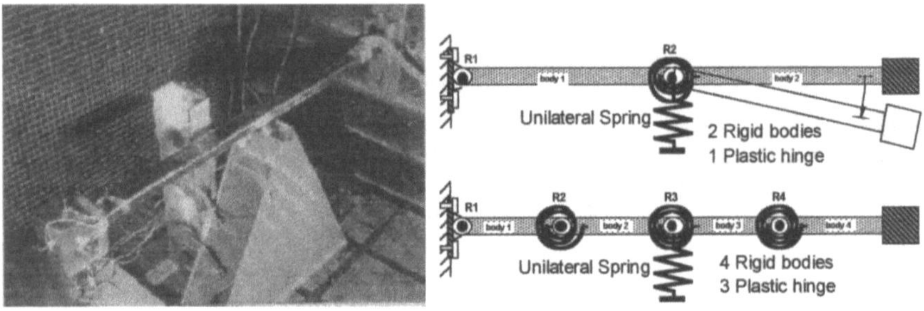

Fig. 3.6. Experimental configuration and multibody models for the impacting beam

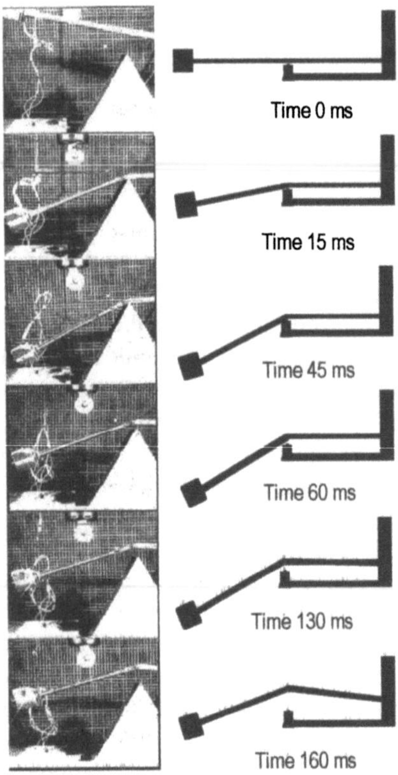

Fig. 3.7. Evolution of the bending angle: experiment and simulation

The measured and simulated accelerations of the ballasting beam are shown in Fig. 3.8. It is observed that the acceleration levels for the experimental and numerical results are similar, in particular during the contact stage.

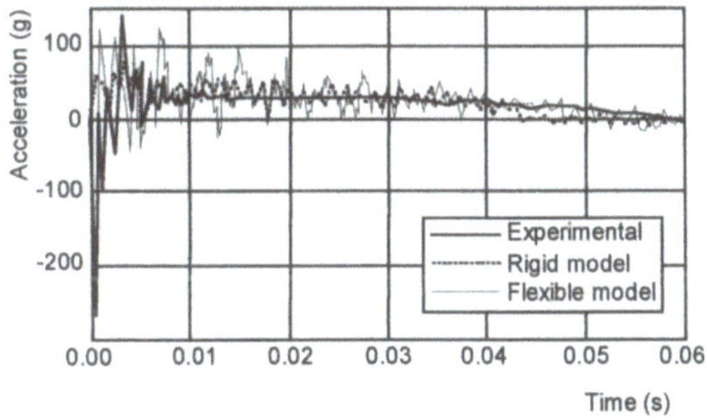

Fig. 3.8. Acceleration of the ballasting beam

4. Optimization of Multibody Systems for Impact Analysis

The methodologies discussed in the previous section are particularly suitable for optimization of flexible multibody systems. This is due to the simplicity of the equations of motion of the multibody system and the reduced set of parameters controlling the lumped stiffness element properties. This corresponds to a small number of design parameters in the optimization process making easier the job of defining their sensitivities. Although some optimization methodologies for multibody systems are reported in the literature, these have not been extensively applied to systems experiencing large deformations (Song 1986; Lust 1992; Dias and Pereira 1994).

4.1. Design Sensitivity Analysis and Optimization Procedures

The general design problem associated with the optimization of multibody systems can be stated in the following form

$$\text{Minimize } \Psi^o(\mathbf{b}) \tag{13}$$

subject to the constraints

$$\Psi_i^c(\mathbf{b}) \Rightarrow \begin{cases} = 0 & i = 1, k \\ \leq 0 & i = k+1, ncon \end{cases} \tag{14}$$

$$\mathbf{b}_l \leq \mathbf{b} \leq \mathbf{b}_u$$

The technological lower and upper bounds on the design variables, represented by $\mathbf{b}_l$ and $\mathbf{b}_u$ respectively, may also reflect acceptable structural designs for the elastic structure under normal loading conditions. The objective function Ψ^0 ($\mathbf{b}$) and the *ncon* constraints $\Psi^c(\mathbf{b})$ are functions of the *nd* design variables. The vector of the design variables $\mathbf{b}$ is also function of the state variables, which include positions, velocities, accelerations and Lagrange multipliers.

During the optimization process the objective and the constraint functionals must be evaluated as well as the corresponding gradients with respect to the design variables, written here as

$$\Psi_{\mathbf{b}}^0 = \frac{\partial \Psi^0}{\partial \mathbf{b}} \quad ; \quad \Psi_{\mathbf{b}}^c = \frac{\partial \Psi^c}{\partial \mathbf{b}} \tag{15}$$

Various types of constraints may be considered corresponding to different design criteria for the dynamic response problem (Haug and Arora 1979). These may be functionals of the particular type with the form

$$\Psi = \max f(\mathbf{q}, \mathbf{v}, \dot{\mathbf{v}}, \boldsymbol{\lambda}, t, \mathbf{b}) \quad \text{for} \quad t^0 \leq t_j \leq t^f \tag{16}$$

representing, for example, the maximum acceleration that occurs during the impact, in a particular point of interest in the structure at distinct instant t_j. In equation (16) t^0 and t^f the initial and final instants of the analysis interval, respectively. Also functionals of the integral type may be considered

$$\Psi_i = \int_{t^0}^{t^f} f_i(\mathbf{q}, \mathbf{v}, \dot{\mathbf{v}}, \boldsymbol{\lambda}, t, \mathbf{b})\, dt \tag{17}$$

which represent, for instance, the total energy absorbed or a standard measure of vehicle crash severity index, in this case given by (Song 1986)

$$SI = \frac{1}{t^f - t^0} \int_{t^0}^{t^f} a(t)^2 \, dt \tag{18}$$

where $a(t)$ is the deceleration of the passenger compartment or the passenger itself when striking an obstacle such as the vehicle interior trimming.

Design sensitivities can be efficiently obtained using the direct differentiation method or standard finite difference techniques. The latter are easier to implement but normally more expensive in terms of computer resources.

Analytical design sensitivities are used in this work, for details in their derivation the interested reader is directed to (Dias and Pereira 1994).

Several optimization algorithms, such as the feasible-direction method, the gradient-projection method, and the linearization method are coded in a multi-purpose program ADS - Advanced Design Synthesis (Vanderplaats 1987), which is used in the present work. A modified feasible-direction method is the selected algorithm to be used in the following example (Vanderplaats and Moses 1973).

4.3. Application to a Railcar End-Underframe Optimization

A railway vehicle is basically a tubular structure or, more precisely a self supported box beam with several openings in the webs, such as windows and doors, as shown in Fig. 4.1. From the manufacturing point of view the carshell is an assembly of the following main structural components: one underframe with two end-underframes; two side panels; one roof and two end frames

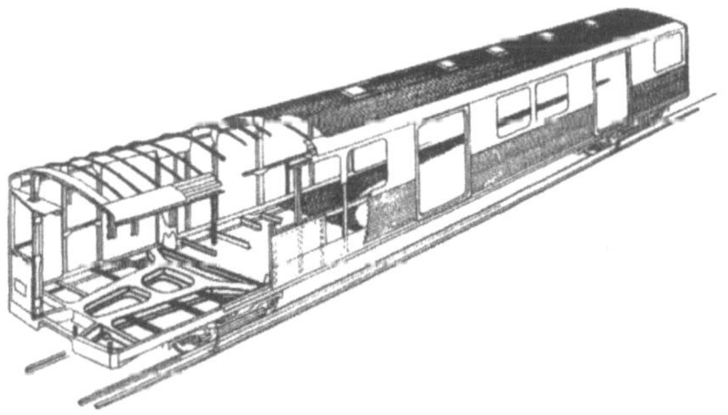

Fig. 4.1. Main structural components of a stainless steel railway vehicle

The two end-underframes located at the vehicle extremities, accounting for about 20% of the total mass of the vehicle structure, have a crucial influence in the carshell crashworthiness behavior. Consequently, the main design efforts focused on these substructures and end-underframes are developed, regarding the crush plastic mechanisms prediction and the energy absorption levels. The design aims to ensure that the passenger compartment remains undeformed in cases of a crash situation. The main design objective is to minimize the vehicle and passengers accelerations during the impact period, subject to constraints on the maximum allowable deformations. To achieve this some structural collapse mechanisms have been explored to be used at the carbody extremities.

In the initial design, plastic hinges will develop initiating the first collapse mechanism. A maximum displacement of 0.4 m, is defined before a second collapse mechanisms is initiated by crushing a *Y* shaped substructure in front of the bolster beam. The total crush displacement at the extremity is a function of the impacting speed and must be from 0.35 m to 1.00 m. 2D multibody models are used to access and explore the different collapse mechanisms of the end-underframe structures that deform essentially in a plane. Fig. 4.2 illustrates a rigid body model of an end-underframe used in a train that was tested for a 30 km/h frontal impact (Ambrósio, Pereira and Dias 1996).

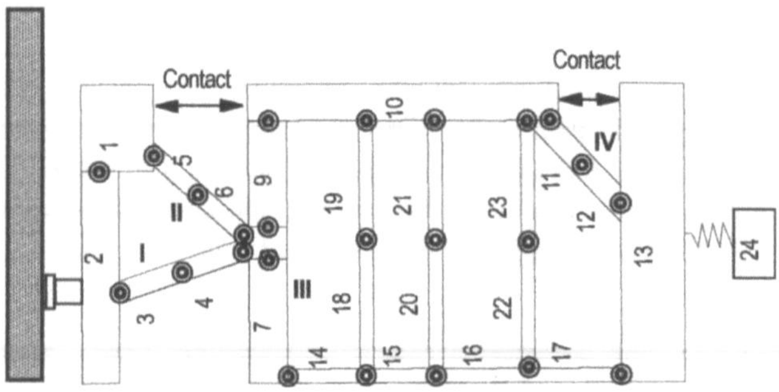

Fig. 4.2 Discrete model of half of the end-underframe

The model is composed of 24 bodies, connected by 24 plastic hinges. The plastic hinge characteristics are obtained from the elastic-plastic behavior of the cross sections and including a dynamic correction factor in order to take into account the strain rate sensitivity of the stainless steel. In the model, body 13 accounts for the remaining mass of the entire carbody shell while body 24 models sand bags, placed in the floor during experimental test, to take into account the mass of the equipment and interior trimming. The contact between the buffer located in front of the end-underframe and a wall is modeled by the continuous contact force model, described in section 2.2.

The simulation of the train crash is carried out from an instant before the carbody touches the wall until the train stops as a result of the impact. During the contact period, that lasts about 120 ms, the end-underframe of the carbody dissipates all the kinetic energy of the train as a result of its deformation. The deformation of the structure, obtained with the discrete deformation model composed of rigid bodies, is described in Fig. 4.3. It is evident that the mechanism of deformation is a result of successive plastic hinges that are formed as the crash progresses. The assumption for the plastic hinge locations used in the discrete deformation models are consistent with the observed deformations in an experimental text.

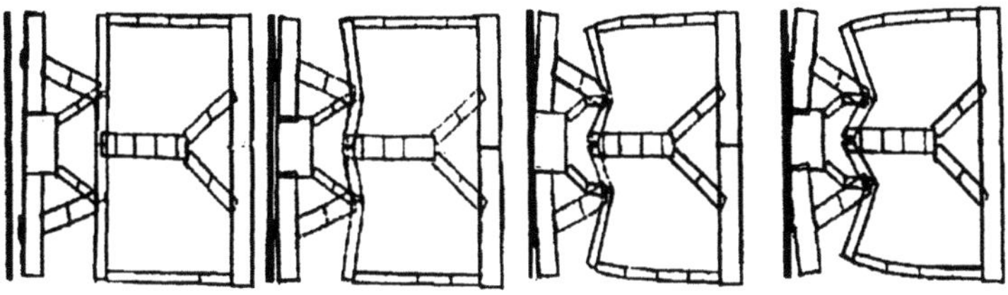

Fig. 4.3. Deformation of the end-underframe with the plastic hinge model

Due to symmetry, only the left side of this model is considered in the simulation of the frontal impact against a rigid wall. The cross-sections of the members, denoted by I, II, III and IV in the end-underframe model of Fig. 4.2, supply the set of design variables used in the optimization process. Two colliding velocities have been considered in the simulation: V=30 Km/h (8.33 m/s) and V=50 Km/h (13.833 m/s). The nonlinear response of the torsional springs is described in terms of a quasi-static moment-angle relationship based on the curves for elastic-plastic behavior, where the bending stiffness of the elastic part is obtained from elementary beam theory as

$$K = EI/l \tag{19}$$

where E is the modulus of elasticity, I is the cross sectional moment of inertia and l is assumed to be the flange width. The limit plastic moment is obtained for each section. For sections I to IV, shown in Fig. 4.4, IPE profiles (European Profile) are considered in the optimization process. A set of design variables is considered corresponding to the different beam heights H, of the cross sections. Other geometric parameters e, a, and ℓ, are obtained as functions of H, considering the range of cross sections commercially available. The plastic moment for this type of section is calculated as

$$M_p = \sigma_y \left[\frac{1}{2} e l^2 + \frac{1}{4} \mathrm{a}^2 (\mathrm{H} - 2\mathrm{e}) \right] \tag{20}$$

where σ_y is the yield stress, and the other constants are the cross-section dimensions. The material is a cold-rolled steel with $\sigma_y = 500$ MPa.

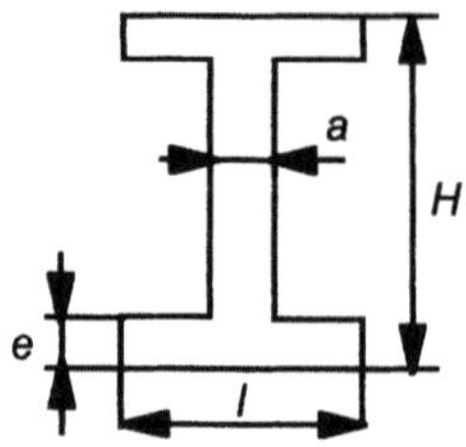

Fig. 4.4 Beam cross sections used in the optimization procedure

The objective function is defined as the acceleration peak experienced at the end-underframe in the compartment zone, i.e., body 13 of Fig. 4.2. Because the acceleration levels are related with human injury tolerance levels, its minimization is directly related to the reduction of the human injuries. Longitudinal constraints are also defined to impose limits on the maximum structural crash deformation in order to avoid failure of the cross sections and overlapping of the structural components. The objective function and constraints are summarized as:

$$\text{Minimize } \max \left|\ddot{x}_{13}(t,\mathbf{b})\right| \tag{21}$$

$$\begin{aligned} \text{Subjected to } \max &\left| \left(x_{13} - x_1\right) - \left(x_{13}^0 - x_1^0\right)\right| - \delta_1 \leq 0 \\ &\left| \left(x_{13} - x_{10}\right) - \left(x_{13}^0 - x_{10}^0\right)\right| - \delta_2 \leq 0 \end{aligned} \tag{22}$$

The subscripts indicate the body number, x is the displacement of the body in the longitudinal direction and the superscript indicates the initial position. The maximum deformations allowed are respectively, $\delta_1 = 0.937$ m and $\delta_2 = 0.254$ m. The purpose of the second constraint is to avoid the overlapping of the structure members in the zone of sections 1, 2 and 3. The design variables considered, corresponding to the sections represented in Fig. 4.2, are $\mathbf{b} = \{H_I, H_{II}, H_{III}, H_{IV}\}$. The initial design vector is $\mathbf{b} = \{0.3, 0.3, 0.4, 0.5\}$ while the lower and upper bounds are, $\mathbf{b}_l = \{0.06, 0.06, 0.06, 0.06\}$ and $\mathbf{b}_u = \{0.6, 0.6, 0.6, 0.6\}$, corresponding to the minimum and maximum depth of the sections available commercially, respectively. The results obtained for the two colliding velocities are summarized in Table 2.1.

For both initial velocities, the acceleration levels decrease as a result of the optimization process. For the impact at 30 Km/h, the acceleration results, presented in Fig. 4.5a, show that a decrease in the maximum acceleration is obtained at the cost of a larger deformation. Globally, a reduction of the section size is obtained with the optimized structure.

Table 2.1. Optimization of the end-underframe of the train

Initial velocity	30 Km/h		50 Km/h	
	Initial	Optimal	Initial	Optimal
Objective function (g)	27.98	17.79	130.8	31.18
Violations (m)				
Constraint 1	-3.77×10^{-1}	1.71×10^{-4}	1.64×10^{-1}	-5.7×10^{-3}
Constraint 2	-3.11×10^{-2}	-8.04×10^{-4}	-2.02×10^{-2}	-2.2×10^{-2}
$b_I = H_I$ (mm)	300.0	99.3	300.0	352.6
$b_{II} = H_{II}$ (mm)	300.0	82.0	300.0	319.0
$b_{III} = H_{III}$ (mm)	400.0	120.9	400.0	469.0
$b_{IV} = H_{IV}$ (mm)	500.0	459.0	500.0	600.0
CPU Time (s)	6 h 46 m		10 h 4 m	

The acceleration results for the impact at V=50 Km/h are shown in Fig. 4.5b . In this case the global deformation constraint 1 is violated in the initial design. For this simulation the initial design yields a very large deformation of the structure as a result of the strong collision of the lower extremity of central beam (body 10) with the lower beam (body 13). Another result of this contact is a large acceleration peak. The optimal design is associated with an increase of the dimensions of all sections, corresponding to higher levels of energy absorption in sections I, II and III. A subsequent decrease in the acceleration peak that occurs when bodies 10 and 13 collide is observed in the simulation.

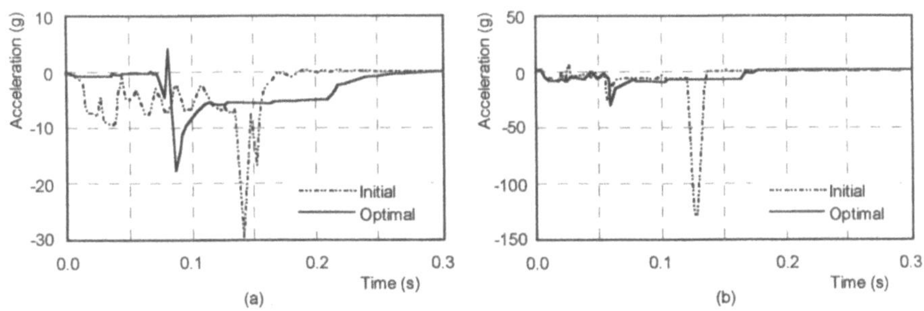

Fig. 4.5. Acceleration for the impact with: (a) V=30 Km/h; (b) V=50 Km/h

5. Distributed Flexibility in Multibody Systems

The methodologies referred to above include the flexibility effects in a multibody system assuming that the patterns of deformation and the collapse mechanisms are known beforehand. In many modeling situations this is not possible or not desirable. Using the finite element method it is possible to describe the elastodynamics of flexible multibody systems including both linear and nonlinear deformations (Shabana 1989; Ambrósio and Nikravesh 1992; Pereira and Proença 1991). A finite element procedure to describe the distributed deformation of flexible components of multibody systems is reviewed here and applied to a model experiencing only elastic deformations and to a case where large nonlinear plastic deformations occur.

5.1. Nonlinear Deformations of a Flexible Body

The most general motion of a flexible body involves a continuous change of its shape while undergoing large rotations and displacements, as shown in Fig. 5.1. The equations of motion of the flexible body are obtained using the principle of virtual work followed by an updated Lagrangian formulation (Ambrósio and Nikravesh 1992). Let the finite element method be used to represent the equations of motion of the flexible body. Referring to Fig. 5.1, the assembly of all finite elements used in the discretization of the flexible body results in the equation of motion for a single flexible body written as:

$$\begin{bmatrix} \mathbf{M}_{rr} & \mathbf{M}_{rf} & \mathbf{M}_{rf} \\ \mathbf{M}_{\phi r} & \mathbf{M}_{\phi\phi} & \mathbf{M}_{\phi f} \\ \mathbf{M}_{fr} & \mathbf{M}_{f\phi} & \mathbf{M}_{ff} \end{bmatrix} \begin{bmatrix} \ddot{\mathbf{r}} \\ \dot{\boldsymbol{\omega}}' \\ \ddot{\mathbf{u}}' \end{bmatrix} = \begin{bmatrix} \mathbf{g}_r \\ \mathbf{g}'_\phi \\ \mathbf{g}'_f \end{bmatrix} - \begin{bmatrix} \mathbf{s}_r \\ \mathbf{s}'_\phi \\ \mathbf{s}'_f \end{bmatrix} - \begin{bmatrix} \mathbf{0} \\ \mathbf{0} \\ {}^{t}_{t'}\mathbf{F} \end{bmatrix} - \begin{bmatrix} \mathbf{0} & \mathbf{0} & \mathbf{0} \\ \mathbf{0} & \mathbf{0} & \mathbf{0} \\ \mathbf{0} & \mathbf{0} & {}^{t}_{t'}\mathbf{K}_L + {}^{t}_{t'}\mathbf{K}_{NL} \end{bmatrix} \begin{bmatrix} \mathbf{0} \\ \mathbf{0} \\ \mathbf{u}' \end{bmatrix} \tag{23}$$

where $\ddot{\mathbf{r}}$ and $\dot{\boldsymbol{\omega}}'$ are, respectively, the translational and angular accelerations of the body-fixed reference frame and $\ddot{\mathbf{u}}'$ denotes the nodal accelerations measured in the body-fixed coordinate system. Note that the local coordinate frame $\xi\eta\zeta$ attached to a point of the flexible body is used to represent the body gross motion and to describe the deformation of the body. Vector $\mathbf{u}'$ denotes the increments of displacements from the previous equilibrium configuration to the current configuration measured in the body-fixed coordinate system.

Eq.(23) describes thoroughly the motion of the flexible body. However, this equation is highly nonlinear, even if only small elastic deformations of the system components are present. The sources of this nonlinear behavior are: a configuration dependent mass matrix; changing external applied forces; gyroscopic and centrifugal forces; nonconstant stiffness matrices. The configuration dependent mass matrix for the flexible body results from the assembly of the individual contributions of each finite element. The contribution of a single finite element for the mass matrix is described by the sub-matrices

$$\begin{aligned} \mathbf{M}_{rr} &= \mathbf{I}\int_{{}^0V} {}^0\rho\, {}^0dv & \mathbf{M}_{r\phi} &= -\mathbf{A}\int_{{}^0V} {}^0\rho\tilde{\mathbf{b}}'\, {}^0dv \\ \mathbf{M}_{rf} &= \mathbf{A}\int_{{}^0V} {}^0\rho\mathbf{N}\, {}^0dv & \mathbf{M}_{\phi\phi} &= -\int_{{}^0V} {}^0\rho\tilde{\mathbf{b}}'\tilde{\mathbf{b}}'\, {}^0dv \\ \mathbf{M}_{\phi f} &= \int_{{}^0V} {}^0\rho\tilde{\mathbf{b}}'\mathbf{N}\, {}^0dv & \mathbf{M}_{ff} &= \int_{{}^0V} {}^0\rho\mathbf{N}^T\mathbf{N}\, {}^0dv \end{aligned} \tag{24}$$

Here, $\mathbf{N}$ is the matrix of shape functions of element j. Submatrices $\mathbf{M}_{rr}$ and $\mathbf{M}_{ff}$ are constant, representing the mass of the entire element and the standard finite element mass matrix, respectively. $\mathbf{M}_{rf}$ and $\mathbf{M}_{\phi f}$ are the nonconstant matrices responsible for the inertia coupling between gross motion and body deformations. The submatrix $\mathbf{M}_{\phi\phi}$, representing the inertia tensor of the flexible body is approximately constant if the body deformations are small; otherwise, its time-variance cannot be neglected. All other submatrices in eq.(23) are either zero or constant for a proper choice of the location and orientation of the body-fixed coordinate frame.

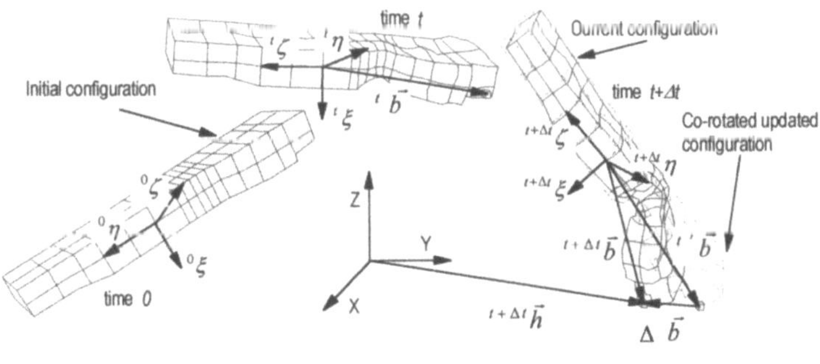

Fig. 5.1. General motion of a flexible body

The right-hand side of eq.(23) contains a time variying vector of gyroscopic and centrifugal forces which are evaluated for each element as

$$\begin{bmatrix} \mathbf{s}_r \\ \mathbf{s}'_\phi \\ \mathbf{s}'_f \end{bmatrix} = \begin{bmatrix} \mathbf{A}\tilde{\boldsymbol{\omega}}'\tilde{\boldsymbol{\omega}}'\int_{{}^0V} {}^0\rho\mathbf{b}'\, {}^0dv \\ \int_{{}^0V} {}^0\rho\tilde{\mathbf{b}}'\tilde{\boldsymbol{\omega}}'\tilde{\boldsymbol{\omega}}'\mathbf{b}'\, {}^0dv \\ \int_{{}^0V} {}^0\rho\mathbf{N}^T\tilde{\boldsymbol{\omega}}'\tilde{\boldsymbol{\omega}}'\mathbf{b}'\, {}^0dv \end{bmatrix} + 2\begin{bmatrix} \mathbf{A}\tilde{\boldsymbol{\omega}}'\int_{{}^0V} {}^0\rho\mathbf{N}\, {}^0dv \\ \int_{{}^0V} {}^0\rho\tilde{\mathbf{b}}'\tilde{\boldsymbol{\omega}}'\mathbf{N}\, {}^0dv \\ \int_{{}^0V} {}^0\rho\mathbf{N}^T\tilde{\boldsymbol{\omega}}'\mathbf{N}\, {}^0dv \end{bmatrix}\dot{\mathbf{u}}' \tag{25}$$

The vector of the external generalized applied forces $\mathbf{g}_f$ for each element is:

$$\begin{bmatrix} \mathbf{g}_r \\ \mathbf{g}'_\phi \\ \mathbf{g}'_f \end{bmatrix} = \begin{bmatrix} \int_{t'A} {}^{t+\Delta t}_{t'}\mathbf{f}_s \, {}^{t'}da \\ \int_{t'A} \tilde{\mathbf{b}}'\mathbf{A}^T \, {}^{t+\Delta t}_{t'}\mathbf{f}_s \, {}^{t'}da \\ \int_{t'A} \mathbf{N}^T\mathbf{A}^T \, {}^{t+\Delta t}_{t'}\mathbf{f}_s \, {}^{t'}da \end{bmatrix} + \begin{bmatrix} \int_{t'V} {}^{t'}\rho \, {}^{t+\Delta t}_{t'}\mathbf{f}_b \, {}^{t'}dv \\ \int_{t'V} {}^{t'}\rho\tilde{\mathbf{b}}'\mathbf{A}^T \, {}^{t+\Delta t}_{t'}\mathbf{f}_b \, {}^{t'}dv \\ \int_{t'V} {}^{t'}\rho\mathbf{N}^T\mathbf{A}^T \, {}^{t+\Delta t}_{t'}\mathbf{f}_b \, {}^{t'}dv \end{bmatrix} \tag{26}$$

where $\mathbf{f}_b$ and $\mathbf{f}_s$ are, respectively, the body and the surface forces. Matrices ${}^t_{t'}\mathbf{K}_L + {}^t_{t'}\mathbf{K}_{NL}$ are the linear and nonlinear stiffness matrices, respectively, and $\mathbf{f}$ is the vector of equivalent nodal forces due to the actual state of stress, given by

$$\begin{aligned} {}^t_{t'}\mathbf{K}_L &= \int_{{}^{t'}V} {}^t_{t'}\mathbf{B}_L^T \, {}_{t'}\mathbf{C} \, {}^t_{t'}\mathbf{B}_L \, {}^{t'}dv \\ {}^t_{t'}\mathbf{K}_{NL} &= \int_{{}^{t'}V} {}^t_{t'}\mathbf{B}_{NL}^T \, {}^t_{t'}\boldsymbol{\tau}' \, {}^t_{t'}\mathbf{B}_{NL} \, {}^{t'}dv \\ \mathbf{f} &= \int_{{}^{t'}V} {}^t_{t'}\mathbf{B}_L^T \, {}^t_{t'}\hat{\boldsymbol{\tau}}' \, {}^{t'}dv \end{aligned} \tag{27}$$

In eqs.(27) $\mathbf{B}_L$ and $\mathbf{B}_{NL}$ denote, respectively, the linear and nonlinear strain matrices and $\boldsymbol{\tau}'$ is the Cauchy stress tensor for the updated configuration. The linearity of the stiffness matrices refers to their relation with the displacements. For a nonconstant constitutive tensor ${}_{t'}\mathbf{C}$ both stiffness matrices are nonlinear, as in the case of a multibody system experiencing elasto-plastic deformations of one or more of its components.

5.2. Partially Rigid-Flexible Multibody Components

Eq.(23) is not efficient for numerical implementation due to the variance of the mass and stiffness matrices. During the integration process the mass matrix is inverted every time step in order to solve the equations explicitly for the accelerations. A simpler form of the equations of motion for a flexible body is obtained using a lumped mass formulation and substituting the local nodal accelerations $\ddot{\mathbf{u}}'$ by the nodal accelerations relative to the inertial frame $\ddot{\mathbf{q}}_f$.

Let the vectors of nodal accelerations be partitioned into translational and angular accelerations as:

$$\ddot{\mathbf{u}}' = \lfloor \ddot{\delta}'^T \quad \ddot{\theta}'^T \rfloor^T \tag{28}$$

The relation between the relative and absolute nodal accelerations for a node k is

$$\ddot{\mathbf{q}}'_{f\,k} \equiv \begin{bmatrix} \ddot{\mathbf{d}}' \\ \ddot{\alpha}' \end{bmatrix}_k = \ddot{\mathbf{u}}'_k + \begin{bmatrix} \mathbf{A}^T & -\left(\widetilde{\mathbf{x}}_k + \widetilde{\delta}_k\right)' \\ \mathbf{0} & \mathbf{I} \end{bmatrix} \begin{bmatrix} \ddot{\mathbf{r}} \\ \dot{\widetilde{\omega}}' \end{bmatrix} + \begin{bmatrix} \widetilde{\omega}'\widetilde{\omega}'\left(\mathbf{x}_k + \delta_k\right)' + 2\widetilde{\omega}'\dot{\delta}'_k \\ \widetilde{\omega}'\dot{\theta}'_k \end{bmatrix} \tag{29}$$

where $\mathbf{x}_k$ is the vector containing the node position in the reference configuration. Upon assembling eqs.(29) for all n nodes and substituting them into eq.(23), one obtains

$$\begin{aligned} &\sum_{k=1}^{n} m_k \ddot{\mathbf{d}}'_k = \mathbf{g}_r \\ &\sum_{k=1}^{n} \left[m\left(\widetilde{\mathbf{x}}' + \widetilde{\omega}'\right)\ddot{\mathbf{d}}'\right]_k = \mathbf{g}'_\theta \\ &\mathbf{M}_{ff}\ddot{\mathbf{q}}'_f = \mathbf{g}'_f - \mathbf{f} - \left({}^t_{t'}\mathbf{K}_L + {}^t_{t'}\mathbf{K}_{NL}\right)\mathbf{u}' \end{aligned} \tag{30}$$

Eqs.(30a) and (30b) describe the motion for the center of mass of a system of particles (Greenwood 1965). If the origin of the body-fixed coordinate system is coincident with the center of mass of the flexible body, eqs.(30a) and (30b) describe the motion of the origin of the $\xi\eta\zeta$ frame. Eq.(30c) is the equation of motion for the nodes of the flexible body, expressed in the body-fixed coordinate system. Due to the use of the lumped mass formulation the mass matrix $\mathbf{M}_{ff}$ is diagonal, i.e.,

$$\mathbf{M}_{ff} = \mathrm{Diag}\left(\mathrm{m}_1\,\mathbf{I}, \mathbf{0}, \cdots, \mathrm{m}_k\,\mathbf{I}, \mathbf{0}, \cdots, \mathrm{m}_n\,\mathbf{I}, \mathbf{0}\right) \tag{31}$$

where m_k is the mass of node k, and $\mathbf{I}$ and $\mathbf{0}$ are 3×3 identity and zero matrices associated with the translational and rotational degrees of freedom, respectively.

Very often it is useful to locate the origin of the body-fixed coordinate system in a point of the flexible body different from its center of mass. For this purpose assume that the flexible body has a rigid part and a flexible part with the body-fixed coordinate frame being attached to the center of mass of the rigid part, as shown in Fig. 5.2. The flexible part is attached to the rigid part by the nodes, belonging to the boundary ψ, that share the body-fixed coordinate frame with the rigid part. For a rigid body, the Newton-Euler equations of motion are written as

$$\begin{aligned} &m\ddot{\mathbf{r}} = \mathbf{f} \\ &\mathbf{J}'\dot{\omega}' = \mathbf{n}' - \widetilde{\omega}'\mathbf{J}'\omega' \end{aligned} \tag{32}$$

where $\mathbf{f}$ and $\mathbf{n}'$ are the external forces applied over the center of mass of the rigid part and moments applied over the body and $\mathbf{J}'$ is the inertia tensor. Eqs.(32) can be used instead of eqs.(28a) and (28b) provided that proper kinematic constraints are introduced between the flexible and rigid parts of the body. These kinematic constraints enforce the nodes in the boundary ψ to have zero displacements, velocities and accelerations with respect to the body-fixed coordinate frame, ie,

$$\mathbf{u}' = \dot{\mathbf{u}}' = \ddot{\mathbf{u}}' = \mathbf{0} \tag{33}$$

The constraint equations can be applied to eq.(30c) for each one of the boundary nodes using the Lagrange multiplier technique. At a later step these multipliers are eliminated from the equations of motion resulting in the dynamic equations (Ambrósio and Nikravesh 1992)

$$\begin{bmatrix} m\mathbf{I} + \overline{\mathbf{A}}\underline{\mathbf{M}}^*\overline{\mathbf{A}}^T & -\overline{\mathbf{A}}\underline{\mathbf{M}}^*\mathbf{S} & \mathbf{0} \\ -\left(\overline{\mathbf{A}}\underline{\mathbf{M}}^*\mathbf{S}\right)^T & \mathbf{J}' + \mathbf{S}^T\underline{\mathbf{M}}^*\mathbf{S} & \mathbf{0} \\ \mathbf{0} & \mathbf{0} & \mathbf{M}_{ff} \end{bmatrix} \begin{bmatrix} \ddot{\mathbf{r}} \\ \dot{\boldsymbol{\omega}}' \\ \ddot{\mathbf{q}}'_f \end{bmatrix} = \begin{bmatrix} \mathbf{f} + \overline{\mathbf{A}}\mathbf{C}'_\delta \\ \mathbf{n}' - \tilde{\boldsymbol{\omega}}'\mathbf{J}'\boldsymbol{\omega}' - \mathbf{S}^T\mathbf{C}'_\delta - \overline{\mathbf{I}}^T\mathbf{C}'_\theta \\ \mathbf{g}'_f - \mathbf{f} - \left({}^t_{t'}\mathbf{K}_L + {}^t_{t'}\mathbf{K}_{NL}\right)\mathbf{u}' \end{bmatrix} \tag{34}$$

where $\underline{\mathbf{M}}^*$ is a diagonal mass matrix containing the mass of the $\underline{n}$ boundary nodes. Matrices $\overline{\mathbf{A}}^T$, $\mathbf{S}$ and $\overline{\mathbf{I}}$ are made from (3x3) matrices as:

$$\overline{\mathbf{A}} = [\mathbf{A} \;\cdots\; \mathbf{A}] \;;\; \mathbf{S} = \left[\left(\tilde{\mathbf{x}}'_1 + \tilde{\boldsymbol{\delta}}'_1\right)^T \;\cdots\; \left(\tilde{\mathbf{x}}'_{\underline{n}} + \tilde{\boldsymbol{\delta}}'_{\underline{n}}\right)^T\right]^T \;;\; \overline{\mathbf{I}} = [\mathbf{I} \;\cdots\; \mathbf{I}]^T$$

Vectors $\mathbf{C}'_\delta$ and $\mathbf{C}'_\theta$ represent, respectively, the reaction force and moment of the flexible part of the body over the rigid part. These reaction forces and moments are written as

$$\begin{aligned} \mathbf{C}'_\delta &= \mathbf{g}'_{\underline{\delta}} - \mathbf{f}_{\underline{\delta}} - \left({}^t_{t'}\mathbf{K}_L + {}^t_{t'}\mathbf{K}_{NL}\right)_{\underline{\delta}\delta}\boldsymbol{\delta}' + \left({}^t_{t'}\mathbf{K}_L + {}^t_{t'}\mathbf{K}_{NL}\right)_{\underline{\delta}\theta}\boldsymbol{\theta}' \\ \mathbf{C}'_\delta &= \mathbf{g}'_{\underline{\theta}} - {}^t_{t'}\mathbf{F}_{\underline{\theta}} - \left({}^t_{t'}\mathbf{K}_L + {}^t_{t'}\mathbf{K}_{NL}\right)_{\underline{\theta}\delta}\boldsymbol{\delta}' + \left({}^t_{t'}\mathbf{K}_L + {}^t_{t'}\mathbf{K}_{NL}\right)_{\underline{\theta}\theta}\boldsymbol{\theta}' \end{aligned} \tag{35}$$

In these equations, the subscripts δ and θ refer to the partition of the vectors and matrices with respect to the translational and rotational nodal degrees of freedom. The underlined subscripts are referred to the boundary nodal displacements between the rigid and flexible parts.

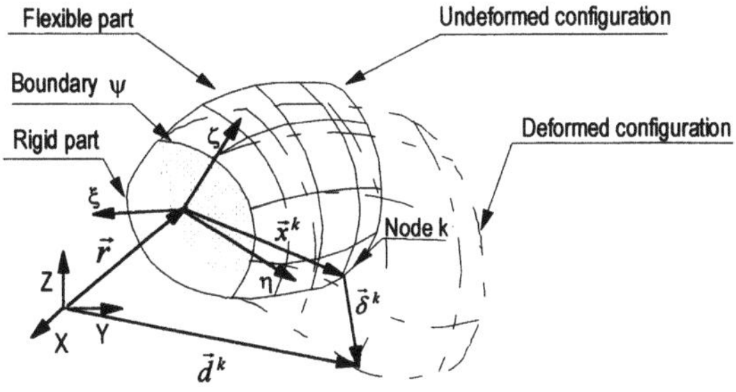

Fig. 5.2. Flexible body with a rigid part and a deformable part

Eq.(34) is the finite element equation of motion for a flexible body. When finite elements with rotational degrees of freedom are used to describe the flexible body, some zero elements appear in the diagonal of the mass submatrix $\mathbf{M}_{ff}$ and eq.(34) cannot be solved explicitly for the accelerations. This problem is avoided using rotational inertias obtained by lumping the off-diagonal terms of the consistent mass matrix $\mathbf{M}_{ff}$ to replace the zero coefficients.

5.3. Application to the Impact of a Railway Vehicle

The spatial flexible multibody models are used to access the overall behavior of a railway car extremity. This model has an assemblage of rigid bodies and partially flexible members experiencing material and geometric nonlinearities. The 3D train carbody model, shown in Fig. 5.3a, has a total weight of 39 tons, including the bogies, equipment and interior trimming In Fig. 5.3b the experimental setup for the train carbody as used in a crash test is displayed (SOREFAME 1993). Each bogie is composed of two wheel sets and two levels of suspension systems between the wheels and train carbody. The bogies are connected to the carbody by a vertical shaft in an elastic bushing allowing the relative position adjustment with a rolling motion. This roll motion is constrained to be relatively small by jounce stops in each side of the carbody.

Considering that crushing takes place in the extremity, it is assumed in the finite element model of the carbody that the end-underframe is deformable. The rest of the carbody between the bolster beams is the rigid part of a partially flexible body. The computer model of the train carbody used in the simulation is composed of 13 moving bodies and 20 sets of spring-dampers corresponding to the suspension systems. The interaction between wheels and rails is modeled by nonlinear spring-damper elements, high lateral sliding friction and low rolling friction coefficients. The initial conditions of the simulation correspond to a train carbody moving with a speed of 30 Km/h towards a rigid wall.

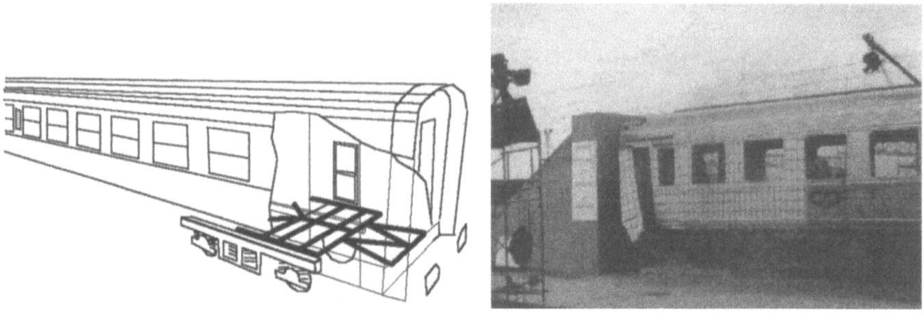

(a) (b)

Fig. 5.3. Train carbody; (a) finite element model of the end-underframe; (b) crash test

The finite element model of the end-underframe, shown in Fig. 5.3, is made of 54 nodal points, corresponding to 150 degrees of freedom for the flexible part and 62 nonlinear beam elements. The material used in this model is a stainless steel with a plastic modulus that is one tenth of its elastic modulus.

During the contact period, that lasts about 175 ms, the end-underframe of the carbody dissipates all the kinetic energy of the train by deformation, as illustrated in Fig. 5.4. The deformation patterns clearly indicate that the end-underframe collapse mechanism is a result of the development of successive plastic hinges as the crash progresses, as assumed before.

Fig. 5.4. Sequence of the train carbody end-underframe deformation

The kinematics of the end-underframe deformation is confirmed using a detailed nonlinear finite element model and a nonlinear finite element code (ABAQUS 1993). A nonlinear static analysis of this model is carried on with the front nodes of the end-underframe loaded until the structure exhausts its capability to deform any further. The progress of the deformation obtained, shown in Fig. 5.5, is similar to the one predicted by the model using the partially flexible body. Furthermore, the deformation of the end-underframe is confirmed in the train crash test performed (SOREFAME 1993), as depicted in Fig. 5.6.

The limit load obtained in both analyses is about 120 tons, for each half of the end-underframe, when the first plastic hinges are formed, as shown in Fig. 5.7b.

However, the prediction by the detailed finite element model for the limit load corresponding to a second plastic hinge, appearing close to the bolster beam, is about 30% higher than what is calculated in the current application. In the experimental test the maximum load carried by each buffer of the vehicle was measured to be 120 tons. Neither a second peak of load was measured nor a second plastic hinge closer to the bolster beam was formed. That is because in the experimental test a collision of the top of the carbody with the rigid wall occurred forcing the kinetic energy to be absorbed by a different region of the carbody. Consequently, the deformation of the end-underframe did not progress any further. The maximum displacement obtained for the deformation of the end-underframe in the experimental test was 0.309 m, which corresponds to the deformation described in the fourth frame of Fig. 5.4. In the numerical model the total displacement of the front of the carbody with respect to its center of mass is 0.875 m. After the first 90 ms of the impact the numerical models behavior diverges from that of the experimental test. These results are consistent with the observation that the second part of the impact observed is not modeled.

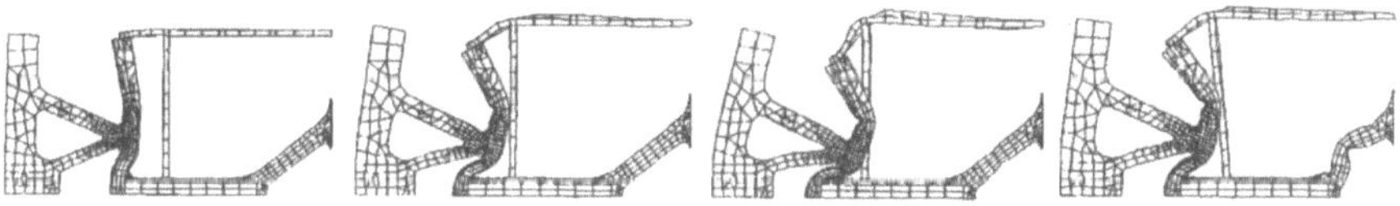

Fig. 5.5. Progress of the end-undeframe deformation predicted by ABAQUS

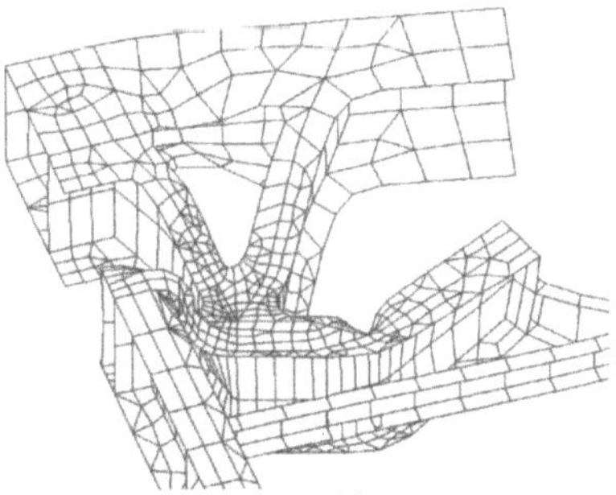

(a)

(b)

Fig. 5.6. Final deformation of the end-undeframe: (a) ABAQUS; (b) experimental testing

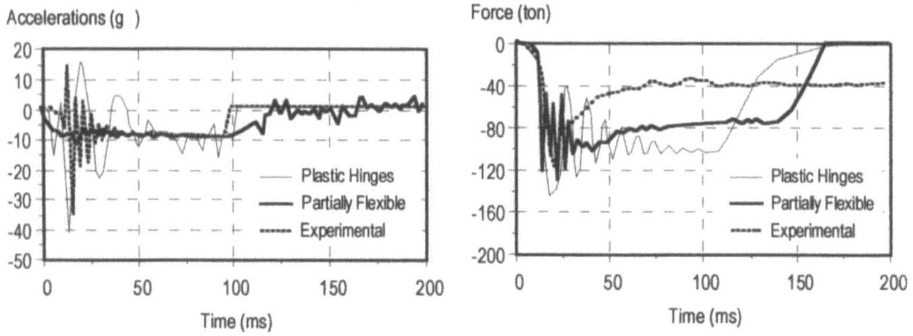

Fig. 5.7 Numerical and experimental results: (a) longitudinal acceleration of the bolster beam; (b) reaction force over the wall

5.4. Linear Elastic Deformation of the Flexible Components

Let the flexible part of the multibody component have a linear elastic behavior. The equation of motion for a flexible body given by eq.(34) is rewritten as

$$\begin{bmatrix} \left(\mathbf{M}_n + \overline{\mathbf{B}}^{*T} \underline{\mathbf{M}}^* \overline{\mathbf{B}}^*\right) & \mathbf{0} \\ \mathbf{0} & \mathbf{M}_{ff} \end{bmatrix} \begin{bmatrix} \ddot{\mathbf{q}}_n \\ \ddot{\mathbf{q}}'_f \end{bmatrix} = \begin{bmatrix} \mathbf{Q} - \overline{\mathbf{B}}^{*T} \mathbf{C}'_\delta - \overline{\mathbf{T}}^{*T} \mathbf{C}'_\theta \\ \mathbf{g}'_f - \mathbf{K}\mathbf{u}' \end{bmatrix} \tag{36}$$

where $\underline{\mathbf{C}}_\delta$ and $\underline{\mathbf{C}}_\theta$, the reaction forces of the flexible part of the body over the rigid part described by eqs.(33), are now written in terms of the nodal displacements rather than in terms of the nodal displacement and rotation increments. All finite element terms used there are readily available from any commercial finite element package. Any type of finite elements with a linear elastic material can be used in the mesh, except for higher-order finite elements that, due to the use of a lumped mass formulation, may lead to incorrect finite element models.

In order to reduce the large number of generalized coordinates to a smaller number of coordinates, the component-mode synthesis technique is used. Only the modes of vibration are used here. Other modes, such as the static-correction modes have proven to be important to achieve numerical efficiency and good numerical precision (Yoo and Haug 1986). Let the nodal displacements of the flexible part of the body be described by a weighted sum of the modes of vibration associated with the lower natural frequencies

$$\mathbf{u}' = \mathbf{X}\mathbf{w} \tag{37}$$

where vector $\mathbf{w}$ represents the amplitudes of the vibration mode contributions towards the flexible part displacements, and $\mathbf{X}$ is the modal matrix. Note that, due to the constraints imposed between the flexible and rigid parts of the body, the modes of vibration considered in this formulation are constrained modes.

In order for the modal superposition technique to be applied it is necessary to describe eq.(36) in terms of local nodal accelerations. Using eq.(29) in eq.(36) and rearranging terms leads to (Gonçalves and Ambrósio 1996)

$$\begin{bmatrix} \underline{\mathbf{M}}_n & \mathbf{0} \\ \mathbf{M}_{ff}\mathbf{H} & \mathbf{M}_{ff} \end{bmatrix} \begin{Bmatrix} \ddot{\mathbf{q}}_n \\ \ddot{\mathbf{u}}' \end{Bmatrix} = \begin{Bmatrix} \mathbf{f}_n \\ \mathbf{f}_f - \mathbf{M}_{ff}\psi \end{Bmatrix} \tag{38}$$

The double time derivative of eq.(37) is substituted in eq.(36) and the second row multiplied by $\mathbf{X}^T$. Using the orthonormality of the modes of vibration with respect to the mass matrix, the result is a much smaller number of equations to describe the flexible body

$$\begin{bmatrix} \underline{\mathbf{M}}_n & \mathbf{0} \\ \mathbf{X}^T\mathbf{M}_{ff}\mathbf{H} & \mathbf{I} \end{bmatrix} \begin{Bmatrix} \ddot{\mathbf{q}}_n \\ \ddot{\mathbf{w}} \end{Bmatrix} = \begin{Bmatrix} \mathbf{f}_n \\ \mathbf{X}^T(\mathbf{g}'_f - \mathbf{M}_{ff}\psi) - \Lambda\mathbf{w} \end{Bmatrix} \tag{39}$$

Note that as a result of the term $\mathbf{X}^T\mathbf{M}_{ff}\mathbf{H}$, the mass matrix is no longer constant, because the transformation matrix $\mathbf{A}$ is included in matrix $\mathbf{H}$. Still, no new finite element terms are involved in this transformation, besides the modes of vibration and the natural frequencies obtained from any chosen FEM package.

5.5. Application to a Sports Car Vehicle Dynamics

The formulation proposed in this paper is applied to model the dynamics of the sports car Lancia Stratos presented in Fig. 5.8. The vehicle is a replica of the original Lancia that was assembled in-house at IDMEC. This provided an opportunity to measure directly the structural and dynamic characteristics of all car components used in the model.

Two models of the vehicle, composed of 17 bodies, are considered here. The system components include elements of the front double A-arm suspension system, the rear McPherson suspension system, wheels and a flexible chassis for the first model and a rigid chassis for the second, as depicted by Fig. 5.9. The tire interaction with the ground is described by an analytical tire model with comprehensive slip (Gim 1988). The interested reader is referred to (Ambrósio and Gonçalves 1996) for a description of the pertinent data required.

Fig. 5.8. Prototype of the sports car

In the flexible model of the chassis, the finite element method is used to describe the deformation, according to the methodology now proposed. The structural components of the chassis are modeled using 262 beam elements, 121 plate elements, and 18 concentrated mass elements, all available in the finite element program ANSYS and shown in Fig. 5.10. The model has 1680 nodal degrees of freedom. The shell of the chassis is not considered in the model due to its small influence in the vehicle structural dynamics.

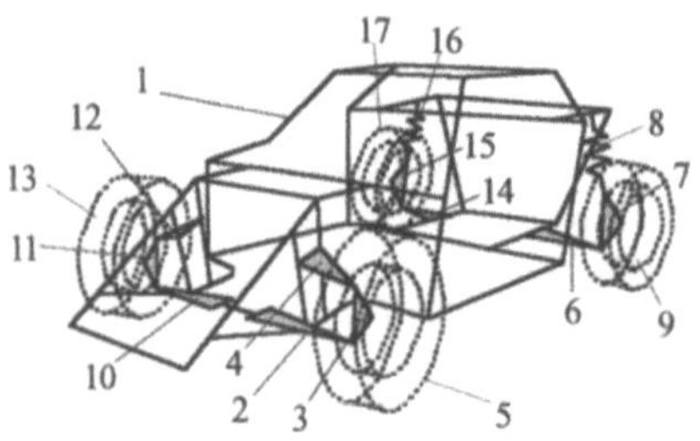

Fig. 5.9. Multibody model of the sports vehicle

When the chassis is defined as a flexible body, it is necessary to include kinematic constraints between the nodes of the flexible chassis and other bodies, namely, the suspension system components.

In the flexible model of the chassis it is assumed that the rigid part of the flexible body, that has the body-fixed frame attached, is a node close to the center of mass of the model. The natural frequencies and corresponding modes of vibration of the chassis are evaluated considering the reference conditions corresponding to a fixed node of the finite element mesh. The finite element code ANSYS is used to find the 10 lower natural frequencies and corresponding modes of vibration of the model. The results are described in Table 5.1 and displayed in Figs. 5.11, 5.12 and 5.13.

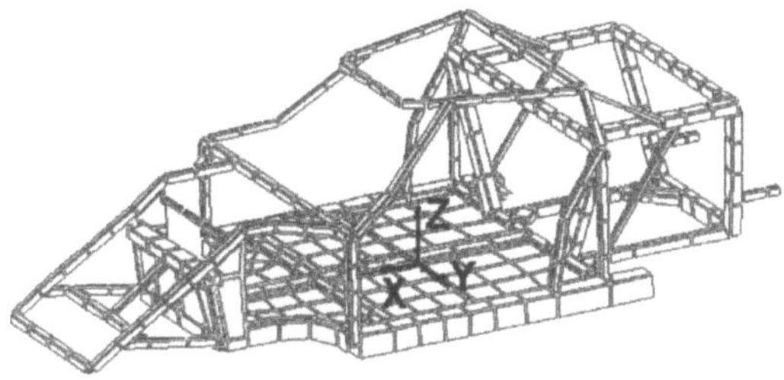

Fig. 5.10. Finite element model of the flexible chassis

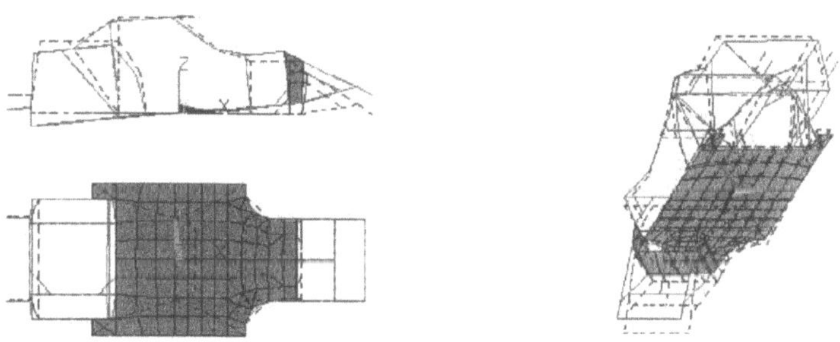

Fig. 5.11 Quasi-rigid rotation about YY (ω_1=6.97 Hz)

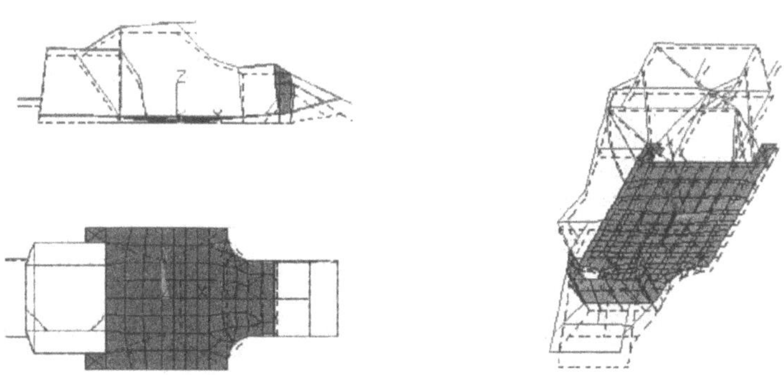

Fig. 5.12. Bending about YY (ω_4=13.81 Hz)

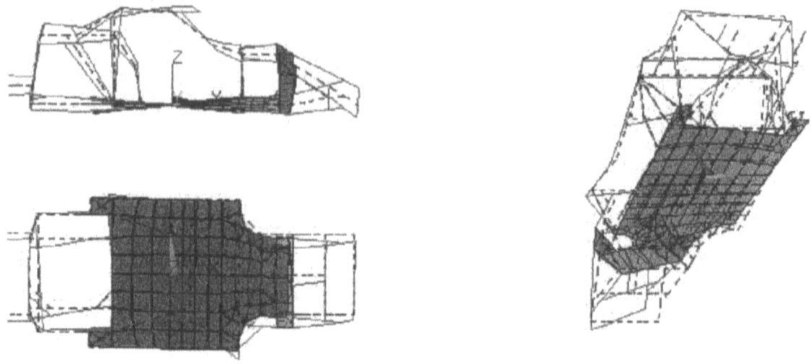

Fig. 5.13 Torsion about XX (ω_5=25.26 Hz)

Both rigid and flexible models of the sports car are simulated for the case of the vehicle riding over two nonsymmetric bumps with a height of 0.1 m on the road, as depicted in Fig. 5.14. The car is moving with an initial velocity of 50 Km/h in the forward direction. These simulations are used to evaluate the stability and maneuverability of the vehicle and the influence of the flexibility of the chassis on these characteristics.

The simulations are carried out for the models of the vehicle with the rigid and the flexible chassis.. The Z coordinate of the center of the chassis for both simulations is displayed in Fig. 5.15a. Note that the initial positions for both models do not correspond to static equilibrium conditions. Furthermore, the final Z coordinate of the center of mass is lower for the flexible chassis relative to the rigid case. This is due to the flexibility of the vehicle.

Table 5.1 Natural frequencies and modes of vibration

Mode	Frequency [Hz]	Mode type
1	6.97	Quasi-rigid rotation mode about YY
2	9.41	Quasi-rigid rotation mode about XX
3	10.96	Quasi-rigid rotation mode about ZZ
4	13.81	Bending about YY
5	25.26	Torsion about XX
6	32.41	Bending about ZZ
7	38.55	Mixed bending and torsion
8	45.95	Torsion about YY
9	50.27	Local mode
10	50.31	Local mode

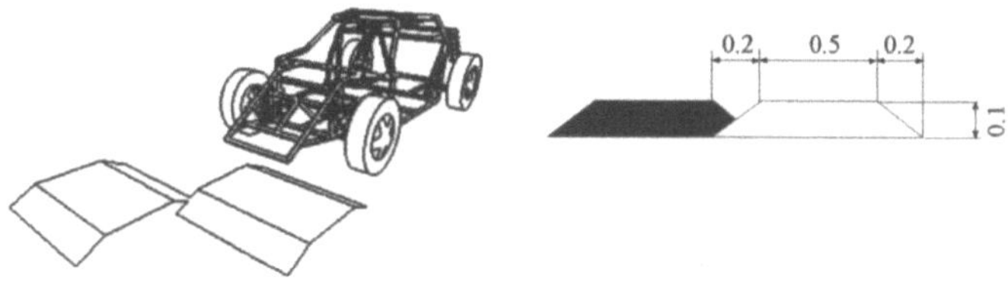

Fig. 5.14 Initial conditions for the simulations and road geometry

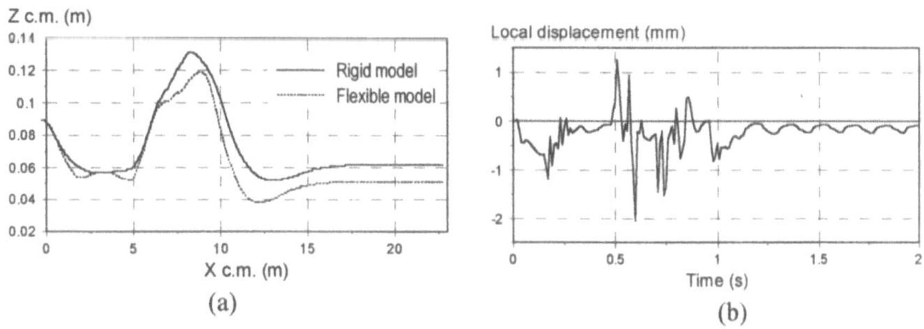

Fig. 5.15 (a) Vertical displacement and (b) suspension support local Z displacement

It is noticeable from Fig 5.15b that displacement peaks of 2 mm of the support point of the upper component of the rear McPherson suspension system are obtained when the vehicle rides over the road bumps. After the vehicle gets back on the flat road the relative displacement oscillates about an equilibrium position that corresponds to a deformed position of the node of 0.25 mm.

6. Conclusions

Different formulations for the dynamic analysis of rigid and flexible multibody systems suitable for crashworthiness and impact conditions have been reviewed here. These formulations range from methodologies where the deformation is lumped at linear or nonlinear spring-damper elements to finite element formulations where the distributed flexibility of the system components may experience linear or nonlinear geometric and material behavior. In the process, a basic form of localized flexibility in multibody dynamics was discussed with the purpose of describing contact-impact between system components and surrounding objects.

Localized plastic deformations represented by plastic hinges are modeled by revolute joints and revolute actuators with appropriate constitutive relationships depending on the cross-sections of the crushed members, obtained experimentally, or using refined finite element representations of such components with codes commercially available. Though leading to numerically efficient codes, the location of plastic hinges needs to be known beforehand, which may pose serious problems. This approach to flexibility showed to be a good analysis tool to support the optimization of multibody systems applied to vehicle impact. This procedure has been developed within the framework of mathematical programming methods involving an efficient and inexpensive dynamic analysis module. The difficulties of these applications are related with the definition of suitable objective functions, efficient evaluation of the sensitivities and, specially, to the adequacy of the optimization procedures to multibody systems. Such complex design efforts using conventional structural optimization techniques are still far from being economical with the power available in current computer platforms. The application to the impact of a railway vehicle demonstrated the feasibility of the use of optimization methods in the crashworthiness dynamics.

Formulations allowing for the combined multibody and nonlinear finite element description of the flexible parts have also been developed and presented. Based on a general description of the continuum mechanics equations and using an updated Lagrangian formulation, the method proposed has no restrictions on the type of deformations appearing in a multibody model. All terms in the equations are related with standard finite elements or with standard rigid multibodies. Consequently, it is not needed to derive any special coupling terms when new finite elements are used. The advantage of this formulation is that no a priori knowledge of the flexible behavior is necessary. However, the computational costs of this procedure are high.

A natural specialization of the method proposed to systems experiencing linear elastodynamics was obtained. Using the mode-component synthesis, the number of generalized coordinates necessary to describe the system model was greatly reduced while the inertial coupling between the system gross motion and its elastodynamics was still preserved. The methodology proposed uses strictly the finite element data obtained from commercial finite element codes and does not require the evaluation of any extra flexible term for the equations of motion.

Further research efforts are still required in order to improve the efficiency of multibody dynamics as applied to flexible systems. These are: adequate modeling of nonlinear flexible bodies with formulations using only structural stiffness and mass terms obtained in standard finite element formulations; to allow a greater interaction with commercially available codes with the possibility to introduce more complex crushable elements; introduction of piecewise analysis techniques based on momentum balance methods for an efficient evaluation of velocity jumps during impact; development of new methodologies for constraint addition and deletion capabilities to deal in a more general manner with the contact problem, and finally, the exploitation of more efficient integration algorithms.

Acknowledgements

The authors gratefully acknowledge the support of: AGARD - Advisory Group for Aerospace Research and Development, project P124; JNICT - Junta Nacional de Investigação Científica e Tecnológica, Project PBICT/P/CEG/2348/95. Some of the work reported herein, namely, the experimental tests and design tools development and enhancement, was carried out within the framework of project BRITE/EURAM BE-3385-89 - TRAINCOL.

References

Ambrósio, J.A.C., Nikravesh, P.E., Pereira, M.S. (1990) Crashworthiness analysis of a truck. J. Math. Comp. Modeling **14**, 959-964

Ambrósio, J.A.C., Nikravesh, P.E. (1992) Elastic-plastic deformations in multibody dynamics. Nonlinear Dynamics **3**, 85-104

Ambrósio, J.A.C., Pereira, M.S., Dias, J. (1996) Distributed and discrete nonlinear deformations on multibody systems. Nonlinear Dynamics **10**(4), 359-379

Ambrósio, J.A.C. Silva, M.P.T., and Gonçalves, J.P. (1996) Development of energy absorbing devices using a kinetostatic multibody dynamics methodology. Int. J. Crashworthiness **1**(2), 127-143

Ambrósio, J.A.C., Gonçalves, J.P.C. (1996) Dynamics of complex flexible multibody systems using natural coordinates. In: ESA International Workshop on Advanced Mathematical Methods in the Dynamics of Flexible Bodies. Proceedings, Noordwijk, The Netherlands, June 3-5

Ambrósio, J.A.C., Pereira, M.S. (1997) Multibody dynamic tools for crashworthiness and impact. In: J.A.C. Ambrósio, M.S. Pereira, F.P Silva (eds.) Crashworthiness of transportation systems: structural impact and occupant protection. NATO ASI Series E. Vol. 332, 475-521. Dordrecht: Kluwer Academic Publishers

Anceau, J.H., Drazetic, P., Ravalard, I. (1992) Plastic hinges behaviour in the multibody systems. Mécanique Matériaux Électricité. n° 444

Baumgarte, J. (1972) Stabilization of constraints and integrals of motion. Computer Methods in Applied Mechanics Engineering. **1**, 1-16

Cardona, A., Geradin, M. (1988) A beam finite element nonlinear theory with finite rotations. Int. J. Num. Meth. In Engng. **26**, 2403-2438

Dias, J.P., Pereira, M.S. (1994) Design for vehicle crashworthiness using multibody dynamics. Int. J. of Vehicle Design **15**(6), 563-577

Erdman, A.G., Sandor, G.N. (1972) Kineto-elastodynamics – a review of the state of the art and trends. Mech. Mach. Theory **7**, 19-33

García de Jalón, J., Bayo, E. (1993) Kinematic And Dynamic Simulation Of Multibody Systems - The Real Time Challenge. New York: Springer-Verlag

Geradin, M., Cardona, A., Doan, D.B., Duysens, J. (1994) Finite element modeling concepts in multibody dynamics. In: M. Pereira, J. Ambrósio (eds.) Computer aided analysis of rigid and flexible mechanical systems. NATO ASI Series E. Vol. 268, 233-284. Dordrecht: Kluwer Academic Publishers

Gim, G. (1988) Vehicle dynamic simulation with a comprehensive model for pneumatic tyres. Ph.D. Dissertation. University of Arizona

Gonçalves, J.P.C., Ambrósio, J.A.C. (1996) Estudo de elementos deformáveis em veículos utilizando sistemas de corpos múltiplos (Study of deformable elements in vehicles using multibody system dynamics). Technical Report IDMEC/CPM-96/004, IDMEC Instituto Superior Técnico

Haug, E.J., Arora, J.S. (1979) Applied optimal design. New York: John Wiley & Sons

Haug, E.J. (1989) Computer-aided kinematics and dynamics of mechanical systems Volume 1: basic methods. Boston: Allyn and Bacon

Huston, R.L. (1990) Multibody dynamics. Stoneham: Butterworth

Huston, R.L., Wang, Y. (1994) Flexibility effects in multibody systems. In: M. Pereira, J. Ambrósio (eds.) Computer aided analysis of rigid and flexible mechanical systems. NATO ASI Series E. Vol. 268, 351-376. Dordrecht: Kluwer Academic Publishers

Kamal, M. M. (1970) Analysis and simulation of vehicle to barrier impact. SAE Paper No. 700414.

Kane, T.R., Ryan, R.R., Banerjee, A.K. (1987) Comprehensive theory for the dynamics of a general beam attached to a moving rigid base. J. of Guidance, Control and dynamics **10**(2), 139-151

Kecman, D. (1983) Bending collapse of rectangular and square section tubes. Int. J. of Mech. Sci. **25**(9-10), 623-636

Kindervater, C.M. (1997) Aircraft and helicopter crashworthiness: design and simulation. In: J.A.C. Ambrósio, M.S. Pereira, F.P Silva (eds.) Crashworthiness of transportation systems: structural impact and occupant protection. NATO ASI Series E. Vol. 332, 525-577. Dordrecht: Kluwer Academic Publishers

Lankarani, H.M., Nikravesh, P.E. (1994) Continuous contact force models for impact analysis in multibody systems. Nonlinear Dynamics **5**(2), 193-207

Lust, R. (1992) Structural optimization with crashworthiness constraints. Structural Optimization. **4**, 85-89

Matolcsy, M. (1997) Crashworthiness of bus structures and rollover protection. In: J.A.C. Ambrósio, M.S. Pereira, F.P Silva (eds.) Crashworthiness of transportation systems: structural impact and occupant protection. NATO ASI Series E. Vol. 332, 321-360. Dordrecht: Kluwer Academic Publishers

Murray, N.W. (1983) The static approach to plastic collapse and energy dissipation in some thin-walled steel structures. In: N. Jones, T. Wierzbicki (eds.) Structural Crashworthiness. 44-65. London: Butterworths

Nikravesh, P.E. ,Chung, I.S., Benedict, R.L. (1983) Plastic hinge approach to vehicle simulation using a plastic hinge technique. J. Comp. Struct. **16**, 385-400

Nikravesh, P.E. (1988) Computer aided analysis of mechanical systems. Englewood Cliffs: Prentice-Hall.

Nikravesh, P.E. (1990) Systematic reduction of multibody equations of motion to a minimal set. Int. J. Non-Linear Mechanics. **25**, 143-151

Pereira, M.S., Proença, P.L. (1991) Dynamic analysis of spatial flexible multibody systems using joint coordinates. Int. J. Num. Meth. in Engng. **32**, 1799-1812.

Pereira, M.S., Drazetic P., Ravalard Y. (1993) A hybrid method for impact analysis of rigid-flexible structural systems undergoing gross motion. J. Impact Engng. (Submmited)

Pereira M.S., Ambrósio J.A.C. (1994) Computer-aided analysis of rigid and flexible mechanical systems. NATO-ASI Series E Vol. 268. Dordrecht: Kluwer Academic Publishers

Pereira, M.S., Ambrósio, J.A.C., Dias, J.P. (1997) Crashworthiness analysis and design using rigid-flexible multibody dynamics with application to train vehicles. Int. J. Num. Meth. in Engng. **40**, 655-687

Roberson , R.E. and Schwertassek, R. (1988) Dynamics of multibody systems. Berlin: Springer-Verlag.

Shabana, A. , Wehage, R. (1983) Variable degree of freedom component mode analysis of inertia variant flexible mechanical systems. ASME J. of Mechanisms, Transmissions and Automation in Design **105**, 371-378

Shabana, A. (1989) Dynamics of multibody systems. New York: John Wiley & Sons

Shampine, L.F., Gordon, M.K. (1975) Computer solution of ordinary differential equations: the initial value problem. San Francisco: V.H. Freeman and Co.

Silva, M.P.T, Ambrósio, J.A.C., Pereira, M.S. (1997) A multibody approach to the vehicle and occupant integrated simulation. Int. J. of Crashworthiness **2**(1), 73-90

Song, J.O., Haug, E.J. (1980) Dynamic analysis of planar flexible mechanisms. Comp. Meth. In Appl. Mech. And Engng. **24**, 359-381

Song, J.O. (1986) An optimization method for crashworthiness design. In: 6th Int. Conf. Vehicle Structural Mechanics, Proceedings. Detroit 1986, 39-46

SOREFAME (1993) Dynamic crash tests: dynamic test of a vehicle against an obstacle. Technical Report n. TRAINCOL/T7.2-F, SOREFAME, Portugal

Thompson, B.S., Sung, C.K. (1986) Survey of finite element techniques for mechanism design. Mech. Mach. Theory **21**(4), 351-359

Vanderplaats, G. (1987) ADS- A fortran program for automated design synthesis, Version 2.01, Eng. Design Optimization Inc., Calif.

Vanderplaats, G., Moses, F. (1973) Structural optimization by methods of feasible directions. J. Comp. & Struct. **3**, 739-755

Winmer. A. (1977) Einfluß der Belastungsgeschwindigkeit auf das Festigkeits- und Verformungsverhalten am Beispiel von Kraftfarhzeugen. ATZ **77**(10), 281-286.

Yoo, W.S., Haug, E.J. (1986) Dynamics of flexible mechanical systems using vibration and static correction modes. ASME J. of Mechanisms, Transmissions and Automation in Design **108**, 315-322

Index

NATO ASI Series F

Including Special Programmes on Sensory Systems for Robotic Control (ROB) and on Advanced Educational Technology (AET)

Vol. 76: Multimedia Interface Design in Education. Edited by A. D. N. Edwards and S. Holland. XIV, 216 pages. 1992. *(AET)*

Vol. 77: Computer Algorithms for Solving Linear Algebraic Equations. The State of the Art. Edited by E. Spedicato. VIII, 352 pages. 1991.

Vol. 78: Integrating Advanced Technology into Technology Education. Edited by M. Hacker, A. Gordon and M. de Vries. VIII, 185 pages. 1991. *(AET)*

Vol. 79: Logic, Algebra, and Computation. Edited by F. L. Bauer. VII, 485 pages. 1991.

Vol. 80: Intelligent Tutoring Systems for Foreign Language Learning. Edited by M. L. Swartz and M. Yazdani. IX, 347 pages. 1992. *(AET)*

Vol. 81: Cognitive Tools for Learning. Edited by P. A. M. Kommers, D. H. Jonassen, and J. T. Mayes. X, 278 pages. 1992. *(AET)*

Vol. 82: Combinatorial Optimization. New Frontiers in Theory and Practice. Edited by M. Akgül, H. W. Hamacher, and S. Tüfekçi. XI, 334 pages. 1992.

Vol. 83: Active Perception and Robot Vision. Edited by A. K. Sood and H. Wechsler. IX, 756 pages. 1992.

Vol. 84: Computer-Based Learning Environments and Problem Solving. Edited by E. De Corte, M. C. Linn, H. Mandl, and L. Verschaffel. XVI, 488 pages. 1992. *(AET)*

Vol. 85: Adaptive Learning Environments. Foundations and Frontiers. Edited by M. Jones and P. H. Winne. VIII, 408 pages. 1992. *(AET)*

Vol. 86: Intelligent Learning Environments and Knowledge Acquisition in Physics. Edited by A. Tiberghien and H. Mandl. VIII, 285 pages. 1992. *(AET)*

Vol. 87: Cognitive Modelling and Interactive Environments. With demo diskettes (Apple and IBM compatible). Edited by F. L. Engel, D. G. Bouwhuis, T. Bösser, and G. d'Ydewalle. IX, 311 pages. 1992. *(AET)*

Vol. 88: Programming and Mathematical Method. Edited by M. Broy. VIII, 428 pages. 1992.

Vol. 89: Mathematical Problem Solving and New Information Technologies. Edited by J. P. Ponte, J. F. Matos, J. M. Matos, and D. Fernandes. XV, 346 pages. 1992. *(AET)*

Vol. 90: Collaborative Learning Through Computer Conferencing. Edited by A. R. Kaye. X, 260 pages. 1992. *(AET)*

Vol. 91: New Directions for Intelligent Tutoring Systems. Edited by E. Costa. X, 296 pages. 1992. *(AET)*

Vol. 92: Hypermedia Courseware: Structures of Communication and Intelligent Help. Edited by A. Oliveira. X, 241 pages. 1992. *(AET)*

Vol. 93: Interactive Multimedia Learning Environments. Human Factors and Technical Considerations on Design Issues. Edited by M. Giardina. VIII, 254 pages. 1992. *(AET)*

Vol. 94: Logic and Algebra of Specification. Edited by F. L. Bauer, W. Brauer, and H. Schwichtenberg. VII, 442 pages. 1993.

Vol. 95: Comprehensive Systems Design: A New Educational Technology. Edited by C. M. Reigeluth, B. H. Banathy, and J. R. Olson. IX, 437 pages. 1993. *(AET)*

Vol. 96: New Directions in Educational Technology. Edited by E. Scanlon and T. O'Shea. VIII, 251 pages. 1992. *(AET)*

Vol. 97: Advanced Models of Cognition for Medical Training and Practice. Edited by D. A. Evans and V. L. Patel. XI, 372 pages. 1992. *(AET)*

NATO ASI Series F

Including Special Programmes on Sensory Systems for Robotic Control (ROB) and on Advanced Educational Technology (AET)

Vol. 98: Medical Images: Formation, Handling and Evaluation. Edited by A. E. Todd-Pokropek and M. A. Viergever. IX, 700 pages. 1992.

Vol. 99: Multisensor Fusion for Computer Vision. Edited by J. K. Aggarwal. XI, 456 pages. 1993. *(ROB)*

Vol. 100: Communication from an Artificial Intelligence Perspective. Theoretical and Applied Issues. Edited by A. Ortony, J. Slack and O. Stock. XII, 260 pages. 1992.

Vol. 101: Recent Developments in Decision Support Systems. Edited by C. W. Holsapple and A. B. Whinston. XI, 618 pages. 1993.

Vol. 102: Robots and Biological Systems: Towards a New Bionics? Edited by P. Dario, G. Sandini and P. Aebischer. XII, 786 pages. 1993.

Vol. 103: Parallel Computing on Distributed Memory Multiprocessors. Edited by F. Özgüner and F. Erçal. VIII, 332 pages. 1993.

Vol. 104: Instructional Models in Computer-Based Learning Environments. Edited by S. Dijkstra, H. P. M. Krammer and J. J. G. van Merriënboer. X, 510 pages. 1993. *(AET)*

Vol. 105: Designing Environments for Constructive Learning. Edited by T. M. Duffy, J. Lowyck and D. H. Jonassen. VIII, 374 pages. 1993. *(AET)*

Vol. 106: Software for Parallel Computation. Edited by J. S. Kowalik and L. Grandinetti. IX, 363 pages. 1993.

Vol. 107: Advanced Educational Technologies for Mathematics and Science. Edited by D. L. Ferguson. XII, 749 pages. 1993. *(AET)*

Vol. 108: Concurrent Engineering: Tools and Technologies for Mechanical System Design. Edited by E. J. Haug. XIII, 998 pages. 1993.

Vol. 109: Advanced Educational Technology in Technology Education. Edited by A. Gordon, M. Hacker and M. de Vries. VIII, 253 pages. 1993. *(AET)*

Vol. 110: Verification and Validation of Complex Systems: Human Factors Issues. Edited by J. A. Wise, V. D. Hopkin and P. Stager. XIII, 704 pages. 1993.

Vol. 111: Cognitive Models and Intelligent Environments for Learning Programming. Edited by E. Lemut, B. du Boulay and G. Dettori. VIII, 305 pages. 1993. *(AET)*

Vol. 112: Item Banking: Interactive Testing and Self-Assessment. Edited by D. A. Leclercq and J. E. Bruno. VIII, 261 pages. 1993. *(AET)*

Vol. 113: Interactive Learning Technology for the Deaf. Edited by B. A. G. Elsendoorn and F. Coninx. XIII, 285 pages. 1993. *(AET)*

Vol. 114: Intelligent Systems: Safety, Reliability and Maintainability Issues. Edited by O. Kaynak, G. Honderd and E. Grant. XI, 340 pages. 1993.

Vol. 115: Learning Electricity and Electronics with Advanced Educational Technology. Edited by M. Caillot. VII, 329 pages. 1993. *(AET)*

Vol. 116: Control Technology in Elementary Education. Edited by B. Denis. IX, 311 pages. 1993. *(AET)*

Vol. 117: Intelligent Learning Environments: The Case of Geometry. Edited by J.-M. Laborde. VIII, 267 pages. 1996. *(AET)*

Vol. 118: Program Design Calculi. Edited by M. Broy. VIII, 409 pages. 1993.

Vol. 119: Automating Instructional Design, Development, and Delivery. Edited by. R. D. Tennyson. VIII, 266 pages. 1994. *(AET)*

Vol. 120: Reliability and Safety Assessment of Dynamic Process Systems. Edited by T. Aldemir, N. O. Siu, A. Mosleh, P. C. Cacciabue and B. G. Göktepe. X, 242 pages. 1994.

NATO ASI Series F

Including Special Programmes on Sensory Systems for Robotic Control (ROB) and on Advanced Educational Technology (AET)

Vol. 121: Learning from Computers: Mathematics Education and Technology. Edited by C. Keitel and K. Ruthven. XIII, 332 pages. 1993. *(AET)*

Vol. 122: Simulation-Based Experiential Learning. Edited by D. M. Towne, T. de Jong and H. Spada. XIV, 274 pages. 1993. *(AET)*

Vol. 123: User-Centred Requirements for Software Engineering Environments. Edited by D. J. Gilmore, R. L. Winder and F. Détienne. VII, 377 pages. 1994.

Vol. 124: Fundamentals in Handwriting Recognition. Edited by S. Impedovo. IX, 496 pages. 1994.

Vol. 125: Student Modelling: The Key to Individualized Knowledge-Based Instruction. Edited by J. E. Greer and G. I. McCalla. X, 383 pages. 1994. *(AET)*

Vol. 126: Shape in Picture. Mathematical Description of Shape in Grey-level Images. Edited by Y.-L. O, A. Toet, D. Foster, H. J. A. M. Heijmans and P. Meer. XI, 676 pages. 1994.

Vol. 127: Real Time Computing. Edited by W. A. Halang and A. D. Stoyenko. XXII, 762 pages. 1994.

Vol. 128: Computer Supported Collaborative Learning. Edited by C. O'Malley. X, 303 pages. 1994. *(AET)*

Vol. 129: Human-Machine Communication for Educational Systems Design. Edited by M. D. Brouwer-Janse and T. L. Harrington. X, 342 pages. 1994. *(AET)*

Vol. 130: Advances in Object-Oriented Database Systems. Edited by A. Dogac, M. T. Özsu, A. Biliris and T. Sellis. XI, 515 pages. 1994.

Vol. 131: Constraint Programming. Edited by B. Mayoh, E. Tyugu and J. Penjam. VII, 452 pages. 1994.

Vol. 132: Mathematical Modelling Courses for Engineering Education. Edited by Y. Ersoy and A. O. Moscardini. X, 246 pages. 1994. *(AET)*

Vol. 133: Collaborative Dialogue Technologies in Distance Learning. Edited by M. F. Verdejo and S. A. Cerri. XIV, 296 pages. 1994. *(AET)*

Vol. 134: Computer Integrated Production Systems and Organizations. The Human-Centred Approach. Edited by F. Schmid, S. Evans, A. W. S. Ainger and R. J. Grieve. X, 347 pages. 1994.

Vol. 135: Technology Education in School and Industry. Emerging Didactics for Human Resource Development. Edited by D. Blandow and M. J. Dyrenfurth. XI, 367 pages. 1994. *(AET)*

Vol. 136: From Statistics to Neural Networks. Theory and Pattern Recognition Applications. Edited by V. Cherkassky, J. H. Friedman and H. Wechsler. XII, 394 pages. 1994.

Vol. 137: Technology-Based Learning Environments. Psychological and Educational Foundations.

Vol. 138: Exploiting Mental Imagery with Computers in Mathematics Education. Edited by R. Sutherland and J. Mason. VIII, 326 pages. 1995. *(AET)*

Vol. 139: Proof and Computation. Edited by H. Schwichtenberg. VII, 470 pages. 1995.

Vol. 140: Automating Instructional Design: Computer-Based Development and Delivery Tools. Edited by R. D. Tennyson and A. E. Barron. IX, 618 pages. 1995. *(AET)*

Vol. 141: Organizational Learning and Technological Change. Edited by C. Zucchermaglio, S. Bagnara and S. U. Stucky. X, 368 pages. 1995. *(AET)*

Vol. 142: Dialogue and Instruction. Modeling Interaction in Intelligent Tutoring Systems. Edited by R.-J. Beun, M. Baker and M. Reiner. IX, 368 pages. 1995. *(AET)*

Vol. 143: Batch Processing Systems Engineering. Fundamentals of Chemical Engineering. Edited by G. V. Reklaitis, A. K. Sunol, D. W. T. Rippin, and Ö. Hortaçsu. XIV, 868 pages. 1996.

NATO ASI Series F

Including Special Programmes on Sensory Systems for Robotic Control (ROB) and on Advanced Educational Technology (AET)

Vol. 144: The Biology and Technology of Intelligent Autonomous Agents. Edited by Luc Steels. VIII, 517 pages. 1995.

Vol. 145: Advanced Educational Technology: Research Issues and Future Potential. Edited by T. T. Liao. VIII, 219 pages. 1996. *(AET)*

Vol. 146: Computers and Exploratory Learning. Edited by A. A. diSessa, C. Hoyles and R. Noss. VIII, 482 pages. 1995. *(AET)*

Vol. 147: Speech Recognition and Coding. New Advances and Trends. Edited by A. J. Rubio Ayuso and J. M. López Soler. XI, 505 pages. 1995.

Vol. 148: Knowledge Acquisition, Organization, and Use in Biology. Edited by K. M. Fisher and M. R. Kibby. X, 246 pages. 1996. *(AET)*

Vol. 149: Emergent Computing Methods in Engineering Design. Applications of Genetic Algorithms and Neural Networks. Edited by D.E. Grierson and P. Hajela. VIII, 350 pages. 1996.

Vol. 150: Speechreading by Humans and Machines. Edited by D. G. Stork and M. E. Hennecke. XV, 686 pages. 1996.

Vol. 151: Computational and Conversational Discourse. Burning Issues – An Interdisciplinary Account. Edited by E. H. Hovy and D. R. Scott. XII, 202 pages. 1996.

Vol. 152: Deductive Program Design. Edited by M. Broy. IX, 467 pages. 1996.

Vol. 153: Identification, Adaptation, Learning. Edited by S. Bittanti and G. Picci. XIV, 553 pages. 1996.

Vol. 154: Reliability and Maintenance of Complex Systems. Edited by S. Özekici. XI, 589 pages. 1996.

Vol. 155: Cooperation: Game-Theoretic Approaches. Edited by S. Hart and A. Mas-Colell. VIII, 328 pages. 1997.

Vol. 156: Microcomputer-Based Labs: Educational Research and Standards. Edited by R.F. Tinker. XIV, 398 pages. 1996. *(AET)*

Vol. 157: Logic of Computation. Edited by H. Schwichtenberg. VII, 396 pages. 1997.

Vol. 158: Mathematical Methods in Program Development. Edited by M. Broy and B. Schieder. VIII, 528 pages. 1997.

Vol. 159: Fractal Image Encoding and Analysis. Edited by Y. Fisher. XIX, 362 pages. 1998

Vol. 160: Discourse, Tools, and Reasoning: Essays on Situated Cognition. Edited by L.B. Resnick, R. Säljö, C. Pontecorvo and B. Bunge. XII, 474 pages. 1997 *(AET)*

Vol. 161: Computational Methods in Mechanical Systems. Edited by J. Angeles and E. Zakhariev. X, 425 pages. 1998.

Druck: **Strauss Offsetdruck, Mörlenbach**
Verarbeitung: **Schäffer, Grünstadt**

MIX
Papier aus verantwortungsvollen Quellen
Paper from responsible sources
FSC® C105338

If you have any concerns about our products,
you can contact us on
ProductSafety@springernature.com

In case Publisher is established outside the EU,
the EU authorized representative is:
Springer Nature Customer Service Center GmbH
Europaplatz 3, 69115 Heidelberg, Germany

Printed by Libri Plureos GmbH
in Hamburg, Germany